Earth System Analysis: Integrating Science for Sustainability

Springer
Berlin
Heidelberg
New York
Barcelona
Budapest
Hong Kong
London
Milan
Paris
Singapore
Tokyo

H.-J. Schellnhuber, V. Wenzel

Earth System Analysis

Integrating Science for Sustainability

Complemented Results of a Symposium
Organized by the Potsdam Institute (PIK)

With 92 Figures, 32 in Color and 16 Tables

Springer

Editors:

Professor Dr. Hans-Joachim Schellnhuber
Dr. Volker Wenzel

Potsdam Institute for Climate Impact Research (PIK)
P.O. Box 60 12 03
D-14412 Potsdam / Germany

ISBN 978-3-642-52356-4 ISBN 978-3-642-52354-0 (eBook)
DOI 10.1007/978-3-642-52354-0

Library of Congress Cataloging-in-Publication Data
Earth systems analysis: integrating science for sustainability: complemented results of a symposium organized by the Potsdam Institute (PIK) / Hans-Joachim Schellnhuber, Volker Wenzel (eds.). p. m.
Includes bibliographical references and index.

1. Environmental sciences - Congresses. 2. System analysis - Congresses. 3. Environmental management - Congresses. 4. Human ecology - Congresses. 5. Sustainable development - Congresses. I. Schellnhuber, Hans-Joachim. II. Wenzel, Volker. III. Potsdam-Institut für Klimafolgenforschung.
GE 105.E18 1998 363.7--dc21 98-24066 CIP

Softcover reprint of the hardcover 1st edition 1998

Typesetting: Camera ready by author
Cover design: Design & production, Heidelberg
SPIN: 10062509 32/3020 - 5 4 3 2 1 0 - Printed on acid -free paper

Opening Words by Steffen Reiche, Minister for Science, Research and Culture of the Federal State of Brandenburg

We all know that our earth is a highly sensitive, complex–structured organism, which has developed astounding mechanisms for resistance and regeneration in response to interference of various sorts. However, as the destruction of the ozone layer increases, so the threat to our environment grows. We think also of the intensification of the greenhouse effect, climatic changes, damage to forests, floods, the pollution of waters, trash mountains and far-reaching anthropogenic disturbance of the landscape. These world-wide environmental changes can have devastating consequences for humanity and the natural world.

The most recent example of this is the "flood of the century" in Brandenburg, Poland and the Czech Republic. This summer, it brought ecological and economic damage on a huge scale to the affected regions; but also a great wave of solidarity. Here it becomes apparent how important basic preventative and environmental research and strategies towards the sustainable, environmentally sound development of our society are. Such research results are urgently needed and anticipated - and not only then when we are literally up to our necks in water.

We have in the State of Brandenburg non–university research institutions which have taken on and are engaged in this important subject. One of the internationally best known and most committed institutions in the area of climate change research and environmental systems analysis is the Potsdam Institute for Climate Impact Research (PIK), founded in 1992. PIK, in close collaboration with research institutions within and beyond Germany, pursues environmental research using new approaches in the natural and social sciences. Its scientists describe and analyse the state of the environment and environmental changes, and develop solutions to resulting problems.

The results of the workshop on "Earth System Analysis" show that a scientifically comprehensive analysis of the interactions between mankind and the environment demands an interdisciplinary approach. Not only the natural but also the human and societal dimensions should be considered. Computer simulation and integrated modelling are in this respect very suitable scientific instruments in their ability to depict complex, non–linear environmental systems and their changes.

Great endeavours are needed in order to correct undesirable developments of the past. We have a duty to future generations to maintain for them an environment in which they can live. In its efforts to bring about economic and ecological regeneration in Brandenburg, the regional Government has to rely on the assistance of science and research. The potential is considerable, as the results of the workshop show. I would be very glad if this publication could help to give an additional impetus to research in this sense also.

Potsdam, June 1997

Preface

Earth System Analysis (ESA) is a science *in statu nascendi.*

It is a science as it has

1. a genuine *subject*,
   namely the total Earth in the sense of a fragile and "gullible" dynamic system,
2. a genuine *methodology*,
   namely transdisciplinary systems analysis based on, i.a., planetary monitoring, global modelling and simulation,
3. a genuine *purpose*,
   namely the satisfactory (or at least tolerable) coevolution of the ecosphere and the anthroposphere (*vulgo*: Sustainable Development) in the times of Global Change and beyond.

Note that the third item is not a strictly necessary ingredient for the establishment of a scientific field – certain branches of contemporary mathematics seem to lack any raison d'être that transcends pure intellectual curiosity. History tells us, however, that almost all scientific developments are driven by direct or subtle societal forces, and the potential applicability of a research enterprise – such as helping to identify ways to secure humanity's life-support systems – might be viewed as a nice bit of added value, after all.

Note also that the notion "Earth System Analysis" constitutes a quite natural amalgamation of the terms "Earth System" and "Systems Analysis", reflecting the essence of ingredients (1) and (2). Now the sheer use of this wording may induce a multitude of negative reactions from experts as well as lay persons: is it not complete hubris to even strive to analyse a system of such overwhelming complexity which, ultimately, also embraces our own material functioning and intellectual activity? A major objective of this book is precisely to answer that fundamental question. We may anticipate here, though, the argumentation that the organs of a locust are at least as complex as the geophysical atmosphere; the attempt of a psychologist to ameliorate his/her own mental condition is no less self-referential in character than the investigation of "correct" environmental behaviour as conducted by ESA; the rather large size of ESA's specimen will hardly impress cosmologists who routinely juggle with entire galaxy clusters. It is the combination of all those aspects, however, that make Earth System Analysis a rather difficult task, a *Grand Scientific Challenge* indeed ...

ESA is obviously still *in statu nascendi*, but who or what "engendered" this emerging scientific entity? Employing T. Kuhn's all too popular epistemological theory, we have to search for a generating *paradigm shift* as triggered by some "*experimentis crucis*", a major historical event, certain revolutionary technological developments, or the like. If that theory applies to ESA at all, then the co-operation of at least three crucial factors is responsible for its coming into being:

- The race for the Moon created the opportunity to observe planet Earth from space with sophisticated equipment and, even more important, to contemplate it as a finite semi-open system from outside. Thus ESA became *conceivable.*
- The advent of electronic super-computers established the technological platform for sufficiently fast and comprehensive global simulation modelling based on adequate management of the plethora of now available monitoring data. Thus ESA became *feasible.*
- The discovery of the ozone hole over Antarctica ignited a cognitive shockwave, confronting the international community with the evidence that humanity can and, in fact, is about to transform the character of the global environment. Thus ESA became *mandatory.*

One might even contend that (environmental) science will never be the same as it was in pre-ozone hole times. Thus the measurements carried out by *Farman* et al. (1985) [67] mark a modern Copernican revolution ultimately emanating, i.a., from billions of mundane hairsprays ....

The making of a science generally proceeds both through the pioneering contributions of individual researchers and through illuminating and exciting meetings. The latter bring together the dedicated followers of the new trend in order to summarize what has already been achieved and to trace out the lines along which the field will or should further develop. In the case of ESA, such a meeting took place in Potsdam in 1994. This symposium had both a formal purpose, namely the inauguration of the Scientific Advisory Board of the Potsdam Institute for Climate Impact Research (PIK), and an essential cause, namely the advancement of Earth System Analysis as an integrated (i.e., transdisciplinary) scientific effort to help ensure "Sustainable Development".

The Potsdam meeting convened an eminent group of researchers active in various fields pertinent to ESA such as climatology, global biosphere dynamics, impacts assessment, agro-landscape ecology, catchment hydrology, international political economy, energy systems analysis, and development sociology. This group was complemented by leading "generalists" representing the unique epistemological and methodological aspects of ESA as evoked by catchwords like "integrated modelling", "game theory", "stochastic optimization", "non-linear dynamics", "cognitive psychology" and "philosophy of science".

The Symposium was structured in the following way: about a dozen key lectures, mostly delivered by members of PIK's Scientific Advisory Board, tried to span the range of ESA's topical complexity. Each of these lectures were commented upon by a mixed tandem of researchers active within and outside, respectively, the specific field in question. The resulting thematic blocks were wrapped up by intense plenary discussions. In a final round, the attempt was made to summarize the results of the meeting, and to identify a canonical research programme for ESA in the years ahead.

From the beginning of this planning process, there were intentions to publish the oral contributions to the Potsdam Symposium in some sort of "proceedings" monograph. According to the unanimous assessment of the participants, however, the meeting was so successful and ground-breaking that only a much extended and complemented written account seemed to be adequate. Thus the editors set out on the difficult task of assembling the elements for an ESA book, which are based on the oral contributions yet considerably transcend them in length and content a posteriori. The result of this effort – a volume of some 500 pages – is now at the reader's disposal.

Owing to a number of reasons and circumstances, the recruitment and composition process for this monograph was rather slow. The delayed appearance of our ESA book, which has to be blamed entirely on the editors, is on the one hand certainly deplorable. Take for instance R. Rockwell's paper unfolding the concept of the regional foundation for global integrated modelling: the corresponding manuscript was received in autumn 1994 and had to endure several years on the shelf, while similar ideas were developed independently and published by other authors in the meantime. Nuisances of that type were also inflicted on a few other authors who delivered well in time.

On the other hand, the editors are firmly convinced that the delay period did not really diminish the value and originality of any of the submitted pieces. Some of the contributions – and, in fact, the book as a whole – have even significantly benefited from the prolonged maturation time, i.e. from the "Sauternes effect": the sweet white wines of that "Appellation" in south-western France owe their unrivalled character to a deliberate extension of the ripening time, which allows a certain micro-organism (*Botrytis cinerea*) to concentrate and transform the essence of the grapes. Although microorganisms did not play a crucial role in the making of this book, the delayed harvesting allowed for a number of alternative enrichment processes.

The material of the resulting monograph is organized as follows: Part 1 consists of a discourse by one of the editors (H.J.S.) that tries to work out in some detail what ESA is all about. The essay is supposed to set the stage for the more specific contributions below, and to create a general *conceptual and formal framework* for the heretofore often hazy debate about Global Change, Sustainable Development, and Intergenerational Equity. This piece of work has tremendously benefited from the symposium and particularly from the various discussions that took place there. Part 2 of the book is dedicated to the *natural dimensions* of ESA, i.e. to the description of some of the most pertinent and fragile elements of the biogeophysical "metabolism" of planet Earth. Part 3 deals with the *human dimensions* of ESA, a topical complex which is much harder to define and to structure than the "natural" counterpart contemplated in Part 2. The intellectual, ethical and material elements of the "anthroposphere" are so diverse and subtle that anything other than an eclectic approach seems unfeasible at present. The editors are confident, though, that some really crucial issues have been selected for this monograph. Part 4 concludes our book by epitomizing various *roads to integrated modelling* of the coupled dynamics of ecosphere and anthroposphere. These efforts - and, in fact, the overall ESA enterprise - are finally scrutinized from the point of view of "moral epistemology" by H.P. Dürr. For our readers' convenience, we now provide the condensed abstracts of the individual contributions.

In **Part 1** of the book a basic methodological approach towards Earth System Analysis has been outlined by Hans-Joachim Schellnhuber. The "scope of the challenge" is reflected by a calculus describing geo-cybernetics, i.e. the managed coevolution of nature and humanity. The pictorial quality of this description is continually assisted by examples and analogies taken primarily from the art of navigation, i.e. the original cybernetics of the ancient Greeks, which thus achieves the status of a metaphor.

The calculus - as a reference to Leonardo da Vinci, for whom a science was only a science if it contained mathematics - has four desired consequences:

- Few basic principles lead to a quasi-axiomatic foundation.
- The principles and a cybernetic type of conclusion give rise to a structural ordering of the scientific contents.
- Each basic principle induces the incorporation of useful mathematical theorems and their intrinsic formalism.
- Thus, the present limits of knowledge become visible together with the most urgent questions. This provides a potential tool for future research.

Using the example of Sustainable Development the value of the approach becomes evident. In contrast to the often stated vagueness of this term, the concept of *Sustainable Development* can be expressed quite concretely according to the five fundamental paradigms *Standardization, Optimization, Pessimization, Equitization* and *Stabilization.* All five paradigms explain Sustainable Development in terms of concrete operations which are accessible to assessment and mathematical treatment.

Another example is represented by the development of control strategies in case of uncertain or fragmentary information, which is a quite typical situation in geo-cybernetics. Here, *Fuzzy Control* is cultivated as a fundamental operational principle for a stepwise reaching of certain target states to be realized by perpetual readjusting, which includes also the strategy itself.

Part 2 deals with the natural dimensions of ESA. In **Chapter 1**, Colin Prentice gives a state-of-the-art report on the potential effects of changes in climate on the Earth's biosphere. Emphasis is laid on natural and semi-natural ecosystems. These ecosystems are represented by the pattern of biomes which are derived from so-called plant functional types. Net primary production (NPP) and the coupling of biosphere and atmosphere through biogeophysical and biogeochemical feedbacks are described as key processes. In equilibrium, these processes realize certain optimality principles – particularly in the case of NPP. If this equilibrium is disturbed by climate change, a transient phase comes into operation in which processes adapt to a new equilibrium of optimal conditions.

The idealized models of theoretical ecology which are tuned to describe realistically the present equilibrium state will not be capable of meeting this challenge. New models with higher predictive power based on greater physical consistency have to be developed. New tools have to be acquired – from the requirements of scale transition to the incorporation of fuzzy logic methods. The status report illuminates the ongoing development of methods as well as future requirements and expectations.

The next two chapters deal with the effect of climate change on agro-ecosystems. In **Chapter 2**, Martin Parry et al. describe how land use can change as a result of climate change in a specific region (England and Wales). For this integrated regional study Parry and coworkers had to apply a whole series of avant-garde models in a co-ordinated approach. Various interim results were derived through well directed scenario analyses as follows: Firstly, a restriction to particular climate change scenarios is made. Then a future scenario for the year 2060 is defined and two levels of possible adaptation to altered yields are introduced. For each scenario, the global cereal yield corresponding to the respective conditions, the change in global price of foodstuffs and the regional change in yield are determined. Finally, an integrated analysis of the regional changes in land use resulting from the anticipated climate, yield and price changes is performed.

In **Chapter 3**, Heidrun Mühle illuminates the dual role played by agriculture with regard to Global Change, as both a cause of it and as suffering/benefitting from it. This suggests the necessity of regarding future agro-ecological concepts as integral parts of the shaping of the landscape in general. They should find acceptance in national and international development programmes.

To understand the flows of material and energy in and between the ecosystems of a landscape, physiologically detailed plant models and the coupling of these with soil models are required. Scenario exercises with these models can help us obtain the necessary understanding for effective decision-making in accordance with a sustainable development of agrarian landscapes.

Part 3 of the book is dedicated to the human dimensions (HD) of the Earth System. In **Chapter 1**, Harold Jacobson presents an analysis of what models for HD have to perform in order to overcome the inadequacies of the pioneer-phase models as regards the "limits to growth". The difficulties above all are due to the fact that changes in behaviour and trends have to be depicted whose key variables are scarcely possible to measure. Behaviour is determined by norms, rules, traditions and expectations, the disregard of which to a large extent explains the failures of environment protection policy hitherto.

Models of a strictly modular construction offer an alternative. The content of the module is determined by the economic, cultural and social institutions relevant to the behavioural aspect being modelled. This includes firms and markets, since these are institutions which perform a structural function as regards behaviour and changes in behaviour. Thus regional variations should also be included.

It seems possible to overcome the anticipated increase of expenditure involved in focusing on many regions, as investigations can often be confined to the respective most important or typical regions, as examples show.

In **Chapter 2**, Udo E. Simonis evaluates the potential and chances of international environment policy instruments, in particular those for the reduction of greenhouse gas emissions. A basic idea in this respect is the creation of a new market – for trade in emission certificates. The *conditio sine qua non* here, however, is the fulfilment of boundary conditions in the form of three criteria, namely, economic efficiency, development policy equity and the potential for decision-making under uncertainty. The second criterion here is receiving the most attention at the present time.

In this connection, a "joint implementation" of taxes and obligations for the reduction of emissions appears to be an important instrumental step on the road towards trade with certificates, since it will be easier to find a consensus on bilateral contracts. It not only causes reducing costs and increasing effectiveness, but also represents a stimulation to the acceleration of technology transfer.

The subsequent implementation of flexible trade conditions for certificates requires a balance between market-orientated, price-linked instruments on the one hand and regulating, quantity-related instruments on the other.

A global projection of "northern" habits of consumption indicates disastrous consequences for the Earth System. Ernst U. von Weizsäcker takes this as a basis for his call in **Chapter 3** for new thinking in this regard. This could be expressed in new models of wealth. Calculations have shown that reducing intensity of resource use by a factor of four is not only necessary but – through the introduction of stimulants and tax structures (from an ecological reform of taxes to changes in lifestyle) – also possible.

In **Chapter 4**, James K. Mitchell focuses on the increasing vulnerability of the socio-technical systems of urban regions. The intensification of the urban metabolism and the danger for sustainable development arising from this has both social and biophysical aspects. The growth of huge conurbations, in particular, carries with it a considerable potential for catastrophe.

The internal dynamics of these urban processes forces, i.a., the increasing settlement of population in endangered areas – often involving people uprooted from other areas and generally still lacking a firm basis of support. This affects stability, intensifies risks and creates specific new risks – for example through synergy effects. However, new opportunities, e.g. for learning processes to limit natural catastrophes or their consequences, also result.

In **Part 4**, approaches for integrated models will be presented and the epistemological value of such models will be estimated in a very general way. In **Chapter 1**, Klaus and Susanne Hasselmann expand their "Structurally Integrated Assessment Model" for the cost optimization of CO_2 emissions reduction in the case of a single actor in relationship to the accounting of many actors. "Structurally integrated" means the coupling of different components. Firstly, an impulse-response relationship is depicted which allocates a climate response to an annual emission rate for greenhouse gases integrated over a period of time. Secondly, the impulse-response climate model is then coupled to a cost model. The total costs appear here as the sum of the emission reduction costs and the damage costs caused by climate change. In different scenarios, the actors are either completely autonomous or they have trading relationships with one another.

The numerical applications imply, for example, that individual climate protection measures of a single or of a small number of developed industrialized countries can be very effective. In addition, they are in the actors' own interests, since their effects are already of the magnitude of those of international concerted actions.

In **Chapter 2**, Jan Rotmans outlines the intentions, structure, and possibilities for application of the global integrated assessment model TARGETS. The model is one of the first comprehensive instruments for a simulation-based study of Global Change and sustainability. The modular construction with both horizontal and vertical structure and the multiplicity of incorporated branch modules permit even the very complicated interactions between nature and society to be studied. The intention is to shorten the causal chain to an operable size without any essential losses, through means of a hierarchical system of indicators and the definition of aggregated indices.

The term sustainability can only be defined in connection with a basic system of values, which can also be understood as a subjective component. Scenario calculations are therefore determined and carried out in dependence on particular points of view or from a subjective perspective, borrowed from the cultural theory of Thompson. This is of assistance in realizing one of the essential purposes of the model, i.e. support of the communication between politics and science.

Richard C. Rockwell's intention in **Chapter 3** is to prove that the concept of a uniform globally integrated model cannot achieve its aim. Natural processes can certainly be described in a globally uniform way, but social processes are subject to huge regional variation, so that no undifferentiated global model of the social sphere can exist.

The concept of a global model must be replaced by a global system of local and regional models, which have to be skilfully coupled. The most difficult question, which cannot yet finally be answered, concerns the appropriate political, economic and geographical units for modelling. In many cases this unit is that of the nation state. It is the role of human ecology to study the interactions between man and the environment, and thus also to develop criteria for the appropriate units for modelling. Four key elements form the framework for the description of human ecosystems: Population, Organization, Environment and Technology (POET).

Hans-Peter Dürr's paper appears predestined to constitute the reflective closing **Chapter 4** to the book, thanks to his relativizing perspective as a quantum physicist. From this viewpoint the Earth System is categorized first and foremost as a macroscopic system with very specific characteristics observable nowhere else, in other words the existence of life and of humankind with all its potential. The original question about the feasibility of global modelling is refined to the question of whether the global models we are able to produce, now and in the near future, are good enough and appropriate enough. To judge their ability, they need to be measured against some objective. This objective is the safeguarding of planetary engineering for the achievement of particular goals.

The feasibility of such engineering is very limited and is marked by threshold values or critical points from phase transition to catastrophes. To avoid these or to draw the Earth System back from the brink of such critical phases is the major aim of global modelling and the justification for planetary engineering. At the same time the modeller is compelled (for reason of the drastic consequences) to be careful not to overestimate his own abilities.

* * *

Taken all in all, the editors feel that this first tentative state-of-the-art report on ESA was clearly well worth the effort involved. Members of the active research community, (environmental) policy makers, interested readers with no specific scientific training - all may find instructive and thought-provoking parts and pieces in this book. We do not recommend a particular digestive scheme – a "linear" consumption of the volume from front end to back end certainly fits the dialectic composition of the material, but a "random-walk" investigation of this material may help to reveal unexpected intellectual insights and associations.

A preface traditionally ends with an acknowledgement of all those persons and institutions that have made the publication in question possible. It seems to us that such an acknowledgement is particularly appropriate here, for the production of this book turned out to resemble the taming of a wild mustang.

First and foremost the editors would like to thank the authors and participants of the Symposium who, also through their contribution to the discussions, generated the scientific substance of the book.

Among the scientists at PIK, we are greatly indebted to Jürgen Kropp for his valuable comments and the extended work in the final technical redaction of the book mastering the complications with Part 1. We thank particularly also Arthur Block,who designed and conducted extensive calculations for the realization of a number of figures for Part 1, further, Werner von Bloh, Martin Cassel-Gintz, Markus Erhard, Dietmar Gibietz-Rheinbay, Oliver Moldenhauer, Gerhard Petschel-Held and Manfred Stock for their contributions.

Thanks for their outstanding commitment is due to the following members of staff at PIK (in alphabetical order): Sulja Bergmann, Johann Grüneweg, Alison Schlums, Bärbel Uffrecht, Klaus Urban, Ursula Werner, and Christina Wienken.

We received support also from external contributors: Jaqueline Budendorf, and, to a great extent, Corinna and Volker Raab, who we would also like to thank in this place.

Our thanks go also to Springer-Verlag Heidelberg for giving us the possibility to publish this book and for their friendly co-operation on a number of details.

* * *

For a multitude of reasons, this book is dedicated to Petra Schellnhuber.

Potsdam, Germany
June, 1997

Hans-Joachim Schellnhuber
Volker Wenzel

Contents

Part 3. Human Dimensions

Part 4. Integration for Sustainability

Commentary:
Global Environment and Society (GES) Models, a Problem of Multiple Control Theory

Chapter 2:
Global Change and Sustainable Development: Towards an Integrated Conceptual Model

Commentary on
"Global Change and Sustainable Development"

List of Contributors

Siegfried Bauer
Regional Planning Dept.
Justus-Liebig-University
Diezstr. 15
D-35390 Gießen
Germany

Lennart Bengtsson
Director of the Max-Planck-Institute for Meteorology
Bundesstr. 55
D-20146 Hamburg
Germany

Hans-Jürgen Bolle
Institute of Meteorology
Geoscience Dept.
Free University of Berlin
Carl-Heinrich-Becker-Weg 6-10
D-12165 Berlin
Germany

Gyula Bora
Economic Geography Dept.
Budapest University of
Economic Sciences
Fovám tér 8
H-1093 Budapest
Hungary

Martin Claussen
Max-Planck Institute
for Meteorology
Bundesstr. 55
D-20146 Hamburg
Germany

Hans-Peter Dürr
Director of the Max-Planck
Institute for Physics
Werner-Heisenberg Institute
Föhringer Ring 6
D-80805 München
Germany

Wolf-Dieter Eberwein
Science Center Berlin (WZB)
Reichpietschufer 50 D
D-10587 Berlin
Germany

Otto Fränzle
Geography Dept.
Kiel University
Ludewig-Meyn-Str. 14
D-24098 Kiel
Germany

G. S. Golitsyn
Institute for Atmospheric Physics
University of Moscow
Pyzhevsky, 3
109017 Moscow
Russia

Klaus Hasselmann
Susanne Hasselmann
Director/Senior Scientist
of the Max-Planck Institute
for Meteorology
Bundesstr. 55
D-20146 Hamburg
Germany

Carsten Helm
Potsdam Institute for Climate Impact Research
P.O. Box 60 12 03
D-14412 Potsdam
Germany

Harold Jacobson
Center for Political Studies
University of Michigan
Ann Arbor
Michigan 48106-1248
USA

Felix Kienast
Swiss Federal Institute for Forest,
Snow and Landscape Research
Züricher Str. 111
CH-8903 Birmensdorf
Switzerland

Gundolf H. Kohlmaier
Institute for Physical and
Theoretical Chemistry
Johann Wolfgang Goethe University
Marie-Curie-Str. 11
D-60439 Frankfurt
Germany

Lenelis Kruse-Graumann
Ecological Psychology Dept.
University of Hagen (GHS)
P.O. Box 9 40
D-58084 Hagen
Germany

Urs Luterbacher
Ellen Wiegandt
The Graduate Institute of
International Studies
11A, Avenue de la Paix
CH-1202 Geneva
Switzerland

Arkadii Maltsev
Russian Academy of Sciences
Leninskii Prospect, 14
117901 Moscow
Russia

James K. Mitchell
Dept. of Geography
The State University of New Jersey
RUTGERS
Lucy Stone Hall
Livingston Campus
P.O. Box 5080
New Brunswick
New Jersey 08903-5080
USA

Tillmann Mohr
President of the Central Office
German Meteorological Service
Frankfurter Str. 135
D-63067 Offenbach
Germany

Heidrun Mühle
Head of Section
Agricultural Landscapes
Center for Environmental Research
Leipzig-Halle (UFZ)
Permoser Str. 15
D-04318 Leipzig
Germany

Peter Nijkamp
Department of Economics
Free University of Amsterdam
De Boelelaan 1105
NL-1081 HV Amsterdam
The Netherlands

Martin L. Parry et al.
Dept. of Geography, and Director,
Environmental Studies Programme
University College London
26 Bedford Way
London WC1H OAP
UK

I. Colin Prentice
Plant Ecology Dept.
Lund University
Östra Vallgatan 14
S-22361 Lund
Sweden

Richard C. Rockwell
Executive Director
Inter-university Consortium for
Political and Social Research
University of Michigan
P.O. Box 1248
Ann Arbor, Michigan 48106-1248
USA

Jan Rotmans
International Centre
for Integrative Studies
University of Maastricht
P.O. Box 616
NL-6200 MD Maastricht
The Netherlands

Hans-Joachim Schellnhuber
Director of the Potsdam Institute
for Climate Impact Research (PIK)
P.O. Box 60 12 03
D-14412 Potsdam
Germany

Udo E. Simonis
Science Center Berlin (WZB)
Reichpietschufer 50
D-10785 Berlin
Germany

Max Tilzer
Director of the
Alfred Wegener Institute for
Polar and Marine Research (AWI)
Postfach 12 01 61
D-27515 Bremerhaven
Germany

Ernst U. von Weizsäcker
President of the Wuppertal Institute
for Climate, Environment
and Energy
Döppersberg 19
D-42103 Wuppertal
Germany

Volker Wenzel
Potsdam Institute for Climate Impact Research (PIK)
P.O. Box 60 12 03
D-14412 Potsdam
Germany

Wilhelm Windhorst
Ecosystems Dept.
Christian-Albrechts-University
Olshausenstr. 40
D-24118 Kiel
Germany

Part 1

Earth System Analysis – The Concept

Discourse: Earth System Analysis – The Scope of the Challenge

H. J. Schellnhuber

List of Frequently Used Mathematical Symbols

$\mathcal{A}$ Anthroposphere
$\mathbf{A}$ Macro-state of the Anthroposphere $\mathcal{A}$
$\mathfrak{B}$ Bundle of Paths
$\mathcal{B}$ Global Brain
$\mathbf{C}$ Coevolution Space
C Class of paths ($C[\mathbf{S}]$)
$\mathfrak{C}(\mathbf{P}_1 \curvearrowright \mathbf{P}_2)$ Geo-cybernetic Corridor from $\mathbf{P}_1$ to $\mathbf{P}_2$
$\mathfrak{D}, \mathfrak{K}, \mathfrak{N}, \mathfrak{r}, \mathfrak{S}, \mathfrak{f}$ Domains and Subdomains
$\mathcal{E}$ Overall Earth System
Γ Manifold (Separating Domains in $\mathbf{C}$)
$\mathcal{H}$ Human Factor
L Lagrange Function
$\mathfrak{M}, \mathfrak{M}^*, \mathfrak{M}^{**}$ Pool of Control Options
$\mathbf{M}, \tilde{\mathbf{M}}, \mathbf{M}^*, \ldots$ Management Sequences ($\mathbf{M} \in \mathfrak{M}$)
$\mathcal{M}$ Executive (Management) Component of Global Subject
$\mathcal{N}$ Ecosphere (Nature)
$\mathbf{N}$ Macro-state of the Ecosphere $\mathcal{N}$
$\mathcal{P}_i$ Paradigm
$\mathbf{P}(\ldots)$ Path in $\mathbf{C}$
$\mathbf{P}$ Point in Coevolution Space ($\mathbf{P} \in \mathbf{C}$)
Π Trajectory in $\mathbf{C}$
Q Scalar Quality Functional
$\mathcal{S}$ Global Subject
$\mathbf{S}$ Generalized Equilibrium
t, τ Time
$\mathfrak{U}$ Accessible Universe in $\mathbf{C}$
$\mathcal{V}$ Value System of Global Subject

1 Prologue

Imagination is indispensable to scientific achievement, even more so to the emerging field of "Earth System Analysis" (ESA). Imagine ... that astronomers were warning us of a huge asteroid heading towards our planet. The collision was supposed to occur in some twenty years from now, but neither the date of the impact nor its site could be predicted with satisfactory precision at this point in time. From the already available approximate knowledge of the celestial maverick's mass and orbital parameters it could be inferred, however, that the collision energy would correspond to an explosion of at least 10 gigatons of TNT.

Let us pause for a second to emphasize two points.

First, the probability of such a strange encounter of the x-th kind is *by no means negligible*: *Lewis* [134] estimates that the mean time interval between 10-gigaton events in the history of the Earth is 70,000 years. As a matter of fact, our planet is studded with impact scars and impact-induced phenomena, such as the formation of thousands of tonnes of diamonds discovered beneath the Bavarian town of Nördlingen (*Hough* et al., 1995 [106]). This town lies within the Ries Crater, which is 24 km across and was gouged out of the rock 14.7 million years ago by a meteorite of presumably 1 km in diameter. NASA takes the threat of comet and asteroid bombardment so seriously that it recently announced the launch of a research programme costing hundreds of millions of dollars to investigate humanity's chances of coping with such a threat.

Secondly, a 10-gigaton collision would release about half as much destructive energy as an all-out exchange involving the world's entire arsenal of nuclear weapons! Clearly, such an event has to be classified as a *global perturbation*, which would shake the foundations of life and civilization on this planet to the core. The famous Cretaceous extinction some 65 million years ago, which wiped out probably more than 60% of the then-existing species, was almost certainly caused by a gigantic asteroid hitting the Yucatan peninsula (*Budyko* and *Golitsyn*, 1988 [35]). The Alfred Wegener Institute for Polar and Marine Research is currently investigating geological evidence of another major impact (*Kyte* et al., 1988 [126]; *Gersonde* et al., 1997 [86]), which took place in the south-eastern Pacific 2.4 million years ago. This 12-gigaton collision generated spring tides of more than 200 metres in height and so much dust and water vapour in the atmosphere that the Earth sank into darkness for millennia ...

Returning to our *Gedankenexperiment* we ask: how would society react to such terrifying news? The shock might initially wreak havoc almost comparable to that of the announced impact itself – stock market crashes, political unrest, a surge of religious movements and sects, etc. But finally, humankind would recover its senses and ask the scientific community to work out a comprehensive "impact analysis" which was capable of answering persistent questions like the following ones:

- What does the collision probability distribution look like, i.e. the estimated likelihood of a given region being bombarded with a given amount of energy?
- What will be the most crucial consequences of the impact? In particular, which areas and systems (natural or civilizatory) are most vulnerable?
- What options for protection, adaptation or rehabilitation are available? In particular, are there any precautionary steps (of technical, economic or political character) that should be taken immediately?
- What options for mitigating or even preventing the collision are available? In particular, is it possible to set up an international rocket programme to deflect or disintegrate the bolide?

There would be no way of answering such questions through the means of reductionist studies conducted within the narrow bounds of disciplinary eruditeness, local technocratic wisdom or the self-serving interpretations of pressure groups. The asteroid impact would trigger an entire cascade of effects that would not be stopped by any geographical, sectoral, cultural, political or social frontier. If Siberia was struck, then the stratospheric dust generated in this event would also deprive the tropical rainforests of their sunlight; if Japan's heartland around Tokyo were pulverized, then the world's economy would be heading straight for disaster.

This means that the required impact analysis would be nothing less than a genuine *Earth System Analysis* – the scientific investigation of how the infinitely interrelated complex of ecosphere and anthroposphere responds as a whole to major perturbations, and how these reactions might be favourably influenced by wise global management. With respect to its aims and scope, the so-defined research programme would be unprecedented in history. Even the notorious Manhattan Project (*Stoff* et al., 1991 [224]), which also set out to save the world (not from cosmic assault but from fascist apocalypse) and ended in opening Pandora's box, could not compare with it in any way.

The asteroid-related Earth System Analysis clearly ought to be

(i) inter-disciplinary,
(ii) inter-national, and
(iii) inter-objective.

The last qualification is of paramount importance and refers to the intricate tangle of partially conflicting needs, intentions and interests of the actors involved (ranging from individuals to multi-national coalitions). Let us give a few examples:

One plausible strategy for "de-sensitizing" the global civilizatory system with respect to bolide impact might be the world-wide dispersion of settlements, infrastructures, and industrial and agricultural production zones – as opposed to the present unabated trends of urbanization, concentration and specialization. If feasible at all, such a sweeping concept would generate a brutal geographic pattern of medium-term winners and losers and, in particular, open up new gulfs between the affluent countries of the North and the nations of the South, which crave for development.

Or consider the problem of burden sharing in creating a rocket system capable of disintegrating the asteroid just before it enters the Earth's atmosphere. Is there any "equitable" contribution scheme that takes into account all the complex disparities between the regions, cultures and social strata of this world and, nevertheless, has a chance of meeting with unanimous approval?

Finally, when it comes to weighing the option "adaptation" against the option "mitigation", are there any criteria powerful enough to warrant the right decision – a decision that is crucial to the subsistence of human civilization? Probably some bold macro-economists would be ready to determine the optimal portfolio of measures by using, e.g., a one-dimensional global damage measure as a function of the size of the largest impacting fragment. However, who will have the political mandate to fix the conversion factors needed for monetizing the damage, i.e., for collapsing all affected items of the planetary inventory – human lives, species, landscapes, power-plants, traffic systems, historic monuments, golf courses, rice paddies, etc. – onto the common dollar axis? So can we find "first ethical principles" from which these factors might be derived in a straightforward manner?

There is another element adding to the uniqueness of our fictional analysis, namely *uncertainty* about the concrete challenge to be met in the not-too-distant future. Since not even the most powerful super-computers will be able to locate in advance the precise spots where the bolide or its scattered debris will hit the surface of planet Earth, the global analysts will resort to *scenarios* such as:

- the "Pacific Crash": sea-quakes and tsunamis of unparalleled ferocity devastate the coastal zones of all riparian nations,

- the "China Syndrome": a direct hit on the Shanghai region wipes out millions of people immediately and destroys the life-support systems of the most populated country in the world, or
- the "California Split": impact-induced tectonic processes in the vicinity of the San Andreas fault separate off vital parts of the "Golden State", triggering an avalanche of world-wide economic and cultural repercussions (such as the elimination of Hollywood's fantasy machine).

Hundreds of scenarios of comparable likelihood will have to be considered, and each of them requires a tedious integrated *if-then* assessment on the continental and planetary scale. The so-defined scientific challenge is of agonizing scope, yet it might be mastered for two reasons:

First, most of these assessments might be constructed from a universal module system, whose building blocks are the answers to fundamental questions about the operation of the ecosphere-anthroposphere complex. Typical questions might concern, for instance, the stability of climate patterns and ocean currents, the vulnerability of ecosystems and coastal zones, the sensitive dependence of the global food production system on specific regions, the robustness of infrastructures with respect to natural hazards, the effectiveness of international institutions in managing trans-boundary problems, and so on.

Second, the advent of sophisticated parallel computer hard- and software (*Akl*, 1997 [2]) in combination with recent progress made in scientific modelling of complex systems might allow the establishment of *virtual impact laboratories*. Renewable artificial earth systems could be exposed there to various simulated crash scenarios in order to study the potential consequences. As a matter of fact, these cyberspace experiments should be the most powerful tool for generating entire ensembles of assessments within a reasonable stretch of time.

* * *

Keeping in mind that any sort of Earth System Analysis will have to rely heavily on the two techniques just described, let us now turn back to the real world at the threshold of the third millennium of the Christian era. Although there is no immediate threat by cosmic tramps, the future of the planet looks almost as black as in our collision fantasy: *Global Change* is all about us – a breathless frenzy of expansion of western industrial patterns, lifestyles and technological knowledge to transmute the very character of the Earth System. The crisis thus induced between nature and humanity expresses itself in a number of fully-developed or emerging phenomena such as:

- the depletion of stratospheric ozone as opposed to the regional enrichment of tropospheric ozone,

- the diffusive spread of physical, chemical and biological pollutants (e.g. mercury) all over the planet,
- the accelerated degradation of humanity's soil resources,
- the global reduction of biodiversity and ecological functions as a consequence of anthropogenic transformations of forests, range- and wetlands,
- the ruining of pristine as well as cultivated landscapes by urban sprawl and rural exodus, and above all,
- the civilizatory modification of the global climate that will trigger an avalanche of environmental impacts.

All these trends are facets of a non-fictional collision, namely the self-generated clash between the biogeophysical Earth System and its own evolutionary offspring – a thermodynamic singularity sometimes called *Homo sapiens sapiens.* The mega-syndrome "Global Change", which will be described in some more detail below, is the very real threat we will have to cope with! This threat, of course, differs in various fundamental respects from the asteroid scenario. We name just a few of them.

Firstly, Global Change is not a predicament inflicted upon humanity by obscure divine or cosmic forces – it is the planetary emanation of trillions of individually expedient decisions of billions of deceased or living actors at all levels of social organization. Therefore, in principle, Global Change may be retarded, modified, or steered according to certain paradigms if the collective will for doing so can be organized. Formally, this boils down to a (very hard) problem of *multiple self-referential control.*

Secondly, Global Change is not hitting humanity in the form of a sudden shock – it is rather a highly interrelated bundle of gradual co-developments. Only occasionally may certain elements of this dynamic pattern cross critical thresholds or points of no return, thereby unleashing abrupt and disastrous impacts on natural or civilizatory systems. It is well known that such insidious processes are most difficult to handle.

Thirdly, Global Change does not have to be as detrimental as an asteroid impact – taken all in all, the long-term benefits for humankind may very well dominate. Even anthropogenic global warming might be turned into positive effects if the right adaptation strategies (e.g. optimal food production and distribution) were chosen and implemented. But note that our civilization is not even passably adapted to the present climate!

Earth System Analysis is generally one of the "Grand Challenges" to scientific endeavour in the decades to come. Its main task will be to provide tools for *managing Global Change* in order to secure an *acceptable long-term coevolution of nature and civilization.* In other (more fashionable) words, Earth System Analysis should yield a good deal of the information and methods required for defining and materializing "*Sustainable Development*".

However, isn't this ambition – or even the very notion – of top-down control of the planetary coevolution hubris of the worst kind? Isn't it precisely this type of thinking, still coveting the intellectual heritage of the Age of Enlightenment, that paved the way to the deep crisis we are at present facing? The answer is Yes – and yet there is no real alternative to judicious management of the global commons! Humankind is modifying the Earth System anyway at a breathtaking pace - the crucial question is whether this transformation should continue in the familiar, effectively planless way or whether we should at least strive to influence the course of the overall process according to common trans-national objectives. (In fact, the first and most important test regarding the feasibility of Earth System Analysis and control will be the materialization of the AGENDA 21 of the Earth Summit in Rio (*UNCED*, 1992 [238]).

W. *Clark* [42], who is one of the early visionaries regarding global environmental management, has introduced a somewhat naive yet powerful allegory: planet Earth as a vast garden cultivated by humankind. Without doubt this garden cannot be shaped and maintained "à la française", i.e., according to detailed plans based on rigorous scientific principles like Euclidean geometry. The global garden has to be managed rather like an English park, which is allowed to develop "naturally" at the micro- and meso-levels within the boundary conditions of a carefully designed macro-structure.

We have to accept that the "environment" or even "the wilderness" are not just human conceptions, but to some extent actually social constructions, at least at the local and regional scale. This intricate issue is thoroughly discussed in a recent book entitled "Uncommon Ground. Towards Reinventing Nature" (*Cronon*, 1995 [48]). Now the time has finally come to extend the gardening to the planetary scale – if only to counteract anthropogenic global despoliation that, ironically, results in part from the measures taken to protect limited-area environments (e.g., constructing high smoke-stacks or trading toxic waste).

But what about the daunting complexity of the system to be managed, what about its chaotic, i.e., unpredictable behaviour as a consequence of ubiquitous non-linear processes and interactions? Is it, therefore, not highly probable that all efforts to control the ecosphere at large will only exacerbate the present crisis between nature and civilization?

A serious effort to respond to this sweeping argument against any type of global environmental management, put forward in particular by conservationist organizations, will be undertaken in one of the later sections. Let us make here just two preliminary comments. First, people have been successfully managing entire landscapes (like Tuscany or Punjab) for millennia, although these regional systems behave in many physical, biological or cultural respects in no less intricate a way than the planetary one. Second, the ignoristic argumentation may very well be turned around: all the predictions about environmental catastrophes lurking around the corner or the statements about the dynamic malevolence of the ecosphere have to be exposed to the uncertainty considerations introduced above as well. Thus *positive* surprises like the existence of as yet unknown self-stabilizing mechanisms within the ecosphere, which may greatly facilitate Earth System control, cannot be ruled out. As a matter of fact, Lovelock's geophysiological theory (better known under the somewhat misleading label "GAIA Hypothesis") is reasoning precisely along these lines (*Lovelock*, 1991 [140]).

* * *

The rest of my essay is organized as follows: in Sect. 2 the present anthropogenic environmental crisis is briefly reviewed with special emphasis on its novel *character* due to globality. In the following section this crisis is conceived as a *cybernetic task* for the emerging "Global Subject". Plausible *guiding principles* for the so-defined control problem are formulated in Sect. 4. In Sect. 5 *simulation modelling* is featured as the indispensable prerequisite for global environmental management. Sect. 6 is devoted to the question of how control schemes for complex systems like the ecosphere can be efficiently *implemented* in spite of major cognitive deficits. A short Epilogue will conclude our *tour d'horizon*.

As indicated by its title, my contribution primarily attempts to provide a very rough outline of what Earth System Analysis is all about: what are the main topics, the – few – appropriate methods, the pertinent questions and the possible answers.

My specific hope is that the precursor of a consistent and comprehensive programmatics for this intellectual enterprise might transpire from the discourse. This programmatics should help to enrich the slowly materializing research agenda for the science of planet Earth (see, for instance, *NASA*, 1988 [161]; *Pickering* and *Owen*, 1994 [182]; *Moore* et al., 1996 [155]; *Munn* et al., 1996 [158]; *Williamson* and *Liss*, 1996 [249]; *Turco*, 1996 [233]) and to guide the further development of the latter toward a science of Global Sustainability.

2 Global Change: Quantity Turns into Quality

Environmental crises are ancient concomitants of the history of humankind. As a rule, crises of this type were caused by "natural" spatiotemporal fluctuations in the dynamic equilibrium of the ecosphere, especially by astrophysical disturbances, tectonic events, climatic excursions or biological disasters such as new types of pest plagues, for example. These fluctuations had a direct effect on the quality, quantity and distribution of the natural media that were indispensable for civilization, and they therefore indirectly acted as history- and culture-forming factors. The scientific investigations on the historical significance of weather and climate antics are especially varied (see e.g. *Brown*, 1992 [34]; *Hole*, 1994a [103]; *Hole*, 1994b [104]) – there is, in fact, hardly any bygone, advanced civilization the downfall of which had not been or could not be connected in some way with these factors.

A series of environmental crises were, however, not due to the whims of sovereign nature, but instead to the goal-oriented, yet in the long run disastrous activities of human societies themselves. There is also a great deal of speculation on this, though rarely of substance. Did, for example, the American Indian tribal nation of Anasazi really ruin, with lasting effect, its cultural landscape surronding Mesa Verde (Colorado) through overuse of the available resources in the form of biomass, soils and fresh water supplies? In contrast, other historical examples are scientifically assured, for instance the deforestation of the states of the former Roman Empire, which was driven by ship building and which spread outwards from the coasts, with its spectacular ecological consequences. Or the great environmental crisis in Western Europe after the middle ages (especially in Germany, France and Great Britain), where the forests disappeared for the most part by the mid-18th century as a result of charcoal production for the steel and iron industry and where the soils were exhausted by the traditional three-field farming (see e.g. *Montanari*, 1995 [154]; *Bork*, 1988 [28]). Remarkably, the innovative thrust of the Industrial Revolution delivered the way out of the crisis; among other things, this revolution gave rise to the systematic use of coal as a substitute for wood and the large-scale deployment of artificial nitrogen fertilizer in agriculture.

What distinguishes these cases of historical environmental destruction from the present-day situation that is complained about with force and verbosity in the official documents of the *UNCED* (1992) [238]? Many factors could be named here, but we want to concentrate on the most elementary of all these differences, namely the "order of magnitude" of the effect on nature: earlier crises were, in the final analysis, always local (i.e. non-planetary) phenomena that were not capable of triggering a drastic and lasting change in the *basic character of the Earth System.* As far as the *time dimension* is concerned, the subsequent "renaturalization" of anthropogenically degraded areas through exchange movements with the "outside world" that has remained intact has frequently been possible up to now. This process is sketched in Fig. 1.

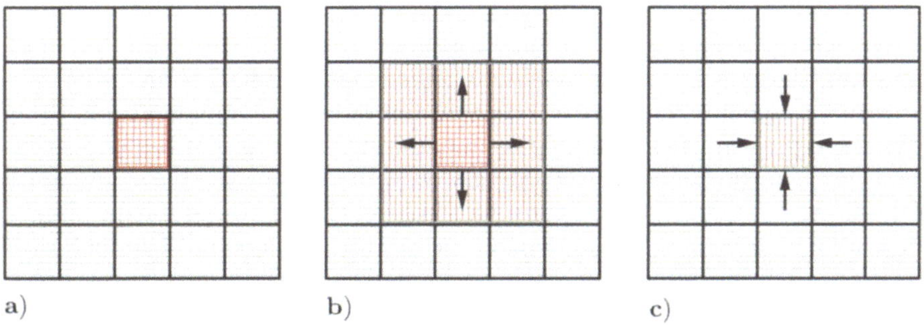

Figure 1. Successive stages in the "Healing Process" (in German: "Renaturierung") of a degraded environmental cell. **(a)** Local degradation. **(b)** Outward migration of actors, polluting substances, etc. **(c)** Inward migration of ecosystems, unpolluted media, etc.

The scheme described is similar to the natural healing of a minor wound in the tissue of the human body, in the process of which a permanent scar may remain. The time constants for the "healing" of heavily-damaged ecotopes are very different and depend on the respective geographic conditions. In most of the historical cases, however, the "fallow time" enforced by nature was significantly longer than the planning horizon of the degraded cultural communities, so a direct resettlement of the renaturalized areas did not take place.

On the other hand, there is also a series of examples of anthropogenic environmental destruction that has led in the past to nearly irreversible damage, locally or regionally. This applies above all to instances of intervention by civilizations that brought about the complete loss of the fertile soil layers and, with that, consequential effects such as karstification and desertification (*WBGU*, 1995 [82]; *Blaikie* and *Brookfield*, 1987 [26]; *Bork*, 1988 [28]). If the spatial order of magnitude of historical events of this type is put into proportion with the magnitude of the overall Earth System, however, it is then clear that these limited environmental crises cannot interfere in any crucial way with the qualitative functions of the ecosphere – such as large-scale atmospheric and oceanic circulation, the global water cycle, planetary biogeochemical cycles, macro-dynamics of the biomes, etc. In particular, the average values of the key planetary parameters (for instance the CO_2 content of the atmosphere or the pH value of the oceans) remain almost unchanged by anthropogenic disturbances of this type: the Earth System behaves vis-à-vis that intervention like a thermodynamic reservoir with quasi-infinite inertia (see e.g. *Callen*, 1985 [37]).

If we want to stay within our "geo-physiological" picture introduced above, then limited, irreversible, historical environmental degradations are comparable to the loss of non-vital body parts (for instance, a fingertip or a tuft of hair) – in other words with damage that can be functionally compensated for nearly completely.

We will try to summarize in this discourse, over and over again, the most important findings in an *extremely simplified* mathematical form. Just as a caricature frequently has a greater recognition value than a photograph, these symbolic representations can help to maintain the overview of the important connections in the analysis. Accordingly, we will describe the basic dynamics of the Earth System in general through a system of only two coupled ordinary differential equations that can depend on various parameters! The historical situation discussed above can then be formalized with regard to its global aspects as follows:

$$\begin{aligned} \dot{\mathbf{N}}(t) \equiv \frac{d\mathbf{N}}{dt}(t) &= F_0(\mathbf{N};t), \\ \dot{\mathbf{A}}(t) \equiv \frac{d\mathbf{A}}{dt}(t) &= G_0(\mathbf{N},\mathbf{A}). \end{aligned} \tag{1}$$

Before we interpret the actual message of this system of equations, we have to first explain the symbols used. t naturally stands for the *time variable*; a suitable origin $t = 0$ has to be chosen. $\mathbf{N}$ designates the *macro-state* of the *ecosphere* $\mathcal{N}$, thus the total natural environment of significance for humanity consisting of atmosphere, hydrosphere, lithosphere and biosphere. Along the lines of *Lovelock* [140], $\mathbf{N}$ is therefore simply to be identified with the state of the planetary "super-organism" GAIA.

$\mathbf{A}$ designates the *macro-state* of the *anthroposphere* $\mathcal{A}$, thus all of humanity along with all of the products of its civilization such as settlements, infrastructures and factories.

Note: Within the framework of this classification, cultivated ecosystems such as a wheat field, for example, are part of $\mathcal{N}$. A constituent of $\mathcal{A}$, for instance a stone house, arises through a specific processing of components of $\mathcal{N}$; this part will usually disintegrate again in the course of the centuries into mineral components and thereby return to the domain of the ecosphere. In other words, no *separate* conservation principles apply for $\mathcal{N}$ and $\mathcal{A}$ – this fact, recognized by Anaximander of Miletus 2600 years ago (see e.g. *Schmitz*, 1988 [209]), cannot be better expressed than through the biblical words referring to human beings themselves: "you have come from the earth, you will become earth once again ... ".

Eq. 1 describes the temporal development of the Earth System; as with the simplest dynamic systems, the first time derivatives of the individual macro-variables can be essentially calculated as functions of the momentary state of all variables. The corresponding mathematical operations are designated here as F_0 and G_0.

It must be emphasized at this point that our system of equations does not represent an outrageous and absurd simplification of the real dynamics, but instead an *extremely condensed description* of the same: **N** and **A** do not necessarily have to be equated with well-defined, real variables; those symbols may very well represent multi-dimensional, geographically explicit fields or even probability distributions. In a similar way, deterministic or probabilistic tensor operators can be concealed behind the symbols d/dt, F_0 and G_0 without further ado. We always have trust in Occam's Razor in the following, however, and will keep our symbols as simple as permitted for the transport of the content.

Let us now come to the actual message for the historical environmental situation as coded in Eq. 1: The temporal evolution of human civilization first depends on its actual, current state, but also directly on the respective macro-state of the ecosphere (which may pass through a glacial as opposed to a warm period, for instance) with all of its local consequences – the historical human is at the mercy of GAIA to a great extent. By way of contrast, the anthroposphere has *no* influence on the dynamics of the ecosphere on the planetary or continental scale. The latter dynamics first depend on GAIA's current macro-state, but also on external disturbances such as the perpetual modulations of the Earth-Sun configuration (*Milankovitch*, 1930 [152]), or the passing of cosmic dust fields through our planetary system (*Muller* and *MacDonald*, 1997 [157]). On the grounds of these "externalities", F_0 is an explicit function of time and (1) is not an autonomous dynamic system. Note that the dependence of the anthroposphere on extraterrestrial influences is mediated by the ecosphere, because only the excursions of $\mathcal{N}$ are directly perceptible for people. The main message of Eq. 1 is, though: The global development of GAIA is *decoupled* from the development of civilization in the pre-industrial age; the historical human had no influence on the overall operational process of the Earth System.

If one wants to illustrate this situation with a simple physical system, then the parametric double pendulum outlined in Fig. 2, with a heavily asymmetric mass ratio, comes under consideration. The behaviour of such a system is best characterized by its "phase space portrait", which reflects the dynamics of the canonically-conjugated variables as defined by Hamiltonian theory

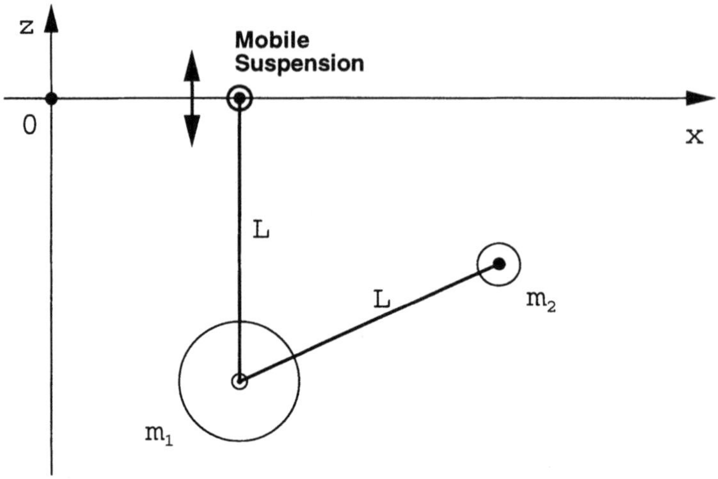

Figure 2. Planar double pendulum, moving under the combined influence of the gravitational forces $\mathbf{F}_i = -m_i g\hat{z}, i = 1, 2$, and a temporal modulation of the pivotal suspension. g denotes the constant acceleration due to the Earth's gravitational field, the two rigid links have identical length L. We initially assume that the first mass is clearly dominating the dynamics, i.e., $m_1/m_2 \gg 1$.

(see e.g. *Lieberman* and *Lichtenberg*, 1983 [135], *Abraham* and *Marsden*, 1977 [1]). Under the assumption that the parametric drive at the suspension point (symbolizing the variation of the external influence on the Earth System), as well as the mass m_2 (symbolizing the influence of civilization on the ecosphere), is practically negligible, the following picture results for the corresponding quantities q_1 ("position") and p_1 ("momentum") of the first pendulum (Fig. 3):

* * *

Let us consider, in contrast, the present-day relationship between nature and civilization: The totality of anthropogenic perturbations of the ecosphere can no longer be compared in any way with the historical influence as regards *intensity* and *geographic extent.* The corresponding current state of the global environment is reviewed in detail in thousands of articles and monographs, although occasionally in a contradictory way; we merely make reference in connection with this to the annual reports of the World Resources Institute (*WRI*, 1986 [255]), the Worldwatch Institute (*WI*, 1984 [257]) and the German Advisory Council on Global Change (*WBGU*, 1993 [81]), and to the book series of the Scientific Committee on Problems of the Environment (*SCOPE* [214]). Humanity has now long since outgrown the role of a free rider in the Earth System and has become a *global actor.*

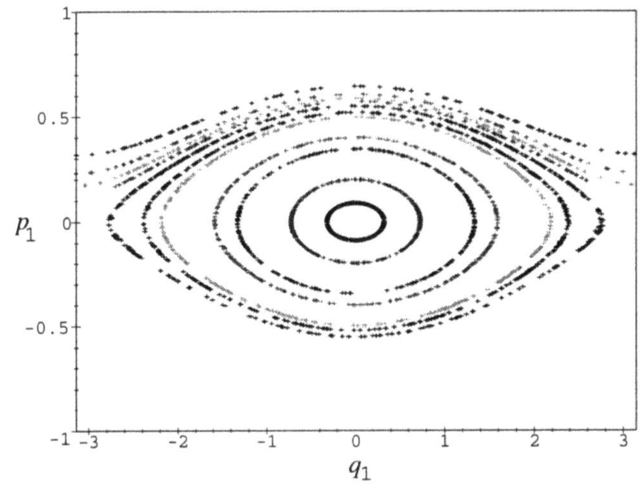

Figure 3. Phase space portrait for the motion of the (almost) unperturbed dominating mass in the double pendulum system depicted in Fig. 2. Each contour corresponds to a given (practically conserved) initial energy of the system component considered. The analyticity of the graph indicates the perfect regularity of the dynamic behaviour observed.

As evidence for this assertion, a series of rather spectacular figures can be given that have meanwhile become a part of the environmental folklore. We will quote a few of these, without dealing more closely with the relevant sources and their reliability.

- Since the start of the Industrial Revolution, the atmospheric CO_2 content has increased by more than 25% because of human activities.
- In the course of the last Spring seasons in the southern hemisphere, episodes of practically complete disappearance of the stratospheric ozone layer over the Antarctic came about.
- The so-called arctic haze, a specific photochemical smog, reigns over a range of up to 9% of the Earth's surface.
- Approx. 8% of the global freshwater flow is used for purposes of civilization.
- Approx. 11% of the ice-free areas of the continents are farmland.
- Since World War II, close to 30% of the fertile soils have been lost through mostly anthropogenic erosion.
- Movements of minerals due to civilization already exceed today, in volume, the global mass movements due to natural forces.
- Humanity now manipulates more than 50% of the net primary terrestrial production caused by green plants.

- The surface of the tropical forests is reduced by 1% annually because of forestry and agriculture.
- Human activities lead, intentionally or indirectly, to the extinction of about 50 species per day.

A limitation or an end to this "list of successes" cannot be foreseen for the time being, since the driving factors are more likely to still be gaining in momentum: for example, it can be assumed that the world's population will approximately *double* by the year 2050 vis-à-vis today's status. The world's economy could even increase its extent *five-fold* by then, if the average growth rate since 1930 of 3% per year is maintained. This would – ceteris paribus – mean an increase of the use of fossil fuels by around 20% per decade world-wide. Such a development almost has the character of a force of nature, especially if one considers that only one fifth of the world's population lives today in relative affluence (guaranteed satisfaction of the basic needs with regard to nutrition, drinking water, accommodation, health, education and mobility). Only a gigantic global campaign on the scale of AGENDA 21 (*UNCED*, 1992 [238]) would be able to perceptibly change this development; realistically, one should instead try to reduce its environmental impact as effectively as possible – but more of that later.

In summary, humanity has intervened *on the scale of the system* in the operation of planet Earth, i.e., the modern anthropogenic modifications no longer vanish – other than the historical ones – in the global limit. The changes in the ecosphere due to civilization can be roughly classified according to:

(i) *Content* (physico-chemical fabric of the atmosphere, bio-chemical state of soils and bodies of water, etc.);
(ii) *Form* (texture of landscapes, topology or connectivity of ecosystems, etc.);
(iii) *Composition* (make up of species, balance of terrestrial biomes, etc.);
(iv) *Artificiality* (distribution and state of minerals, etc.).

Categories *(i)* – *(iv)* naturally do not describe any strictly disjunctive quantities, but instead multiply overlapping problem areas that coalesce to bring about *"Global Change"*.

Our classification simultaneously supplies, in a certain way, a ranking with regard to the degree of severity of the human intervention. All actions of substitution or sealing up of natural areas with settlements and/or infrastructure are, as an example, to be subsumed under the most serious Category *(iv)*. The relative share of such areas of "artificial nature" is rapidly increasing world-wide with shanty-town building, urban sprawl and landscape fragmentation. The progressive conversion of the rain forest ecosystem into savannas, steppes or farmland in practically all tropical zones, for instance, falls into

Category *(iii)*. The planet is thereby robbed of an "organ" that is important, among other things, for the hydrological cycle; as a consequence, the "geo-physiology" significantly changes. Category *(ii)* contains, in contrast, seemingly milder forms of environmental modifications, for instance measures for the regional development of rural areas (in Germany, this has become notorious under the name "Flurbereinigung", i.e., land consolidation). Interestingly, the anthropomorphism of landscapes can be directly determined through the fractal dimension derived from aerial photographs (*May* and *Sugihara*, 1990 [146]). Finally, the intervention of civilization frequently viewed as being harmless is part of Category *(i)*. This intervention in general only leads to slight shifts in the average values of certain environmental parameters, e.g. of the average lead content of coastal sediments.

In fact, modifications of all four categories can contribute to significantly transforming the dynamic character of the Earth System – its *"operating mode"*. We want to discuss this briefly, using examples for the less obvious Categories *(i)* and *(ii)*. The *first Category* includes especially changes in large-scale key parameters such as the global share of CO_2 in the atmosphere, which only represents a marginal admixture with regard to volume. The 3-atom CO_2 molecule has quite specific quantum-mechanical resonance characteristics in the infrared range, however, which make it a potent heat reflector. The progressive anthropogenic increase of the CO_2 content is therefore leading to a continuous revision of the radiation budget of planet Earth and thereby – with a high probability – to a qualitative modification of the complex atmospheric circulation pattern, which in the end represents a sun-driven, dissipative state far from thermodynamic equilibrium (on this, see especially *IPCC*, 1996 [107]).

A different geo-physiological control parameter is the salinity of the North Atlantic. This variable has a decisive significance for the formation of heavy deep water to the south of Greenland, which is mainly responsible for driving the "Conveyor Belt" of world-wide thermohaline oceanic circulation (*Broecker*, 1995 [32]). The saline content, in turn, can in principle be influenced by anthropogenic changes of the entry of fresh water into the North-Atlantic basin (e.g. through direct modification of the continental outlets or through climate-caused modification of the precipitation conditions). A series of intensive computer experiments (see, among others, *Rahmstorf*, 1995 [188]; *Manabe* and *Stouffer*, 1993 [143]) supply clear indications that even relatively slight manipulations of the freshwater valve could lead to new types of patterns of thermohaline circulation. In particular, it is not difficult to bring about a complete shutdown of the Conveyor Belt in the simulation – an event that has otherwise been realized many times in the earlier history of Earth and that has always had natural causes up to now.

A related computer experiment may also be interesting, in which the hydrological cycle is *not* changed, but the salt in the world's oceans is completely removed. The Conveyor Belt does not then come to a standstill, remarkably;

it reorganizes itself, though, in an Atlantic and a – practically decoupled – Pacific vortex (*Ganopolski* and *Schellnhuber*, 1997 [80]).

The *second Category* of globally-effective environmental changes due to civilization includes above all the sum of the topographic rearrangements of the Earth's surface. This process has, for instance in Central Europe, now transformed nearly all of the original natural biogeotopes – a fact that can be easily verified by inspection during an appropriate business flight. A cultural landscape with intermixtures of small areas of forest, meadows, wetlands and farmland fulfills other ecological functions, however, than a macro-structured area – even if the relative total weights of the various landscape components are identical. This problem has been researched very little and is even less understood (*Prentice* et al., 1993 [187]; *Baker*, 1995 [13]; *Mosier* et al., 1997 [156]). Nevertheless, it can be assumed that the texture of the "planetary garden" has a significant influence on the hydrological, climatological and above all biological characteristics of the overall system: geometry and interlinking of the habitats presumably determine the abundance pattern of the regional species and thereby also the ecological elasticity or resilience of the respective domains. Within the framework of an extended "Daisy-World Model" (*Watson* and *Lovelock*, 1983 [245]) , the connection between landscape fragmentation, biological diversity and self-regulation can be explicitly studied (*von Bloh* et al., 1997 [242]).

How the strength of key parameters as well as geometry control the dynamic mode of a dissipative system can be visually demonstrated through the legendary Bénard convection experiment (see Fig. 4). This system now represents one of the icons of modern non-linear dynamics, because it serves as a model case for a series of convective phenomena in nature (see e.g. *Ebeling* et al., 1990 [62]; *Lanius*, 1995 [128]), on the one hand, and has formed the starting point for the Lorenz equations and thereby for the discovery of the first "strange attractor", on the other hand (see, e.g., Chapt. 4).

In the extensive literature on this physical phenomenon it has been laid out in detail how the convection pattern ("software") is determined by the system parameters ("hardware") such as temperature gradient and preparation of the experimental arrangement. The overall spectrum of patterns and dynamic states that can be generated here by proper tuning is extraordinarily rich (see e.g. *Bergé*, 1990 [20]).

The Bénard experiment does not just show how sensitively the motion of a complex system reacts to small, but macroscopic, variations of the boundary conditions. It further conveys the insight that sharply-defined "critical" values can exist for the key parameters of such a system; the character of the dynamic pattern – thus the operating mode – abruptly changes when they are exceeded: *quantity then suddenly changes into quality!* For example, when the fundamental system quantities are known, the temperature gradients at

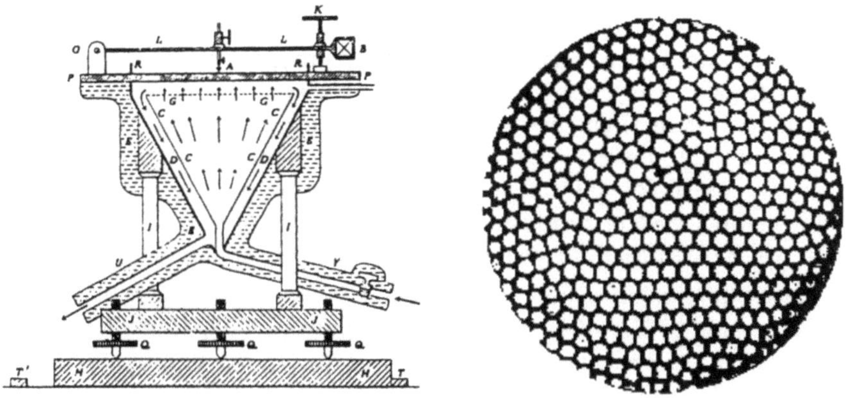

a) b)

Figure 4. The famous Bénard experiment demonstrating pattern formation in a dissipative dynamic system. **(a)** Typical experimental set-up. A layer of viscous fluid confined by two planar slabs is heated uniformly from below. As long as the resulting temperature gradient across the layer is sufficiently small, the heat is transported exclusively by diffusion processes at the molecular level. **(b)** If the temperature gradient transgresses a critical threshold, then a hexagonal convective pattern emerges as a self-organized macroscopic model of heat transport (pictures taken from *Wittig*, 1983 [252]).

which the cylindrical convectional motion sets in and when it is finally replaced by a truly complicated flow pattern can be exactly calculated. The emergence of structural transformations or phase transitions as a result of the continuous variation of certain system parameters is generally described by the so-called catastrophe theory (*Saunders*, 1980 [204]).

One of the most burning questions of modern environmental research is therefore whether there is a possibility that *critical thresholds* likewise exist in the much more complex Earth System for certain global conditions that can be modified by nature, God, humanity or whatever. The knowledge of such threshold values would be all the more important if so-called *"points of no return"* were involved – exceeding them would induce practically irreversible transformations of the ecosphere's character. The scientific discussion on this question has unfortunately been speculative to a great extent up to now, because the available information regarding data and processes are just as inadequate for a satisfactory answer as today's computer resources for running simulation models. We have, without doubt, here run up against one of the main research fronts in future Earth System Analysis.

Various environmentally relevant threshold values have already been postulated or discussed, especially in connection with the estimation of the potential consequences of anthropogenic climate change: we just mention the presumed minimum temperature of 27 °C for the top ocean layer as a necessary condition for the formation of tropical hurricanes, or the temperature value of approx. 15 °C as a switching point between the methane-generating and the almost methane-free reaction path during the microbial decomposition of organic material in soils. These two examples belong to the class of process thresholds determined by the laws of nature, however, not to the really relevant *system thresholds.* A genuine criticality analysis for planet Earth will look instead for answers to questions of the following type:

- Above which value of the atmospheric CO_2-concentration will the ENSO phenomenon (*Philander*, 1989 [181]; *Latif* et al., 1994 [129]) become permanent (or, conversely, shut down)? (*Meehl* et al., 1993 [150]; *Knutson* and *Manabe*, 1994 [122])?
- At which level of anthropogenic biotope disintegration in the temperate latitudes do the annual, transcontinental migrations of certain animal species collapse?
- What minimum extent (diameter) does the tropical rain forest need to subsist as an ecosystem in its remaining refuges, and what consequences would its disappearance have on the global water cycle?
- What percentage of the Earth's surface may be sealed up, without destroying the homeostatic (self-stabilizing) functions of GAIA (Lovelock, 1991 [140])?

These questions have been asked quite naively here and will presumably not find sensible answers in this form. More sophisticated versions of questions of this type will play a dominant role in future global environmental research, though!

If we formally summarize the present-day dynamic relationship between nature and civilization on the planetary scale similarly to the way this has been done in the historical case (see above), we obtain the following system of equations:

$$\begin{aligned} \dot{\mathbf{N}}(t) &= F_1(\mathbf{N}, \mathbf{A}; t), \\ \dot{\mathbf{A}}(t) &= G_1(\mathbf{N}, \mathbf{A}). \end{aligned} \tag{2}$$

All symbols have the same meaning as in (1); only the index of the functions F and G was replaced, in order to indicate the new type of situation. The structural difference to Eq. 1 is the *dynamic quasi-symmetry between the global quantities* **N** *and* **A**: the anthroposphere has now become an active component also at the system scale and therefore represents a significant driving force for coevolution.

In the mechanical example of the parametric double pendulum, this would mean that the mass m_2 of the second pendulum is no longer negligible vis-à-vis the mass m_1 of the first one, thus $m_1 > m_2$ or even $m_1 \approx m_2$. If we draw up again a phase space portrait of the non-linear model system under this modified prerequisite as a counterpart to Fig. 3, a qualitatively clearly different picture results (Fig. 5).

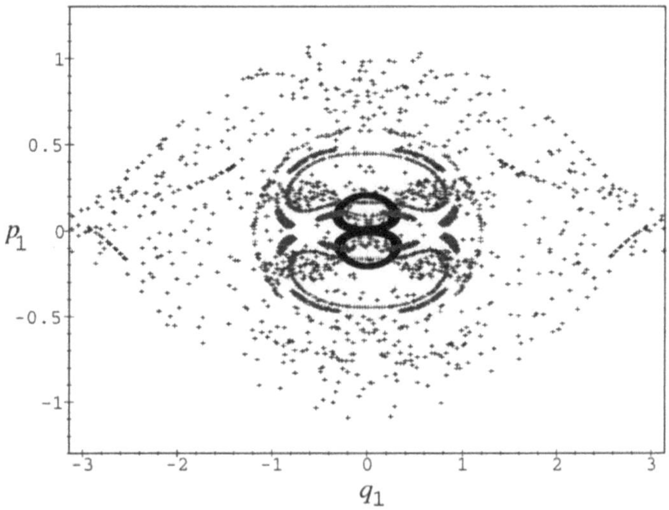

Figure 5. Phase space portrait for the motion of the first mass in the double pendulum system of Fig. 2 for the mass ratio $m_1/m_2 = 1$. Note that the qualitative character of the graph has changed completely in comparison with Fig. 3: the non-analytic texture now heralds chaotic dynamics.

Because of the fact that the second pendulum itself now has an influence on the overall movement, the dynamic character of the system is transformed: with the loss of the analytic nature of the phase space texture, the door is opened for deterministic chaos (*Arnold* and *Avez*, 1968 [8]).

In a similar way, the dynamics of the Earth System have become significantly more complicated since humanity emerged as a global system component – even if comparisons with models that show signs of chaos appear to have no substance at first. We will return again to this point later. This section will be concluded, however, with a few thoughts on the geographically explicit realization of Global Change as characterized above, and on the implications of this spatial phenomenology.

The modern anthropogenic redesign of our environment achieves its global effectiveness through two types of spreading mechanisms that are basically different, namely through *short-range diffusion processes* and *long-range induction processes.* All forms of direct, physical or biogeochemical transport of ecologically relevant substances or entities from sources associated with civilization are part of the *first type:* gases that enhance the natural greenhouse effect or that affect the ozone abundance, aerosols and dusts, heavy metals and organic pollutants, radioactive substances, artificial genetic material, etc. These items spread out in a front-like way through strictly local processes that achieve most diverse expansion velocities and horizons. As an example, particles infiltrating the stratosphere are transported more quickly, by several orders of magnitude, than toxic herbicides in groundwater.

The possibility of diffusive dilution of local emissions of pollutants formed the basis in the 1960s and 1970s for a regionally effective, but globally disastrous environmental-policy principle that can be rewritten as the "strategy of the high smokestacks". (For this, see also the current discussion of the "High-Smokestacks Syndrome", *WBGU*, 1997a [84]). The finite nature of the Earth System necessarily implies that the substances "thrown away" remain within the aura of the anthroposphere and, in fact, often even return to the direct emitters through complicated paths within the food webs. The attempt at the "global homogenization" of the environmental load by civilization may even be substantially more damaging for humanity as a whole than a regional "confinement" at the expense of the party causing it.

The reason for that is twofold. On the one hand, the "homeopathic dilution" of environmental modifications reduces their individual and social *perceptibility* enormously, and the illusion of an unspoiled nature is prolonged, particularly in the geographic centres of the disturbance. Until direct perception of the continually-developing global changes finally comes about, several decades will be lost in general for an effective, international mitigation policy with clear-cut regional specifications.

On the other hand, it is precisely the uniform distribution of originally concentrated pollution that can significantly increase its negative *overall impact.* As an example, the homogenization of landscape fragmentation due to urban sprawl, thoroughly damaging the diversity and vitality of the self-organized balance of species, could be named. Thus the "migration resistance" of these landscapes is significantly increased, or to put it the other way round, the biological mobility as a prerequisite for reactive adaptation movements is dramatically lowered.

We want to demonstrate these elementary effects with a simple electrotechnical example and, in the process, also throw some light on the popular, but infrequently explained statement "*The whole is more than the sum of its parts*". Our object for demonstration is a circuit that essentially consists of N parallel branches (Fig. 6 a). Each of these parallel, conductive segments is supposed to have an internal resistance of the magnitude ε/N. According to Kirchhoff's rules, then the total resistance R_0 amounts to

$$R_0 = \frac{\varepsilon}{N^2}, \tag{3}$$

and, with a given applied voltage U, the following current flows:

$$I_0 = \frac{U}{R_0} = U\frac{N^2}{\varepsilon}. \tag{4}$$

Now an additional resistance of the fixed overall magnitude ρ is to be introduced to the circuit; the portioning and distribution of the "perturbative" entity still remains to be determined. Resistor arrangements that allow for strong (or even maximum) currents are especially sought after. We consider a perfectly homogeneous distribution of ρ on the parallel branches (Fig. 6 b) and an extremely non-homogeneous distribution loading down a single branch (Fig. 6 c) as extreme cases.
In the *homogeneous case*, we find that the total resistance R_h is given by

$$R_h = \frac{\varepsilon + \rho}{N^2}, \tag{5}$$

so the electric current amounts to

$$I_h = \frac{UN^2}{\varepsilon + \rho}. \tag{6}$$

In the *inhomogeneous case*, in contrast, the following applies:

$$R_i = \left[\frac{N(N-1)}{\varepsilon} + \frac{N}{\varepsilon + N\rho}\right]^{-1}, \tag{7}$$

$$I_i = \frac{UN(N-1)}{\varepsilon} + \frac{UN}{\varepsilon + N\rho}. \tag{8}$$

Thus for the ratio between I_i and I_h, the inequality

$$\begin{aligned}\frac{I_i}{I_h} &\geq \left[\frac{UN(N-1)}{\varepsilon}\right] \Big/ \left[\frac{UN^2}{\rho}\right] \\ &= \left(\frac{N-1}{N}\right)\frac{\rho}{\varepsilon} \approx \frac{\rho}{\varepsilon} \quad \text{for large} \quad N\end{aligned} \tag{9}$$

holds. So, under the prerequisite $\rho \gg \varepsilon$ already used for this estimation, we observe that

$$I_i \gg I_h. \tag{10}$$

Therefore, if one wants to keep the current as high as possible, it makes sense to distribute the additional *resistance as non-homogeneously as possible!*
Let us now make two further assumptions:

- The reduction of current by the arbitrarily-distributed incremental resistance may only be realized *locally*, i.e., through observations at the parallel branches of the network. The *perception threshold* is reached when the current in the individual branch drops below the value $\frac{1}{10}\left(\frac{UN}{\varepsilon}\right)$.

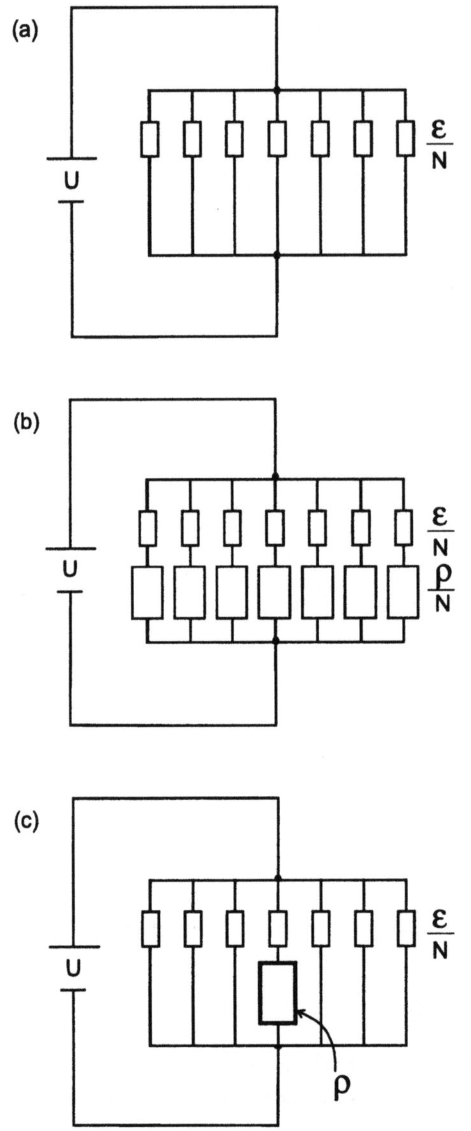

Figure 6. Variants of an electric network for demonstrating the dependence of overall system performance on distributional topology. **(a)** Basic configuration. **(b)** Homogeneous allocation of total incremental resistance ρ to all parallel links. **(c)** Inhomogeneous distribution as a result of allocating ρ to one single link.

– The circuit serves to drive another system, the function of which is no longer fulfilled if the *total current* falls short of the *critical value* of $\frac{1}{2}\left(\frac{UN^2}{\varepsilon}\right)$.

Select $\rho = 4\varepsilon$, for instance. In the *homogeneous* case, we then have

$$I_h^{local} = \frac{1}{5}\left(\frac{UN}{\varepsilon}\right), \tag{11}$$

$$I_h = \frac{1}{5}\left(\frac{UN^2}{\varepsilon}\right), \tag{12}$$

i.e. the flow disturbance caused by the additional resistance is not noticed locally, although the total current is already sub-critical!

In the *inhomogeneous* case, things are precisely the other way around. For the individual branch loaded with the additional resistance, we obtain, under the prerequisite $N > 2$,

$$I_i^{local} = \frac{1}{4N+1}\left(\frac{UN}{\varepsilon}\right) < \frac{1}{10}\left(\frac{UN}{\varepsilon}\right) , \tag{13}$$

while the total current amounts to

$$\begin{aligned} I_i &= \left(N - 1 + \frac{1}{4N+1}\right)\left(\frac{UN}{\varepsilon}\right) \\ &\geq (N-1)\left(\frac{UN}{\varepsilon}\right) > \frac{1}{2}\left(\frac{UN^2}{\varepsilon}\right) . \end{aligned} \tag{14}$$

Thus the perturbation can already be verified locally, although the overall system is still super-critical!

Conclusion:
The homogeneous distribution of the load not only leads to poorer system performance in general, but moreover veils the crossing of critical thresholds.

* * *

Let us return to the general implementation mechanisms of Global Change. By no means all of the elements of ecosphere change are caused by the diffusion processes described above: note, for example, that the world-wide phenomenon of accelerated soil degradation (see, *WBGU*, 1995 [82], and the references therein) is hardly progressing through the physical expansion of fronts. Environmental changes of this type – and also those driven by diffusion in a certain way – are more likely the result of profound "Global Change of the Anthroposphere", which is essentially mediated through *long-range effects*. This "action at a distance" is particularly realized by the tele-communication (in the broadest sense) of information, technologies, economic incentives, lifestyles, political convictions and value systems, as well as by the physical tele-transport of persons and materials within the framework of the world-wide flow pattern of goods, services and traffic. Because of this it is possible, for example, that booming East Asian regions will move into direct economic competition with traditional industrial territories of Europe, that the campesinos of the drought-stricken Brazilian Northeast move to the coastal shantytowns in droves, allured by the elusive prospects of an upper-class life as conveyed by the ubiquitous "telenovelas", or that in South Jordan fossil groundwater under the blazing desert is squandered for a bizarre non-sustainable plantation culture, which supplies table grapes to London in the winter via air freight. As a whole, the various long-range effects induce innumerable local instances of intervention in the natural budget, which bring about *global patch structures* of environmental change or of environmental degradation in the end.

If one wants to symbolize this form of ecosphere transformation through a statistical phase transistion, the phenomenon of *nucleation* comes into consideration, whereas the diffusion-driven environmental changes can more likely be categorized as *infection* processes. Fig. 7 outlines the different phenomenologies of these two basic mechanisms of stochastic state substitution.

a) 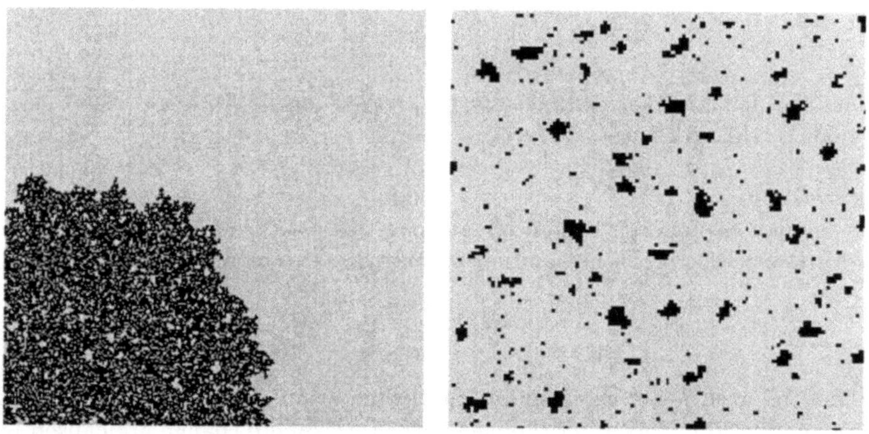 b)

Figure 7. Fundamentally different spatial patterns resulting from distinct ways of performing a phase transition. **(a)** In infection processes, the two phases are separated by connected invasion fronts. **(b)** In nucleation processes, one of the phases appears as a patchy structure embedded within the other phase.

The nucleation of a new type of phase can only take place, though, if the conditions for existence of this phase have already been fulfilled in a sufficiently large domain. Only sporadic impulses from outside are then required

– e.g., the introduction of condensation nuclei for the artificial generation of rain – in order to put the polycentric transformation into operation. In the case of Global Change, the preconditions for the world-wide transition to the modern technical-commercial society and for the accompanying ubiquitous rearrangement of the ecosphere have developed over the long term, in part over centuries – but the catalyst of tele-interaction has now converted this potential into an extensive conflagration in a short period of time!

This process cannot, by the way, be studied better in any region than in the marginalized Northwest of Spain, where the ancient cultural landscapes of Cantabria, Asturia and Galicia have been virtually pulverized since the end of the Franco regime and the entry of the nation state into the European Union. The remains of traditional land-use areas still jut out like sinking islands from an ocean of rapidly-growing, profoundly incoherent commerce and settlement structures.

The insight into the topological nature of Global Change does not just have academic value, it supplies hints and indicators, among other things, on how this change could be controlled or how its most crass effects could be reduced. Whereas diffusive spreading processes are typical *surface phenomena* that can be limited or screened off by local or regional measures, in the case of the main mechanisms of Global Change (as described above) *bulk phenomena* are involved. The latter may only be significantly influenced by the well-specified preparation of the global – economic, socio-cultural, political – conditions. Particularly the instruments of tele-interaction come under consideration, which helped to orchestrate the modern crisis in the Earth System. The elaboration of this thought represents one of the main aspects of the following section.

3 Global Environmental Management: The Physics and the Metaphysics

Isn't it strange that both the state of affairs and the cause of the stratospheric ozone loss were discovered "just in time" to be able to (presumably) ward off a really threatening development for the majority of the Earth's population? Imagine that the chemists of the Industrial Revolution were already in a position two centuries ago to manufacture substances similar to CFCs on a large scale and to subsequently dump them into the atmosphere through various types of uses. How would one have tried to explain at that time, however, the unavoidable impacts of this act – destruction of the ozone layer, even over the temperate latitudes, significant increase in the UV-B radiation, vast amounts of physical, genetic and economic damage to the societies that have been exposed? Similar to the case of the greatest natural catastrophe of Western civilization up to now, the plague of 1347 - 50, the scholars presumably would have offered an entire spectrum of obscure causes – from a mysterious "miasma" to the apocalyptic punishment by God ...

Today, humanity is actually changing its habitat on a global scale, as described above – not just the reflection properties of the stratosphere. However, to the same degree that the redesign of the planetary environment progresses, science and technology are generating better and better preconditions for channeling this development along acceptable paths. In particular, modern civilization is in a position to contemplate its own activities of unleashing the forces of "nature" against itself and to comprehend the mechanisms and consequences of these processes – to a great extent. Because of this, the global actor gains a new type of identity and becomes a "*Global Subject*", who is starting to define its (her/his) self-conscious role in the Earth System. These fundamental observations are to be briefly discussed in the following.

The prerequisite that is perhaps the most important for the emerging self-consciousness of the Global Subject is its rapidly-growing ability *to conceive the planetary system in its entirety.* According to the great philosopher (and natural scientist) I. Kant, nature is just what a human mind composes from consensual affections according to its own rules (which, in turn, arise from the "transcendental synthesis of apperception"; see, for instance, *Kant*, 1991 [114]). Thus the design of that nature is necessarily shaped by the characteristics of our perception apparatus. This apparatus defines, like an optical lens system, the construction laws for the image of the outside world to emerge from innumerable particular impressions.

The Kantian approach has its merits, even though the realist will meet it with a great deal of scepticism. As an example, note that the phenomenon of "climate" is without any doubt an ("apperceptive") construct from the observations of the atmospheric states actually prevailing in space and time. We do not want to discuss here whether a "social construction" could even be involved in the case of climate (*Stehr* and *v. Storch*, 1997 [221]).

However, the senses of an individual person are in no way sufficient to perceive, for example, the status quo and development of the *world* climate, or generally any global changes in the environment. Who has ever observed "the ozone hole" or even watched the "extinction of species"? Nevertheless, terms of this type have now become standard elements of everyday conversation, and children portray these phenomena with crayons, just as they drew flowers and butterflies in those good old days. For humanity as a Global Subject perceives the Earth System with the senses of the world-wide "scientific-medial complex". Part of the main modules of this complex are all types of monitoring equipment (satellites, aeroplanes, stratospheric balloons, deep-sea vehicles, drilling devices, weather stations, etc.), the planetary hierarchy of computers and data storage, the explosively growing electronic networks, such as the World Wide Web, all public and commercial means of communication (television, radio, newspapers, magazines, books, etc.), and naturally the agents of this complex who may occasionally appear like the abbots, monks or lay brothers and sisters of a bizarre order – hundreds of thousands of co-operating members of the scientific system who attempt to fit together, interpret and translate for the public the results of the global fact-generating machine to the best of their knowledge and (usually) belief, as well as crowds of media representatives who further popularize the products of the scientists, convert them into "news" and frequently degenerate them into canapés for infotainment. The so-produced flood of correct or distorted information on the state of the globe invades the most obscure corner of human civilization; it regenerates in the daily rhythm and it seems to expand and accelerate incessantly during this process.

There is already a wealth of practical applications for those new types of artificial sensory organs of humanity that penetrate deeply into the area of everyday affairs: just think of the optimization of fertilizer use in agriculture through precision orientation with the aid of the Pentagon's "Global Positioning System", or the control of environmental criminality via remote sensing.

Fully immersed in the electronic ether of the scientific-medial complex, the individuum may further develop into "l'homme symbiotique" (*de Rosnay*, 1995 [55]), into part of a planetary super-organism that will never attain the horrific maturity of the fictional Solaris ocean (*Lem*, 1991 [132]), though. The emergence of this form of collective consciousness is accompanied by a series of negative phenomena, however, in particular "*global hallucinations*". Some of the notorious highlights of ecological folklore belong to this class of collective "disturbances in consciousness", for instance the unshakable expectation of the biological "collapse" of certain oceanic basins (like the North Sea), the invocation of the "imminent climatic catastrophe" or the almost hysterical fear of anthropogenic ground-level ozone, the damaging effect of which cannot compete with that of cigarette consumption in any way. By way of contrast, truly terrifying developments such as the soil degradation in the agricultural centres of the Earth or the over-exploitation of the most productive fishing grounds world-wide are only gradually moving into the consciousness of the Global Subject.

The contemplation of the Earth System in its totality is best done from a real or an artificial *distance*, which lets the planet shrink to an object of manageable dimensions for the purpose of investigation. Today's geographic information systems (GIS) which allow all essential system aspects to be illuminated via thematic maps are most instrumental in this shrinking process. As an example, we sketch in Fig. 8 the texture of land use and land degradation that humanity has impressed on the terrestrial biosphere since the last Ice Age.

Fig. 9, in contrast, conveys a real (although GIS-processed) vista of a segment of the Earth's surface: the interweaving and mutual penetration of ecosphere $\mathcal{N}$ and anthroposphere $\mathcal{A}$ can be clearly recognized here.

In fact, though, not only the regional dovetailing of the structures of $\mathcal{N}$ and $\mathcal{A}$ is visually evident. As already expressed by Eq. 2, *the metabolisms of nature and civilization are dynamically coupled today at the scale of the Earth System.* The anthroposphere maintains its "metabolism" by perpetually sucking up raw materials from the ecosphere, consuming them and expelling them again in a mostly devalued or distorted form (*Harrison* and *Jeffries*, 1976 [97]; *Baccini* and *Brunner* (1991) [9]). Because of this, the metabolism of the "host organism" $\mathcal{N}$ is significantly perturbed.

The main topic of Earth System Analysis is the medium-term to long-term *coevolution of* $\mathcal{N}$ *and* $\mathcal{A}$ on a global scale (see also, *Pitt* and *Samson*, 1997 [183]). This coupled development can be formally represented by a path in the physical *coevolution space* $\mathbf{C}$, which is spanned by the state variables $\mathbf{N} = (N_1, N_2 \ldots)$ and $\mathbf{A} = (A_1, A_2 \ldots)$ of $\mathcal{N}$ and $\mathcal{A}$, respectively (see Fig. 10).

Thus, the coevolution path $\mathbf{P}(t) \equiv (\mathbf{N}(t), \mathbf{A}(t))$ describes the future succession of evolutionary stages of the global $\mathcal{N} - \mathcal{A}$ tandem as initiated by the current state $\mathbf{P}_0 \equiv \mathbf{P}(0) \equiv (\mathbf{N}(0), \mathbf{A}(0))$. We denote by $\mathbf{P}_T(t)$ the *path segment* that unfolds until the time $T > 0$, thus

$$\mathbf{P}_T(t) \equiv \mathbf{P}(t) \mid_{[0,T]} . \tag{15}$$

We now come to a decisive point in our analysis: despite the extreme formal reductions that have been carried out, the representation of the dynamics

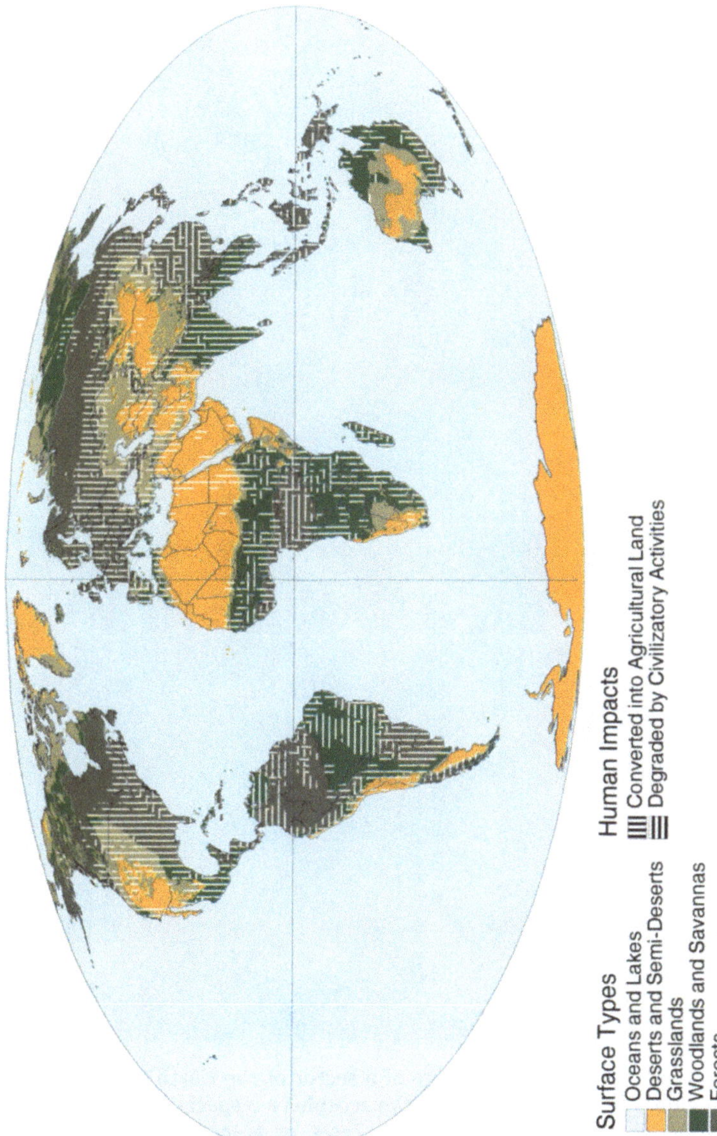

Figure 8. Anthropomorphicity of the Earth's terrestrial surface (M. Plöchl, M. Lüdeke and M. Cassel-Gintz, Potsdam Institute for Climate Impact Research, unpublished).

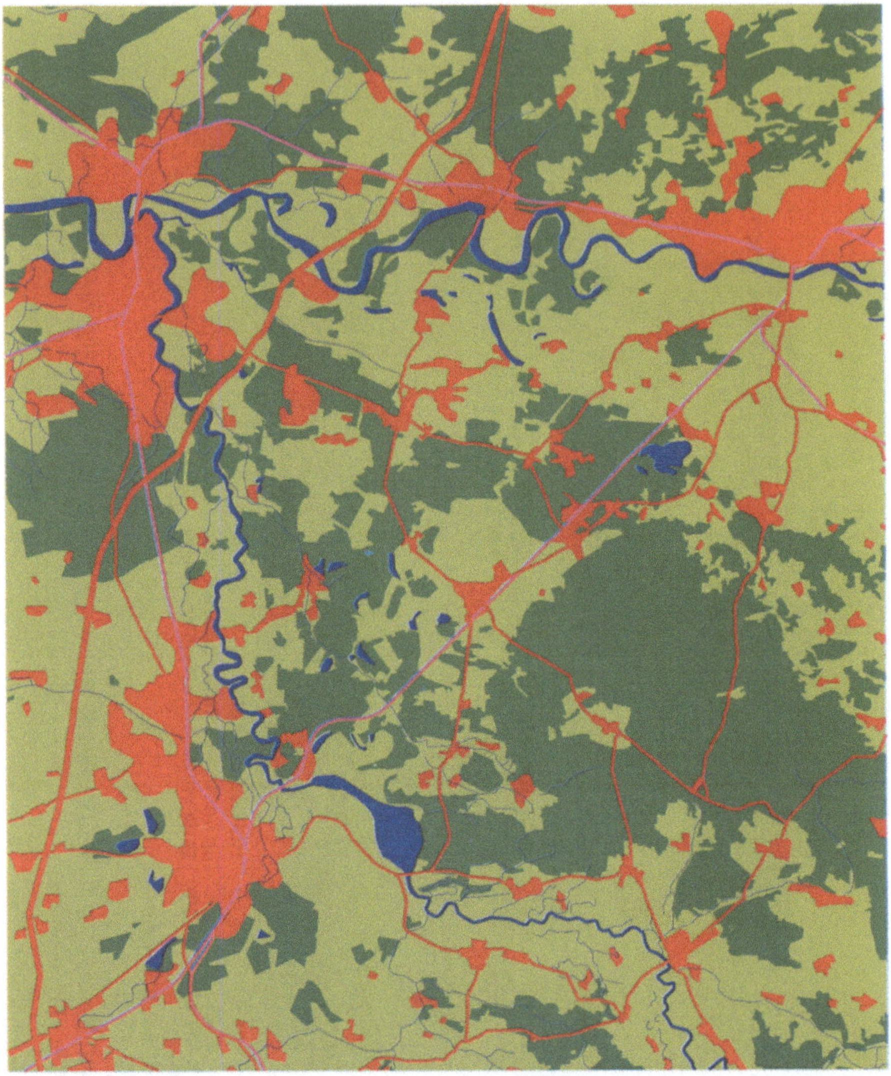

Figure 9. GIS-processed satellite image of a sector of the Earth's surface (Dessau-Wittenberg region). The components of the ecosphere (especially forests, meadows, fields and bodies of water) are coloured in *green* or *blue*; the components of the anthroposphere (especially settlements and traffic paths) in *red*.

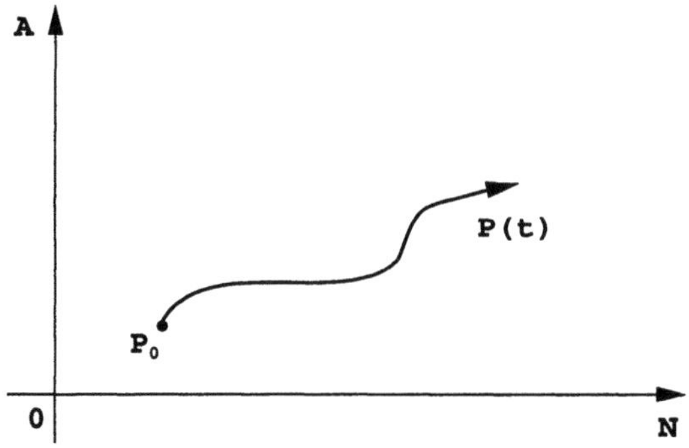

Figure 10. Two-dimensional caricature of the coevolution space **C**, the points of which represent possible states of the global human-environment system. For illustration, a segment of a hypothetical coevolution path $\mathbf{P}(t)$ is drawn in.

of the human-environment system up to now basically seems to be correct. But this representation suggests the conclusion – entirely in the spirit of a physics-oriented view of the world – that the future coevolution $\mathbf{P}(t), t > 0$, is *determined* (in principle at least) by the "initial condition" $\mathbf{P}_0$ and the "dynamic equation" (2).

This statement would only have to be qualified in so far as the time development described by Eq. 2 does not solely have to comply with the deterministic rules of the natural laws involved, but is also influenced by *genuinely non-deterministic* elements. We are not refering here to stochastic aspects of the dynamics, which result from information deficits regarding quantities and processes that are principally deterministic. What is rather meant is the *a-priori indefiniteness* of the future intentional potpourri of billions of humans, which induces the overall behaviour of the anthroposphere. On the other hand, it is conceivable that this ensemble of individual actors and volitions can be taken as a thermodynamic system, the average behaviour of which can be precisely described by means of suitable macro-variables. Macroeconomics, like no other scientific discipline, illustrates the glory and misery of such an approach, in particular with its mathematical pricing and welfare theory à la Walras and Pareto (*Walras*, 1874 [244]; *Pareto*, 1917 [171]; *Samuelson* and *Nordhaus*, 1995 [203]). Consequently, Eq. 2 could fully prove its validity as a time-development rule for global, thus macroscopic, quantities, although in the sense of a probabilistic description.

Under these conditions of "predestined coevolution", however, only the role of prospective contemplation and retrospective documentation would remain for science. The task of organizing the best possible local adaptation to an inexorable development – in a way that is similar to how organizers of open-air events react to the weather forecast – would fall to politics. Such a fatalistic attitude is naturally blatantly inconsistent with the mentality that has brought about the desire for global environment and development (E & D) policies and has thereby installed, among other things, the system of international environmental conventions (*WBGU*, 1996 [83]). The fundamental axiom of this gigantic initiative is the conviction that the coevolution of the $\mathcal{N}$ - $\mathcal{A}$ system does *not* unfold in a globally deterministic way, but can instead be "positively" influenced, may even be *controlled.* An autonomous willpower is required for this, though, which intervenes *from "outside"* in the physical human-environment metabolism according to a top-down strategy.

This intention for E & D design will be brought about by the very "Global Subject" whose basis of existence has been outlined at the start of this section. The Global Subject *transcends* the sum of the physical-individual desires and impulses of all elements of $\mathcal{A}$ as a result of a self-referential process. The collective target structure emerges through million-fold communication, perception and evaluation of personal value-systems as a synergistic control quantity (for physical metaphors like mean-field theory see e.g. *Negele*, 1982 [163]; *Haken*, 1983 [95]; *Baxter*, 1990 [16]; *Landau*, 1996 [127]). One element of this target structure might, for instance, be the intention of limiting anthropogenic warming of the Earth's atmosphere to a maximum of $2\,°C$ – a project that would profoundly shatter and revise the respective manoeuvring spaces for individual action regarding energy consumption, mobility, etc. in every respect.

This means, however, that in the Earth System – besides $\mathcal{N}$ and $\mathcal{A}$ – yet another entity exists, which manifests itself in a "*metaphysical dimension*", so to speak. Fig. 11 provides a synopsis of the overall situation in cartoon form.

Traditionally, the metaphysical space is occupied by "God" or a comparable, poorly-defined trans-human subject of volition, to whom a controlling or just a maliciously-interfering influence on $\mathcal{N}$ and $\mathcal{A}$ is ascribed. We shall not commit ourselves on the existence of such an entity and will restrict ourselves instead to the observation that humanity is today confronted with itself in the form of a seemingly exogenous force unleashing the powers of nature. Thus, if in the analysis one considers the Global Subject as an operator of international E & D strategies, then the composition of the Earth System can be – purely formally – represented as follows (the used letter type refers to *entities*, not state variables):

$$
\begin{array}{ccccc}
 & \mathcal{E} = (\mathcal{N} & , & \mathcal{H}) & \\
 & \downarrow & & \downarrow & \\
(a, b, c, \ldots) = & \mathcal{N} & & \mathcal{H} = (\mathcal{A}, & \mathcal{S}) \\
 & & & & \downarrow \\
 & & & & \mathcal{S} = (\mathcal{B}, \mathcal{V}, \mathcal{M}).
\end{array}
\qquad (16)
$$

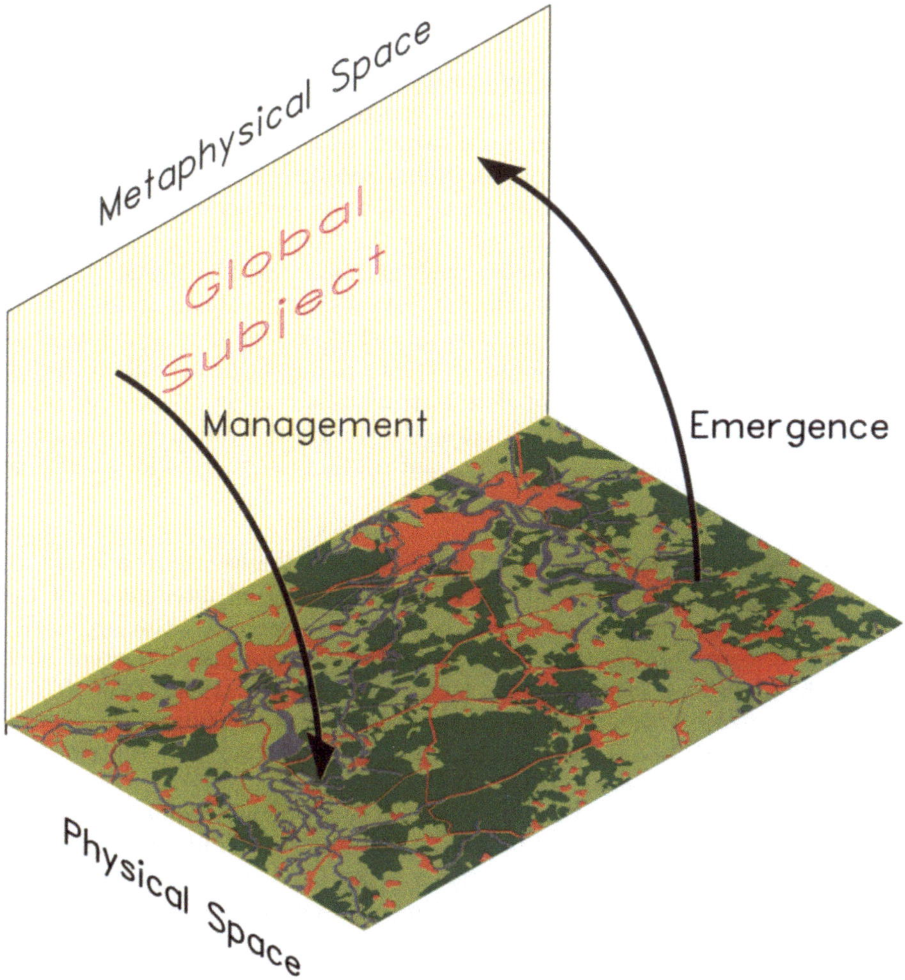

Figure 11. The physical and the metaphysical dimensions of the Earth System, which are "orthogonal" but mutually constitutive.

This "formula" is to be read like this: The overall Earth System $\mathcal{E}$ consists of two main components, namely the ecosphere $\mathcal{N}$ and the *human factor* $\mathcal{H}$. $\mathcal{N}$ can be quite conventionally "spelled" according to the sub-spheres $\mathfrak{a}$ ("atmosphere"), $\mathfrak{b}$ ("biosphere"), $\mathfrak{c}$ ("cryosphere"), etc. In contrast, the human factor consists of the "physical" sub-component $\mathcal{A}$ (the anthroposphere as the aggregate of all *individual* lives, actions and products) and the "metaphysical" sub-component $\mathcal{S}$ (the Global Subject as a *collective* E & D factor). The latter can, in turn, be decomposed into the triple ($\mathcal{B}$, $\mathcal{V}$, $\mathcal{M}$), where – borrowing from the structure of human beings – $\mathcal{B}$ stands for the "*brain*", $\mathcal{V}$ for the "*value system*" (or even more vaguely expressed, for the "soul") and $\mathcal{M}$ for the "*executive organs*" of the Global Subject.

Note that $\mathcal{S}$ has no *independent* material basis whatsoever in the sense of a "body": its physical elements, including the "organs", are all components of the anthroposphere. The specific combination of the modules of $\mathcal{A}$, which have developed as a result of evolution, establishes the dynamic identity of $\mathcal{S}$ as an immaterial construction, however, which is just as real or unreal as "the life" of an organism composed of billions of molecules or "the music" of a symphony orchestra with complete instrumental casting. We should warn the reader, not to take our little anatomy of the Earth System too literally – it has a primarily allegorical character, but it also proves useful for analytical purposes.

The collective brain $\mathcal{B}$ of the Global Subject is essentially a result of the trans-individual senses and the computer-supported intelligence of the scientific-medial complex described above. In this context, the big, international research programmes for the analysis of the Earth System such as WCRP, IGBP or IHDP play a crucial role (*WCRP*, 1996 [253]; *IGBP*, 1994 [108]; *HDP*, 1996 [101]), as do the activities of the Intergovernmental Panel on Climate Change (IPCC), which represents a gigantic "truth tribunal" of unprecedented character (*IPCC*, 1996 [107]).

The collective ethics $\mathcal{V}$ of $\mathcal{S}$ is created in an extremely complex and less transparent interaction process between groups and individuals, officials and non-governmental organizations, decision makers and those impacted by decisions ("decision takers"), where certain social classes and sectors, mainly of the industrialized nations, dominate on the grounds of their privileged access to the media. The world-wide supply of the masses with ecologically correct elements of attitude by governments and environmental-protection organizations is rapidly taking on professional characteristics in the process and reflects the formation of a generally-accepted, although contradictory canon of moral guiding principles for Global Change. The desiderata on the corresponding list extend from the "preservation of Creation" to "intergenerational equity", from the protection of the whales to setting fair prices for the plantation products of "indigenous cultures".

The collective executive organs $\mathcal{M}$ of the Global Subject (the symbolism is motivated by the Latin word "manus" for hand) are primarily supranational institutions for E & D management, which belong to two distinct categories within the taxonomic system of political science. There are, on the one hand, binding agreements between independent states, together with their reporting, verification and sanction mechanisms. At present, a great deal of hope is placed on international conventions and protocols for the protection of the global environment (climate stabilization, preservation of biological diversity, control of desertification, etc.) (*WBGU*, 1996 [83]), but a certain sobering-up with regard to the effectiveness of these instruments will probably set in before long. On the other hand, concrete institutions and agencies can be named that are either supposed to form the material backbone of the mentioned conventions (such as the office for the Framework Convention on Climate Change that moved to Bonn in 1996, for instance) or that are the appropriate organs of relevant state alliances (e.g. UN or OECD). The international spectrum is extraordinarily heterogeneous and extends from capital-flow regulators (World Bank, GEF) to cultural care agencies with extremely idealistic programmes (UNESCO). The Global Subject materializes through the activities of institutions of these types, but also through the development-aid initiatives of individual countries or the boycott campaigns of non-governmental organizations operating world-wide. As far as that is concerned, the inflatable rafts with which Greenpeace boarded the scrapped oil-drilling platform "Brent Spar" in 1995 are also elements of $\mathcal{M}$ – quite irrespective of the sense or nonsense of this spectacle.

On the basis of its self-consciousness, which is established and further developed in the interplay of the components $\mathcal{B}$, $\mathcal{V}$ and $\mathcal{M}$, the Global Subject $\mathcal{S}$ can semi-autonomously *select and implement global strategies of E & D management* from the available pool of measures that we want to describe through the formal set $\mathfrak{M}$ embracing the elements $\mathbf{M}(t)$, $t \geq 0$. Each $\mathbf{M}(t)$ represents a certain time sequence of management modules that can be activated. With such elements, direct intervention into the biogeophysical metabolism of $\mathcal{N}$ may be involved in those extreme cases that are generally propagated under the heading of "*geo-engineering*": specific injections of propane into the stratosphere for the neutralization of chlorine molecules that endanger the ozone layer, stimulation of the marine "carbon pump" by iron-fattening of the plankton carpets in the South Seas, reforestation of huge areas of savanna and steppe for sequestration of the anthropogenic CO_2 surplus, doping of aeroplane fuel with sulphur to generate a world-wide homogeneous aerosol haze for correction of the planetary radiation balance, massive increase in the continental storage of fresh water with the help of retaining dams to avoid an increase in sea level – this is only a small selection of mostly eccentric proposals for direct technical repair or control of the Earth System.

As a rule, however, $\mathfrak{M}$ stands for strategies that aim primarily at $\mathcal{A}$ and, because of this, indirectly – through avoidance, structuring or adaptation activities – have an effect on the ecosphere. Regulatory-law measures (e.g. the establishment of environmental standards of all types), economic instruments (e.g. tax incentives or the granting of certificates), political campaigns (e.g. public propaganda to improve voluntary family planning), educational and instructional programmes (e.g. advancement of resource-saving individual behaviour), internationally-organized search surveys for the minimization of industrial greenhouse-gas emissions (e.g. joint implementation), specific externally-financed E & D projects with a model character such as technology transfer or capacity building (especially in the developing countries), consumer initiatives (e.g. shopping recommendations to influence the tropical wood or trophy markets) or international agreements on the regulation of trans-boundary problems (e.g. immigration, water consumption or trade treaties) are among the most important management elements of this type. In particular, consistent long-term strategies for the implementation of nation-specific E & D conventions of the AGENDA-21 caliber can be composed out of those elements. For detailed information, we make reference to the extensive recent literature on this topic (see e.g. *Victor*, 1996 [241]; *Simonis*, 1996 [216]; *Sprinz* and *Luterbacher*, 1996 [220]; *Bergesen* and *Parmann*, 1996 [22]; *Efinger* and *Breitmeier*, 1992 [63]; *Breitmeier*, 1992 [31]). In addition, we should emphasize that the intervention options of the Global Subject consist less in massive interference with material processes (like industrial production or agriculture), than in the creation of "intellectual ethers" or collective public moods ("morphogenetic fields") that impress orientation or design instructions for "sustainable" self-organization on the anthroposphere.

Whatever the pool of instruments of $\mathcal{S}$ may look like, our fundamental tenet is that the Global Subject is basically in a position to *control* the coevolution process, and it has already started to do so – although in a groping and still clumsy way. That is why the equation of motion (2) for the "physical" state variables of the Earth System has to be replaced by the following dynamic rule:

$$\begin{aligned} \dot{\mathbf{N}} &= F_2(\mathbf{N}, \mathbf{A}; t; \mathbf{M}(t)) \ , \\ \dot{\mathbf{A}} &= G_2(\mathbf{N}, \mathbf{A}; \mathbf{M}(t)) \ , \\ &\text{with} \ \ \mathbf{M}(t) \in \mathfrak{M} \ . \end{aligned} \tag{17}$$

Eq. 17 describes a dynamics *controlled by arbitrary selection of the exogenous strategy* $\mathbf{M}(t)$, whereas Eq. 2 represents the "undisturbed flow" in coevolution space, as it would develop without intervention of the Global

Subject. As a matter of fact, the coevolution state can be forced through the activities of $\mathcal{S}$ onto a path deviating from the "natural development".

If we want to again make use of a physical example for illustration of these circumstances, then *the control of a boat in a prescribed flow field* seems to be a most useful allegory (Fig. 12).

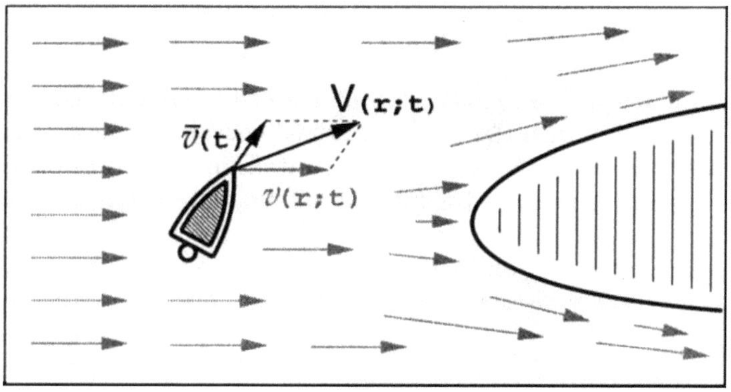

Figure 12. Motion of a propelled vessel on a river. $v(r;t)$ denotes the spatiotemporally varying velocity field of the river relative to a fixed riparian coordinate system. $\overline{v}(t)$ symbolizes the instantaneous boat velocity relative to the supporting fluid as generated by the vessel's driving mechanism, and $V(r;t)$ denotes the resulting absolute velocity of the boat with respect to the river bank.

The movement of the boat can be controlled through a motor, which exerts, for a given steering wheel (and rudder) position, the vector force $m(t)$ on the vehicle at time t. Under certain idealized assumptions – negligible boat mass, scalar effective coefficient of friction γ – the process can be described purely kinetically as defined by the Galilean transformation. The boat velocity in the fixed coordinate system as defined by the shoreline then turns out to be

$$\begin{aligned} V(r;t) &= v(r;t) + \overline{v}(t), \\ \text{with } & \overline{v}(t) = m(t)/\gamma. \end{aligned} \tag{18}$$

The relative velocity $\overline{v}(t)$ is therefore the instantaneous result of the control variable $m(t)$. Note that we have *not* generally assumed stationarity, i.e., $v \equiv v(r)$, for the natural velocity field $v(r;t)$ here: the fluid may, for instance, change its velocity pattern with the seasonal water level.

Despite its extreme simplicity, this mechanical system is indeed suitable to illustrate the essence of the formal dynamic equation (17) for the $\mathcal{N}$-$\mathcal{A}$ complex and to identify the fundamental features of suitable control strategies (see above all Sect. 5). The velocity field $v(r;t)$ represents in this allegory the uncontrolled coevolution as established by Eq. 2. Depending on whether this development is autonomous or not (according to the existence of time-modulated extraterrestrial perturbations), v can be specified as a stationary or as an unsteady flow. The driving force $m(t)$ corresponds to the management strategy of the Global Subject – in spite of the fact that the influence of $\mathbf{M}(t) \in \mathfrak{M}$ on the coevolution will, in general, hardly be representable as an additive quantity in the equation of motion. The resulting absolute velocity $V(r;t)$ finally symbolizes the real development of the coupled metabolisms of the ecosphere and the anthroposphere under the joint effect of micro- and macroscopic forces.

Because of the self-conscious intervention of the Global Subject, we are therefore not dealing with the *one* predestined coevolution path, but instead with the *path bundle* $\{\mathbf{P}(t \mid \mathbf{M}) \mid \mathbf{M} \in \mathfrak{M}\}$ of optional futures of the physical Earth System, which in principle can all emanate from the initial state (Fig. 13).

For the sake of clarity and comprehensiveness of the analyses conducted below, it is neccessary to specify and to complete our notation a bit.

Definition 1: Let $\hat{t} \geq 0$ be an arbitrary point in time, $\hat{\mathbf{P}} \in \mathbf{C}$ an arbitrary state in coevolution space, and $\mathbf{M} \in \mathfrak{M}$ a given management sequence. We then denote by $\mathbf{P}(\hat{\mathbf{P}}, \hat{t}; t \mid \mathbf{M})$ the *coevolution path* that develops from the state $\hat{\mathbf{P}}$ at the time $\hat{t}$ under the influence of $\mathbf{M}$ for $t > \hat{t}$.

For each $T > 0$, the corresponding *path segment* $\mathbf{P}(\hat{\mathbf{P}}, \hat{t}; t \mid \mathbf{M}) \mid_{[\hat{t}, \hat{t}+T]}$ will be symbolized by the expression $\mathbf{P}_T(\hat{\mathbf{P}}, \hat{t}; t \mid \mathbf{M})$.

We designate the accompanying *trajectories* in coevolution space with $\boldsymbol{\Pi}(\hat{\mathbf{P}}, \hat{t} \mid \mathbf{M})$ or $\boldsymbol{\Pi}_T(\hat{\mathbf{P}}, \hat{t} \mid \mathbf{M})$, i.e.,

$$\boldsymbol{\Pi}(\hat{\mathbf{P}}, \hat{t} \mid \mathbf{M}) := \{\mathbf{P}(\hat{\mathbf{P}}, \hat{t}; t \mid \mathbf{M}) \mid t \geq \hat{t}\} \ , \tag{19}$$

$$\begin{aligned} \boldsymbol{\Pi}_T(\hat{\mathbf{P}}, \hat{t} \mid \mathbf{M}) &:= \{\mathbf{P}_T(\hat{\mathbf{P}}, \hat{t}; t \mid \mathbf{M}) \mid t \geq \hat{t}\} \\ &\equiv \{\mathbf{P}(\hat{\mathbf{P}}, \hat{t}; t \mid \mathbf{M}) \mid t \in [\hat{t}, \hat{t} + T]\} \ . \end{aligned} \tag{20}$$

If $\hat{t} = 0$, we obviously have

$$\mathbf{P}(\hat{\mathbf{P}}, \hat{t}; \hat{t} \mid \mathbf{M}) = \hat{\mathbf{P}} = \mathbf{P}_0 \ , \tag{21}$$

in agreement with the original notation. ■

We should point out here that we have explicitly exogenized the state of the Global Subject via the decomposition of the Earth System into "physical" and "metaphysical" components according to Eq. 16. As a consequence, we have actually banished the selection act carried out by $\mathcal{S}$ among the available strategies $\mathbf{M}$ into the realm of *a-priori* incalculability, for the time being. This approach does not of necessity imply, however, that the pool $\mathfrak{M}$ of management sequences is an exogenous quantity that does not depend on the development of $\mathcal{N}$ and $\mathcal{A}$ in any way.

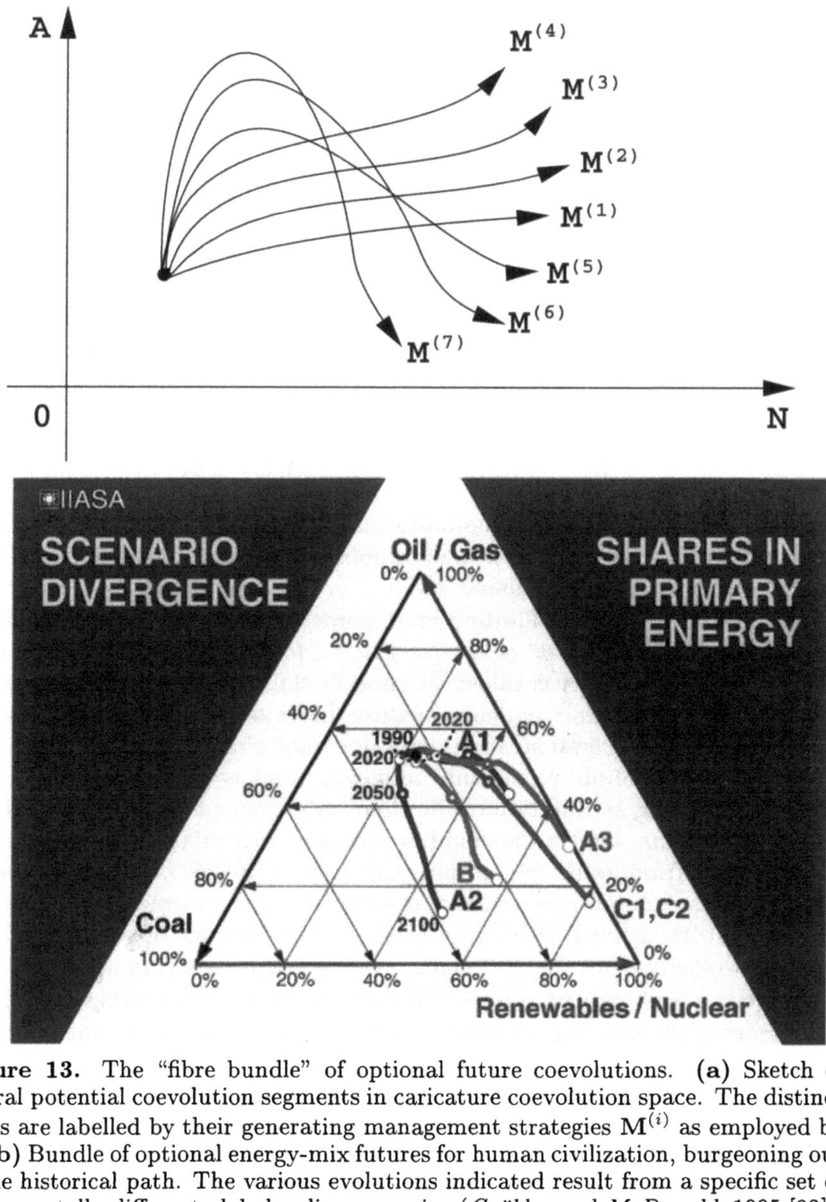

a)

b)

Figure 13. The "fibre bundle" of optional future coevolutions. **(a)** Sketch of several potential coevolution segments in caricature coevolution space. The distinct paths are labelled by their generating management strategies $\mathbf{M}^{(i)}$ as employed by $\mathcal{S}$. **(b)** Bundle of optional energy-mix futures for human civilization, burgeoning out of the historical path. The various evolutions indicated result from a specific set of fundamentally different global policy scenarios (*Grübler* and *McDonald*, 1995 [93]). (Courtesy of N. Nakićenović)

Vigorous world-wide economic growth, improved general conditions of international political cooperation or qualitative technological progress could, for instance, significantly expand the global control options. So we are not fundamentally ruling out the possibility of the "coevolution" of $\mathfrak{M}$ with the $\mathcal{N}$-$\mathcal{A}$ complex, but we will not discuss it here in more detail because of the associated mathematical complications. In fact, the dynamism of control is captured in Eq. 17 to a great extent through the observation that a given strategy $\mathbf{M}$ is coupled with the respective states $\mathbf{N}$ and $\mathbf{A}$ through the time-development functions F_2 and G_2, respectively. This accounts for the fact that one and the same management element – e.g. an energy tax of x percent over a time period of τ years - can have different effects in accordance with the factual state of coevolution.

In order to keep the analysis in the following sections reasonably simple, we even assume, as a rule, that the strategy pool $\mathfrak{M}$ is *invariant in time*. This precisely means that from every point in time $\hat{t} \geq 0$ on, the same set of optional management strategies is available:

$$\{\mathbf{M}(t - \hat{t})\ |_{t \geq \hat{t}} |\ \mathbf{M} \in \mathfrak{M}\} = \{\mathbf{M}(t)\} \equiv \mathfrak{M}. \tag{22}$$

Such an invariance of control possibilities is, for example, evident if measures from a finite list can be chosen and combined in discrete time steps. If we symbolize this list by an alphabet $\{\alpha, \beta, \dots, \omega\}$, then $\mathfrak{M}$ can be represented as the set of all possible, infinitely long words that can be formed from the given alphabet. The "word" $\alpha\beta\omega\alpha\beta\omega\alpha\beta\omega\dots$, for instance, then means that at time t_1 the measure α is taken, at time t_2 the measure β, at time t_3 the measure ω, and that this sequence of three basic units periodically repeats. It is clear that the overall strategy pool does not change at all if the first n letters are cut out of all words and the initial point is rescaled.

The simplifying assumption of the time invariance of $\mathfrak{M}$ is justified, above all, for two reasons. On the one hand, we remain with this prerequisite within the structural framework of conventional control theory (see the references in Sect. 5, for example), where the amount of control elements is generally considered to be a fixed, stipulated pool. In our allegory of steering a boat through a velocity field, the optional control activities are completely defined by the technical apparatus; a possibility of free choice only exists with regard to the *order* of the steering measures. On the other hand, management strategies are long-term in character, as a rule. Thus the selection and succession of measures take place at a certain point in time for a usually wide planning horizon, and it is generally not feasible to take the potential development of the control mechanisms themselves into consideration in this planning at this point. Yet mid-way correction or even perpetual readjustment of long-term strategies are by no means ruled out as a result of this, as we will explain in more detail in Sect. 6. A completely new breed of control analysis will then be required, however.

All of the difficulties associated with the potential variability of $\mathfrak{M}$ could possibly be remedied within the framework of a *"super-deterministic approach"*, which makes the Global Subject and, as a consequence, the selection and development of $\mathbf{M}(t)$ *endogenous*. An approach of this type would have to be captured in a formal system of equations with the following structure:

$$\begin{aligned} \dot{\mathbf{N}} &= F_3(\mathbf{N}, \mathbf{A}; t; \mathbf{M}), \\ \dot{\mathbf{A}} &= G_3(\mathbf{N}, \mathbf{A}; \mathbf{M}), \\ \dot{\mathbf{M}} &= H_3(\mathbf{N}, \mathbf{A}; \mathbf{M}), \end{aligned} \tag{23}$$

where H_3 is a suitable time-development function that correctly calculates the variation of the management strategy $\mathbf{M}$ of $\mathcal{S}$ from the extended current coevolution triplet ($\mathbf{N}$, $\mathbf{A}$, $\mathbf{M}$).

However, even if a description of this type appeared to be feasible with regard to content and methodology, one would not thereby escape the trap of *"infinite regression"*. On the basis of the conclusions resulting from the solution of the dynamic equation (23) by the agents of the Global Subject (read: scientists and politicians), $\mathcal{S}$ would presumably try to realize a different strategy $\tilde{\mathbf{M}}(t)$ than the one that is traced out for it – which leads to the problem of "self-avoiding prophecy". This self-referential process might possibly converge towards a strategy that is both stable and acceptable over the long term, but the epistemological complications arising here have at least the same quality as the difficulties obscuring the quantum-mechanical measurement process (see, e.g., *Schrödinger*, 1935 [210]; *Bohm*, 1951 [27]; *d'Espagnat*, 1979 [57]; *Mermin*, 1985 [151]; *Davies*, 1986 [54]; *Griffiths*, 1989 [92]; *Gribbin*, 1991 [91]; *Rosen*, 1996 [198]).

After these observations, we can formulate the main message of this section as follows: E & D policy is not primarily a forecasting problem, but a *control task on the scientific basis of Earth System Analysis.* We may denote this activity of $\mathcal{S}$ as *"geo-cybernetics"* – a process that only has very little in common with the simple technical repair exercises dreamed up by "geo-engineering". Geo-cybernetics is already being pursued on a large scale (via environmental conferences, shareholders' meetings, scientific symposia, media campaigns, grass-root movements and educational programmes), but it is high time to transform this bundle of activities, which has been rather erratic up to now, into systematic international management. It will be the main task of geo-cybernetics to create *now* the prerequisites for a coevolution that is acceptable over the long term within the leeway available in $\mathbf{C}$ in compliance with the natural laws. This requires above all that we *make decisions*, because we can only forecast to a limited extent what we are going to want!

Within the other mechanical allegory made use of several times above – the Earth System as a parametric double pendulum – geo-cybernetics corresponds to the deliberate attempt to control the inherent chaotic dynamics of the construct and, if possible, to put it on the desired track. We will deal with the general *feasibility* of such a project in some depth and detail in Sect. 6.

We now turn to the *key questions* of global E & D management, the answers to which depend heavily on the results of Earth System Analysis (*Clark*, 1989 [42]; *Blackburn*, 1991 [25]):

1. *What kind of world do we have?*
2. *What kind of world do we want?*
3. *What must we do to get there?*

These questions sound naive and presumptuous at the same time; the sheer act of formulating them provokes protest (especially from social scientists), which occasionally increases to hostility. The three key questions nevertheless outline in an adequate way the setting of tasks for the Global Subject, which does not differ *in structure* from the typical everyday management problems of an individual person. The actions of workaday routine are also taken on the basis of uncertain knowledge, and despite the fact that demands and realizability generally gape far apart from each other, nobody is willing to give up the option of attempting, at least, to influence the course of events for the better.

Key Questions 2 and 3 emphasize a fundamental aspect of geo-cybernetics that we have not yet discussed in detail up to now, namely the *targets issue*: Which guiding model or "paradigm" shows the "right" way and by which means can this path best be taken and kept to? Global E & D management is thereby defined as a problem of *optimal control* in the broadest sense.

It is therefore formally necessary to propose or impose deliberately a *paradigm* $\mathcal{P}$. In the simplest case, $\mathcal{P}$ defines a *scalar quality functional* $Q_{\mathcal{P}}[\mathbf{P}(\mathbf{M})]$ for the evaluation of all of the coevolution paths $\mathbf{P}(\mathbf{M}) \equiv \mathbf{P}(\mathbf{P}_0, 0; t \mid \mathbf{M})$ that can be generated by corresponding management sequences $\mathbf{M} \in \mathfrak{M}$. One may then search for *"best strategies"* $\mathbf{M}^*$, which satisfy the following inequality:

$$Q_{\mathcal{P}}[\mathbf{P}(\mathbf{M}^*)] \quad \geq \quad Q_{\mathcal{P}}[\mathbf{P}(\mathbf{M})] \quad \forall\, \mathbf{M} \in \mathfrak{M} \ . \tag{24}$$

The best strategy is unique under certain circumstances; in general, a larger sub-set of equivalent management sequences from $\mathfrak{M}$ will satisfy Eq. 24. Depending on the design of $\mathcal{P}$, though, it can also be the case that no one-dimensional quality measure for the ranking of possible coevolution paths can be constructed. Then the evaluation of $\mathbf{P}(\mathbf{M})$ – and therefore of the control entity $\mathbf{M}$ – will be achieved through more complex and possibly even qualitative or fuzzy rules (see below). For more than ten years now, a heated debate on the E & D paradigm issue has been dragging on among the "agents" of the

Global Subject, whereas very little thought has so far been given to effective procedures or algorithms for determining the best strategies with respect to a given paradigm (with the exception of the rather broad "indicator discussion"; see, e.g., *WB*, 1995 [246]; *FPOB*, 1995 [68]; *UN*, 1996 [237]; *Moldan* and *Billharz*, 1997 [153], and further references in Sect. 4). However, even the scientific analysis of the right models for future coevolution management has remained remarkably vague and is mainly oriented towards the fashionable term "Sustainable Development" (*WCED*, 1989 [254]). This term, which we will abbreviate as SD, has by now been interpreted in hundreds of different ways (Daly's equilibristic definitions standing out here (*Daly*, 1990 [49]), thanks to their intellectual clarity) and has occasionally brought about some peculiar forms of political implementation proposals (such as the plans for "sustainable development" of individual communities or tiny economic segments, testifying a complete failure to appreciate the global interdependencies just complained about).

Despite all of the shortcomings and ideological distortions, the debate about "Sustainable Development" shows the right way for a determination of possible models of geo-cybernetics. We will therefore not avoid this notion, but will rather attempt to put it into concrete terms in the next section. The result will not be a single and binding definition, but instead a collection of *precise optional paradigms of coevolution*, which emphasize different fundamental motives of human action. At the same time, we will attempt to investigate more closely the methods for the selection of the best strategies for realizing the respective ideal.

4 Sustainable Development: One + Four Paradigms

If one tries to analyse the term "Sustainable Development" according to the rules of linguistic logic, one comes to the conclusion that it represents a slightly masked synonym for "*continuable progress*": this interpretation both accounts for the explicitly optional nature of the word "sustainable" and the implicitly positive meaning of the word "development". At the same time, our translation explains, at least in part, the attractiveness of a notion that latently and in technical jargon invokes the naive dream of unlimited improvability of the human conditions of existence. The objection that the realization of such a dream could be shattered already by rigid natural-law restrictions (such as, for example, thermodynamic efficiencies, finite reaction speeds or upper limits for exploitable raw materials supplies) is often casually dismissed with the reference to "qualitative" instead of "quantitative" development.

In general, the term "Sustainable Development" is used, though, with a meaning that mixes the utopia of perpetual progress with fundamental spatiotemporal equity claims. *Clark* (1989) [42] provides a prototypical example, when he quotes the classic formulation of the so-called Brundtland Report (*WCED*, 1989 [254]) as follows: "The WCED, chaired by Prime Minister Brundtland, characterizes Sustainable Development as paths of social, economic and political progress that 'meet the needs of the present without compromising the ability of future generations to meet their own needs.' "

In recent years, a positive definition is generally no longer attempted at all; by way of contrast, certain development trends on all scales of the Earth System are frequently classified as "*non-sustainable*". This tendency may thoroughly reflect a laudable insight into the inadequacy of previous interpretation attempts or even into the fundamental indefinability of "sustainable development". It is also reminiscent, though, of the unfortunate practice of anti-liberal societies to linguistically exclude suspicious or disagreeable activities and movements as "non-christian", "non-German", "non-socialistic", etc. In a similar way, it can happen that the target object with the rating "non-sustainable" will be fundamentally denied "ecological correctness", and therefore the right to objective treatment. Thus by attempting to wangle one's way around a binding clarification of the expression, full license is given for misuse, more than ever! The basically understandable claim for ending the academic debate on "Sustainable Development" by a concrete *operationalizing* of the "concept" can have comparably negative effects: if one wants to take the second (or third) step before the first one in this way, then the danger arises that every player involved in the "sustainability game" will pass off his/her respective recipe (as characterized by individual interests) for local, regional or global E & D management as the exclusive implementation scheme for an exceedingly vague paradigm. Thus the debate would finally deteriorate to the level of a political contest where the strongly promiscuous body of voters is above all won over through the principle of selling (preferably in the media) opportunistic or even unavoidable particular measures as elements of a thoughtful general strategy.

Therefore a *structural analysis* has to precede the operationalizing effort, and this requires in turn a minimum level of formal precision. The rest of this section is devoted to this analysis, which will be restricted to conceptual considerations and nevertheless attempts to create the foundation for a positive definition of the term "Sustainable Development". We will *not* try, however, to review or even to comment on the relevant literature on the topic. We merely make reference to a few important sources within the deluge of more recent contributions (*Pearce* et al., 1989 [176]; *Rees*, 1989 [193]; *Pearce* and *Turner*, 1990 [174]; *Simonis*, 1990 [215]; *Daly*, 1990 [49]; *Brown* et al., 1991 [33]; *Pearce*, 1991 [175]; *Lélé*, 1991 [131]; *Goodland*, 1991 [89]; *Solow*, 1992 [218]; *Buitenkamp* et al., 1992 [36]; *EIAR*, 1992 [64]; *ReVelle* and *ReVelle*, 1992 [194]; *WRI*, 1992 [256]; *Darmstaedter*, 1992 [53]; *Carley*, 1993 [38]; *Pearce*, 1993 [173]; *Daly* and *Townsend*, 1993 [51]; *Lundgren*, 1993 [141]; *Redclift*, 1994 [192]; *McKenzie-Mohr* and *Marien*, 1994 [147]; *WI*, 1994 [258]; *RSU*, 1994 [189]; *Fritsch* et al., 1994 [75]; *Kastenholz* et al., 1996 [116]; *Loske*, 1996 [138]). Otherwise, we will take up certain lines of thought or proposals of other authors if necessary in the course of the analysis.

The pictographic-intuitive approach has proven to be rather successful in the analysis of complex dynamic systems. The following discussion of the potential SD models is therefore mainly based on the illustration of coevolution as a controlled motion in space **C**. We will make a practice of condensing the coevolution space down to *two dimensions* for these illustrative purposes, as has already taken place in Figs. 10 and 13. We may, for instance, interpret the **N** axis as the scale for the *state of the global climate*, and the **A** axis as the scale for the *degree of development of human civilization*. All of the fundamental geo-cybernetic options can be graphically worked out and compactly formalized within the framework of this heavily simplified representation.

The structure of our stylized coevolution space is characterized by the existence of an "*ecological niche*" for the biosphere in general and humanity in particular. On the grounds of favourable astrophysical and geophysical conditions (*Kasting* et al., 1988 [117]), possibly ameliorated and consolidated by "geo-physiological" self-organization processes (*Lovelock*, 1991 [140]; *Krumbein* and *Schellnhuber*, 1990 [125]) our planet offers ideal subsistence conditions to life. For our further illustrative analysis, we will make the simplifying assumption that the (physiogenic or anthropogenic) change of certain planetary master parameters (CO_2 content of the atmosphere, surface albedo, oceanic salinity, etc.) could destabilize the current dynamic equilibrium of the Earth System and – depending on the type of disturbance – might put a "runaway greenhouse process" or a "runaway cooling chamber process" into operation (*Kasting* et al., 1988 [117]). These processes would finally boil down to robust, stationary operating modes again, that are either characteristic for our neighbouring planet Venus, close to the Sun, or for our neighbouring planet Mars, on the far side of the Sun. The ecological niche of life on Earth is thus defined in this context as a window between the Venerian and the Martian regime, i.e. between the basins of attraction of the alternative planetary modes (see Fig. 14).

The domain characterized by the "terrestrial operating mode" embraces a continuum of conceivable states of the coupled ecosphere-anthroposphere system, but which do not all have to be reachable from the current starting point $\mathbf{P}_0 = (\mathbf{N}_0, \mathbf{A}_0)$. As an example, it is difficult to imagine that humanity could accomplish an extremely high level of civilization in the direct vicinity of the Martian regime. Nevertheless, a multitude of development options, which are determined in turn by the set $\mathfrak{M}$ of strategy options, are available to the Global Subject S within the "habitable zone". The accessible **N**-**A** states can be described concisely with the help of the following formalization.

Definition 2: Consider an arbitrary initial time $\hat{t}$, an arbitrary successive instant $T \geq \hat{t}$ and an arbitrary initial point $\hat{\mathbf{P}} \in \mathbf{C}$. Then the set of coevolutionary states

$$\mathfrak{U}(\hat{\mathbf{P}}, \hat{t}; T) := \bigcup_{\mathbf{M} \in \mathfrak{M}} \mathbf{P}(\hat{\mathbf{P}}, \hat{t}; t = T \mid \mathbf{M}) \tag{25}$$

is called a *geo-cybernetic front.*

Similarly, the set

$$\mathfrak{U}_T(\hat{\mathbf{P}}, \hat{t}) := \bigcup_{\mathbf{M} \in \mathfrak{M}} \Pi_T(\hat{\mathbf{P}}, \hat{t} \mid \mathbf{M}) \tag{26}$$

is called a *geo-cybernetic plume.*

Finally the set

$$\mathfrak{U}(\hat{\mathbf{P}}, \hat{t}) := \bigcup_{\mathbf{M} \in \mathfrak{M}} \Pi(\hat{\mathbf{P}}, \hat{t} \mid \mathbf{M}) \tag{27}$$

is called the *geo-cybernetic universe* or the *accessible universe* with respect to the initializing pair $(\hat{t}, \hat{\mathbf{P}})$.

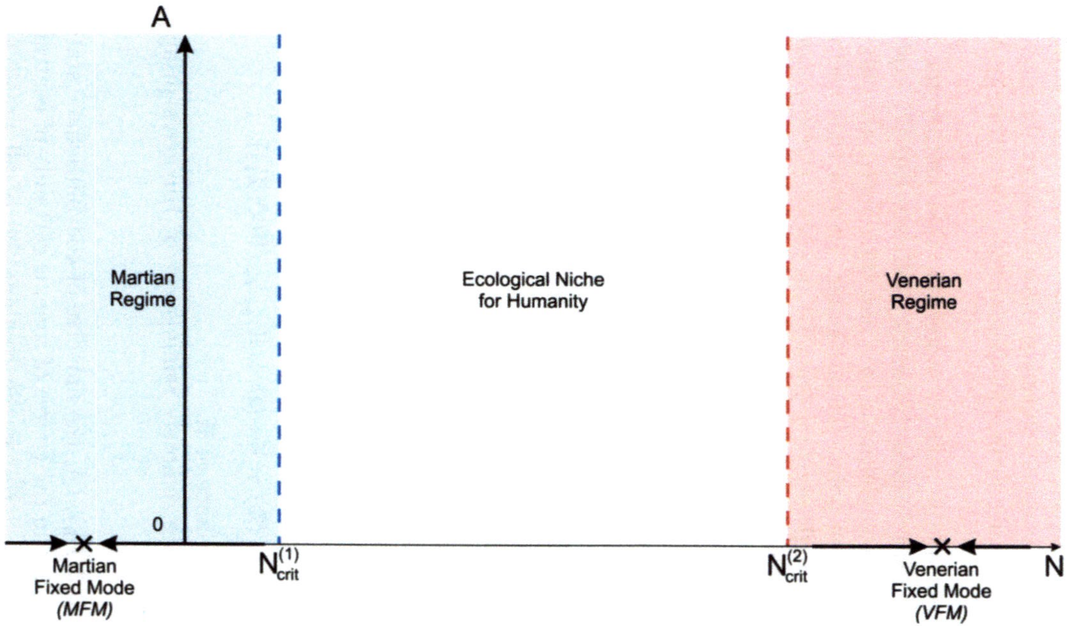

Figure 14. The "habitable zone" in our toy coevolution space is bounded by the vertical lines $\mathbf{N} = \mathbf{N}^{(1)}_{\mathrm{crit}}$ and $\mathbf{N} = \mathbf{N}^{(2)}_{\mathrm{crit}}$, respectively.

Evidently, we have the inclusion

$$\mathfrak{U}(\hat{\mathbf{P}},\hat{t};T) \subset \mathfrak{U}_T(\hat{\mathbf{P}},\hat{t}) \subset \mathfrak{U}(\hat{\mathbf{P}},\hat{t}) = \lim_{T\to\infty} \mathfrak{U}_T(\hat{\mathbf{P}},\hat{t}) \ . \tag{28}$$

■

The accessible universe $\mathfrak{U}(\hat{\mathbf{P}},\hat{t})$ is a particularly useful entity here, as it comprises all the states in coevolution space which might be reached sometime from $\hat{\mathbf{P}}$ by implementing some available management strategy $\mathbf{M}$ at time $\hat{t}$.

If we assume, for the sake of simplicity, that neither the management pool $\mathfrak{M}$ nor the Earth-System dynamics itself (see Eq. 17) *explicitly* depend on time, then the geo-cybernetic sets just introduced are only functions of the respective starting point. In particular, we can then easily extend the definition from a single starting point to an entire *starting domain* $\mathfrak{D} \subset \mathbf{C}$. We therefore also consider the quantities

$$\mathfrak{U}(\mathfrak{D};T) := \bigcup_{\hat{\mathbf{P}}\in\mathfrak{D}} \bigcup_{\mathbf{M}\in\mathfrak{M}} \mathbf{P}(\hat{\mathbf{P}},0;t=T \mid \mathbf{M}) \ , \tag{29}$$

$$\mathfrak{U}_T(\mathfrak{D}) := \bigcup_{\hat{\mathbf{P}}\in\mathfrak{D}} \bigcup_{\mathbf{M}\in\mathfrak{M}} \Pi_T(\hat{\mathbf{P}},0 \mid \mathbf{M}) \ , \tag{30}$$

$$\mathfrak{U}(\mathfrak{D}) := \bigcup_{\hat{\mathbf{P}}\in\mathfrak{D}} \bigcup_{\mathbf{M}\in\mathfrak{M}} \Pi(\hat{\mathbf{P}},0 \mid \mathbf{M}) \ , \tag{31}$$

where $T \geq 0$ is an arbitrary stretch of time.

A series of relations between these geo-cybernetic quantities can be derived. Assuming $T_2 \geq T_1 \geq 0$, we find for instance that

$$\mathfrak{U}(\hat{\mathbf{P}};T_2) = \mathfrak{U}(\mathfrak{U}(\hat{\mathbf{P}};T_1);T_2 - T_1) \tag{32}$$

and

$$\mathfrak{U}_{T_2}(\mathfrak{D}) = \mathfrak{U}_{T_1}(\mathfrak{D}) \cup \mathfrak{U}_{T_2-T_1}(\mathfrak{U}(\mathfrak{D};T_1)) \ , \tag{33}$$

due to the time invariance of $\mathfrak{M}$.

These relations are remotely reminiscent of Huygen's principle of wave mechanics.

The quantity $\mathfrak{U}(\mathbf{P}_0)$, i.e. the universe of coevolutionary states accessible from the current one in the course of time, is certainly of central importance for geo-cybernetics. $\mathfrak{U}(\mathbf{P}_0)$ is the combination of all possible trajectories $\Pi(\mathbf{P}_0, 0 \mid \mathbf{M})$ that can be realized through control – the entire tangle of optional "coevolution spaghetti", so to speak. One should note, though, that two arbitrarily chosen specimens of this tangle may have long segments in common, so they can "stick together" over sections. A conceivable structure of the universe accessible for geo-cybernetics within the framework of our two-dimensional illustration is presented in Fig. 15.

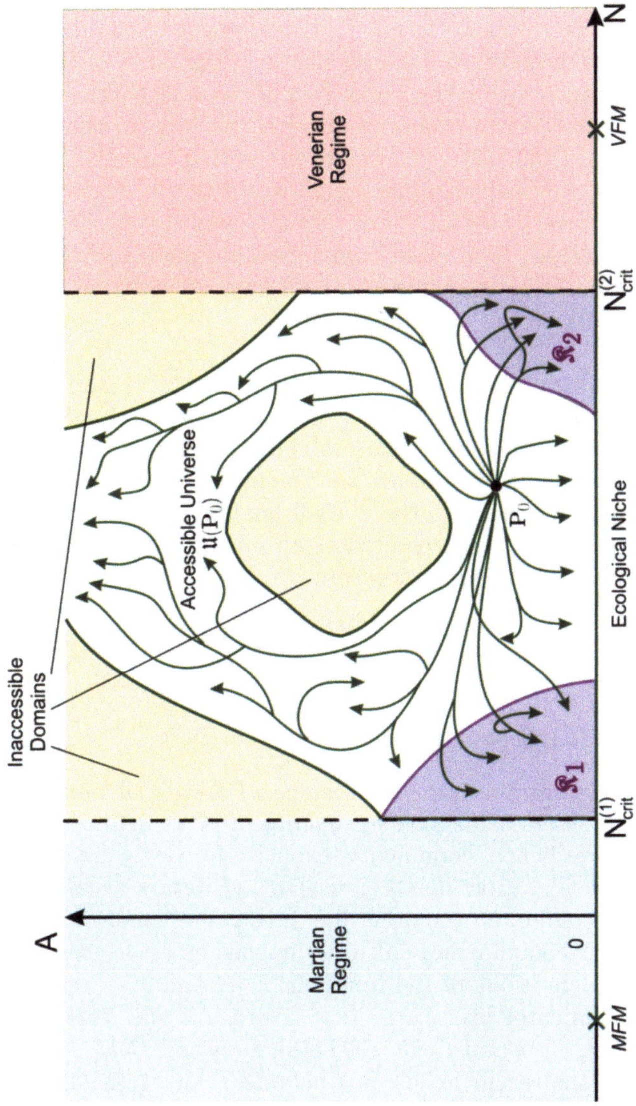

Figure 15. Set of all coevolution states which can be reached from $\mathbf{P}_0$ by appropriate management (*green contour*). The *violet* sub-sets $\mathfrak{R}_1$, $\mathfrak{R}_2$ of $\mathfrak{U}(\mathbf{P}_0)$ are "catastrophe domains" (see explanation in the text below).

In our example, $\mathfrak{U}(\mathbf{P}_0)$ is not a simply-connected quantity. The specific shape may be interpreted as follows: in order to achieve a higher level of civilizatory development (among other things, industrialization of the "Third World"), humanity has to accept either a warming of the Earth's climate (through unabated consumption of fossil fuels) or a cooling of it (because of an increase of the planetary albedo accompanying massive land use changes, for instance) *over the medium term.* The attainment of the higher development stage permits, however, a subsequent throttling down of the "ecological overload" (for instance within the framework of a complete transition to some form of regenerative energy economy or with the help of the "New Green Revolution" (*FAO*, 1996 [72]). A climate-preserving upwards movement is, in contrast, no more possible in our model world than a vertical backwards drop from a high level of civilization – that is why we observe a "hole" in our accessible universe. The other non-accessible domains result from the plausible assumption that an extremely high level of civilization can only be realized in proximity to the optimal climate.

In Fig. 15, we have also indicated specific sub-sets $\mathfrak{K}_1$ and $\mathfrak{K}_2$ of $\mathfrak{U}(\mathbf{P}_0)$. These represent exemplary ***catastrophe domains***, which should be avoided at all costs in the geo-cybernetic process! For the precise definition of such domains it is necessary to introduce an "*environmental-economic Lagrange function*" $L(\mathbf{P})$, which assesses the overall quality of the coevolution state $\mathbf{P} = (\mathbf{N}, \mathbf{A})$. L can also be interpreted as an integrated utility function and will typically have the following structure:

$$L(\mathbf{N}, \mathbf{A}) = l(w(\mathbf{N}), q(\mathbf{A})) \quad , \tag{34}$$

with

$$w(\mathbf{N}) = \omega\left(w_A(\mathbf{N}), w_S(\mathbf{N}), w_N(\mathbf{N})\right) \quad . \tag{35}$$

Here, w_A represents the immediate ***value of the global natural resources*** for humanity, w_S assesses ***the systems quality of*** $\mathbf{N}$ as defined by ecological theory (stability, resilience, complexity, capacity for development, etc.) and w_N takes into consideration ***the "eigenvalue" of nature*** according to some determination scheme. q, on the other hand, is a compound indicator for the quality of the socioeconomic and cultural dimensions associated with $\mathbf{A}$. For possible concrete definitions of the functions l, w, and q we refer the reader to the relevant "indicator literature" (e.g., *Nordhaus* and *Tobin*, 1972 [165]; *Zolotas*, 1981 [263]; *Cobb* and *Cobb*, 1994 [46]; *Daly* and *Cobb*, 1994 [50]; *WB*, 1995 [246] and to the recent article of *Costanza* et al., 1997 [47]). We point out here that the "Lagrange function" L can certainly also depend on the time derivatives of $\mathbf{N}$ or $\mathbf{A}$ and, in addition, explicitly on t, i.e.

$$L \equiv L(\mathbf{N}, \dot{\mathbf{N}}; \mathbf{A}, \dot{\mathbf{A}}; t) \tag{36}$$

or even

$$L \equiv L(\mathbf{N}, \dot{\mathbf{N}}, \ddot{\mathbf{N}}, \ldots; \mathbf{A}, \dot{\mathbf{A}}, \ddot{\mathbf{A}}, \ldots; t) \quad . \tag{37}$$

In these cases, it would be useful to switch from the representation of coevolution in "configuration space" **C** to *phase-space representation* (*Goldstein*, 1982 [88]; *Desloge*, 1982 [56]). A complication of this type is not necessary, however, for the main part of our exploration of fundamental SD paradigms as carried out in this section. Unless stated otherwise, we will also assume in our illustrative analysis that the $\mathcal{N}$-$\mathcal{A}$ dynamics is autonomous in the sense that it is controlled by $\mathcal{S}$ but not influenced by varying external forces.

As a rule, a given coevolution state $\mathbf{P} = (\mathbf{N}, \mathbf{A})$ will be acceptable – or at least tolerable – if its overall quality $L(\mathbf{N}, \mathbf{A})$ has reached a certain *minimum level* L_{min}. We can now exactly describe what we mean by catastrophe domains.

Definition 3: A non-void sub-set $\mathfrak{K}$ of **C** is called a *catastrophe domain*, if the following conditions are met.

(i) $\mathfrak{K}$ is *accessible*, i.e., $\mathfrak{K} \subset \mathfrak{U}(\mathbf{P}_0)$.
(ii) All states in $\mathfrak{K}$ are *intolerable*, i.e., $L(\mathbf{P}) < L_{min}$ for all $\mathbf{P} \in \mathfrak{K}$.
(iii) $\mathfrak{K}$ is *inescapable*, i.e., $\mathfrak{U}(\mathfrak{K}, T) \subset \mathfrak{K}$ for all $T \geq 0$, therefore $\mathfrak{U}(\mathfrak{K}) = \mathfrak{K}$.

■

Expressed in less technical terms, we are considering here irreversible coevolutionary collapses brought about by mismanagement! One should note that the Martian and Venerian regimes are not catastrophe domains taken as a whole, because they represent largely inaccessible areas. The abscissa sections between $N_{crit}^{(1)}$ and MFM resp. $N_{crit}^{(2)}$ and VFM in the example of Fig. 15 belong to $\mathfrak{K}_1$ or $\mathfrak{K}_2$, though. One can, for instance, imagine here that all these catastrophe domain states will be attracted, in the long run and independently of the management sequence **M** applied, by the Mars mode or the Venus mode, i.e.,

$$\lim_{T \to \infty} \mathfrak{U}(\mathfrak{K}_1; T) \equiv \mathfrak{U}(\mathfrak{K}_1; \infty) = MFM \quad , \tag{38}$$

$$\lim_{T \to \infty} \mathfrak{U}(\mathfrak{K}_2; T) \equiv \mathfrak{U}(\mathfrak{K}_2; \infty) = VFM \quad . \tag{39}$$

Within the picture of the boat allegory (Fig. 12) for managed coevolution, the asymptotic states MFM or VFM consequently represent pointlike maelstroms that cannot be escaped from once one has dared to get too close to them.

Here we may, of course, ask ourselves where the coevolution would move to if the Global Subject $\mathcal{S}$ entirely refrained from geo-cybernetic activities. We identify this "non-strategy" with the element zero in $\mathfrak{M}$, i.e. $\mathbf{M} = 0$. The resulting coevolution path will be denoted as the "business-as-usual scenario" (BAU), symbolized by $\mathbf{P}^{(0)}(t)$:

$$\mathbf{P}^{(0)}(t):=\mathbf{P}(\mathbf{P}_0, 0; t \mid 0) \ . \tag{40}$$

$\mathbf{P}^{(0)}(t)$ is consequently the properly initialized solution of Eq. 2, which has to turn into the corresponding solution of Eq. 17 for $\mathbf{M} = 0$. In our exemplary model world, we can make the assumption that this BAU path will enter the catastrophe domain $\mathfrak{K}_2$ in a finite time and, thus, terminate in VFM (Fig. 16).

We have now created the formal and substantial prerequisites for the identification and description of the fundamental (or "pure") SD paradigms in the further course of this section. These paradigms can be assigned to elementary control principles, which in turn correspond to archetypical *human ideals* (positive interpretation) or, likewise archetypical, *human vices* (negative interpretation), respectively. The resulting little taxonomy (Tab. 1) concludes our preliminary remarks.

Table 1.
Classification of the pure SD paradigms which will be explained in detail below.

Symbol	Name of Paradigm	Positive Goal	Negative Motive
$\mathcal{P}_0$	***Standardization***	***Order***	***Despotism***
$\mathcal{P}_1$	***Optimization***	***Prosperity***	***Greed***
$\mathcal{P}_2$	***Pessimization***	***Security***	***Cowardice***
$\mathcal{P}_3$	***Equitization***	***Fairness***	***Jaundice***
$\mathcal{P}_4$	***Stabilization***	***Reliability***	***Indolence***

4.1 Standardization

We begin our survey with a geo-cybernetic paradigm, which is not rooted in some deeper theoretical systems principle, and that therefore only represents a *pseudo-paradigm*. What we mean by this is the direct prescription of norm values, development corridors, target domains, etc. for controlled coevolution – the *establishment of absolute E & D standards*, in brief. The coevolution criteria involved here are not stringently derived – or even derivable – from the

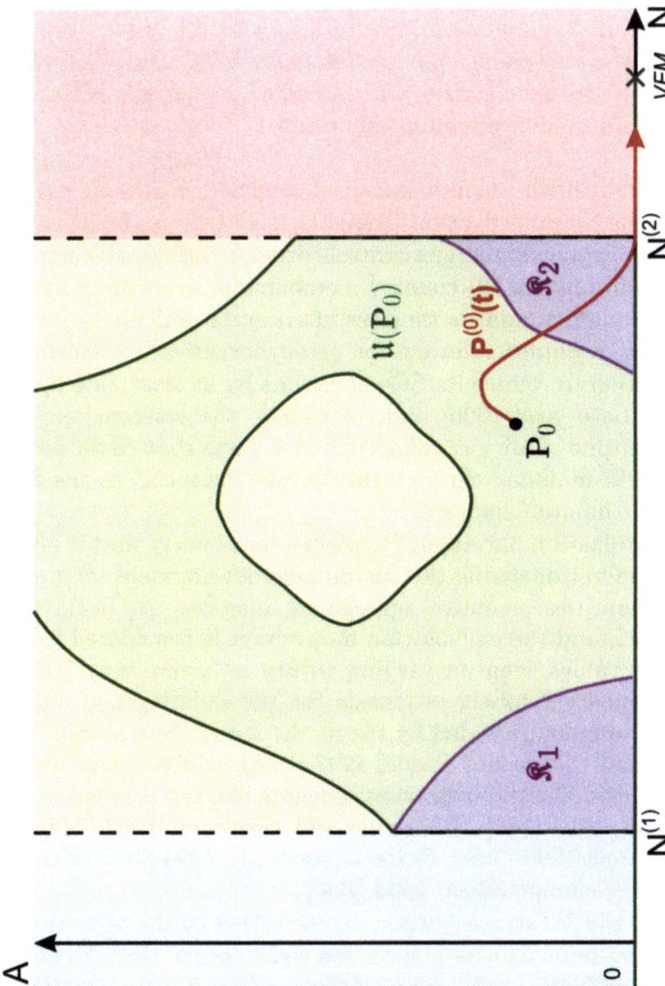

Figure 16. The disastrous coevolution path as generated by global "Business As Usual".

internal dynamics of the system under consideration, but from external normative settings of the Global Subject instead. The actual systems dynamics only comes into play via the question concerning the attainability in principle of the standards that have been set.

We label this paradigm, which seems to be trivial at first glance, with the symbol $\mathcal{P}_0$ and call it "*Standardization*" (see Tab. 1). The underlying philosophy has, on the one hand, much to do with the naive arbitrariness of the Stalinist five-years plans, but on the other hand also with the sense of responsibility of eco-toxicologists who attempt to establish adequate critical loads for hazardous environmental pollutants.

The real public life in the industrialized countries is already more strongly influenced by the Standardization principle than one would assume: norms, quota, criteria – from exhaust gas control through "affirmative action" to the Maastricht conditions for EU country membership, everything and everyone is regulated through standards that are often quite difficult to justify!

Illuminating examples relevant for geo-cybernetics are Germany's self-imposed obligation to reduce its CO_2 emissions by at least 25% by 2005 with respect to the base year 1990, and, of course, the international reduction quota for the output of all sorts of greenhouse gases that could be stipulated at the end of 1997 in Japan through the "Kyoto Protocol" to the Framework Convention on Climate Change.

The Standardization paradigm $\mathcal{P}_0$ offers a seductively simple and concrete possibility for operationalizing the sustainable-development idea: certain environmental quantities or entire aggregated functions are declared *sustainability indicators*, and the coevolution in question is considered to be correct if the indicator values keep on varying within a "*green range*" that is likewise to be defined. Explicit proposals for the definition of such a green range are, for example, provided by the model study "Sustainable Germany" (*Loske*, 1996 [138]; *Loske* and *Sachs*, 1997 [139]), which constitutes a rather impressive analysis of standardization concepts like the ecocapacity principle (*Opschoor* et al., 1991 [167]; *Weterings* and *Opschoor*, 1992 [248]; *Opschoor* and *v. d. Straaten*, 1993 [168]; *Buitenkamp* et al., 1994 [36]) or the so-called MIPS approach (*Schmidt-Bleek*, 1994 [208]).

In principle, the SD strategy $\mathcal{P}_0$ is characterized by the hope that complex environmental systems can be put on the right course through some sort of "technical instructions" in the sense of German regulatory law. This hope is generally justified, in turn, by the assumption that the *non-sustainable* value ranges of the key indicators are evident to a certain extent. In other words, the respective system is supposed to be steerable, more or less "by sight", i.e. via the largely statistical, perpetual evaluation of spatiotemporal close-range information. Fig. 17 illustrates this fundamental strategy and, at the same time, sketches some of the problems associated with it.

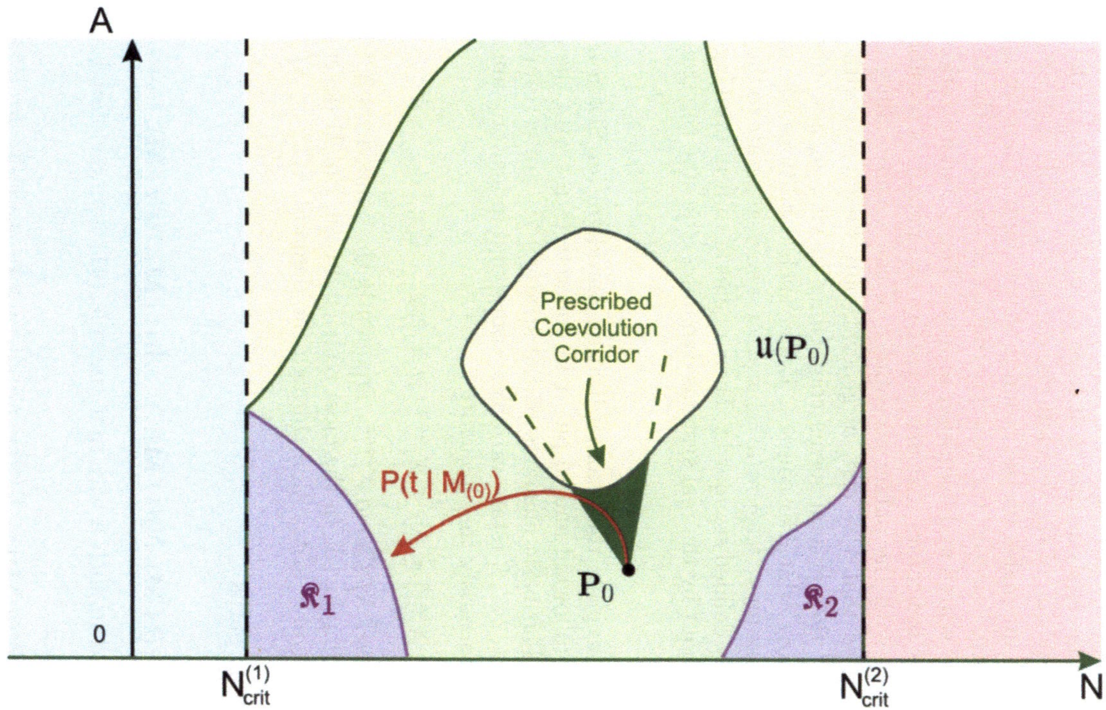

Figure 17. Exemplary realization of the Standardization paradigm in our toy world. The SD corridor (indicated in *green*) is defined according to certain standards, which do not take into account the general accessibility conditions. A coevolution path is shown, which is generated by the management sequence $\mathbf{M}_{(0)}$ complying with the normative principle $\mathcal{P}_0$. Note that $\mathbf{P}(t \mid \mathbf{M}_{(0)})$ has to leave the prescribed corridor and ultimately even risks entering the catastrophe domain $\mathfrak{K}_1$.

The fundamental arbitrariness in the stipulation of coevolution norms for the Standardization paradigm involves the great danger that essential interactions, repercussions and side-effects in the non-linear system to be controlled are not taken into consideration. In the most favourable case, the standards that are to be set are at least intuitively or proto-analytically derived from one of the "*teleological*" paradigms still to be described, i.e., on the basis of a systems analysis that is at least rudimentary. Under the latter circumstances, a revision of the selected standards is possible if those norms should prove to be untenable on the grounds of new findings and insights.

In the worst case, however, the geo-cybernetic standards are the result of a policy-dominated search and definition process. The Standardization then tends to favour key indicators that are as catchy (= simple) as possible and hastens to comply with the public acceptance of the corresponding norms. Standards generated that way can, in general, only be revised with much difficulty – even if the revision seems compelling in view of new scientific insights or pioneering ethical discourses. Under such conditions, the control strategy $\mathcal{P}_0$ may take on a "*teleocratic*" character, i.e., degenerate to the ritual fulfillment of rigidly defined standards. Then a general shift in thinking could only be brought about by the real systems dynamics exposing the standards as an illusion.

We will try to express the Standardization paradigm for SD in a more formal way in the following and to put it in concrete terms with the aid of a few examples. The basic elements of the strategy $\mathcal{P}_0$ are certain E & D quantities X_l; $l = 1, 2 \ldots L$ (state variables, compound indicators, functions, etc.), which have to satisfy the *inequalities*

$$X_l^{min} \leq X_l \leq X_l^{max} \quad ; l = 1, 2 \ldots L \ , \tag{41}$$

with fixed *limit values* X_l^{min} and X_l^{max}. Frequently the boundaries $X_l^{min} = 0$ or $X_l^{max} = \infty$ are employed. In the most restricted case, Eq. 41 degenerates to an exact target-value instruction through the demand $X_l^{min} = X_l^{max}$ for certain l.

All coevolution paths

$$\mathbf{P}(t \mid \mathbf{M}) \equiv (N_1(t \mid \mathbf{M}), N_2(t \mid \mathbf{M}), \ldots ; A_1(t \mid \mathbf{M}), A_2(t \mid \mathbf{M}), \ldots) \tag{42}$$

are to be controlled in such a way that the rule (41) is not violated at any point in time, i.e.

$$X_l^{min} \leq X_l(\mathbf{P}(t \mid \mathbf{M})) \leq X_l^{max} \quad \text{for all} \quad t \geq 0 \quad \text{and all} \quad l \ . \tag{43}$$

Example 1 ("Eco-Centrism"):

In this case, the E & D rules exclusively focus on the *"natural components"* of the Earth System; the socio-economic standards are completely ignored. The requirement is, for example, quite simply that

$$N_l^{min} \leq N_l \leq N_l^{max} \quad ; l = 1, 2 \ldots L \ . \tag{44}$$

A somewhat more subtle, normative strategy would prescribe, e.g.,

$$w_{min} \leq w(\mathbf{N}) \ , \tag{45}$$

where $w(\mathbf{N})$ is the quality index for the ecosphere introduced through Eq. 35.

Example 2 ("Tolerable E & D Window"):
Here a fixed admissible range in coevolution space is declared within which each possible development is permitted. In contrast to the eco-centric version of $\mathcal{P}_0$, the *human dimensions* are explicitly taken into account by requiring:

$$N_i^{min} \leq N_i \leq N_i^{max} \ ; i = 1, 2 \ldots I \ , \tag{46}$$

$$\text{and} \quad A_j^{min} \leq A_j \leq A_j^{max} \ ; j = 1, 2 \ldots J \ , \tag{47}$$

where $\{N_1, \ldots, N_I; A_1, \ldots, A_J\}$ represents a carefully-selected sub-set of the total ensemble of state variables.

Example 3 ("Living Standard Maintenance"):
By way of contrast to the previous examples, where detailed *separate* standards were set for the individual components of the Earth System, the tactical variant of $\mathcal{P}_0$ considered now refers to a *highly-aggregated coevolution indicator*. An obvious choice for the latter would be the "Lagrange function" $L(\mathbf{P})$ defined in Eq. 34. Then the management sequence $\mathbf{M} \in \mathfrak{M}$ has to be selected in such a way that the so-created coevolution path $\mathbf{P}(t \mid \mathbf{M})$ fulfills the inequality

$$L_{min} \leq L(\mathbf{P}(t \mid \mathbf{M})) \tag{48}$$

for all times $t \geq 0$. The establishment of the minimum standard L_{min} is, of course, a complex and highly political matter!

The corresponding SD paradigm may be refined through additional rules, which will *guarantee the geographic homogeneity* of the coevolution quality measured by L, or *limit the time variation* of $L(t) \equiv L(\mathbf{P}(t \mid \mathbf{M}))$ for example. The "variation" of a scalar function is a non-trivial mathematical entity, though; we refer the reader to the relevant monographies on functional analysis for its precise definition (see e.g. *Kreyszig*, 1978 [123]).

We wish to point out that the concrete realization of $\mathcal{P}_0$ sketched in this example explicitly permits trade-offs between natural and civilizatory qualities, i.e., between "ecology" and "economy". A coevolution path can therefore be classified as "sustainable" if a transient degradation of environmental standards is *overcompensated* through massive socio-economic advantages. A corresponding development philosophy laid the foundations for the "Green Revolution" of the 1960s and 1970s (see *Barrow*, 1995 [15] and the references therein), which was more or less fixated on food-production rates.

Example 4 ("Continuous Progress"):

We emphasized at the beginning of this section that the term "Sustainable Development" is so attractive because it suggests the fundamental possibility of "continuable progress" in a slightly cryptic way. The following operationalization of $\mathcal{P}_0$ comes closest to this extreme interpretation.

The Standardization paradigm is based here on *dynamic quality functionals*

$$Q_{\mathcal{P}_0}[\mathbf{P}(t \mid \mathbf{M})] \equiv Q_0(t \mid \mathbf{M}) \ , \tag{49}$$

which continuously assess the "progress" associated with the controlled coevolution $\mathbf{P}(t \mid \mathbf{M})$ (see also Sect. 3). A simple choice for such a functional is, for example,

$$Q_0(t \mid \mathbf{M}) := \frac{d}{dt} \{L(\mathbf{P}(\mathbf{P}_0, 0; t \mid \mathbf{M}))\} \ , \tag{50}$$

i.e. Q_0 represents the *time variation of the generalized utility function.* According to this prescription, a sequence $\mathbf{M}$ would then warrant permanent progress if, and only if,

$$0 < Q_{min} \leq Q_0(t \mid \mathbf{M}) \quad \text{for all} \quad t \geq 0 \ . \tag{51}$$

More complex progress measures are also conceivable, however, for instance

$$Q_0(t \mid \mathbf{M}) := \frac{1}{T} \{L(\mathbf{P}(\mathbf{P}_0, 0; t+T \mid \mathbf{M})) - L(\mathbf{P}(\mathbf{P}_0, 0; t \mid \mathbf{M}))\} \ , \tag{52}$$

with suitable averaging time $T \in (0, \infty)$.

This measure defines, in conjunction with condition (51), the requirement of perpetual *mean* coevolutionary progress.

For the general construction of Q_0 the "quality factors" $w(\mathbf{N})$ and $q(\mathbf{A})$ (see Eq. 34) should be combined in a reasonable (maybe non-linear) way and the resulting aggregate should be then subject to appropriate mathematical operations (possibly including higher time derivatives). The concrete form merely depends on society's preferences that are to be reflected in the notion "progress". For the state of Malaysia, for example (*WN* (1996) [247]), coevolutionary progress is synonymous with "sustained and self-supporting economic growth", while preserving the stock of natural resources of the country. The Standardization paradigm obviously determines the government's structural policy here: Malaysia is to be "a fully developed industrialized nation" by the year 2020, but in the course of this development, "only three of the total of 19 million hectares of forest area will be released for human use", for instance.

* * *

The basic question remains as to whether the fundamental strategy reflected in the Standardization paradigm can be successful *over the long term.* In particular, we have to ask: Is the "continuous progress" as defined through a largely arbitrary indicator Q_0 also "*continuable*" – as required by the SD ideal? Does the observance of certain norms for short-term E & D management furnish any guarantee at all for the fulfillment of the requirement? We will try to shed some light on the structure of this dilemma through an examplary, yet formal discussion.

The Standardization strategy will, in reality, focus on a limited time period $[0, T]$, $T > 0$, which we want to designate as the "*planning horizon*". The scope of the planning horizon is determined on the one hand by the reliable findings of the natural and social sciences regarding the state and dynamics of the environmental system under consideration (e.g. the entire Earth System), and on the other hand by the distribution of power and further political circumstances that the E & D managers have to observe. Assuming democratic conditions, standards like the duration of legislative periods, re-electability, statutes of limitation on legal affairs (e.g. environmental liability), etc. must be considered. Under authoritarian conditions (in the "most favourable" case, an "eco-dictatorship"), the planning horizon tends to be set much longer; it is, however, directly linked with the stability of the ruling regime, which is practically incalculable (see for instance the still unpredictable ecological consequences of the economically-caused collapse of the Eastern-European E & D management à la Russe).

Let us now consider a given progress measure $Q_0(t \mid \mathbf{M})$, a *minimum standard* $Q_{min} > 0$ and a planning horizon T for some environmental systems to be controlled. We further assume that the progress measure is not determined by the management *future*, but exclusively by the respective coevolution *history.* A well-defined measure of that type is obtainable, for example, if we declare

$$Q_0(t \mid \mathbf{M}) := \frac{1}{\Delta} \{L(\mathbf{P}(t \mid \mathbf{M})) - L(\mathbf{P}(t - \Delta \mid \mathbf{M}))\}, \quad \Delta > 0 \ , \qquad (53)$$

and postulate certain regularity properties of the quantities involved. We generally assume, anyway, that the coevolution path $\mathbf{P}(t \mid \mathbf{M})$ is a *continuous function of time* for every admissable management sequence $\mathbf{M}(t) \in \mathfrak{M}$. If L is not constructed in an irregular way, this property is conveyed to $Q_0(t \mid \mathbf{M})$, the initial value $Q_0(0)$ of which is unambiguously determined as a consequence of the causality principle.

For the further discussion, we will also introduce the notion of a *management episode.* Let T_1, T_2 be two arbitrary points in time, with $0 \leq T_1 \leq T_2$, and $\mathbf{M} \in \mathfrak{M}$ an arbitrary management sequence. The accompanying management episode is then defined as

$$\mathbf{M}_{[T_1,T_2]} := \mathbf{M}(t)\Big|_{[T_1,T_2]} \quad . \tag{54}$$

The symbol $\mathfrak{M}_{[T_1,T_2]}$ then denotes the entire set of all of the management episodes that can be selected in the time period $[T_1, T_2]$, i.e.,

$$\mathfrak{M}_{[T_1,T_2]} := \left\{ \mathbf{M}(t) \mid_{[T_1,T_2]} \Big| \mathbf{M} \in \mathfrak{M} \right\} \quad . \tag{55}$$

In the special case $T_1 = 0$, we can replace T_2 by the symbol T and use the simplified notations

$$\mathbf{M}_T \quad := \quad \mathbf{M}(t)\Big|_{[0,T]} \quad , \tag{56}$$

$$\mathfrak{M}_T \quad := \quad \left\{ \mathbf{M}(t) \mid_{[0,T]} \Big| \mathbf{M} \in \mathfrak{M} \right\} \quad . \tag{57}$$

If we assume time invariance of the management pool $\mathfrak{M}$ (see Eq. 22), then the following identity holds:

$$\mathfrak{M}_{[T_1,T_2]} = \left\{ \mathbf{M}_{T_2-T_1}(t + T_1) \Big| \mathbf{M} \in \mathfrak{M} \right\} \quad . \tag{58}$$

We are now in a position to analyse more closely the planning problems that typically arise with the paradigm $\mathcal{P}_0$, as outlined in Fig. 18.

This picture is particularly supposed to make clear that some management strategy which satisfies all desired standards over the short term can lead nevertheless to an irreversible development which destroys all of the long-term demands of the paradigm. In such a case, the initial progress is *not continuable* or, in other words, the development is not sustainable. There is, however, no systematic procedure for avoiding erroneous developments of this type within the framework of the pseudo-paradigm $\mathcal{P}_0$. The environmental system under consideration can have "*lines of no return*" that are not recorded by the local/instantaneous available indicators and are only detectable by a *teleological analysis.* The most important element of such an analysis is the careful exploration of the potential limitations of future options induced by the selection of a specific management episode.

The considerations necessary for this can be better structured through a precise concept formation. We begin by defining the set $\mathfrak{M}(\hat{\mathbf{M}}_T)$ of all possible management sequences *which can be generated from the episode* $\hat{\mathbf{M}}_T \in \mathfrak{M}_T$, i.e.,

$$\mathfrak{M}(\hat{\mathbf{M}}_T) := \left\{ \mathbf{M} \in \mathfrak{M} \Big| \mathbf{M}(t) = \hat{\mathbf{M}}_T(t) \quad \forall \quad t \in [0, T] \right\} \quad . \tag{59}$$

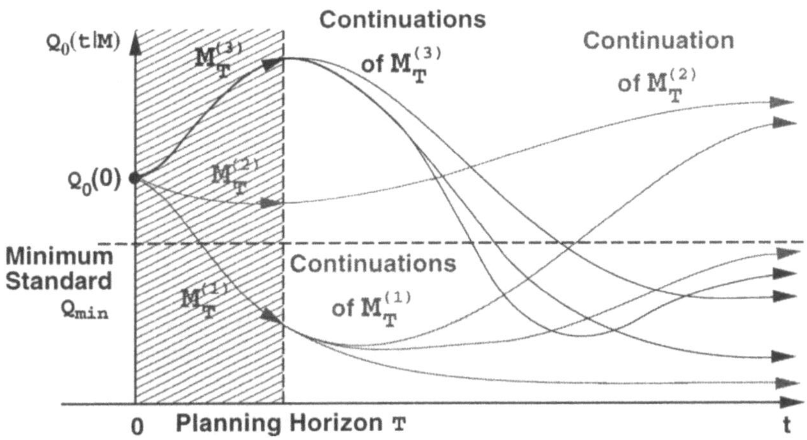

Figure 18. Time development of coevolution progress measure $Q_0(t \mid \mathbf{M})$ for a number of possible management sequences. According to the Standardization paradigm the coevolution segments generated by the management episodes $\mathbf{M}_T^{(2)}$ and $\mathbf{M}_T^{(3)}$, respectively, are admissible – at least with respect to the controllable initial time window. However, note that all available continuations of $\mathbf{M}_T^{(3)}$ ultimately violate the standard, so this management episode leads directly into a dead-end street. The episode $\mathbf{M}_T^{(2)}$, on the other hand, possesses at least one admissible continuation and therefore generates a "sustainable" coevolution segment.

Evidently, we have

$$\mathfrak{M}(\hat{\mathbf{M}}_T) \subset \mathfrak{M} \tag{60}$$

and – again applying the time invariance of $\mathfrak{M}$ -

$$\mathfrak{M}(\hat{\mathbf{M}}_T) = \left\{ \mathbf{M} \in \mathfrak{M} \middle| \mathbf{M}(t) = \left\{ \begin{array}{lll} \hat{\mathbf{M}}_T(t), & 0 & \leq t \leq T \\ \tilde{\mathbf{M}}(t-T), & T & < t, \text{with arbitrary } \tilde{\mathbf{M}} \in \mathfrak{M} \end{array} \right. \right\} . \tag{61}$$

In addition, we introduce the entity $\mathfrak{M}_T^*$ of all *management episodes compatible with* $\mathcal{P}_0$, i.e.,

$$\mathfrak{M}_T^* := \left\{ \mathbf{M}_T \in \mathfrak{M}_T \middle| Q_{min} \leq Q_0[\mathbf{P}(t \mid \mathbf{M}_T)] \quad \forall \quad t \in [0, T] \right\} \quad , \tag{62}$$

and the set $\mathfrak{M}^*$ of all (infinite) *management sequences compatible with* $\mathcal{P}_0$, i.e.,

$$\mathfrak{M}^* := \left\{ \mathbf{M} \in \mathfrak{M} \middle| Q_{min} \leq Q_0[\mathbf{P}(t \mid \mathbf{M})] \quad \forall \quad t \geq 0 \right\} \quad . \tag{63}$$

Then the following obvious observation can be made:

$$\mathfrak{M}^* \subset \bigcup_{\mathbf{M}_T^* \in \mathfrak{M}_T^*} \mathfrak{M}(\mathbf{M}_T^*) \quad . \tag{64}$$

If we finally define the set $\mathfrak{M}^*(\mathbf{M}_T^*)$ of *admissible continuations of management episodes complying with* $\mathcal{P}_0$, i.e.,

$$\mathfrak{M}^*(\mathbf{M}_T^*) := \left\{ \mathbf{M} \in \mathfrak{M}(\mathbf{M}_T^*) \middle| Q_{min} \leq Q_0[\mathbf{P}(t \mid \mathbf{M})] \quad \forall \quad t > T \right\} \quad , \tag{65}$$

then we even have the identity

$$\mathfrak{M}^* = \bigcup_{\mathbf{M}_T^* \in \mathfrak{M}_T^*} \mathfrak{M}^*(\mathbf{M}_T^*) \quad . \tag{66}$$

The last relation reflects the evident fact that admissible (infinite) management sequences can only be generated from admissible management episodes.

We are now ready to formulate precisely the term "sustainability" in the sense of a Standardization paradigm. In doing so, we make a fundamental distinction between a "sustainable path segment" and a "sustained path".

Definition 4: A coevolution path segment $\mathbf{P}_T(t \mid \hat{\mathbf{M}}) \equiv \mathbf{P}_T(\mathbf{P}_0, 0; t \mid \hat{\mathbf{M}})$ is called *sustainable* if it is

(i) generated from an admissible management episode, i.e.,

$$\hat{\mathbf{M}}_T \equiv \hat{\mathbf{M}}\Big|_{[0,T]} \in \mathfrak{M}_T^* \quad , \tag{67}$$

and

(ii) continuable in an admissible way, i. e.,

$$\mathfrak{M}^*(\hat{\mathbf{M}}_T) \neq \emptyset \quad . \tag{68}$$

A coevolution path $\mathbf{P}(t \mid \hat{\mathbf{M}}) \equiv \mathbf{P}(\mathbf{P}_0, 0; t \mid \hat{\mathbf{M}})$ is called *sustained* if it is generated from an admissible management sequence, i.e.,

$$\hat{\mathbf{M}} \in \mathfrak{M}^* \quad . \tag{69}$$

■

When scrutinizing an environmental system to be controlled, it will generally be relatively simple to identify the elements of $\mathfrak{M}_T^*$, i.e., the management episodes admissable over the short or the medium term. The task becomes significantly more difficult, as a rule, for the admissable continuations, and $\mathfrak{M}^*(\mathbf{M}_T^*)$ will probably be void for most of the episodes $\mathbf{M}_T^* \in \mathfrak{M}_T^*$! Only a brute-force long-term analysis can supply really dependable information here, but this requires that the dynamic equations (17) are integrated for all possible continuations of the management episodes from $\mathfrak{M}_T^*$. This is only possible, though, if the "teleological horizon" is extented, based on valid simulation models of coevolution (see Sect. 5), substantially beyond the actual planning horizon.

Definition 5: The path segment $\mathbf{P}_T(t \mid \mathbf{M}_T^*)$ generated by $\mathbf{M}_T^* \in \mathfrak{M}_T^*$ is called *sustainable to degree* x, where $x \in (0, 1]$, if a fraction x of all possible continuations is admissible, i.e., the sub-set $\mathfrak{M}^*(\mathbf{M}_T^*) \subset \mathfrak{M}(\mathbf{M}_T^*)$ has the relative weight x.

$\mathbf{P}_T(t \mid \mathbf{M}_T^*)$ is called *absolutely sustainable* if all possible continuations are admissible, i.e., $\mathfrak{M}^*(\mathbf{M}_T^*) = \mathfrak{M}(\mathbf{M}_T^*)$.
Evidently, an absolutely sustainable path segment is sustainable to degree 1.
■

As a matter of fact the expression "sustainable to degree x" may have a great deal of practical significance, because the management episodes selected by future generations cannot be predicted with certainty.

There is, by the way, a trivial option to guarantee "sustainable coevolution" in the sense of the Standardization paradigm. In order to achieve this one "only" has to push the planning horizon arbitrarily far ahead, i.e., $T \to \infty$, and prescribe an explicit, complete management sequence $\mathbf{M}^*(t)$, which ensures the compliance with the desired standards for all times! *Per constructionem*, each segment of the associated coevolution path is then sustainable.

This extreme version of $\mathcal{P}_0$ may already be qualified as the "*Desideration paradigm*", because it is by no means clear for a complex environmental system – and certainly even less so for the Earth System – whether a suitable $\mathbf{M}^*$ exists or how it is to be determined. However, even if the latter problems could be solved, such a "10,000-years plan" would fundamentally contradict all our ideas of freedom and personal responsibility. The "true" SD paradigms to be introduced in the next subsections undoubtedly also have a greater "ethical value".

4.2 Optimization

The SD paradigm to be discussed in this section represents the classic example of a *fundamental teleological strategy*, which fully accounts for the internal systems dynamics. After all, the seemingly perfect physical world view of the 19th century was founded almost exclusively on the axiom that "nature" indeed singles out motions of the type *that optimize certain functionals over the long term*. The differential equations for the dynamics of the respective mechanical system to be described result in a straightforward manner from this "Hamiltonian Variational Principle" (see, *Goldstein*, 1982 [88]).

In the same spirit, the "*Optimization paradigm*" $\mathcal{P}_1$ requires from controlled coevolution "simply" the *maximization of generalized utility* over the prescribed time periods. The "utility" or "benefit" is hereby defined in a *normative* way from social preferences via innumerable acts and interactions of individual volition. The notion itself, as well as the possibilities for operationalizing it, play an important role in the scientific developments of the modern age – ranging from the philosophically-based "Utilitarianism" of J. Bentham and J. S. Mill to "Game Theory" (*von Neumann*, 1944 [243]) and the modern political-economic theory of "Rational Choice" (see, e.g., *Kirsch*, 1993 [121], and references therein). However, however the generalized utility may be determined, the paradigm $\mathcal{P}_1$ reflects a fundamental constant of human motivation, namely "*wanting the best*".

With respect to our geo-cybernetic task, this means that we have to select *optimal management sequences* $\mathbf{M}^* \in \mathfrak{M}$. They have to maximize some total coevolutionary utility that may be defined via the already familiar "environmental-economic Lagrange function" $L(\mathbf{P})$ (see Eq. 34). Let us begin with a few examples for potential concrete maximization rules:

Example 1 ("Canonical Optimization"):

For an arbitrarily-chosen planning horizon $T > 0$, the management episode $\mathbf{M}_T^*(t)$ is to be selected in order to maximize the *functional*

$$I\left[\mathbf{P}(\mathbf{M}_T)\right] := \int_0^T L(\mathbf{P}(t \mid \mathbf{M}))dt \quad , \tag{70}$$

i.e.,

$$I\left[\mathbf{P}(\mathbf{M}_T)\right] \leq I\left[\mathbf{P}(\mathbf{M}_T^*)\right] \quad \text{for all} \quad \mathbf{M} \in \mathfrak{M}_T \quad . \tag{71}$$

Under certain conditions (see below), this variational task has a unique solution for each T; the limit $T \to \infty$ then generates the optimal management sequence $\mathbf{M}^*(t) \in \mathfrak{M}$, which incorporates all episodes.

We point out that such a canonical optimization, which represents the most simple realization of the general SD paradigm $\mathcal{P}_1$, plays a major role in the current debate confronting environment and economics. This applies above all to the various *integrated assessment* studies on the global climate problem (*Kaya* et al., 1993 [119]; *Nakićenović* et al., 1994 [159] *Nakićenović* et al., 1996 [160]). As an example, *Nordhaus*'s [164] DICE model is based on Ramsay's theory of optimal economic growth (see, e.g., *Intriligator*, 1971 [109]).

Example 2 ("Optimal Outcome"):

As an alternative to the maximization of the integral (= mean) utility over a longer planning period, an optimization of the environmental-economic quality *at the end of this period* might be considered. Temporary drawbacks are accepted within the framework of such a *"per aspera ad astra"* strategy, if only the target utility can be increased.

This is formally achieved by selecting for a given planning horizon $T \gg 0$ a $\mathbf{M}_T^* \in \mathfrak{M}_T$ in such a way that

$$L\left(\mathbf{P}(T \mid \mathbf{M}_T)\right) \leq L\left(\mathbf{P}(T \mid \mathbf{M}_T^*)\right) \quad \text{for all} \quad \mathbf{M}_T \in \mathfrak{M}_T \ . \tag{72}$$

A geo-cybernetic strategy that is optimal in this sense would, for example, do without economic growth rates that are maximal due to the exploitation of non-renewable natural resources in the first decades of the next century, in order to obtain, in return, a coevolution level that is all the higher in the year 2100, say. The opinion prevails among most economists, though, that canonical optimization also ensures the optimal outcome *at all times.* For the justification of this opinion, the so-called "discount rate" (see, for instance, *Samuelson* and *Nordhaus* (1995) [203]) plays a role that is almost as mysterious as the part of Einstein's "cosmological constant" in the physical theory of the development of the Universe ...

Example 3 ("Qualified Optimization"):

The rigorous optimization goals pursued in the above examples can be moderated through a number of *boundary conditions*, that will bring to bear especially the ecological or social aspects involved. We may, for instance, consider the following type of auxiliary conditions:

(i)

$$L_{min} \leq L\left(\mathbf{P}(t \mid \mathbf{M})\right) \quad \text{for all} \quad t \geq 0; \tag{73}$$

(ii)

$$N_{min} \leq N_i(t \mid \mathbf{M}), \quad A_{min} \leq A_j(t \mid \mathbf{M}) \quad \text{for all } t \geq 0 \ , \tag{74}$$

where the indices i and j designate specific variables of the ecosphere and the anthroposphere, respectively.

Because of the introduction of such "constraints", the Optimization paradigm is enriched with Standardization elements. We will discuss the *blending* of fundamental SD strategies, which turns out to be most significant in practice, rather broadly in the concluding Sect. 4.6.

* * *

There exists a highly-developed mathematical theory for the solution of *dynamic optimization tasks* of the type just considered, i.e., for the determination of the "best" management sequences $\mathbf{M}^*(t)$. This theory is associated above all with the names of Lagrange, Bolza, Mayer, Pontryagin and Bellman, and it provides necessary and sufficient conditions, respectively, for the existence and uniqueness of the optimal control functions. The corresponding bulk of wisdom primarily applies to *deterministic* problems only, however, – the theory of *stochastic* optimization has certainly made a lot of progress in recent decades, but is still restricted to the supply of efficient search algorithms, in the best case (see, e.g., *Fleming* and *Rishel*, 1975 [71]; *Papageorgiou*, 1991 [170]). Furthermore, the quantities involved have to satisfy *regularity requirements* that are relatively demanding: continuity, differentiability, convexity, etc. We will briefly compile in the following the *necessary* conditions for the solution of a dynamic optimization task that

(a) fulfills the mathematical requirements just mentioned and
(b) fits the spirit of the geo-cybernetic paradigm $\mathcal{P}_1$, within the framework of our illustrative formalization.

Our description will borrow from *Cesari*'s [40] excellent monography.

Mathematical Excursion - the necessary conditions for Mayer problems of optimal control:
Let t_1, t_2 be successive points in time, i.e., $t_1 < t_2$. Let $x(t) = \left(x^1(t), \ldots, x^n(t)\right)$ be the systems (vector) variable in question and $u(t) = \left(u^1(t), \ldots, u^m(t)\right)$ the control (vector) variable, which can be chosen from a set $U(t)$. x and u have to satisfy the set of *differential equations*

$$\frac{dx^i}{dt} = f_i\left(x(t), u(t), t\right), \quad i = 1, \ldots, n \ , \tag{75}$$

and certain boundary conditions. An *optimal pair* $(x_*(t), u_*(t))$ is searched for which *minimizes the Mayer functional*

$$I[x, u] = g\left(t_1, x(t_1), t_2, x(t_2)\right) \ , \tag{76}$$

where g is an appropriate real-valued function.

In other words,

$$I[x_*, u_*] \leq I[x, u] \tag{77}$$

for all admissible pairs (x, u). (Note that minimization is equivalent to maximization here as g may be replaced by $-g$ without loss of generality.)

Let $\lambda(t) = \left(\lambda^1(t), \ldots, \lambda^n(t)\right)$ be an auxiliary vector (of "*Lagrange multipliers*") and

$$H(x, u, \lambda, t) := \sum_{i=1}^{n} \lambda^i f_i(x, u, t) \tag{78}$$

an auxiliary real-valued function ("*Hamiltonian*").

Finally let R be the *infimum of H* if the latter is considered as a function of u only, i.e.,

$$R(x, \lambda, t) := \inf_{u \in U} H(x, u, \lambda, t) \quad . \tag{79}$$

Then the optimal pair (x_*, u_*) necessarily has the following properties:

1.

$$\frac{d\lambda^i}{dt} = -\frac{\partial H}{\partial x_*^i} (x_*(t), u_*(t), \lambda(t), t) \quad . \tag{80}$$

Note that we evidently also have

$$\frac{dx_*^i}{dt} = \frac{\partial H}{\partial \lambda^i} \quad . \tag{81}$$

The combined set of differential relations form the so-called "*canonical equations*" of the optimization problem.

2.

$$R(x_*, \lambda, t) = H(x_*(t), u_*(t), \lambda(t), t) \quad \text{for all} \quad t \in [t_1, t_2] \quad . \tag{82}$$

3.

$$\frac{dR}{dt}(x_*(t), \lambda(t), t) = \frac{\partial H}{\partial t} (x_*(t), u_*(t), \lambda(t), t) \quad . \tag{83}$$

4. A "*transversality condition*" is satisfied, i.e.,

$$\lambda^0 \, dg + \left[R(t) \, dt - \sum_{i=1}^{n} \lambda^i(t) \, dx_*^i \right]_1^2 = 0 \quad , \tag{84}$$

where $\lambda^0 \geq 0$ is some constant and the variations of t and x_* at the "end events" have to comply with the boundary conditions. ■

In addition to this important result, *sufficient* conditions for the existence and uniqueness of optimal solutions can be formulated in many cases. We refer here to the relevant literature, particularly to the outstanding Russian school of control theory.

Fig. 19 sketches the *ideal situation* for the geo-cybernetic paradigm $\mathcal{P}_1$, namely a situation where the optimal management sequence $\mathbf{M}^*_{(1)}(t)$ – and thus the optimal coevolution path $\mathbf{P}(t \mid \mathbf{M}^*_{(1)})$ – can be determined for all times ahead in an unambiguous way.

At a first glance, the Optimization paradigm appears to be an absolutely convincing fundamental strategy for Sustainable Development; however, we have only touched upon some basic difficulties relating to both *contents and methodology* yet. With respect to this the following comments apply:

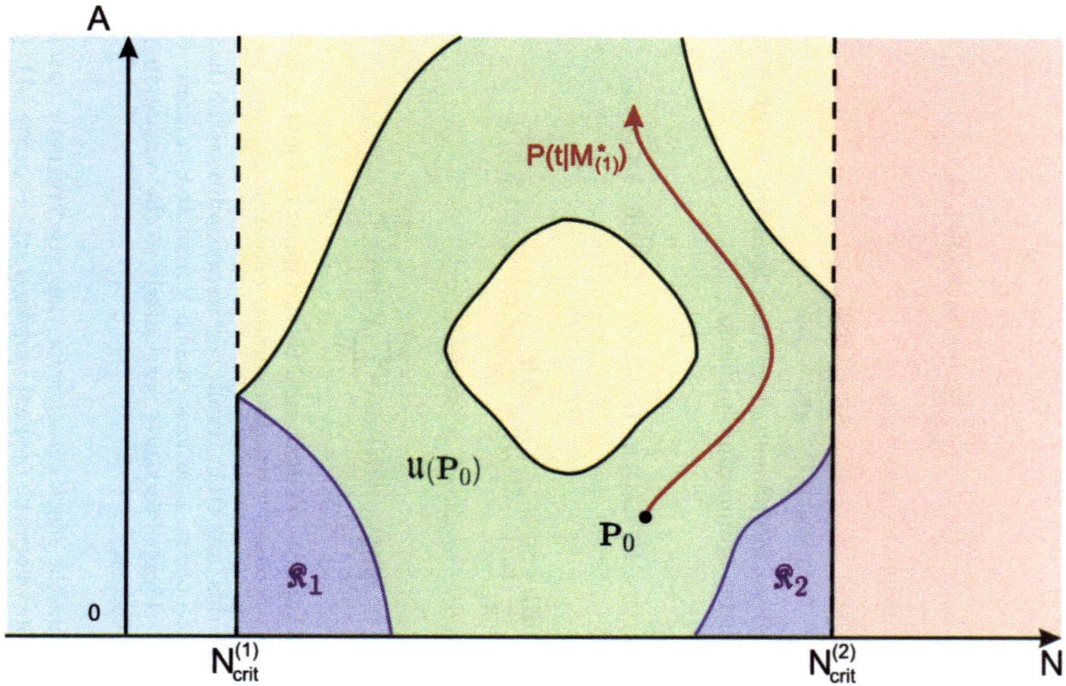

Figure 19. Unique optimal coevolution path $\mathbf{P}(t \mid \mathbf{M}^*_{(1)})$ according to one possible realization of the SD paradigm $\mathcal{P}_1$ within our caricatural Earth System.

(i) Generally speaking, it should be analytically difficult and politically almost impossible to define or impose *scalar* target values for optimization. This point is supported, for example, by the hot debate that developed during the production of the recent IPCC Report (*IPCC*, 1996 [107]) about the monetization of climate impacts: is it possible (or acceptable) to trade off, say, human lives against consumer goods – and if so, to monetize at geographically varying "exchange rates" (*Fankhauser*, 1995 [66])? Geo-cybernetics will therefore not be able to rest on a "simple" environmental-economic Lagrange function; it will have to face within the framework of the paradigm $\mathcal{P}_1$ the complex problem of "*multi-objective optimization*" (see, e.g., *Bell* et al., 1977 [17]). We will look more closely at some of the main aspects of this problem in Sect. 4.6.

(ii) If we nonetheless succeed in defining and providing all the necessary mathematical quantities, then the particular features of the optimization problem under consideration may permit the derivation of *local*, i.e., differential conditional equations for the desired solution (see Mathematical Excursion above). Actually finding the solution can then become an extremely labour-intensive and slow business, however.
Even if this most favourable case should apply, there is no guarantee that the solutions determined from local equations are also *globally* optimal. Here particular attention has to be paid to the sufficient conditions for existence and uniqueness. Otherwise one runs the risk that short-term optimal management episodes are automatically seen as initial segments of the desired management episode for a considerably wider planning horizon. However, for $T_2 > T_1$, the following may apply:

$$\mathbf{M}^*_{T_2}(t) \mid_{[0,T_1]} \not\equiv \mathbf{M}^*_{T_1}(t) \quad . \tag{85}$$

In other words, it may happen that for an optimal coevolution with respect to a large time window, a *completely different* control option should be chosen than for the short-term maximization of utility (in this connection, see also the discussion in the previous section).

We illustrate this problem by an extremely simple *mechanical example*: Let us consider a charged particle of mass m, which is moving in one dimension under the combined influence of a local electrostatic force f and a spatially homogeneous but time-dependent control force u. If $x \equiv x(t)$ denotes the instantaneous position of the particle, then its dynamics is governed by the differential equation

$$m\ddot{x} = f(x) + u(t) \quad . \tag{86}$$

We assume that the control force u is restricted to a finite interval, namely

$$u \in [-1, 1] \quad . \tag{87}$$

The optimization task is to maximize the kinetic energy,

$$E = \frac{m}{2}\dot{x}^2 \ , \tag{88}$$

for a given planning horizon $T > 0$ under the restrictions (initial conditions)

$$x(0) = \dot{x}(0) = 0 \ . \tag{89}$$

Suppose now that $f(x)$ has the following (approximate) shape:

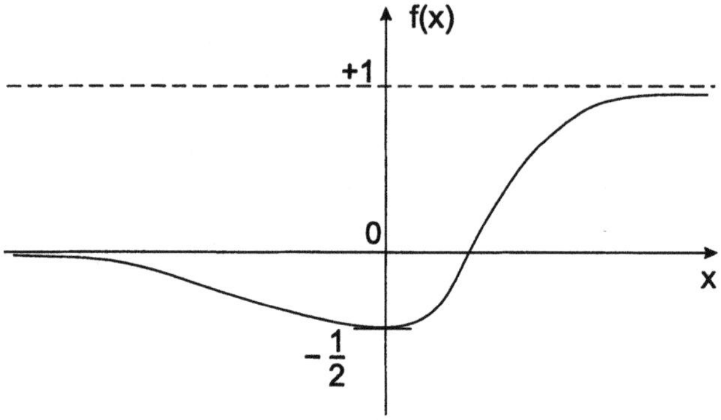

It is then clear, without any explicit calculation, what the optimal control force $u(t)$ has to look like:
For small planning horizon $T = T_1$ we will have to choose

$$u(t) = -1 \quad \text{for all} \quad t \in [0, T_1] \ , \tag{90}$$

while for large $T = T_2 \gg T_1$ we must pick

$$u(t) = +1 \quad \text{for all} \quad t \in [0, T_2] \ . \tag{91}$$

Note that the short-term strategy is obviously not compatible with the long-term one, so at a certain time horizon, $T = T_{crit}$, the optimal control *history* will have to change abruptly!

(iii) Even in the ideal situation sketched in Fig. 19, where the globally optimal coevolution path can be uniquely determined for all times, the fundamental strategy $\mathcal{P}_1$ may not be without problems: an optimal result is namely utterly dependent on perfect control, which means within our geo-cybernetic context above all the long-term *self-control* of the Global Subject $\mathcal{S}$. But what happens in the case of episodic excursions from the ideal control scheme? How far will the Earth System then deviate from the optimal course, which may lead terrifyingly close to inhospitable regions or even catastrophe domains? This problem is one of the main motivations for an alternative paradigm to "Optimization", as described in the next section.

4.3 Pessimization

The Optimization paradigm for Sustainable Development is supported by a fundamentally *optimistic* attitude that assumes that the best possible coevolution can be actually realized under all circumstances. This requires a more or less *perfect base of knowledge* and a *"secular"* discipline, so to speak, of the Global Subject in the form of a coherent, collective volition process extending over many communicating generations. Everybody who tends to be more *pessimistic* will have strong doubts about the realization of such ideal requirements and, in contrast, will lean towards a fundamental strategy that practically represents the opposite position to the Optimization paradigm. The core of this fundamental SD strategy $\mathcal{P}_2$, which we want to call the *Pessimization paradigm*, is the elementary precautionary principle of "*preventing the worst*". Thus, if one wants to guarantee the smallest possible amount of damage instead of the greatest possible benefit, then the "optimal" coevolution path $\mathbf{P}(t \mid \mathbf{M}^*_{(1)})$ in Fig. 19, for example, is taboo: it runs much too close along the catastrophe domain $\mathfrak{K}_2$, and the mismanagement of intermediate generations could lead directly to disaster ...

The pessimization paradigm $\mathcal{P}_2$, in comparison, strives for the greatest possible safety vis-à-vis human failure and whims of nature in every form. In view of Murphy's (semi-humorous) Law "Everything that can go wrong, will go wrong!", $\mathcal{P}_2$ may be qualified as the *"Anti-Murphy Strategy of SD"*. This strategy will attempt under all circumstances to avoid *disastrous coevolution paths*, i.e., paths that lead directly into catastrophe domains. But it will, moreover, shy away from *potentially disastrous coevolution paths*: these are developments that do not themselves terminate in catastrophe domains, yet support sufficiently many *disastrous by-paths* as generated by slight modifications of the management sequence. We will discuss, in the following, several ways of operationalizing $\mathcal{P}_2$ which differ considerably in sophistication.

Example 1 ("Minimax Strategy"):
The main prerequisite here is the existence of a *damage functional* $D[\mathbf{P}(\mathbf{M})]$. The latter calculates the "maximum damage" for each and every $\mathbf{M} \in \mathfrak{M}$, and the associated coevolution path $\mathbf{P}(\mathbf{M})$ generated through this. The computation may involve, for example, the *lowest value of the environmental-economic Lagrange function* L as realized by $\mathbf{P}(\mathbf{M})$ or the *duration of violating the stipulated standards* for certain ecosphere and/or anthroposphere variables. The "pessimal" management sequence $\mathbf{M}^*_{(2)}(t)$ is then defined to be that element of $\mathfrak{M}$ which *minimizes the maximum damage*, i.e.,

$$D[\mathbf{P}(\mathbf{M}^*_{(2)})] \leq D[\mathbf{P}(\mathbf{M})] \quad \text{for all} \quad \mathbf{M} \in \mathfrak{M} \ . \tag{92}$$

This is a well-known strategy from game or decision theory (see, e.g., (*Kreyszig*, 1991 [124]; *Binmore*, 1990 [24]). In the form described here, the scheme is less a "true" variant of the Pessimization paradigm, than rather an optimization strategy in disguise: disastrous paths are, in fact, ruled out with certainty, but not *potentially* disastrous ones. Thus the target damage has been minimized but not the *risk* !

Example 2 ("Strong Anti-Murphy Strategy"):

In the case of the "genuine" Anti-Murphy strategy for risk confinement, a "game" is played against the inability or disinformation of the coming generations. In the most favourable case, a coevolution course can be taken such that *even foolish management will have no chance of wreaking havoc at a later point in time* ... We are explicitly assuming here a perfectly foreseeable and therefore particularly deterministic Earth System, which is only subject to the voluntative uncertainty associated with the will of the Global Subject (see also Sect. 6).

For the illustrative description of the pertinent management principle, we will focus on one specific catastrophe domain $\mathfrak{K} \subset \mathfrak{U}(\mathbf{P}_0)$, e.g., on $\mathfrak{K}_2$ in Fig. 15. We introduce the set $\mathfrak{U}^{-1}(\mathfrak{K})$, the *basin of access to* $\mathfrak{K}$, which will serve as the counterpart to the accessible universe $\mathfrak{U}(\mathfrak{K})$ associated with the set $\mathfrak{K}$:

$$\mathfrak{U}^{-1}(\mathfrak{K}) := \{\mathbf{P} \in \mathfrak{U}(\mathbf{P}_0) \mid \mathfrak{U}(\mathbf{P}) \cap \mathfrak{K} \neq \emptyset\} \quad , \tag{93}$$

i.e., $\mathfrak{U}^{-1}(\mathfrak{K})$, is the set of all points $\mathbf{P}$ in the geo-cybernetic universe $\mathfrak{U}(\mathbf{P}_0)$ from which the catastrophe domain $\mathfrak{K}$ can be actually reached through (mis-)management.

Evidently, we have

$$\mathbf{P}_0 \in \mathfrak{U}^{-1}(\mathfrak{K}) \quad , \tag{94}$$

due to the assumption $\mathfrak{K} \subset \mathfrak{U}(\mathbf{P}_0)$.

This does *not* necessarily mean, though, that $\mathfrak{U}(\mathbf{P}_0)$ is identical with $\mathfrak{U}^{-1}(\mathfrak{K})$. For the sake of feasibility of our strong Anti-Murphy strategy, we will, in fact, presume that

$$\mathfrak{U}^{-1}(\mathfrak{K}) \neq \mathfrak{U}(\mathbf{P}_0) \quad , \tag{95}$$

i.e., $\mathfrak{U}^{-1}(\mathfrak{K})$ is a *true sub-set* of $\mathfrak{U}(\mathbf{P}_0)$.

This is certainly only possible if the geo-cybernetic dynamics ("physical" Earth System dynamics + management) *is not perfectly reversible.* Otherwise, one could return from every point $\hat{\mathbf{P}}$ in $\mathfrak{U}(\mathbf{P}_0)$ to $\mathbf{P}_0$ and subsequently reach $\mathfrak{K}$. Fig. 20 gives a non-trivial example for such a situation within the framework of our hydrographic illustration of geo-cybernetics (see Fig. 12).

Under the conditions just specified, an efficient Anti-Murphy strategy might be designed as follows:

a)

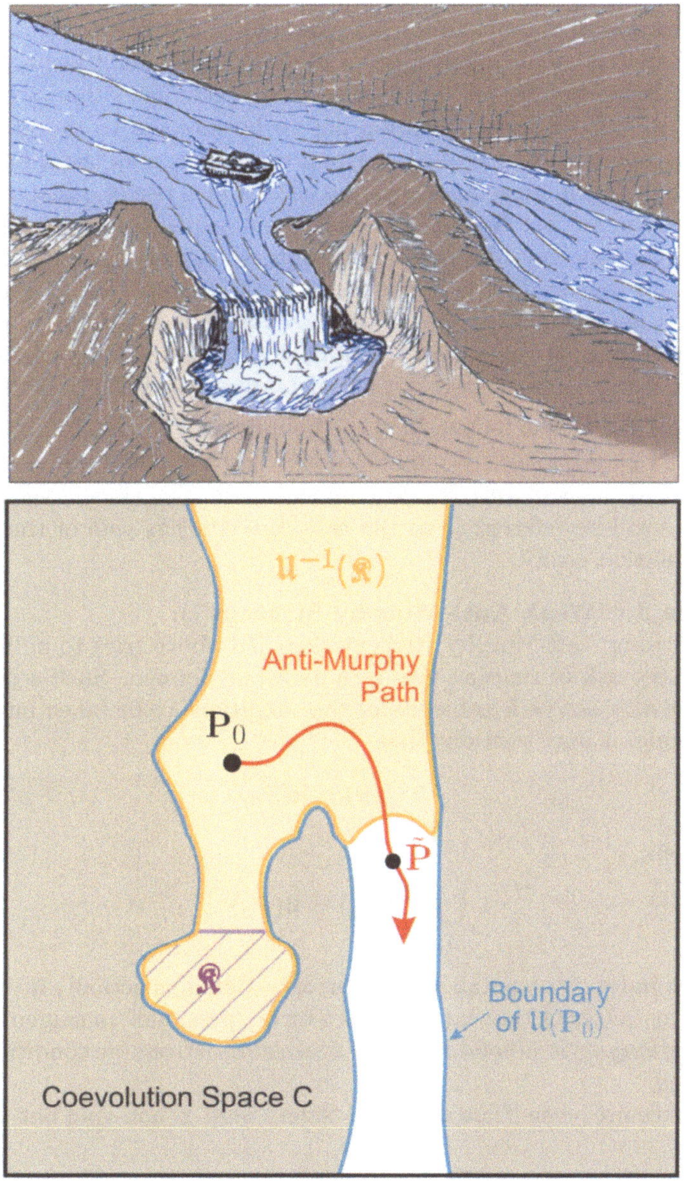

b)

Figure 20. Allegoric representation of a geo-cybernetic situation where the strong Anti-Murphy strategy works. **(a)** Boat situated upstream of a river bifurcation. Depending on the actual steering, the vessel will either enter the disastrous path towards the waterfall, or the tolerable path through the rapids. The propulsion of the boat is assumed to be too weak to cross the rapids against the current. **(b)** Translation of the real constellation into the abstract two-dimensional coevolution space. Note that $\mathfrak{U}^{-1}(\mathfrak{R})$ is a true sub-set of $\mathfrak{U}(\mathbf{P}_0)$, so there are management sequences which will steer the stystem in question forever away from the basin of access to $\mathfrak{R}$ (towards the point $\tilde{\mathbf{P}}$, for instance).

(i) *Determine* by brute-force scanning of $\mathfrak{U}^{-1}(\mathbf{P}_0)$ and $\mathfrak{M}$, through inverse integration of the dynamics (an example in a different context is given by *Kappertz* et al., hjs:Kappertz.1994 [?], via careful inspection of the reversibility conditions, or by alternative techniques the *basin of access* to $\mathfrak{K}, \mathfrak{U}^{-1}(\mathfrak{K})$. Special attention has to be directed to the identification of the boundaries of $\mathfrak{U}^{-1}(\mathfrak{K})$: beyond these lines even the silliest or most irresponsible management episodes cannot give rise to disaster any more!

(ii) *Determine* the set $\mathfrak{M}_{(2)}$ of all management sequences $\mathbf{M}_{(2)}(t) \in \mathfrak{M}$, which will bring the coevolution state ultimately out of the basin of access to $\mathfrak{K}$. Each coevolution path $\mathbf{P}(t \mid \mathbf{M}_{(2)}) \equiv \mathbf{P}(\mathbf{P}_0, 0; t \mid \mathbf{M}_{(2)})$ generated by such a management sequence will be called an *Anti-Murphy path* and is acceptable from the point of view of the Pessimization paradigm.

(iii) As a more sophisticated option, *determine* those $\mathbf{M}^*_{(2)} \in \mathfrak{M}_{(2)}$ which minimize the time for reaching the boundary of $\mathfrak{U}^{-1}(\mathfrak{K})$ from the starting point $\mathbf{P}_0$. $\mathbf{M}^*_{(2)}$ may be unique under certain conditions and will then be called the *best Anti-Murphy management.* The associated entity $\mathbf{P}(t \mid \mathbf{M}^*_{(2)})$ will be referred to as the *best Anti-Murphy path* or the "pessimal coevolution path".

Example 3 ("Weak Anti-Murphy Strategy"):

The strong Anti-Murphy strategy sketched above tries to nullify the coevolutionary risk of running into a catastrophe domain. Such a perfect risk avoidance may not be feasible under the conditions to be taken into account. For example, it may turn out that

$$\mathfrak{U}^{-1}(\mathfrak{K}) = \mathfrak{U}(\mathbf{P}_0) \tag{96}$$

or, at least,

$$\bigcup_j \mathfrak{U}^{-1}(\mathfrak{K}_j) = \mathfrak{U}(\mathbf{P}_0) \quad , \tag{97}$$

where the index j counts all the catastrophe domains actually involved. Under circumstances of that type, a short-term "pessimal" management for establishing *long-term absolute* security against disastrous life conditions *cannot* be realized.

Furthermore, even if the available "safety belt" is non-void but *very small,* i.e.,

$$\mathfrak{U}(\mathbf{P}_0) \setminus \mathfrak{U}^{-1}(\mathfrak{K}) \quad \approx \quad \emptyset \quad , \tag{98}$$

$$\text{resp.} \quad \mathfrak{U}(\mathbf{P}_0) \setminus \bigcup_j \mathfrak{U}^{-1}(\mathfrak{K}_j) \quad \approx \quad \emptyset \quad , \tag{99}$$

then the hope for a more or less riskless shaping of the future may remain an illusion: the *precision control* of coevolution, necessary under these conditions, may not be realizable in practice.

In adverse situations of the types just described, geo-cybernetic actions within the Pessimization paradigm must aim at steering the coevolution as long as possible or necessary through those subdomains of $\mathfrak{U}(\mathbf{P}_0)$ from which disaster ranges might be approached by rather *drastic mismanagement* only. Thus, this kind of *weak or relative Anti-Murphy Strategy* is trying to minimize the *unavoidable* residual risk. Evidently, the underlying principle also applies to the initial component of any strong Anti-Murphy strategy in the sense of a tactical element, especially when the road from the starting point $\mathbf{P}_0$ to the security zone is very long or tortuous.

For the precise description of the weak Anti-Murphy strategy some further formal elements are required.

Definition 6: Let $x \in [0,1]$ and assume that systems dynamics and the management pool do not depend explicitly on time. The *x-risk environment* of $\mathfrak{K}$, symbolized by $\mathfrak{N}_x(\mathfrak{K})$, is defined as follows:

$$\mathfrak{N}_x(\mathfrak{K}) := \{\mathbf{P} \in \mathfrak{U}(\mathbf{P}_0) \mid \text{A fraction } y \geq x \text{ of all possible management sequences } \mathbf{M} \text{ connects } \mathbf{P} \text{ with } \mathfrak{K}\}. \quad (100)$$

The *complement* of $\mathfrak{N}_x(\mathfrak{K})$ in $\mathfrak{U}(\mathbf{P}_0)$ will be called the *x-safety zone with respect to* $\mathfrak{K}$ and symbolized by $\mathfrak{S}_x(\mathfrak{K})$. Thus

$$\mathfrak{S}_x(\mathfrak{K}) := \mathfrak{U}(\mathbf{P}_0) \setminus \mathfrak{N}_x(\mathfrak{K}) \quad . \quad (101)$$

Each point in $\mathfrak{S}_x(\mathfrak{K})$ therefore has the property that only a fraction z of all possible management sequences $\mathbf{M} \in \mathfrak{M}$ connects this point to $\mathfrak{K}$, and z is *smaller* than x. ■

Several *technical remarks* concerning these definitions seem to be in order:

(i) The expression "fraction of all possible management sequences" will only have a precise meaning if a *measure* μ regarding the set $\mathfrak{M}$ has been declared (see for instance the classical treatise by *Halmos* (1974) [96]). Such a measure assigns a non-negative real number $\mu(\Omega)$ to each sub-set $\Omega \subset \mathfrak{M}$. Let $\mathfrak{M}(\tilde{\mathbf{P}} \mid \mathfrak{K})$ denote the sub-set of all management sequences which connect an arbitrary point $\tilde{\mathbf{P}} \in \mathfrak{U}(\mathbf{P}_0)$ to the catastrophe domain $\mathfrak{K}$, i.e., for each $\mathbf{M} \in \mathfrak{M}(\tilde{\mathbf{P}} \mid \mathfrak{K})$ there is a time $t(\mathbf{M}) < \infty$ such that

$$\mathbf{P}(\tilde{\mathbf{P}}, 0; t \mid \mathbf{M}) \in \mathfrak{K} \quad \text{for all} \quad t \geq t(\mathbf{M}) \quad . \quad (102)$$

Then the *fraction* of all possible management sequences contained in $\mathfrak{M}(\tilde{\mathbf{P}} \mid \mathfrak{K})$ is simply defined as

$$Fr(\tilde{\mathbf{P}} \mid \mathfrak{K}) := \frac{\mu(\mathfrak{M}(\tilde{\mathbf{P}} \mid \mathfrak{K}))}{\mu(\mathfrak{M})} \quad , \quad (103)$$

where we assume that $\mu(\mathfrak{M}) < \infty$.
Note that a naive computation of such a fraction by forming an aleatoric sequence $\mathbf{M}^{(1)}, \mathbf{M}^{(2)}, \mathbf{M}^{(3)}, \ldots$ of elements in $\mathfrak{M}$ and calculating the number of hits (= connection of $\tilde{\mathbf{P}}$ to $\mathfrak{K}$ actually achieved) divided by the number of trials is quite dubious here:
Generally $\mathfrak{M}$ is a *non-denumerable* set, so our aleatoric sequence may be absolutely irrepresentative! Therefore we have to resort to more sophisticated techniques like the functional integration methods employed in Feynman's formulation of quantum mechanics (*Feynman* and *Hibbs*, 1965 [70]). This topic, however, cannot be scrutinized in any depth within the scope of this essay.

(ii) In a sense, $Fr(\tilde{\mathbf{P}} \mid \mathfrak{K})$ measures how easily $\mathfrak{K}$ can be approached from $\tilde{\mathbf{P}}$ by (in-)appropriate management. As the latter exclusively depends on the will of the Global Subject $\mathcal{S}$, we may call $Fr(\tilde{\mathbf{P}} \mid \mathfrak{K})$ the strength of the *voluntative connectivity* between $\tilde{\mathbf{P}}$ and $\mathfrak{K}$. An overall topological structure of the coevolution space $\mathbf{C}$ can be induced along these lines by extending the notion of "voluntative connectivity" to arbitrary pairs ($\mathbf{P}_1$, $\mathbf{P}_2$) of points in $\mathbf{C}$. Again, we refrain from investigating this intricate issue here.

(iii) Note that the definition of $\mathfrak{R}_x(\mathfrak{K})$ and $\mathfrak{S}_x(\mathfrak{K})$, respectively, directly implies

$$\mathfrak{R}_0(\mathfrak{K}) = \mathfrak{U}(\mathbf{P}_0) \tag{104}$$

$$\text{and} \quad \mathfrak{S}_0(\mathfrak{K}) = \mathfrak{U}(\mathbf{P}_0) \setminus \mathfrak{U}(\mathbf{P}_0) = \emptyset \ . \tag{105}$$

For $x' > x$ we further have

$$\mathfrak{R}_x(\mathfrak{K}) \supset \mathfrak{R}_{x'}(\mathfrak{K}) \tag{106}$$

$$\text{and} \quad \mathfrak{S}_x(\mathfrak{K}) \subset \mathfrak{S}_{x'}(\mathfrak{K}) \ . \tag{107}$$

If we define

$$\mathfrak{r}_x(\mathfrak{K}) := \{\mathbf{P} \in \mathfrak{U}(\mathbf{P}_0) \mid \text{A fraction } x \text{ of all possible management sequences } \mathbf{M} \text{ connects } \mathbf{P} \text{ to } \mathfrak{K}\} \ , \tag{108}$$

then the following relations obviously hold:

$$\mathfrak{R}_x(\mathfrak{K}) = \bigcup_{x \leq z \leq 1} \mathfrak{r}_z(\mathfrak{K}), \tag{109}$$

$$\mathfrak{r}_x(\mathfrak{K}) = \bigcap_{x < z \leq 1} \mathfrak{R}_x(\mathfrak{K}) \setminus \mathfrak{R}_z(\mathfrak{K}) = \mathfrak{R}_x(\mathfrak{K}) \setminus \bigcup_{x < z \leq 1} \mathfrak{R}_z(\mathfrak{K}). \tag{110}$$

We may say that a risk of degree x is associated with the "marginal set" $\mathfrak{r}_x(\mathfrak{K})$, while the risk associated with the set $\mathfrak{R}_x(\mathfrak{K})$ is *at least* of degree x!

Using the complementary definition just introduced, we further observe:

$$\mathfrak{U}(\mathbf{P}_0) \setminus \mathfrak{U}^{-1}(\mathfrak{K}) \subset \mathfrak{r}_0(\mathfrak{K}) \quad , \tag{111}$$

because no management whatsoever connects the "zone of absolute safety" (which may be void, however) to the catastrophe domain. However, note that (111) is not necessarily an equality, for $\mathfrak{r}_0(\mathfrak{K})$ may embrace additional sub-sets of $\mathfrak{U}(\mathbf{P}_0)$ which are connected to $\mathfrak{K}$ by management sets of measure zero.
We also have

$$\mathfrak{R}_1(\mathfrak{K}) = \mathfrak{r}_1(\mathfrak{K}) \tag{112}$$

$$\text{and} \quad \mathfrak{K} \subset \mathfrak{r}_1(\mathfrak{K}). \tag{113}$$

Here we emphasize that

$$\mathfrak{S}_1(\mathfrak{K}) \equiv \mathfrak{U}(\mathbf{P}_0) \setminus \mathfrak{R}_1(\mathfrak{K}) \tag{114}$$

is the sub-set of $\mathfrak{U}(\mathbf{P}_0)$, where the probability of picking at random a disastrous management sequence is at least less than unity!

The weak Anti-Murphy strategy can now be formulated as follows:
On the basis of societal demand analyses a $\hat{x} \in [0,1]$ is selected, which defines the *maximum tolerable disaster risk.* The associated sub-set $\mathfrak{R}_{\hat{x}}(\mathfrak{K})$ of $\mathfrak{U}(\mathbf{P}_0)$ includes all points of the accessible universe that have a "*voluntative distance*" to $\mathfrak{K}$ of $1-\hat{x}$ or smaller — at least a fraction $\hat{x}$ of all volitions select management sequences which drive the coevolution into $\mathfrak{K}$. $\mathfrak{S}_{\hat{x}}(\mathfrak{K})$ therefore represents the zone of relative security, i.e. the set of all points in $\mathfrak{U}(\mathbf{P}_0)$ from which management of arbitrary incompetence leads to disaster in an "acceptable" fraction of cases only. We assume now that

$$\mathbf{P}_0 \in \mathfrak{S}_{\hat{x}}(\mathfrak{K}) \ . \tag{115}$$

Then the strategic instruction – i.e., the operationalization of $\mathcal{P}_2$ – simply reads: Determine and employ an $\mathbf{M}_{(2|\hat{x})} \in \mathfrak{M}$ such that $\mathbf{P}(t \mid \mathbf{M}_{(2|\hat{x})})$ *develops entirely within the $\hat{x}$-safety zone* $\mathfrak{S}_{\hat{x}}(\mathfrak{K})$. This fundamental strategy provides, after all, a relatively reliable protection against human failure, i.e., against dramatically wrong decisions of future decision makers. The set of all weak Anti-Murphy managements will be symbolized by $\mathfrak{M}_{(2|\hat{x})}$, i.e.,

$$\mathfrak{M}_{(2|\hat{x})} := \left\{ \mathbf{M}_{(2|\hat{x})} \in \mathfrak{M} \mid \Pi(\mathbf{P}_0, 0 \mid \mathbf{M}_{(2|\hat{x})}) \subset \mathfrak{S}_{\hat{x}}(\mathfrak{K}) \right\} \ . \tag{116}$$

The *feasibility* of this variant of the Pessimization paradigm is guaranteed by the following result.

Theorem: *The set* $\mathfrak{M}_{(2|\hat{x})}$ *of weak Anti-Murphy managements is not void if certain (mild) conditions are satisfied.*

The proof of this theorem will be presented elsewhere, since a number of technical considerations have to be worked out in detail. But let us sketch at least the basic lines of argumentation:

Let $\hat{\mathbf{P}} \in \mathfrak{S}_{\hat{x}}(\mathfrak{K})$, for instance $\hat{\mathbf{P}} = \mathbf{P}_0$, as presumed above. Choose an arbitrary time step $T > 0$ and consider the geo-cybernetic front $\mathfrak{U}(\hat{\mathbf{P}};T)$, which is generated from the coevolution centre $\hat{\mathbf{P}}$ through the application of $\mathfrak{M}$. By definition (see Eqs. 25 and 29) we have

$$\tilde{\mathbf{P}} \in \mathfrak{U}(\hat{\mathbf{P}};T) \Leftrightarrow \tilde{\mathbf{P}} = \mathbf{P}(\hat{\mathbf{P}},0;T \mid \tilde{\mathbf{M}}) \quad \text{for some} \quad \tilde{\mathbf{M}} \in \mathfrak{M} \ . \tag{117}$$

Each point $\tilde{\mathbf{P}}$, in turn, is the origin of a bundle

$$\mathfrak{B}(\tilde{\mathbf{P}};T) := \left\{\mathbf{P}(\tilde{\mathbf{P}},T;t \mid \mathbf{M}) \mid \mathbf{M} \in \mathfrak{M}\right\} \tag{118}$$

of continuations of the corresponding initial coevolution segment(s).

This means that the total set of all possible continuations of coevolution after time T is given by the collection

$$\bigcup_{\tilde{\mathbf{P}} \in \mathfrak{U}(\hat{\mathbf{P}};T)} \mathfrak{B}(\tilde{\mathbf{P}};T) \ . \tag{119}$$

Now let $x(\tilde{\mathbf{P}})$ denote the fraction of (management sequences that generate) paths in $\mathfrak{B}(\tilde{\mathbf{P}};T)$ which will ultimately enter $\mathfrak{K}$. Assume that

$$x(\tilde{\mathbf{P}}) \geq \hat{x} \quad \text{for all} \quad \tilde{\mathbf{P}} \in \mathfrak{U}(\tilde{\mathbf{P}};T). \tag{120}$$

This assumption obviously implies that an overall fraction $\tilde{x} \geq \hat{x}$ of *all* possible coevolution paths emanating from $\hat{\mathbf{P}}$ at time 0 – and passing through $\mathfrak{U}(\hat{\mathbf{P}};T)$ at time T – will end in the catastrophe domain $\mathfrak{K}$. However, this is clearly *in contradiction* to the initial presumption that $\hat{\mathbf{P}}$ is an element of $\mathfrak{S}_{\hat{x}}(\mathfrak{K})$!

Hence there must exist at least one point $\hat{\hat{\mathbf{P}}} \in \mathfrak{U}(\hat{\mathbf{P}};T)$ with $x(\hat{\hat{\mathbf{P}}}) < \hat{x}$, i.e., the fraction of disastrous coevolution paths (continuations) emerging from $\hat{\hat{\mathbf{P}}}$ is less than $\hat{x}$. So $\hat{\hat{\mathbf{P}}} \in \mathfrak{S}_{\hat{x}}(\mathfrak{K})$ by definition. Therefore, we can make the general statement that *each* coevolution front originating from *any* point in $\mathfrak{S}_{\hat{x}}(\mathfrak{K})$ contains at least one point in the $\hat{x}$-safety zone:

$$\mathfrak{U}(\hat{\mathbf{P}};T) \cap \mathfrak{S}_{\hat{x}}(\mathfrak{K}) \neq \emptyset \, , \quad \text{if} \quad \hat{\mathbf{P}} \in \mathfrak{S}_{\hat{x}}(\mathfrak{K}) \ . \tag{121}$$

As the time step $T > 0$ can be chosen arbitrarily (small), we conclude that a complete coevolution path within $\mathfrak{S}_{\hat{x}}(\mathfrak{K})$ can be generated from the starting point $\hat{\mathbf{P}} \in \mathfrak{S}_{\hat{x}}(\mathfrak{K})$ by appropriate management. ∎

If $\mathbf{P}_0 \notin \mathfrak{S}_{\hat{x}}(\mathfrak{K})$ then there are two possibilities to rescue the weak Anti-Murphy strategy:

(i) Reduce the safety standard, i.e., increase $\hat{x}$ until $\mathbf{P}_0$ lies within the corresponding safety zone.

(ii) Find and apply the management episode which establishes the minimum risk passage from $\mathbf{P}_0$ to $\mathfrak{S}_{\hat{x}}(\mathfrak{K})$. The "risk" associated with a given passage may be defined, for example, as the (properly normalized) time integral over $x(\mathbf{P}(t))$.

We have thus shown that the weak (or relative) Anti-Murphy strategy can be realized under the specified preconditions and restrictions. As mentioned above, this particular operationalization of $\mathcal{P}_2$ does usually not provide a unique management sequence and a corresponding unambiguous $\hat{x}$-safe coevolution path. Actually all management sequences in $\mathfrak{M}_{(2|\hat{x})}$ are allowed, and they generate the bundle of trajectories

$$\mathfrak{f}_{\hat{x}}(\mathfrak{K}) := \left\{\Pi(\mathbf{P}_0, 0 \mid \mathbf{M}_{(2|\hat{x})}) \mid \mathbf{M}_{(2|\hat{x})} \in \mathfrak{M}_{(2|\hat{x})}\right\} \quad . \tag{122}$$

We refer to $\mathfrak{f}_{\hat{x}}(\mathfrak{K})$ as *the $\hat{x}$-safely explorable universe with respect to $\mathfrak{K}$*, since this sub-set of $\mathfrak{U}(\mathbf{P}_0)$ embraces all those coevolution states which can be (or have to be) visited from $\mathbf{P}_0$ in consideration of the stipulated relative security standards.

We emphasize that

$$\mathfrak{f}_{\hat{x}}(\mathfrak{K}) \subset \mathfrak{S}_{\hat{x}}(\mathfrak{K}) \quad , \tag{123}$$

but the equality would generally not be true. It is easy to construct explicit examples of geo-cybernetic settings where some point in $\mathfrak{f}_{\hat{x}}(\mathfrak{K})$ cannot be reached from $\mathbf{P}_0$ by any managerial efforts – even if $\mathbf{P}_0 \in \mathfrak{S}_{\hat{x}}(\mathfrak{K})$.

The characteristics of the weak Anti-Murphy strategy are sketched in Fig. 21.

We should point once again to the fact that the statements made in this section so far are valid in the specified form only if the "eigen-dynamics" of the Earth System as well as the management pool $\mathfrak{M}$ are *not* explicitly time-dependent. Yet even if these preconditions are not satisfied, an analysis of the Pessimization paradigm is possible along the lines pursued above. The price to be paid, however, will be a significant complication of both the formalism and the deductive process.

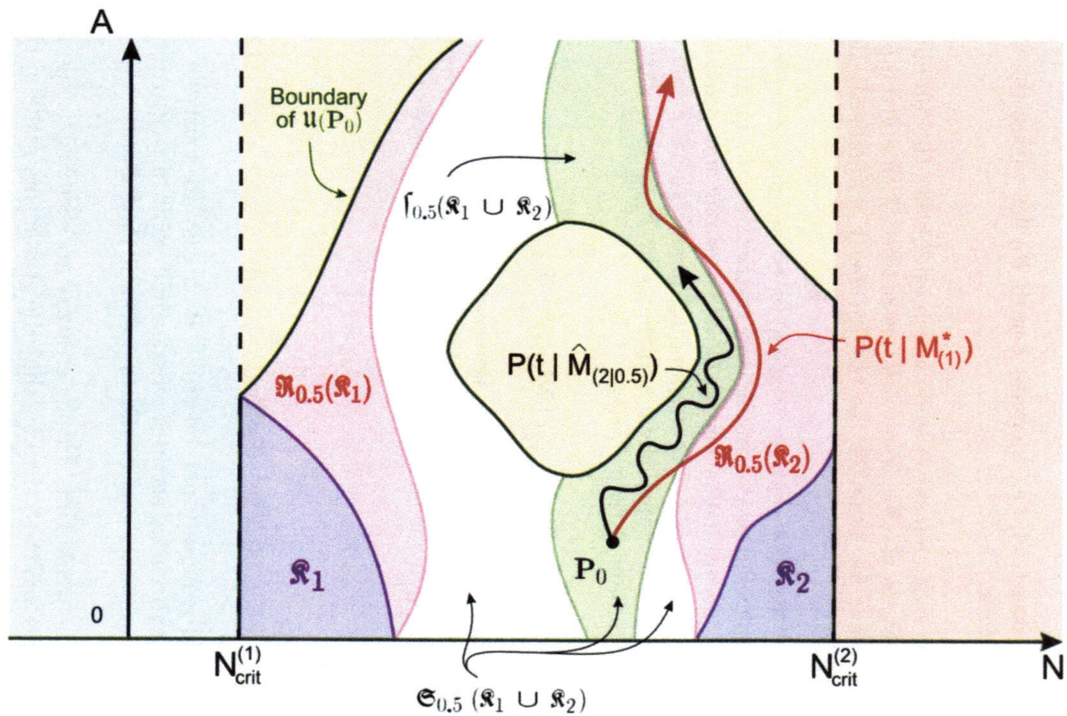

Figure 21. Cartoon of weak Anti-Murphy strategy as one possible variant of the SD paradigm $\mathcal{P}_2$. $\hat{x} = 0.5$ here; the 0.5-safety zone and the 0.5-safely explorable universe, respectively, refer to the union of catastrophe domains $\mathfrak{K}_1 \cup \mathfrak{K}_2$. In this example $\mathfrak{f}_{0.5}$ is a true sub-set of $\mathfrak{S}_{0.5}$. Note that the "optimal" coevolution path according to $\mathcal{P}_1$ from Fig. 19, $\mathbf{P}(t \mid \mathbf{M}^*_{(1)})$, trespasses into the 0.5-risk environment at least twice and is therefore forbidden by the Pessimization paradigm. By way of contrast, the management $\hat{\mathbf{M}}_{(2|0.5)}$ steers the "Earth System" as safely as required through coevolution space. The progress associated with $\mathbf{P}(t \mid \hat{\mathbf{M}}_{(2|0.5)})$ may be rather slow, however.

We should also point out here that especially the weak Anti-Murphy variant of $\mathcal{P}_2$ evokes the risk of "*infinite regression*" (see e.g., *Popper*, 1992 [186], and Sect. 3): for to what degree of relative security can you actually realize a relatively secure path? With respect to the problem structure, we are here in the same situation as someone who has to cross an abyss on a slippery plank and who tries to keep as far as possible to the centre of the plank: how well can this person actually maintain a safe distance from the edges of that miserable bridge?

Therefore, the set $\mathfrak{M}_{(2|\hat{x})}$ of relative Anti-Murphy managements might, in turn, be subject to a relative Anti-Murphy analysis and one could, e.g., select those sequences which may be readjusted rather easily in case of voluntative deviations. However, to what degree can we then be secure that these corrections will be actually be implemented, and so on... As a matter of fact, this is just a theoretical and not really a practical problem: in the individual case, one simply has to decide how many steps in the security hierarchy one wants to take into account. Error tolerability with respect to mismanagement will, as a rule, rapidly grow with the regression level (see for this issue also Sect. 6).

Example 4 ("Probabilistic Pessimization"):

The general setting for the potential operationalizations of the Pessimization paradigm discussed so far involves considerably idealized assumptions regarding determinism, predictability and autonomy of the environmental system to be controlled. If we add some elements of realism to our analysis, then those assumptions have to be modified in at least two respects:

(i) A considerable *cognitive uncertainty* with respect to the system's components, processes and boundary conditions has to be taken into account.
(ii) The strongly inhomogeneous probability distribution for the realization of the various voluntative options associated with the elements in $\mathfrak{M}$ has to be considered.

When both these aspects are integrated into the paradigm $\mathcal{P}_2$ then Pessimization becomes, in essence, "a game against our descendants and against nature". Within the framework of this interpretation, the fundamental guiding principle may be operationalized, for instance, by assigning to each point $\hat{\mathbf{P}} \in \mathfrak{U}(\mathbf{P}_0)$ a probabilistic distance to the pertinent catastrophe domain, i.e. a quantity $d(\hat{\mathbf{P}} \mid \mathfrak{K})$. This distance accounts for the total ensemble $\mathfrak{M}(\hat{\mathbf{P}} \mid \mathfrak{K})$ of eligible managements connecting $\hat{\mathbf{P}}$ to $\mathfrak{K}$, the probabilistic weights of these disastrous control options as derived from a "psychogram" of the Global Subject $\mathcal{S}$, and the uncertainties associated with the dynamics of the "physiologic" components $\mathcal{N}$ and $\mathcal{A}$ of the whole Earth System $\mathcal{E}$ (see Eq. 16). Thus $d(\hat{\mathbf{P}} \mid \mathfrak{K})$ may be qualified as an *integrated risk measure.*

Based on this measure, a *sophisticated minimax strategy* (see above) may be employed. Define for each $\mathbf{M} \in \mathfrak{M}$

$$d_{min}(\mathbf{M} \mid \mathfrak{K}) := \min_{\hat{\mathbf{P}} \in \mathbf{P}(t|\mathbf{M})} d(\hat{\mathbf{P}} \mid \mathfrak{K}) \quad , \tag{124}$$

i.e., $d_{min}(\mathbf{M} \mid \mathfrak{K})$ quantifies the integrated probabilistic proximity to $\mathfrak{K}$ for the coevolution path $\mathbf{P}(t \mid \mathbf{M})$ generated by the management sequence $\mathbf{M}$.

The "pessimal" management $\mathbf{M}^*_{(2)}$ therefore has to satisfy

$$\begin{aligned} d_{min}(\mathbf{M}^*_{(2)} \mid \mathfrak{K}) &= \max_{\mathbf{M} \in \mathfrak{M}} d_{min}(\mathbf{M} \mid \mathfrak{K}) \\ &= \max_{\mathbf{M} \in \mathfrak{M}} \left[\min_{\hat{\mathbf{P}} \in \mathbf{P}(t|\mathbf{M})} d(\hat{\mathbf{P}} \mid \mathfrak{K}) \right] \quad . \end{aligned} \tag{125}$$

This means that $\mathbf{M}^*_{(2)}$ provides the *best worst case scenario* with respect to disastrous coevolution. Note, however, that $\mathbf{M}^*_{(2)}$ does not have to be unique.

The variant of $\mathcal{P}_2$ just discussed may be further refined, of course. One might, for example, include the evaluation of the respective times that are spent by the system in a probabilistic distance from $\mathfrak{K}$ through appropriate integrals along the coevolution path considered. For the sake of brevity, we will not further explore here this rather promising approach to operationalizing the Pessimization paradigm.

* * *

Let us conclude this section devoted to $\mathcal{P}_2$ with two general remarks:

First, we must re-emphasize that this SD paradigm does not primarily determine *the* pessimal management sequence, but rather *excludes non-tolerable control options.* As a consequence, a fairly large manoeuvring space for geocybernetic activities usually remains. The precision control within that "decision leeway" may be pinned down to a unique sequence of steering elements by employing auxiliary SD paradigms according to a "nested strategy" (see Sect. 4.6).

Second, it has to be realized and accepted that the Pessimization paradigm heavily depends on normative elements through the identification of catastrophe domains. From this perspective, $\mathcal{P}_2$ might be perceived as a "qualified Standardization paradigm". The latter approach is advocated by the German Global Change Advisory Board within the context of their "Guardrail Philosophy" (*WBGU*, 1996 [83]): Sustainable Development is interpreted there as a coevolution that, first and foremost, keeps away from the "lethal syndromes" of Global Change. The syndromes themselves are defined as the dominant functional patterns of environmental degradation as a consequence of the global transformation process accelerating since the end of World War II.

4.4 Equitization

The recent E & D debate with its preliminary climax in Rio, 1992, revolves around notions like "fairness", "justice", and "equity". It is therefore structurally characterized by the problems of *comparative endowment* of individuals, groups, classes or nation states with natural resources, development chances, environmental security, agreeable ambiances, etc. As the 20th century draws to a close, this debate actually resumes – on an international (in fact geostrategic) stage – a fundamental conflict, which has generated, under purely political-economic signs, the rise and fall of the great ideologies of that century. This *spatial* extension of the traditional complex of problems towards a genuinely global challenge is complemented, however, by a qualitatively novel aspect: the issue of *inter-generational equity*, i.e., the quest for just allocation of options *in time*.

The yearning for fair spatiotemporal endowment of all the actors involved in the game of life, which is obviously deeply rooted in the human cosmos of values, constitutes the ethical background for a further fundamental SD strategy. We will call it the *Equitization paradigm* and register it as the fundamental strategy $\mathcal{P}_3$. $\mathcal{P}_3$ forms, in a manner that similarly applies to $\mathcal{P}_1$ and $\mathcal{P}_2$, a logical pair together with the Standardization paradigm $\mathcal{P}_0$: the crucial feature exhibited by both paradigms is namely the act of *comparison*. In the case of Standardization, though, it is a matter of comparing the coevolutionary quality with *absolute* standards ("E & D standard metres", so to speak), which are imposed from outside upon the environmental system to be controlled through normative acts of long-term effect. By way of contrast, Equitization aims at the *relative* balancing of the respective interests of the actors in a permanent process, fully taking the intrinsic systems dynamics into consideration.

The distinct consequences of absolute and relative comparison may be well illustrated through the example of the definition of poverty: In the first case, this definition is customarily based on fixed or very slowly changing assessment standards such as canonical market baskets, quality thresholds for social housing development schemes, reimbursement regulations of the public health insurance funds, etc. In the second case, "poverty" is defined on a purely comparative basis – through the 50%-margin of the average available per capita income of a community, for instance. This can lead to the bizarre conclusion that the municipality of, say, Palm Springs accommodates a considerable number of "poor" dollar millionaires.

Now, how can we operationalize the just spatiotemporal allocation of options within the framework of a fundamental SD strategy? The basic prerequisite for this is a handy and robust concept that remains perceptible, communicable and conveyable across the real competition of political interests. In searching for such a concept for the Equitization paradigm, one should try to avoid sophisticated interpretations of the notion of "equity" (see, *Rawls*, 1971 [190]; *Cline*, 1992 [45]; *Rayner*, 1997 [191]; *Tòth*, 1997b [231]), and to factor out the familiar aspects of economic-geographic fair allocation (for this, see Sect. 5). Only then will it be possible to focus on the specific quality of the task, which is entirely due to the evolutionary aspect.

We therefore identify here "equity" essentially with "*equality of E&D options for successive global generations*". By doing so, we are especially capable of providing a solid operational basis for the legendary but evasive Brundtland definition of Sustainable Development – meet the needs of the present generation without compromising the *ability* of future generations to meet their own needs.

We have repeatedly placed strong emphasis on the words "options" and "ability", which are supposed to highlight the importance of self-determination and self-responsibility of each generation within the framework of paradigm $\mathcal{P}_3$. If the latter values are disregarded, then a "teleocratic" prescription for coevolution in the spirit of the fundamental strategy $\mathcal{P}_0$ can readily be formulated. As an example, one might demand that an appropriate geo-cybernetic management $\mathbf{M}^*_{(3)}$ satisfied the requirement

$$\int_{\substack{\text{time span of}\\ \text{generation } n+1}} L(\mathbf{P}(t \mid \mathbf{M}^*_{(3)}))dt = \int_{\substack{\text{time span of}\\ \text{generation } n}} L(\mathbf{P}(t \mid \mathbf{M}^*_{(3)}))dt \quad , \qquad (126)$$

where the index n counts all future generations. Even the dynamic change of needs and the resources available for their satisfaction could possibly be taken into account here through the quality function L; thus $\mathbf{M}^*_{(3)}$ would have to warrant, for instance, a steady *ratio* between these factors over the long term.

Our objectives, however, require more subtle approaches for the operationalization of the Equitization paradigm. In the following discussion of such approaches we will again tacitly assume – if not explicitly specified otherwise – that the geo-cybernetic dynamics is not directly dependent on time.

Example 1 ("Miraculous Equitization"):
We start by considering *radical interpretations of the Brundtland definition* of Sustainable Development, mainly in order to expose the fundamental problems associated with a formula that has become almost ritual. Two examples for coevolutions that leave all of the dynamic options open are outlined in Figs. 22 and 23. For descriptive purposes, the coevolution space $\mathbf{C}$ is collapsed to one dimension only here, and the coevolution paths are properly represented as functions of time.

Both versions of "miraculous Equitization" presented here formalize the naive ideal that one may decide for certain development possibilities without inducing the loss of options at some later point in time. The two variables

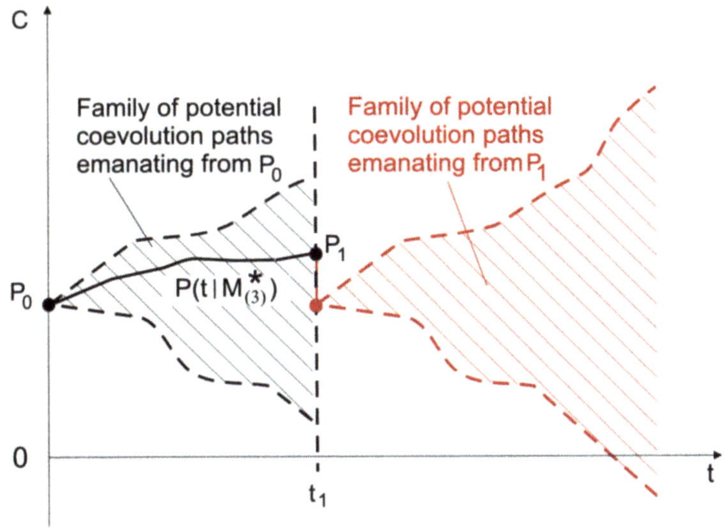

Figure 22. Miraculous Equitization, Variant 1. $\mathbf{M}^*_{(3)}$ generates a coevolution path with the property that at an arbitrary point in time $t_1 > 0$, the same bundle of potential coevolutions as at time 0 is disposable (almost) instantaneously.

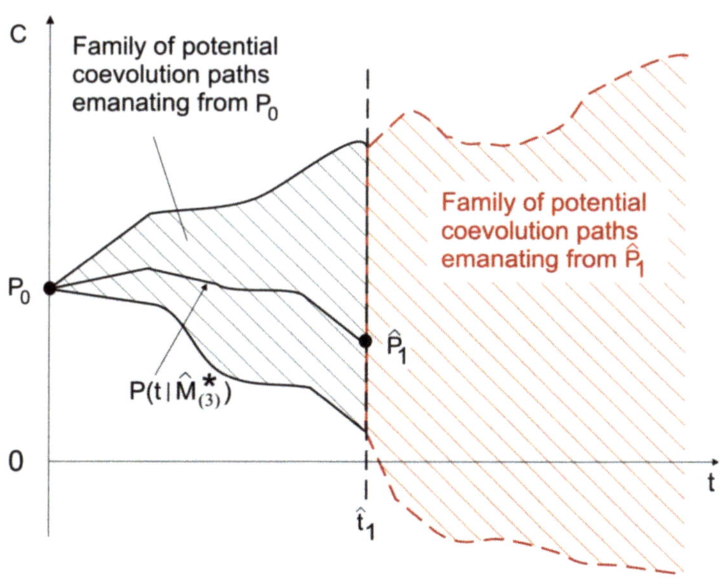

Figure 23. Miraculous Equitization, Variant 2. $\hat{\mathbf{M}}^*_{(3)}$ generates a coevolution path with the property that the bundle of potential coevolutions at some later time $\hat{t}_1 > 0$ is the union of all continuation bundles of all possible path segments in $[0, \hat{t}_1]$.

are structurally different, though: in the first case (Fig. 22), a management sequence $\mathbf{M}^*_{(3)}(t)$ is being searched for *that allows itself to be "undone" quasi-instantaneously at any time.* In the second case (Fig. 23), it is exactly the other way around: by judicious choice of $\hat{\mathbf{M}}^*_{(3)}(t) \in \mathfrak{M}$, the optional management episodes not realized instead of segments of the latter sequence should be *endowed with the capacity of being "done" quasi-instantaneously at any time.* This idea is not altogether absurd if we think of quantum reality, where all potential time developments of a system run virtually in parallel if certain mild conditions are met (see the literature on Schrödinger's fabulous cat, e.g., *Schrödinger*, 1935 [210]; *Bohm*, 1951 [27]; *Davies*, 1986 [54]; *Gribbin*, 1991 [91]). We emphasize, though, that both of the outlined radical versions of the Equitization paradigm fully earn the adjective "miraculous" in the context of geo-cybernetics – they violate, among other things, the requirement for coevolutionary continuity and therefore merely have hermeneutic value with respect to the Brundtland formula.

Under certain circumstances there is, however, a trivial possibility for the realization of Variant 1: if the system is successfully "parked" in the state $\mathbf{P}_0$ through suitable management (see also Sect. 4.5), then the development options will seemingly be identically preserved. In fact, though, this is a matter of "freezing" these options, as one rules out their use by the very fact that one wants to keep them open for all times! This situation is precisely described through the parable of the horse (or the classical dilemma of Buridan's donkey) that starves to death between two piles of hay, because it can't make a decision on either one of the two. The SD debate frequently seems to be characterized by such a horse mentality ...

In the light of the considerations just made, the goal of the Equitization paradigm can be formulated more sharply: *Maximal preservation of E & D options under minimal obstruction of the current coevolution.* If one disregards miracles, this goal will not be achievable. We will outline in the following, however, several approximations to this "Brundtland Ideal" that can be realized in principle.

Example 2 ("Weak Brundtland Strategy"):
The weakest form of inter-generational equity, which is, in return, the version most likely to be operationalized, is the "*preservation of the overall coevolutionary future*". This seemingly paradoxical expression epitomizes a variant of $\mathcal{P}_3$ that conserves the accessible universe, i.e. all *asymptotic E & D options for* $t \to \infty$. The formal recipe for this weak Brundtland strategy is as follows:

Determine $\mathbf{M}^*_{(3)}(t) \in \mathfrak{M}$ in such a way that for arbitrary $\hat{t} \geq 0$ the relation

$$\mathfrak{U}(\mathbf{P}(\hat{t} \mid \mathbf{M}^*_{(3)})) = \mathfrak{U}(\mathbf{P}_0) \tag{127}$$

is satisfied.

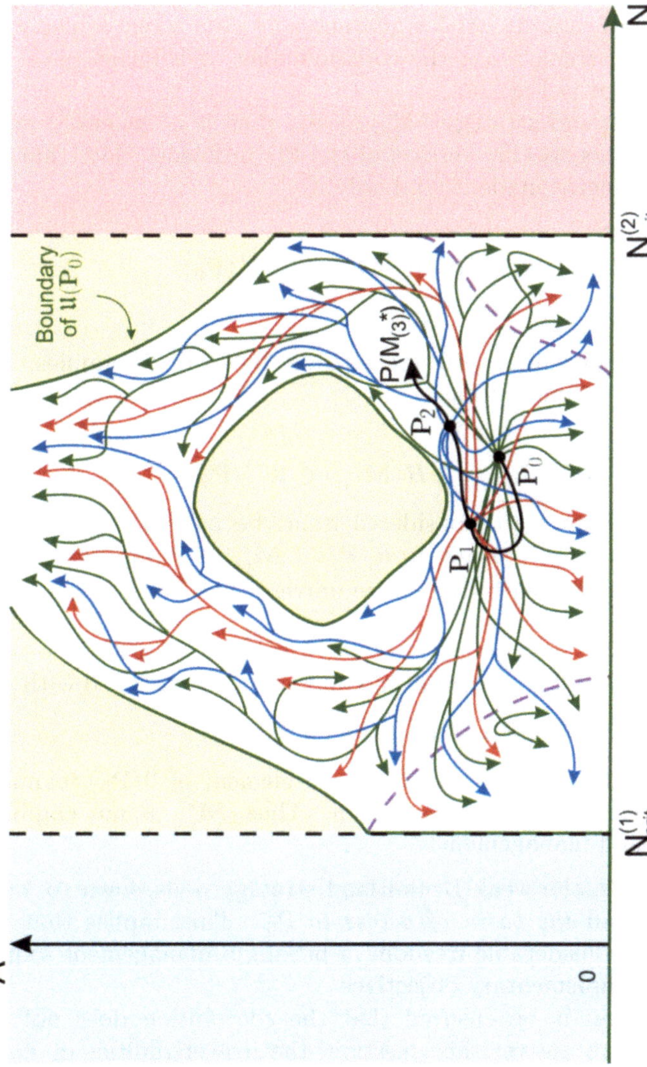

Figure 24. Sketch of weak Brundtland management in our two-dimensional geo-cybernetic toy system. $\mathbf{M}^*_{(3)}$ generates a coevolution path $\mathbf{P}(\hat{t} \mid \mathbf{M}^*_{(3)})$, which preserves all *asymptotic* E & D options. For example, all states in $\mathfrak{U}(\mathbf{P}_0)$ can ultimately be reached from $\mathbf{P}_1 \equiv \mathbf{P}(\hat{t}_1 \mid \mathbf{M}^*_{(3)})$ and $\mathbf{P}_2 \equiv \mathbf{P}(\hat{t}_2 \mid \mathbf{M}^*_{(3)})$, respectively, by appropriate management. However, note that passage type and time may vary considerably with the starting point on $\mathbf{P}(\hat{t} \mid \mathbf{M}^*_{(3)})$.

We emphasize here that, due to *causality*, we have

$$\mathfrak{U}(\mathbf{P}(\hat{t} \mid \mathbf{M})) \subset \mathfrak{U}(\mathbf{P}_0) \tag{128}$$

for arbitrary $\hat{t} \geq 0$ and $\mathbf{M} \in \mathfrak{M}$.

The strategy $\mathbf{M}^*_{(3)}$ stands out for guaranteeing even strict equality of all accessible universes strung along the corresponding coevolution path. This situation is illustrated in Fig. 24.

If a weak Brundtland strategy $\mathbf{M}^*_{(3)}$ exists, then it is generally not the unique solution of this specific version of the SD problem. Note that $\mathbf{M}^*_{(3)}$ satisfies the requirements made if, and only if,

$$\boldsymbol{\Pi}(\mathbf{M}^*_{(3)}) \equiv \boldsymbol{\Pi}(\mathbf{P}_0, 0 \mid \mathbf{M}^*_{(3)}) \subset \mathfrak{U}^{-1}(\mathbf{P}_0) \quad . \tag{129}$$

Proof:

(i) Select an arbitrary time $\hat{t} \geq 0$ and assume that Eq. 129 applies. From this assumption it follows that

$$\mathbf{P}(\hat{t} \mid \mathbf{M}^*_{(3)}) \in \boldsymbol{\Pi}(\mathbf{M}^*_{(3)}) \subset \mathfrak{U}^{-1}(\mathbf{P}_0) \quad , \tag{130}$$

i.e., the coevolution state considered is in the basin of access to $\mathbf{P}_0$. Therefore, $\mathbf{P}_0$ can be reached from $\mathbf{P}(\hat{t} \mid \mathbf{M}^*_{(3)})$ by appropriate management, and subsequently the entire universe $\mathfrak{U}(\mathbf{P}_0)$ may be explored.

(ii) Assume that Eq. 129 is not satisfied, so there is a time $\tilde{t} > 0$ with

$$\mathbf{P}(\tilde{t} \mid \mathbf{M}^*_{(3)}) \notin \mathfrak{U}^{-1}(\mathbf{P}_0) \quad . \tag{131}$$

Then the state $\mathbf{P}_0$, which is definitely an element of $\mathfrak{U}(\mathbf{P}_0)$, cannot be reached from $\mathbf{P}(\tilde{t} \mid \mathbf{M}^*_{(3)})$ by definition. Thus, $\mathbf{M}^*_{(3)}$ is not eligible for weak Brundtland management. ■

As a consequence, the weak Brundtland strategy boils down to *keeping the coevolution within the basin of access to* $\mathbf{P}_0$. This implies that there remains, as a rule, considerable freedom in picking a management sequence which may meet complementary objectives.

However, how can it be ensured that the coevolution does not leave $\mathfrak{U}^{-1}(\mathbf{P}_0)$? In order to answer this question, the irreversibilities in systems control as defined by *critical manifolds*, must be investigated.

Definition 7: Consider a manifold Γ which separates two adjacent domains Ω_1, Ω_2 of coevolution space $\mathbf{C}$.

Γ is called *critical* if for each point $\mathbf{P} \in \Gamma$ we find that the managed coevolution may pass through $\mathbf{P}$ from Ω_1 to Ω_2, but *not* vice versa. ■

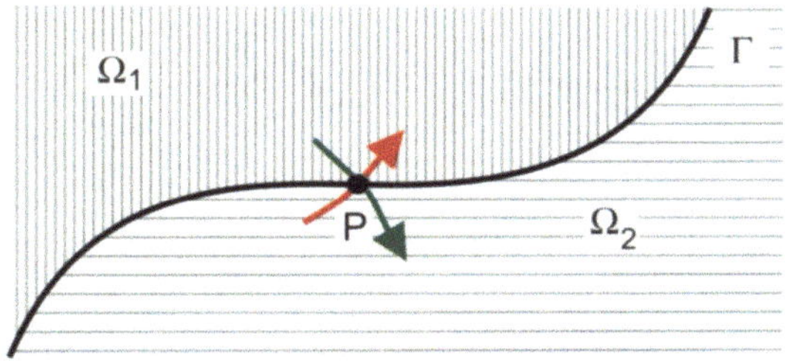

Remarks:

(i) The above definition is made without loss of generality as the labels of the respective domains can be chosen at will.

(ii) The boundary between the domains does not have to be as smooth as the above sketch and the mathematical expression "manifold" suggest. Modern dynamic systems theory tells us that such separating sets may be rather "strange" or "fractal" (*Mandelbrot*, 1991 [144]), even in low-dimensional systems.

(iii) Our definition of "critical" is fairly wide in so far as it identifies this notion with *irreversible* coevolutionary moves, which do not have to be *intolerable* or *discontinuous* as well. Therefore, the boundary of any catastrophe domain is certainly critical, but critical manifolds may also hedge in absolute security zones (see Sect. 4.3).

Existence, form and position of critical manifolds in coevolution space depend on the intrinsic systems dynamics as well as on the peculiarities and limits of the available geo-cybernetic options. In our hydrographic illustration of an Anti-Murphy strategy as depicted in Fig. 20b, for instance, two critical lines were already identified: one of them, the edge of the waterfall as a boundary of the catastrophe domain $\mathfrak{K}$, is quasi-independent of all potential manoeuvres. The borderline which delimits the basin of access to $\mathfrak{K}$ upstream of the rapids, however, is largely determined by the propulsive endowment and the hydrodynamic properties of the boat.

We have to emphasize that the coevolutionary irreversibility defined by a critical manifold has only *local* relevance, if this manifold does not completely isolate an "interior domain" from the rest of the accessible universe $\mathfrak{U}(\mathbf{P}_0)$ within **C**. Consider, for example, the metaphoric hydrographic situation sketched in Fig. 25.

The answer to the question of whether a given critical manifold Γ nevertheless admits *global* reversibility is generally a difficult task. One major prerequisite is a thorough investigation of the manifold's topology with respect to the embedding sets $\mathfrak{U}^{-1}(\mathbf{P}_0)$ and **C**. Such a situation is well known from non-linear Hamiltonian dynamics, where impermeable barriers in phase space

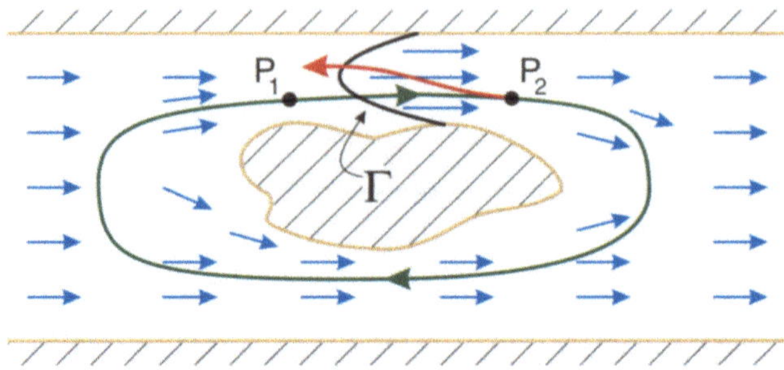

Figure 25. The critical line Γ is determined here by a limit stream velocity which is only exceeded in the upper river arm. The move from $\mathbf{P}_1$ to $\mathbf{P}_2$ is locally irreversible, yet *globally reversible* due to the fact that Γ does not form a *closed loop.*

(the so-called Kolmogorov-Arnold-Moser tori) do not forbid global intrinsic mobility. This phenomenon, which requires sufficiently high dimensionality of the system in question, goes by the name of "Arnold diffusion" (*Arnold*, 1964 [6]) and is responsible for the long-term instability of most arrangements of celestial bodies in the universe (see Sect. 6).

Note that our critical manifolds behave rather like "semi-permeable membranes" with respect to geo-cybernetic action. Let us now assume that the highly non-trivial survey of distributions and properties of these membranes in coevolution space can be performed in a satisfactory way. Let us also assume that this analysis results in some kind of *"navigational chart"* that clearly indicates the set $\{\Gamma(\mathbf{P}_0)\}$ of critical manifolds which delimit $\mathfrak{U}^{-1}(\mathbf{P}_0)$ in the sense of *global irreversibility*. Then weak Brundtland management sequences $\mathbf{M}^*_{(3)}$ characterized by Eq. (129) are implicitly generated by the following operational rule:

Avoid crossing any of the manifolds in the collection $\{\Gamma(\mathbf{P}_0)\}$, i.e., implement only those geo-cybernetic actions which keep the system away from a given $\Gamma(\mathbf{P}_0)$ or make it move tangentially to that critical boundary.

Several *comments* on this innocent-looking instruction have to be made:

(i) As mentioned before, the operational scheme generally does not determine a unique weak Brundtland strategy but leaves the Global Subject with considerable freedom for maneuvring according to complementary E & D criteria. Whether this freedom is sufficiently large – or small – to guarantee the possible satisfaction of all the individual generations to come is by no means evident. We will return to this point below.

(ii) The observance of the boundaries $\{\Gamma(\mathbf{P}_0)\}$ is both a necessary and a sufficient condition for preserving all asymptotic coevolutionary options.

A less efficient operational principle, yet safer and easier to implement, would be the following one:
Just refrain from traversing any *locally* critical manifold $\Gamma_{\text{loc}}(\mathbf{P}_0)$, i.e., any "semi-permeable membrane" for geo-cybernetic motion departing from $\mathbf{P}_0$ (see the general definition of critical manifolds above).
Note that a locally critical manifold with respect to $\mathbf{P}_0$ does not have to be globally critical as well: Fig. 25 features such a situation within our hydrographic allegory. The same allegory also gives us a hint how the entity $\Gamma_{\text{loc}}(\mathbf{P}_0)$ might be identified and avoided in an on-line-fashion, that is, by direct inspection of short-term coevolution dynamics. The steersman "simply" has to perpetually monitor the local velocity of the supporting medium, and to change course if this velocity approaches a maximum value defined by the propulsive equipment.
Note that the "primitive" operational scheme formulated here is clearly *sufficient* in the weak Brundtland sense, as

$$\{\Gamma(\mathbf{P}_0)\} \subset \{\Gamma_{\text{loc}}(\mathbf{P}_0)\} \quad . \tag{132}$$

(iii) Whether a weak Brundtland strategy can be implemented at all – by any operational principle – crucially depends on the topological properties of the family of critical manifolds in coevolution space. These manifolds might, e.g., embrace the starting point $\mathbf{P}_0$ like a dense set of onion skins. In such a case there remains no open domain for geo-cybernetic manoeuvring.

(iv) Things become even more complicated *if the geo-cybernetics depends explicitly on time.* Under such circumstances, the (local and global) critical manifolds may not be stationary, for instance

$$\{\Gamma(\mathbf{P}_0)\} \equiv \{\Gamma(t \mid \mathbf{P}_0)\} \quad . \tag{133}$$

As a consequence, local manoeuvring becomes rather risky and cannot warrant global reversibility of coevolutionary motion any more (just think of steering a vessel through a river delta influenced by strong tidal forces). The formal analysis of this genuinely dynamic control problem requires a fully "teleological" approach and is beyond the scope of this essay.

Example 3 ("Qualified Brundtland Strategies"):
Up to now we have discussed only extreme variants of the Equitization paradigm, which are supposed to secure either the instantaneous or the asymptotic availability of E & D options in the coevolution process. Clearly, the first strategy is not a realistic one, while the second strategy may be intolerable for certain generations along the coevolution line: weak Brundtland management, after all, only guarantees the access to specific pockets of

$\mathbf{C}$ in the long run – but not necessarily within the time span relevant to the generations in question.

So we have to ask ourselves whether *intermediate* variants of $\mathcal{P}_3$ can be devised that warrant "intergenerational equity" in its proper sense. An obvious approach would be to consider time windows of reasonable length ($T = 20$ or 50 years) and the corresponding *geo-cybernetic plumes*. Then the Equitization recipe might be formulated in the following way:

Find a management sequence $\mathbf{M}^*_{(3)}$ with the property that

$$\mathfrak{U}_T(\mathbf{P}(\hat{t} \mid \mathbf{M}^*_{(3)})) = \mathfrak{U}_T(\mathbf{P}_0) \quad \text{for all} \quad \hat{t} \geq 0 \ . \tag{134}$$

Most probably, there is no such element in the management pool $\mathfrak{M}$ which is capable of strictly satisfying the above condition. There may be, however, various ways of relaxing the rigorous requirement without spoiling its essence. We might, for instance, consider two plumes $\mathfrak{U}_T(\mathbf{P}_1)$ and $\mathfrak{U}_T(\mathbf{P}_2)$ to be "nearly equal" if the following statement holds:

Let $d > 0$ be a given (small) distance in the *normed* coevolution space $\mathbf{C}$ (see also Sect. 4.5). Then the d-neighbourhood of any point $\bar{\mathbf{P}}_1 \in \mathfrak{U}_T(\mathbf{P}_1)$ contains at least one point $\bar{\mathbf{P}}_2 \in \mathfrak{U}_T(\mathbf{P}_2)$ and vice versa.

However, note that even such an approximate equality of geo-cybernetic plumes accessible to the generations interlinked through a specific coevolution path might not be attainable. As a consequence, we ultimately may have to resort, within the framework of paradigm $\mathcal{P}_3$, to *qualified* Brundtland strategies which at least preserve the *equivalence* of the co-variant medium-term option space. Let us emphasize that we thereby will introduce, inevitably, patronizing (at best) or gratuitous (at worst) elements into the geo-cybernetic process, so there is a high price to be paid for feasibility.

One operational scheme for generating qualified Brundtland strategies is the *"principle of self-referential positive discrimination"*. We will make no attempt to formalize this principle here, yet we sketch the underlying rationale:

Imagine that the first generation, departing from state $\mathbf{P}_0$, chooses a management episode $\mathbf{M}^*_T$ which produces the path segment $\mathbf{P}_T(\mathbf{P}_0, 0; t \mid \mathbf{M}^*_T)$. This coevolution segment may traverse a number of critical manifolds that will constitute impermeable barriers for the coming generations. The choice of the first generation may nevertheless be accepted as "equitable" *if* these manifolds either

- delimit *miserable* or *dangerous* pockets in coevolution space (like the basins of access to catastrophe domains), or
- encompass optional realms in $\mathbf{C}$ that all future generations supposedly will not wish to reenter *due to the trend-setting initial choice* (like the status quo ante of wetlands transformed into bucolic landscapes or flourishing cities in the Netherlands),

and *if* the manifolds are crossed in the "right" direction, of course.
Then the second generation (or couple of generations) will make their choice, and so on. This interactive process hopefully keeps the optional quality of the respective geo-cybernetic plumes at least at a constant level, although (or because) the future generations are progressively deprived of their "license" to commit fatal mistakes ...

We have to point out, however, that the SD recipe just described only involves "eternal" critical thresholds, so it might be even used for "improving" the weak Brundtland strategy. If we *seriously* want to compare the optional T-plumes available to distinct generations down the coevolution line, we have to deal with a most intricate additional problem: migration along any coevolution path generally rules out certain options for future generations – simply because the Earth System has moved away from initially accessible states in **C** and cannot be driven back there within the limited time span T. On the other hand, new opportunities will emerge, as a rule, due to this drift in coevolution space. Now the difficult task is to *compare* the options lost to the options gained by number and character in order to determine *whether the overall opportunistic quality of the T-plume will be preserved.*

To fulfill such a task, a fully-fledged teleological analysis of all geo-cybernetic paths emanating from $\mathbf{P}_0$ is needed, not to mention the projective value-judgements to be made. As a consequence, a robust and practical implementation scheme for genuinely qualified Brundtland strategies may not exist – a fact that would drastically diminish their political relevance. However, all this is at present scientific "terra incognita" and remains a major research challenge for the future. Fig. 26 tries to give an impression of what the analysis will have to be all about.

4.5 Stabilization

We now come to a principal geo-cybernetic model – the "*Stabilization paradigm*" $\mathcal{P}_4$ – which seems to fit best the popular notions of "Sustainable Development" and which can be expected to yield to operationalization most easily. On closer examination, however, this paradigm proves to be quite problematic in its essence as well as structure.

The entire E & D debate is still – predominantly in the advanced industrialized countries – characterized by *massive fears* of losing the doggedly elaborated level regarding consumption options, mobility, security, educational and recreational offers, etc. The "Limits-to-Growth" philosophy of the Club of Rome continues to resound in these fears, although the public concerns about the future of this planet have recently focused more on the natural "sinks" for the metabolic secretions of the anthroposphere "Beyond the Limits" (*Meadows* et al., 1992 [148]). As a consequence, society and its decision-makers exhibit a strong desire for safety, stability, even *stagnation*

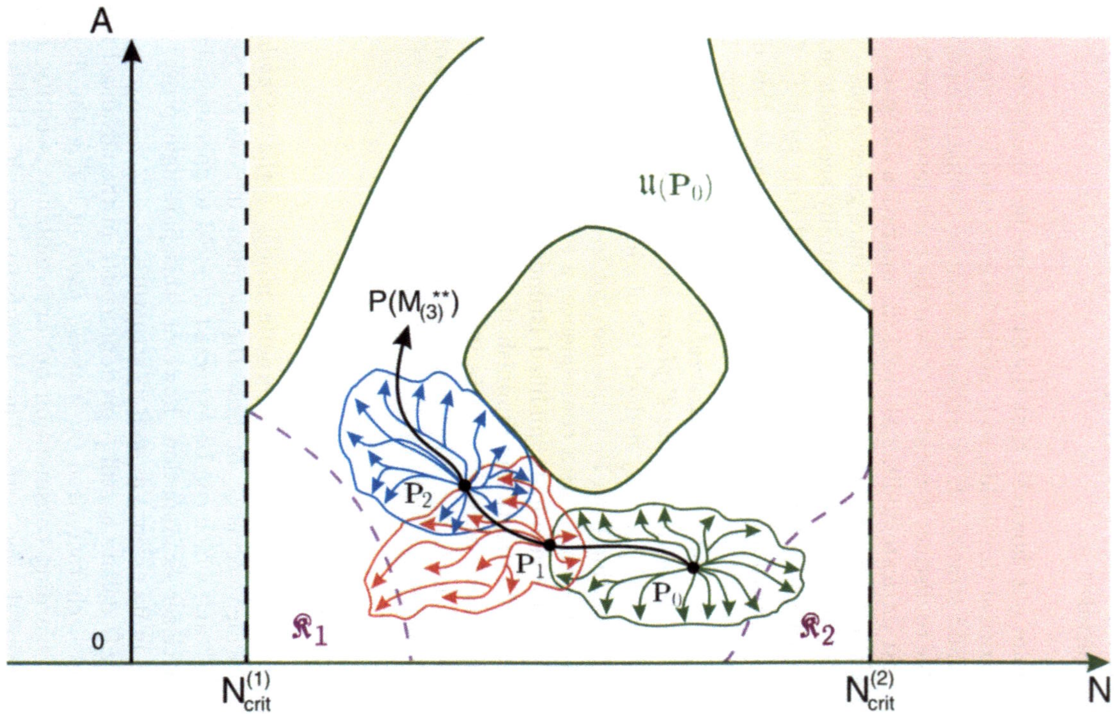

Figure 26. Cartoon of a qualified Brundtland coevolution inducing "equivalent" successive T-plumes in the sense of optional equity. The underlying management sequence is $\mathbf{M}_{(3)}^{**}$, and $\mathbf{P}_1 \equiv \mathbf{P}(T \mid \mathbf{M}_{(3)}^{**})$, $\mathbf{P}_2 \equiv \mathbf{P}(2T \mid \mathbf{M}_{(3)}^{**})$, etc., are the generational stepping stones down the E & D flow. However, note that we may run here again into a tantalizing inherent problem of paradigm $\mathcal{P}_3$, which we already mentioned earlier: in order to preserve (qualified medium-term) options, almost all of them have to be dismissed!

in the relationship between global civilization and nature. Strictly speaking, this is a desire for the *"end of all environmental history"*.

This longing for stable equilibria is expressed in a relatively vague way in various recent publications such as Al Gore's bestseller "Earth in the Balance" (*Gore*, 1992 [90]), but also quite concretely and socio-economically effectively in the stability goals or stabilization mechanisms, respectively, of global environmental treaties such as the "Framework Convention on Climate Change" and the "Montreal Protocol" for protecting the stratospheric ozone layer (*WBGU*, 1996 [83]).

In its most simplified version, the Stabilization paradigm boils down to the deliberate steering towards a selected state $\mathbf{P}^*$ in coevolution space, *where the Earth System can be safely "detained" through suitable geo-cybernetic management.* $\mathcal{P}_4$ is therefore in diametric opposition to the Equitization paradigm $\mathcal{P}_3$, which moves the evolutionary capacity and flexibility of the environmental system under consideration into the limelight. Thus the Stabilization paradigm may be seen as a radical, yet systemic materialization of the quite sweeping Standardization paradigm $\mathcal{P}_0$. On the one hand, $\mathcal{P}_4$ rules out any net long-term coevolutionary progress and therefore replaces the subtle desideratum "Sustainable Development" by the rather blunt objective of "*Sustainability*". On the other hand, the Stabilization paradigm does *not* simply *prescribe* coevolution states or paths, but searches systematically for balancing management options in accordance with the intrinsic dynamics of the Earth System and with the available pool of steering instruments. So we can interpret $\mathcal{P}_4$ as the "true" logical counterpart to paradigm $\mathcal{P}_3$.

One may dream of blending those two SD models. This could be accomplished, in particular, if a coevolution state $\mathbf{P}^{**} \in \mathbf{C}$ were found which united maximum stability and evolutionary capacity – thus constituting a simultaneously optimal defense and attack position, in military jargon. The actual existence of a state of this type, or at least the availability of a corresponding set of states (see below), is a speculative matter that we do not want to pursue further here.

We rather wish to attempt in the following to formulate the Stabilization paradigm more precisely, to discuss its main aspects and to analyse the prerequisites for its operationalization. Let us initially disregard the systemic questions concerning the availability or accessibility of "equilibrium states $\mathbf{P}^*$" and assume that a choice has to be made among many states of this type. We are thus confronted with the second of our fundamental questions formulated above, namely *"What kind of world do we want?"* Or, expressed in more specific terms: Which environmental qualities are to be strived for and to be stabilized without thereby ignoring the pertinent socio-economic dimensions?

The corresponding public and academic debate is generally not characterized by oversized imagination. With very few exceptions the following guiding principles on the road to equilibrium are encountered over and over again:

(i) Preservation.
This strategy aims at practically *freezing the current coevolution state* along with all its natural and civilizatory aspects. This in no way means that the status quo is viewed as being the best possible state of affairs. It possesses, however, the dignity of an unplanned historical product, just like the biological species that it includes. The preservationist attitude certainly lacks sophistication, but it is precisely this shortcoming that makes it a directly convincing and extremely impressive conviction. Everyone who has, for example, argued with governmental curators of historic monuments about the development (or reconstruction) of urban ensembles can tell you a thing or two about that.

(ii) Restoration.
A frequently advanced radical eco-centric alternative to preservation is the *re-establishment of the status quo ante*, i.e. of a "virgin" nature unspoilt by anthropogenic deformations and perturbations. However, what does this primeval reference state look like for a specific biotope to be restored – or for the Earth System as a whole? How many decades or centuries do we have to go back in the flow of time to be able to catch sight of the "pure" state? These questions are especially difficult to answer if anthropomorphic landscapes are to be "renaturalized" – a process that may induce significant losses in biological diversity, ecological performance and aesthetic attractions. Apart from these problems, restoration is illusory as a *comprehensive*, large scale strategy, as it will always be restricted to specific areas within narrow bounds (such as river meadows, for example).

(iii) Segregation.
By way of constrast, an environmental policy that aims at a stepwise *disentanglement of nature and civilization* may be comparatively "successful". The final product of such a strategy would have be a global ensemble of protected areas (ranging from national parks to miniscule wetlands set aside by municipalities) that were largely decoupled from coevolution and left to their own self-organized development. The Federal Environmental Agency of Germany (UBA), e.g., actually defines as the reference state of a "valuable" biogeotope the asymptotic conditions that would form without further human intervention (*Irmer*, 1997 [110]).

Such a "ghettoization strategy" for nature certainly enjoys a relatively high degree of public acceptance or even approval at present, but it is problematic in several respects: what environmental quality standards should apply to the immense rest of the "civilized" areas; what opportunities for experiencing nature would remain for human beings; and, above all, how could the areas to be protected be screened off from large-scale anthropogenic influences (e.g., through the atmosphere)? On closer inspection, a long-term, stable segregation of the ecosphere and the anthroposphere appears to be neither a desirable nor a realizable venture ...

Our brief outline of the various popular, but ill-defined alternatives for the establishment of a coevolutionary target state should already have made it clear that, in addition to the question of social preference, the aspects of *"attainability"* and *"sustainability"* must not be neglected. Thus it is imperative that systems analysis complements the normative considerations. A series of more recent and more subtle approaches to the definition of global or regional environmental-quality goals already meet this demand to a certain extent; i.e. these approaches formulate action principles ("ecological imperatives") that characterize the respective "target state" as an explicit or implicit equilibrium. Let us illustrate this with a few examples:

A number of topical studies on the sustainability of large geographic units like countries or continents (see, e.g., *Buitenkamp* et al., 1992 [36]; *FEE*, 1995 [74]; *Spangenberg*, 1995 [219]; *Loske*, 1996 [138]) rely on the *"environmental space"* concept (*Opschoor*, 1992 [166])) also relevant to the Standardization paradigm (see Sect. 4.1)). That concept is derived from an ecological theory considering the limited "carrying capacity" of the planetary ecosystem to be shared among a growing number of human beings – not to speak of other species. The overall approach therefore envisages not a target coevolution state but rather a *minimum-standard equilibrium* enforced by the laws of nature.

A counterpart of this is provided by the vision of *"Park Earth"*, which evokes the picture of a biosphere that is largely preserved yet optimally designed according to cultural preferences and needs. Within the framework of ecosystems theory, the corresponding equilibrium is rather a *"climax state"* as determined by humanity's abilities as a "gardener" (see, e.g., *Clark* and *Munn*, 1986 [43]). The resulting global environmental patchwork is supposed to guarantee both the maximum mutual permeability and stabilization of nature and civilization, as striven for – at the landscape level – by UNESCO's "Man and the Biosphere Programme" (*UNESCO*, 1988 [236]; *USNC*, 1995 [240]).

Explicit *operational rules* for constructing a coevolutionary *fixed state* are given by the Daly school of environmental economists (*Barbier*, 1989 [14]; *Daly*, 1990 [49]; *Pearce* and *Turner*, 1990 [174]):

(i) The consumption rate of any renewable natural resource must not exceed its regeneration rate.
(ii) The reduction rate of any non-renewable resource must not exceed the rate of substituting it by a fully equivalent renewable resource.
(iii) The emission rate of any anthropogenic substance must not exceed the pertinent absorption and transformation rates by natural sinks.

At first sight, these rules seem to reflect a dull and static concept which simply aims at preserving the very status quo. At second sight, we notice, however, that those rules (in particular, (ii)) refer to *compound variables* which allow for some freedom regarding the values of the individual constituents. As a matter of fact, there are many more subtleties involved here. We have to take into account particularly the fact that the above-mentioned regeneration, absorption or transformation rates of natural systems may be elastic quantities heavily depending on the harvesting or immission intensities. Standard chemical reactions theory certainly supports such a point of view. The "Daly Equilibrium" is therefore a *dynamic* (or homeorhetic) entity in disguise and, as a consequence, is both less illusive and more problematic than it initially appears. An exemplary discussion further below will feature some of the highlights and deficiencies of this concept.

* * *

These first thoughts on the Stabilization paradigm already make clear that the concept of "equilibrium" has to be defined in the SD context much more broadly than usual. It is probably completely futile to strive for a fixed coevolution state with a detailed specification of all micro-variables, even if a social agreement concerning such a desideratum could be reached. Rather, it will be useful, if not necessary, to consider entire *sets of coevolution states* as complex stabilization targets. These sets define *generalized dynamic equilibria* with respect to temporal averaging or compositional aggregation.

So the question "What kind of world do we want?" will have no simple answer, but this is by no means the whole story. Even if a satisfactory *generalized coevolutionary equilibrium* were identified, the question *"How do we get there?"* would still have to be pondered. The answer to the latter question will be complex as well: both the overall *accessibility* of the target equilibrium and the comparative *costs* (in the broadest sense of the word) of the eligible stabilizing avenues have to be investigated in full depth. All these aspects will be dealt with in the rest of this section.

4.5.1 Generalized Equilibria

A. Dynamic Stabilization.

First of all we have to state in precise terms what type of managed coevolution can be classified as a "stable" one.

Definition 8:

(i) As indicated above, let us assume that the coevolution space $\mathbf{C}$ is endowed with a norm $|| \ ||$ that allows us, in particular, to measure the *distance* between any pair of coevolution states. Now let $\varepsilon > 0$ and $\bar{\mathbf{P}}$ be some arbitrary point in $\mathbf{C}$. Then the (open) ε-neighbourhood of $\bar{\mathbf{P}}$ in coevolution space, $\mathfrak{V}_\varepsilon(\bar{\mathbf{P}})$, is defined as follows:

$$\mathfrak{V}_\varepsilon(\bar{\mathbf{P}}) := \{\mathbf{P}' \in \mathbf{C} \mid ||\mathbf{P}' - \bar{\mathbf{P}}|| < \varepsilon\} \ . \tag{135}$$

(ii) Consider an arbitrary coevolution path $\mathbf{P}(\hat{\mathbf{P}}, \hat{t}; t \mid \mathbf{M})$, as generated by the management sequence $\mathbf{M} \in \mathfrak{M}$, and the associated coevolution trajectory $\boldsymbol{\Pi}(\hat{\mathbf{P}}, \hat{t}; \mathbf{M})$ with starting point $\hat{\mathbf{P}}$. For given $\varepsilon > 0$ and coevolution state $\mathbf{P}(\bar{t}) \equiv \mathbf{P}(\hat{\mathbf{P}}, \hat{t}; \bar{t} \mid \mathbf{M})$ the recurrence time $T_\varepsilon(\mathbf{P}(\bar{t}))$ is defined as the minimum time needed by the coevolution to revisit the ε-neighbourhood of $\mathbf{P}(\bar{t})$. So

$$\mathbf{P}(\bar{t} + T_\varepsilon(\mathbf{P}(\bar{t}))) \ \in \ \mathfrak{V}_\varepsilon(\mathbf{P}(\bar{t})) \ , \tag{136}$$

$$\text{i.e.} \quad ||\mathbf{P}(\bar{t} + T_\varepsilon(\mathbf{P}(\bar{t}))) \ - \ \mathbf{P}(\bar{t})|| < \varepsilon \ . \tag{137}$$

Note that, in general, $T_\varepsilon(\mathbf{P}(\bar{t}))$ does not have to be finite, i.e. $\mathfrak{V}_\varepsilon(\mathbf{P}(\bar{t}))$ might not be revisited at all.

(iii) A *stable coevolution path* $\mathbf{P}(t) \equiv \mathbf{P}(\hat{\mathbf{P}}, \hat{t}; t \mid \mathbf{M}^*_{(4)})$ satisfies the following condition:

$$\sup_{t \geq \hat{t}} T_\varepsilon(\mathbf{P}(t)) < \infty \quad \text{for any} \quad \varepsilon > 0 \ . \tag{138}$$

The generating management sequence $\mathbf{M}^*_{(4)}$ is then referred to as a *stabilizing management.* ■

Remark:

The stability condition formulated above guarantees that the corresponding coevolution reapproaches any of its states as closely as desired after a finite waiting time. This implies, in particular, that each point on the trajectory $\boldsymbol{\Pi}(\hat{\mathbf{P}}, \hat{t} \mid \mathbf{M}^*_{(4)})$ is an "accumulation point" of the coevolutionary sequence of states. It does *not* imply, however, that all the distinct points on $\boldsymbol{\Pi}(\hat{\mathbf{P}}, \hat{t} \mid \mathbf{M}^*_{(4)})$ are covered within a finite period.

Examples of Stable Coevolutions.

Let us first emphasize that a stable coevolution path is established by the right choice of the triple $(\hat{\mathbf{P}}, \hat{t}, \mathbf{M}^*_{(4)})$ of control variables, i.e. by appropriate selection of initial state, starting time and management sequence. The triple reduces to the control couple $(\hat{\mathbf{P}}, \mathbf{M}^*_{(4)})$ if the geo-cybernetic process does not explicitly depend on time.

We further note that our stability concept does not directly refer to robustness or resilience properties in the usual sense of stability theory, but primarily to the *recurrent* character of the coevolution path in the sense of *stationarity.* Thus, our definition of stable coevolution accommodates a wide range of dynamic patterns in coevolution space, which will be perceived as generalized equilibria. In the following we present a few illustrating examples.

(i) *Fixed Points:*

$$\mathbf{P}(\hat{\mathbf{P}}, \hat{t}; t \mid \mathbf{M}^{fix}_{(4)}) = \hat{\mathbf{P}} = const. \quad , \tag{139}$$

i.e. the management sequence $\mathbf{M}^{fix}_{(4)} \in \mathfrak{M}$ exactly preserves the initial coevolution state in spite of the intrinsic Earth System dynamics as described by Eq. 2. This is, of course, the trivial case of a generalized equilibrium which coincides with the familiar *static* concept.

$\hat{\mathbf{P}} \equiv (\hat{\mathbf{N}}, \hat{\mathbf{A}})$ is a *fixed coevolution state* from the time $\hat{t}$ on if and only if the following conditions are fulfilled (see Eq. 17):

$$\begin{aligned} \dot{\mathbf{N}} &= F_2(\hat{\mathbf{N}}, \hat{\mathbf{A}}; t; \mathbf{M}^{fix}_{(4)}) = 0 \ , \\ \dot{\mathbf{A}} &= G_2(\hat{\mathbf{N}}, \hat{\mathbf{A}}; \mathbf{M}^{fix}_{(4)}) = 0 \ , \\ &\text{for all} \quad t \geq \hat{t} \ . \end{aligned} \tag{140}$$

If the system is "autonomous", implying that t does not appear in the first line of Eq. 140, then stationary management

$$\mathbf{M}^{fix}_{(4)}(t) \equiv \hat{\mathbf{M}}^{fix}_{(4)} = const. \tag{141}$$

will suffice to do the job. However, note that even under autonomous conditions there might not exist such a management element for any real starting point $\mathbf{P}_0$ of the coevolution. It seems, for instance, impossible to stabilize the present climate by any reduction scenario for the global greenhouse gas emissions because of the inertia of the planetary geophysical circulation system. On the other hand, it cannot be completely ruled out that a judicious addition of certain "geo-engineering ingredients" like afforestation or aerosol injection could meet the preservation goal in spite of that inertia.

(ii) *Periodic Orbits:*

$$\mathbf{P}(\hat{\mathbf{P}}, \hat{t}; t + \hat{T} \mid \mathbf{M}_{(4)}^{per}) = \mathbf{P}(\hat{\mathbf{P}}, \hat{t}; t \mid \mathbf{M}_{(4)}^{per}) \qquad (142)$$

for all $t \geq \hat{t}$ and some *finite period* $\hat{T} > 0$.
The coevolution trajectory $\boldsymbol{\Pi}(\hat{\mathbf{P}}, \hat{t} \mid \mathbf{M}_{(4)}^{per})$ is then a *closed loop* in coevolution space $\mathbf{C}$.

(iii) *Quasiperiodic Orbits:*
Here $\mathbf{P}(\hat{\mathbf{P}}, \hat{t}; t \mid \mathbf{M}_{(4)}^{qp})$ can be expressed as a function (*Bergé* et al., 1984 [21])

$$f(\omega_1 t, \omega_2 t, \ldots, \omega_L t) \; , L \geq 2 \; , \qquad (143)$$

for all $t \geq \hat{t}$.
Quasiperiodicity of such a coevolution is established by the fact that the map $f : \mathbb{R}^L \to \mathbf{C}$ is 1-periodic in each of its arguments, i.e.,

$$f(x_1, \ldots, x_l + 1, \ldots, x_L) = f(x_1, \ldots, x_L) \quad \text{for any} \quad l \in \{1, \ldots, L\} \; , \qquad (144)$$

and the "fundamental frequencies" $\omega_1, \ldots, \omega_L$ are *incommensurate*. The latter simply means that the equation

$$\sum_{l=1}^{L} n_l \omega_l = 0 \qquad (145)$$

cannot be satisfied by a non-trivial L-tuple $(n_1, \ldots, n_L)$ of integers.
We obviously have

$$f(\omega_1 \hat{t}, \ldots, \omega_L \hat{t}) = \hat{\mathbf{P}} \; , \qquad (146)$$

and the coevolution trajectory $\boldsymbol{\Pi}(\hat{\mathbf{P}}, \hat{t} \mid \mathbf{M}_{(4)}^{qp})$ densely covers a *torus* in $\mathbf{C}$, as sketched below for the simplest case $L = 2$ (*Stewart*, 1989 [222]).

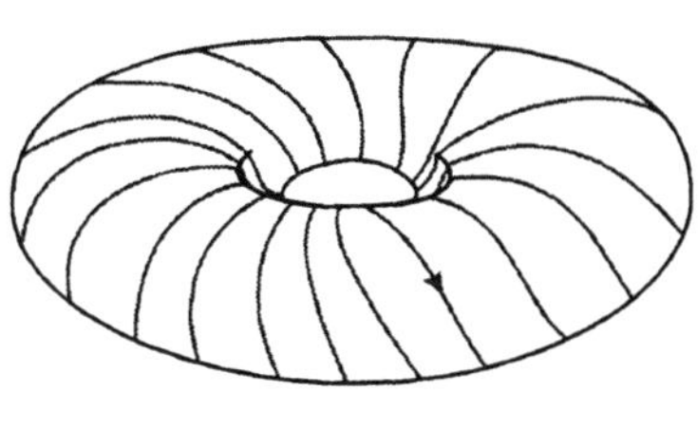

The quasiperiodic functions constitute a specific sub-set of the so-called *almost periodic* functions: as a matter of fact, all of the latter entities may represent stable coevolution patterns according to our definition.

(iv) *Strange Attractors:*
Our qualification of stable coevolutions as recurrent ones also allows for very complicated "equilibrium" motion in **C**, which may produce wild trajectories similar to the pattern shown in Fig. 27.

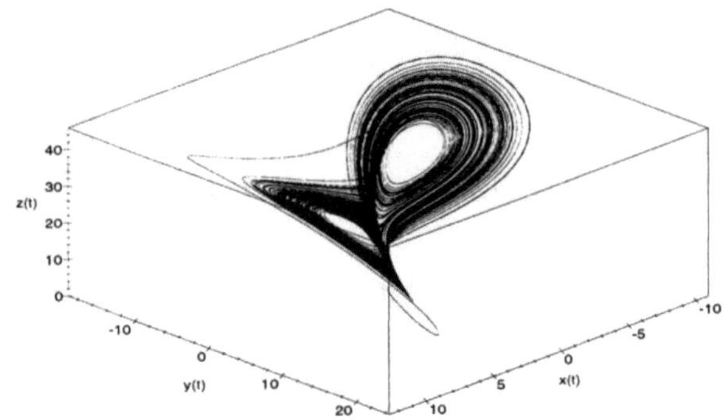

Figure 27. The famous *Lorenz Attractor*, which governs the dynamic behaviour of the solutions of a non-linear reduced-form model for the atmospheric circulation (*Lorenz*, 1963 [137]).

The pattern depicted above is the first example ever recognized by scientists of a so-called "strange attractor" – a complicated invariant set encountered in dissipative dynamic systems that stands out for its *Cantor structure* (see e.g., *Falconer*, 1990 [65]).
It is by no means improbable that certain management sequences $\mathbf{M}_{(4)}^{str}$ are capable of keeping the coevolution, if not in a fixed state, then at least on such bizarrely confined tracks. But contrary to the examples presented in (i) through (iii), we are here not able to describe the explicit form of $\mathbf{P}(\hat{\mathbf{P}}, \hat{t}; t \mid \mathbf{M}_{(4)}^{str})$ in simple mathematical terms.

* * *

All our examples of generalized equilibria in the sense of recurrent motion are taken from the theory of dissipative or symplectic dynamic systems that emerged in the last three decades. This theory is equipped with yet many more patterns according to which a stable coevolution may unfold – like those tenuous fractal objects named "Julia Sets" (*Julia*, 1918 [111]).

However, note that we deal within the geo-cybernetic setting with non-linear dynamic systems which are subject to external control as an additional complication. Thus a thorough analysis might reveal novel and even more complex structures accommodating stable coevolutions.

B. Compound Stabilization.

Due to the extension of the equilibrium concept to recurrent motion, as discussed above, the possibilities for realizing the Stabilization paradigm of Sustainable Development are considerably enhanced. The remaining requirements on geo-cybernetic management are nonetheless high, as virtually *all* sub-variables representing the coevolution state **P** have to be stabilized (dynamically).

As an alternative, the concept of equilibrium may be generalized and thus relaxed to a certain degree, if only *partial aspects* or *aggregated quantities* of the coevolution were considered, stipulating true constancy in time, however. Especially the second option, whose goal is the stabilization of the compound variables, i.e. of the *macro-features* of the environmental system considered, turns out to be quite relevant.

The formal aspects of this fundamental strategy are readily sketched.

Let A, B, C, ... denote pertinent macro-variables which can be derived from the micro-variables $x_1, x_2, \ldots, x_N$ representing the full information on the system in question (for the sake of simplicity, we assume the set of micro-variables to be finite). Thus

$$A \equiv A(x_1, \ldots, x_N) \ , \ B \equiv B(x_1, \ldots, x_N) \ , \text{ etc.} \tag{147}$$

Compound equilibrium then simply means that, e.g.,

$$\dot{A} = 0 \ . \tag{148}$$

The necessary condition for this to be satisfied clearly is

$$\sum_{n=1}^{N} \frac{\partial A}{\partial x_n} \dot{x}_n = 0 \ , \tag{149}$$

but *not* the detailed-balance requirement

$$\dot{x}_1 = \dot{x}_2 = \ldots = \dot{x}_N = 0 \ . \tag{150}$$

Eq. 149 leaves us with a considerable amount of managerial freedom.

This freedom can well be illustrated by a concrete example that is eminently important for geo-cybernetics. If we select as the pertinent compound variable the mean surface temperature T_{global} of the Earth, then this quantity is mainly a function of the spatial distribution of all greenhouse gas and aerosol particles. The equilibrium condition

$$\dot{T}_{global} = 0 \tag{151}$$

can, in principle, be satisfied by most different regional patterns of compensatory simultaneous enrichment of the Earth's atmosphere with climate-effective substances, the distinct residence times of which have to be taken into account meticulously. The *robustness*, however, of such an artificially-balanced geophysical equilibrium state with respect to natural and anthropogenic disturbances is a different cup of tea – humanity is probably just about to manoeuvre itself unintentionally into a rather risky and fragile situation (*Schult* et al., 1997 [211]; *Feichter* et al., 1997 [69]).

Daly's SD rules (see above) certainly rely on the concept of compound equilibrium as they explicitly allow for ***substitution*** between "equivalent" resources and do not enforce a detailed-balance ecological accounting. However, which are the aggregate entities to be preserved, and what substitutes are rated equivalent? The answers to these questions are not trivial and the implications of the operational rules may be rather undesirable, as will be demonstrated now for the *energy issue*.

Let us assume that the energy resources that will be accessible to civilization in the medium-term future are restricted to three main components, namely fossil fuels F, renewable biomass energy B, and photovoltaic energy S harvested directly from the sun. Now a naive application of Daly's second rule would call for the preservation of the compound entity

$$E := F + B + S \quad , \tag{152}$$

i.e. the sum of energy bulks stored in some form somewhere on this planet.

However, note that it appears unfeasible to accumulate amounts of energy from photovoltaic collection which are in any way comparable to the huge quantities tucked away in the prospected layers of coal or other fossil fuels. As a matter of fact, there is no point in doing so because the crucial entity for civilization is not the overall amount of existing energy but its daily or annual allowance of energy *consumption*. In other words, the quantity to be preserved is the *sum total of available power*

$$L_E = L_F + L_B + L_S \quad , \tag{153}$$

where L_F denotes the energy per unit time provided by fossil fuels, and so on. The preservation rule

$$\dot{L}_E = 0 \tag{154}$$

guarantees that the *energy-consumption option* for humanity is – at least quantitatively – held constant in the course of time. We emphasize that this approach to SD does not take into account explicitly the economic costs involved.

It seems plausible to represent the instantaneous power supply in the following form:

$$L_E = w_F f(F) + w_B g(B) + w_S A \quad . \tag{155}$$

Here w_F, w_B, w_S are *efficiency functions* that depend mainly on technological and logistic developments and therefore explicitly on time t. $f(F)$ and $g(B)$ are *availability functions* for fossil fuels and biomass, respectively, which depend particularly on the distribution and accessibility of these energy forms, while A simply measures the overall area of installed solar-cell panels. Our assumption that the geographic patterns of fossil fuels deposits and biomass-production sites are basically fixed implies that the functions f and g should – to a first approximation – not depend on t. These functions may deviate, however, markedly from the linear shape. The possible form of $f(F)$ is sketched below:

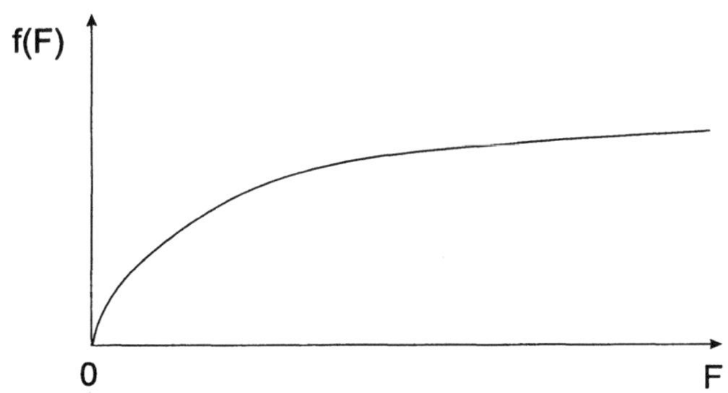

Here the specific feature $f'(0) \gg 1$ is due to the fact that the least accessible deposits will be the last ones to be exploited. Evidently, we have $f(0) = 0$ (as well as $g(0) = 0$).

Let us now scrutinize the stability condition (154), which is equivalent to the equation

$$\dot{w}_F f(F) + w_F f'(F)\dot{F} + \dot{w}_B g(B) + w_B g'(B)\dot{B} + \dot{w}_S A + w_S \dot{A} = 0 \quad , \tag{156}$$

or the even simpler requirement

$$f'(F)\dot{F} + g'(B)\dot{B} + \dot{A} = 0 \quad , \tag{157}$$

if we assume the efficiency functions to have almost constant values close to unity in time.

However,

$$\dot{F} = -\varphi(t) \quad , \tag{158}$$

$$\dot{B} = b(B) - \beta(t) \quad , \tag{159}$$

where $\varphi(t)$ is the current consumption rate of fossil fuels, $b(B)$ is the generation rate of biomass, and $\beta(t)$ is its consumption rate. With all the simplifying hypotheses made, the aggregate stability condition finally boils down to the relation

$$f'(F)\varphi + g'(B)\beta - \dot{A} = g'(B)b(B) \quad . \tag{160}$$

Here the consumption rates φ, β and the substitution rate $\dot{A}$ are the *control variables* for sustained power management.

At a first glance, it seems quite easy to satisfy Eq. 160 in agreement with complementary societal objectives by an appropriate mix of choices for the control variables. This confidence is shattered, however, if we consider, for instance, the limit of fossil fuel depletion, i.e. the process $F \to 0$. Assuming that there is a sub-equilibrium for biomass production and consumption implying

$$\beta = b(B) \quad , \tag{161}$$

we are left with the condition

$$f'(F)\varphi - \dot{A} = 0 \tag{162}$$

to be satisfied in the vicinity of $F = 0$.

However, $f'(F) \gg 1$ for small F as pointed out above. As a consequence, the consumption rate φ has to drop rapidly to zero because the photovoltaic-area expansion rate can obviously not rise to infinity! This is certainly *not* a showcase for inter-generational equity: the full burden has to be borne anyway by the last generations "in power line", if the preceding generations do not refrain from consuming at a finite rate. On the other hand, if the first generations in line actually do renounce consumption options, then the fossil fuels are virtually useless and cannot be invested in economic and technological development around the globe...

This example clearly demonstrates that we may run into serious difficulties with even the more sophisticated recipes for compound equilibrium management. Similar insights could be gained from studying the problem of *stabilizing the overall marine fish stocks* (*Safina*, 1995 [202]). Here it seems obvious that the sum total of wild and aquacultural fish has to be preserved, but a spatiotemporal "fair" implementation of such a plan is all but simple.

C. Dynamic Compound Stabilization.

The notion of "generalized equilibrium" as defined in the context of the Stabilization paradigm is wide enough to allow also for combinations of the interpretations explained in A and B. One might, for instance, only demand from geo-cybernetic management that it organized the *recurrent development of certain aggregate variables.* In doing so, however, we must avoid the transition to a blunt laisser-faire strategy. This can be accomplished, e.g., by stipulating maximum tolerance ranges for the variation of certain core entities (like potable water availability) in the sense of paradigm $\mathcal{P}_0$.

4.5.2 Passage to Equilibrium

The implementation of the Stabilization paradigm $\mathcal{P}_4$ is generally achieved through the following sequence of steps:

1. Definition of the *admissible* set of generalized equilibria.
2. Exploration of the coevolution space $\mathbf{C}$ and management pool $\mathfrak{M}$ in search of available generalized equilibria $\mathbf{P}(\hat{\mathbf{P}}, \hat{t}; t \mid \mathbf{M}^*_{(4)})$ in accordance with Step 1.
3. Selection of the *preferable* generalized equilibrium, in the case that more than one of them exist.
4. Confirmation of the *accessibility* from $\mathbf{P}_0$ for the chosen generalized equilibrium. If our geo-cybernetic system does not depend explicitly on time, then a necessary and sufficient condition is

$$\boldsymbol{\Pi}(\hat{\mathbf{P}} \mid \mathbf{M}^*_{(4)}) \subset \mathfrak{U}(\mathbf{P}_0) \quad . \tag{163}$$

5. Qualification of the optional *passages* to $\mathbf{P}(\hat{\mathbf{P}}, \hat{t}; t \mid \mathbf{M}^*_{(4)})$ from $(\mathbf{P}_0, 0)$ and, perhaps, selection of the optimal one according to additional criteria.

We will not be able to say much about the first three steps here, which mainly require normative considerations. Let us therefore assume that a number of admissible generalized equilibria (of the Clark or Daly type, for instance) have been detected and ranked according to some set of criteria. In the rest of this section we will be concerned with Step 4 and Step 5, respectively.

Let us first illustrate what Step 4 is all about. For that purpose, we return once more to our two-dimensional toy universe and restrict the set of admissible generalized equilibria to coevolution fixed points

$$\tilde{\mathbf{P}}_i \equiv \mathbf{P}(\mathbf{M}^i_{(4)}) \quad , i = 1, 2, \dots \quad , \tag{164}$$

where the $\mathbf{M}^i_{(4)}$ are stabilizing constant managements. A typical constellation is sketched in Fig. 28. In this example, we immediately observe that

$$\tilde{\mathbf{P}}_1 \notin \mathfrak{U}(\mathbf{P}_0), \tilde{\mathbf{P}}_2 \in \mathfrak{U}(\mathbf{P}_0) \quad , \tag{165}$$

so only the second geo-cybernetic fixed point can be approached from the starting point $\mathbf{P}_0$.

When confronted with the full coevolution-management problem, however, the application of the seemingly straightforward criterion (163) may not be feasible or practical any more: why should we trace out the entire accessible universe $\mathfrak{U}(\mathbf{P}_0)$ to confirm the accessibility of a given generalized-equilibrium candidate? Again, it may be more efficient to employ an *inverse approach*, i.e., to start out with the equilibrium in question and to determine its basin of access. Actually the most clever way to prove the attainability of the target state (dynamics) is the *explicit construction* of specific corridors or passages connecting it to $\mathbf{P}_0$. This task is still a highly challenging one, and its solution requires a thorough formal analysis of the structural elements involved.

Such an analysis can be built on previous insights (see, in particular, Sect. 4.3) and is much facilitated by introducing an *"accessibility relation"* between pairs of points in $\mathbf{C}$:

Let us assume in the following that our geo-cybernetics does *not* explicitly depend on time and let $\mathbf{P}_1$, $\mathbf{P}_2$ be arbitrary states in coevolution space. We write

$$\mathbf{P}_1 \curvearrowright \mathbf{P}_2 \tag{166}$$

if $\mathbf{P}_2$ can be reached from $\mathbf{P}_1$ by employing some management episode, i.e. within a finite period of time.

Evidently

$$\begin{aligned} \mathfrak{U}(\mathbf{P}) &= \{\mathbf{P}' \in \mathbf{C} \mid \mathbf{P} \curvearrowright \mathbf{P}'\} \quad (167) \\ \text{and} \quad \mathfrak{U}^{-1}(\mathbf{P}) &= \{\mathbf{P}' \in \mathbf{C} \mid \mathbf{P}' \curvearrowright \mathbf{P}\} \quad . \quad (168) \end{aligned}$$

Note that the symbol "$\curvearrowright$" does *not* describe an equivalence relation (but see below): Obviously, we have

$$\mathbf{P}_1 \curvearrowright \mathbf{P}_1 \quad , \tag{169}$$

but the *"symmetry property"*

$$\mathbf{P}_1 \curvearrowright \mathbf{P}_2 \quad \Rightarrow \quad \mathbf{P}_2 \curvearrowright \mathbf{P}_1 \tag{170}$$

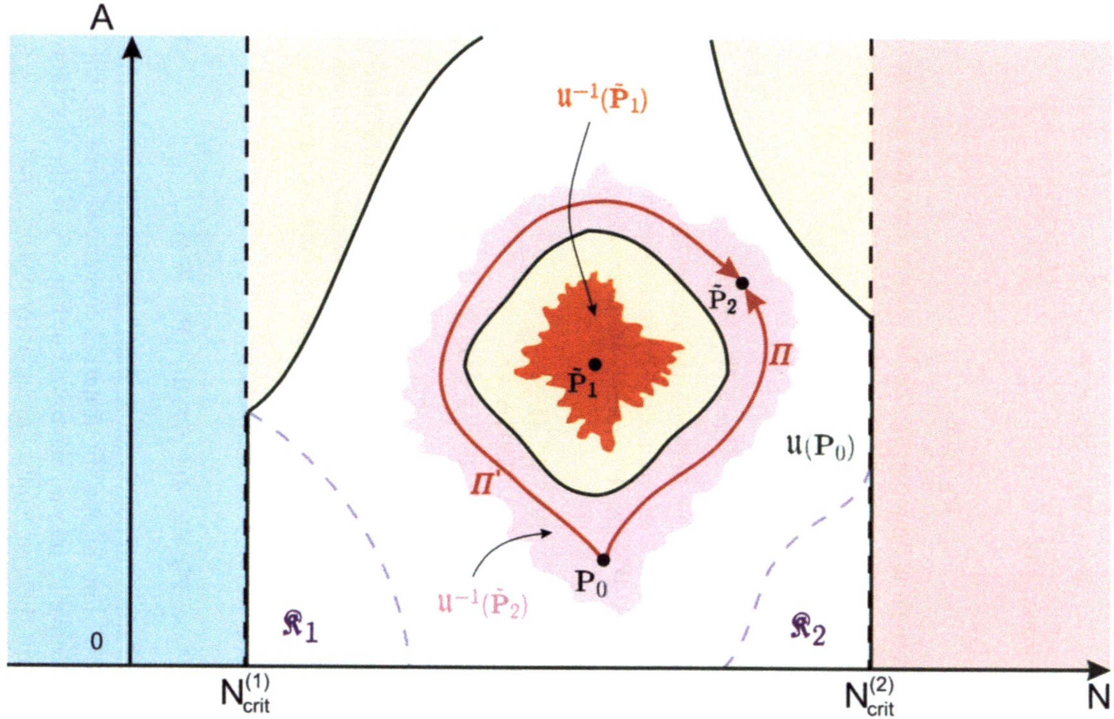

Figure 28. Illustration of pertinent elements involved in the implementation of the Stabilization paradigm. Here $\tilde{\mathbf{P}}_1$, $\tilde{\mathbf{P}}_2$ are admissible coevolution states that can be fixed by appropriate geo-cybernetic management. The preferable state $\tilde{\mathbf{P}}_1$ is unfortunately not accessible from $\mathbf{P}_0$ but the second best choice $\tilde{\mathbf{P}}_2$ is, as indicated by the respective basins of access, $\mathfrak{U}^{-1}(\tilde{\mathbf{P}}_1)$ and $\mathfrak{U}^{-1}(\tilde{\mathbf{P}}_2)$. As a matter of fact, there exist a multitide of coevolution segments which lead from $\mathbf{P}_0$ to $\tilde{\mathbf{P}}_2$. Two rather distinct ones are earmarked through their trajectory segments $\boldsymbol{\Pi}$ and $\boldsymbol{\Pi}'$, respectively.

is generally not fulfilled (due to the possible presence of critical manifolds as discussed in Sect. 4.4).

The *"transitivity property"*, on the other hand, is certainly satisfied: Let $\mathbf{P}_1, \mathbf{P}_2, \mathbf{P}_3 \in \mathbf{C}$ and

$$\mathbf{P}_1 \curvearrowright \mathbf{P}_2, \quad \mathbf{P}_2 \curvearrowright \mathbf{P}_3 \quad . \tag{171}$$

Then we can easily prove that

$$\mathbf{P}_1 \curvearrowright \mathbf{P}_3 \quad . \tag{172}$$

In a similarly straightforward manner it can be shown that the accessibility relations (171) imply

$$\mathfrak{U}(\mathbf{P}_3) \subset \mathfrak{U}(\mathbf{P}_2) \subset \mathfrak{U}(\mathbf{P}_1) \quad , \tag{173}$$

and

$$\mathfrak{U}^{-1}(\mathbf{P}_1) \subset \mathfrak{U}^{-1}(\mathbf{P}_2) \subset \mathfrak{U}^{-1}(\mathbf{P}_3) \quad . \tag{174}$$

More interesting are the relations between the accessible universes and the basins of access of states which are connected by management. Let us assume that $\mathbf{P}_1 \curvearrowright \mathbf{P}_2$ but $\mathbf{P}_2 \not\curvearrowright \mathbf{P}_1$, i.e. $\mathbf{P}_1$ cannot be approached from $\mathbf{P}_2$. As a consequence, we have

$$\mathfrak{U}(\mathbf{P}_1) \neq \mathfrak{U}^{-1}(\mathbf{P}_1) \quad , \tag{175}$$

$$\mathfrak{U}(\mathbf{P}_2) \neq \mathfrak{U}^{-1}(\mathbf{P}_2) \quad , \tag{176}$$

$$\mathfrak{U}(\mathbf{P}_1) \cap \mathfrak{U}^{-1}(\mathbf{P}_2) \neq \emptyset \quad , \tag{177}$$

$$\mathfrak{U}(\mathbf{P}_2) \cap \mathfrak{U}^{-1}(\mathbf{P}_1) = \emptyset \quad . \tag{178}$$

We skip the proof which is again straightforward and not that enlightening. It is most instructive, however, to put together all the topological information contained in Eqs. 173 to 178 within one simple picture as carried out in Fig. 29.

As a matter of fact, the ***general statement***

$$\mathbf{P}_1 \curvearrowright \mathbf{P}_2 \Longleftrightarrow \mathfrak{U}(\mathbf{P}_1) \cap \mathfrak{U}^{-1}(\mathbf{P}_2) \neq \emptyset \tag{179}$$

is true for any pair of states in coevolution space, and this statement directly implies Eqs. 177 and 178 in the specific situation considered here. So the sets $\mathfrak{U}(\mathbf{P}_2)$ and $\mathfrak{U}^{-1}(\mathbf{P}_1)$ have to be well separated in Fig. 29. If these sets would touch or even overlap, then we would immediately observe that

$$\mathfrak{U}(\mathbf{P}_1) = \mathfrak{U}(\mathbf{P}_2), \quad \mathfrak{U}^{-1}(\mathbf{P}_1) = \mathfrak{U}^{-1}(\mathbf{P}_2) \quad , \tag{180}$$

and, consequently, $\mathbf{P}_2 \curvearrowright \mathbf{P}_1$ would hold in contradiction to our assumption.

In this case we might qualify the states $\mathbf{P}_1$ and $\mathbf{P}_2$ as "equivalent" in the sense of geo-cybernetic theory.

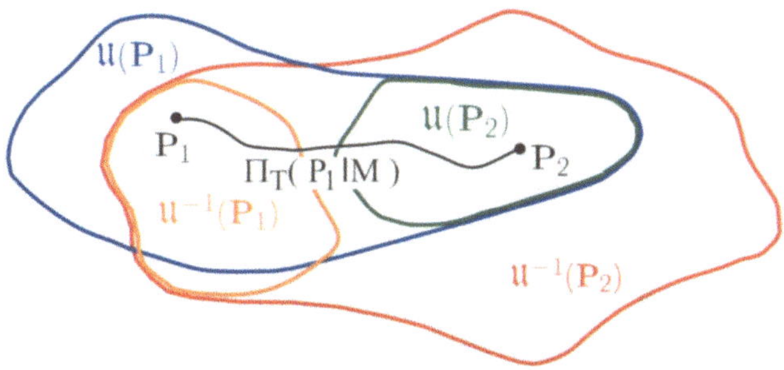

Figure 29. Exemplary topology of pertinent sub-sets of $\mathbf{C}$ in the case $\mathbf{P}_1 \curvearrowright \mathbf{P}_2$, $\mathbf{P}_2 \not\curvearrowright \mathbf{P}_1$. According to our assumption, there exists at least one connecting trajectory segment generated by some management $\mathbf{M}$ within a finite time interval $[0, T]$.

Definition 9: Consider an arbitrary couple of states $\mathbf{P}_1$, $\mathbf{P}_2$ in $\mathbf{C}$.

If

$$\mathbf{P}_1 \curvearrowright \mathbf{P}_2 \quad \text{and} \quad \mathbf{P}_2 \curvearrowright \mathbf{P}_1 \tag{181}$$

then we declare $\mathbf{P}_1$ and $\mathbf{P}_2$ as *geo-cybernetically equivalent* and write

$$\mathbf{P}_1 \sim \mathbf{P}_2 \quad . \tag{182}$$

Note that the symbol "$\sim$" now describes a true equivalence relation in the strict mathematical sense, as the three required properties "reflexivity", "symmetry" and "transitivity" are evidently fulfilled. ■

Actually all coevolution states that are not separated by critical manifolds belong to the same *geo-cybernetic equivalence class.* These classes are most important for the Equitization paradigm $\mathcal{P}_3$ (see Sect. 4.4) but less relevant to the Stabilization paradigm $\mathcal{P}_4$, where the one-way accessibility matters.

Definition 10: Let $\mathbf{P}_1, \ \mathbf{P}_2 \in \mathbf{C}$ and $\mathbf{P}_1 \curvearrowright \mathbf{P}_2$. Then

$$\mathfrak{C}(\mathbf{P}_1 \curvearrowright \mathbf{P}_2) := \mathfrak{U}(\mathbf{P}_1) \cap \mathfrak{U}^{-1}(\mathbf{P}_2) \tag{183}$$

is non-void according to Eq. 179. In fact, each point in this set is a potential intermediate state for geo-cybernetic transition from $\mathbf{P}_1$ to $\mathbf{P}_2$. $\mathfrak{C}(\mathbf{P}_1 \curvearrowright \mathbf{P}_2)$ is therefore called the *geo-cybernetic corridor from* $\mathbf{P}_1$ *to* $\mathbf{P}_2$. ■

In the exemplary situation sketched in Fig. 29 the corridor $\mathfrak{C}(\mathbf{P}_1 \curvearrowright \mathbf{P}_2)$ has the following shape:

We emphasize again that $\mathfrak{C}(\mathbf{P}_1 \curvearrowright \mathbf{P}_2)$ is a one-way street, as a rule. However, if $\mathbf{P}_1 \sim \mathbf{P}_2$ then

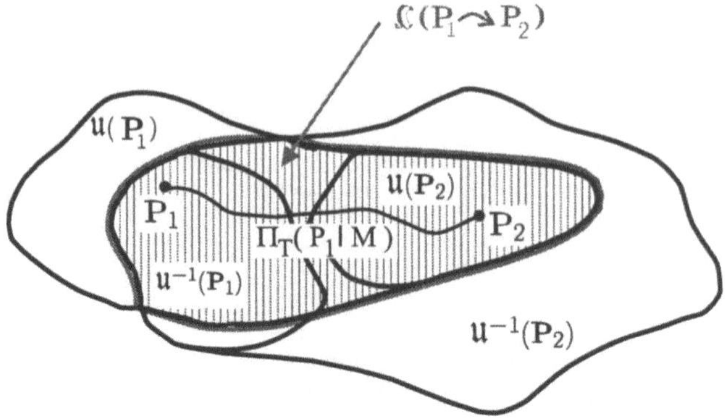

$$\mathfrak{L}(\mathbf{P}_1 \curvearrowright \mathbf{P}_2) = \mathfrak{L}(\mathbf{P}_2 \curvearrowright \mathbf{P}_1) \quad , \tag{184}$$

and we may denote this two-way street (supporting *coevolution loops* visiting both $\mathbf{P}_1$ and $\mathbf{P}_2$) by the symbol $\mathfrak{L}(\mathbf{P}_1 \sim \mathbf{P}_2)$. Let us point out that the geo-cybernetic corridors of any type do not have to be simply-connected sub-sets of $\mathbf{C}$.

The *dynamic* entity behind a static corridor $\mathfrak{L}(\mathbf{P}_1 \curvearrowright \mathbf{P}_2)$ is the family of coevolution episodes leading in finite time from $\mathbf{P}_1$ to $\mathbf{P}_2$.

Definition 11: Let us assume that $\mathbf{P}_2$ is accessible from $\mathbf{P}_1$. Therefore the *set of passing managements* $\tilde{\mathfrak{M}}(\mathbf{P}_1 \curvearrowright \mathbf{P}_2) \subset \mathfrak{M}$, is non-void. $\tilde{\mathfrak{M}}(\mathbf{P}_1 \curvearrowright \mathbf{P}_2)$ is determined by the following requirement:

For any $\tilde{\mathbf{M}} \in \tilde{\mathfrak{M}}(\mathbf{P}_1 \curvearrowright \mathbf{P}_2)$ there exists a minimum time $T(\tilde{\mathbf{M}}) < \infty$ such that

$$\mathbf{P}(\mathbf{P}_1, 0; T(\tilde{\mathbf{M}}) \mid \tilde{\mathbf{M}}) = \mathbf{P}_2 \quad . \tag{185}$$

Then the *geo-cybernetic passage* from $\mathbf{P}_1$ to $\mathbf{P}_2$, $Z(\mathbf{P}_1 \curvearrowright \mathbf{P}_2)$, is defined as

$$Z(\mathbf{P}_1 \curvearrowright \mathbf{P}_2) := \left\{ \mathbf{P}_{T(\tilde{\mathbf{M}})}(\mathbf{P}_1, 0; t \mid \tilde{\mathbf{M}}) \mid \tilde{\mathbf{M}} \in \tilde{\mathfrak{M}}(\mathbf{P}_1 \curvearrowright \mathbf{P}_2) \right\} \quad . \tag{186}$$

As a consequence, we have

$$\mathfrak{L}(\mathbf{P}_1 \curvearrowright \mathbf{P}_2) = \bigcup_{\tilde{\mathbf{M}} \in \tilde{\mathfrak{M}}(\mathbf{P}_1 \curvearrowright \mathbf{P}_2)} \Pi_{T(\tilde{\mathbf{M}})}(\mathbf{P}_1, 0 \mid \tilde{\mathbf{M}}) \quad . \tag{187}$$

(For the sake of clarity, we have abstained from employing the full notation indicating mangement *episodes* $\tilde{\mathbf{M}}_{T(\tilde{\mathbf{M}})}$ in Eqs. 186 and 187.) ■

If our geo-cybernetic problem does not depend explicitly on time and the target generalized equilibrium is simply a "sustainable" coevolution state $\tilde{\mathbf{P}}$, then "all" we have to do is to identify and analyse the passage $Z(\mathbf{P}_0 \curvearrowright \tilde{\mathbf{P}})$. If this set is non-void then it will be called a *stabilizing passage.* If $Z(\mathbf{P}_0 \curvearrowright \tilde{\mathbf{P}})$ contains more than one element, then the question of the right choice for the actual transition path segment $\mathbf{P}_{T(\tilde{\mathbf{M}})}(\mathbf{P}_0, 0; t \mid \tilde{\mathbf{M}})$ arises. Before dealing with these issues in some depth, let us emphasize that the implementation of the Stabilization paradigm generally confronts us with a more complicated situation:

Let

$$\mathbf{S}(t) := \mathbf{P}(\hat{\mathbf{P}}, \hat{t}; t \mid \mathbf{M}^*) \tag{188}$$

be some generalized equilibrium produced by the stabilizing management $\mathbf{M}^* \equiv \mathbf{M}^*_{(4)} \in \mathfrak{M}$. Let $\tau \geq 0$ and

$$\begin{aligned} \mathbf{S}_\tau(t) \quad &:= \quad \mathbf{S}(t)\Big|_{t \geq \hat{t}+\tau} \\ &= \quad \mathbf{P}(\hat{\mathbf{P}}_\tau, \hat{t}+\tau; t \mid \mathbf{M}^*) \quad , \end{aligned} \tag{189}$$

where

$$\hat{\mathbf{P}}_\tau := \mathbf{P}(\hat{\mathbf{P}}, \hat{t}, \hat{t}+\tau \mid \mathbf{M}^*) \equiv \mathbf{S}(\hat{t}+\tau) \quad . \tag{190}$$

According to our previous definitions, $\mathbf{S}_\tau(t)$ is again a generalized equilibrium for any choice of τ. Therefore $\mathbf{S}(t)$ defines a *class*

$$C[\mathbf{S}] := \{\mathbf{S}_\tau(t) \mid 0 \leq \tau < \infty\} \tag{191}$$

of stable, i.e. *recurrent* coevolution paths, and it makes sense *to identify the generalized equilibrium with the whole class.* Note that the closed hulls of all the associated coevolution trajectories are identical:

$$\overline{\boldsymbol{\Pi}(\mathbf{S}_\tau)} = \overline{\boldsymbol{\Pi}(\mathbf{S}_0)} \quad , \tag{192}$$

where $\mathbf{S}_0 \equiv \mathbf{S}$. This statement is trivial for simple stable coevolutions like fixed states or periodic orbits but less so for "strange" recurrent motion in $\mathbf{C}$.

The parameter τ may be conceived here as a control variable which allows for determining *when* (and where) the environmental system enters the target equilibrium coevolution. In order to exercise wise control, however, we have to identify and analyse the overall *stabilizing passage* $Z(\mathbf{P}_0 \curvearrowright C[\mathbf{S}])$, which connects the initial state $\mathbf{P}_0$ to the generalized equilibrium $C[\mathbf{S}]$. $Z(\mathbf{P}_0 \curvearrowright C[\mathbf{S}])$ consists of coevolution path segments $\mathbf{P}_{\hat{t}+\tau(\tilde{\mathbf{M}})}(\mathbf{P}_0, 0; t \mid \tilde{\mathbf{M}})$, where $\tilde{\mathbf{M}} \in \tilde{\mathfrak{M}}(\mathbf{P}_0 \curvearrowright C[\mathbf{S}])$ is an appropriate management sequence such that for $\tau(\tilde{\mathbf{M}}) < \infty$ we have

$$\mathbf{P}_{\hat{t}+\tau(\tilde{\mathbf{M}})}(\mathbf{P}_0, 0; \hat{t} + \tau(\tilde{\mathbf{M}}) \mid \tilde{\mathbf{M}}) = \mathbf{S}(\hat{t} + \tau(\tilde{\mathbf{M}})) \ . \tag{193}$$

Thus the entire management sequences $\mathbf{M}^{**}$ that bring about transition to and maintenance of stable coevolutionary motion are composed in the following way:

$$\mathbf{M}^{**}(t) = \begin{cases} \tilde{\mathbf{M}}(t) & , \quad 0 \le t \le \hat{t} + \tau(\tilde{\mathbf{M}}) \\ \mathbf{M}^{*}(t) & , \quad \hat{t} + \tau(\tilde{\mathbf{M}}) \le t \end{cases} \quad , \tag{194}$$

$$\text{assuming} \quad \tilde{\mathbf{M}}(\hat{t} + \tau(\tilde{\mathbf{M}})) = \mathbf{M}^{*}(\hat{t} + \tau(\tilde{\mathbf{M}})) \ . \tag{195}$$

The investigation of the stabilizing passage $Z(\mathbf{P}_0 \curvearrowright C[\mathbf{S}])$ may become rather involved, especially when our geo-cybernetics depends explicitly on time and, consequently, there are no static corridors $\mathfrak{C}(\mathbf{P}_0 \curvearrowright C[\mathbf{S}])$. For descriptive reasons we will therefore restrict ourselves mainly to simple generalized equilibria like fixed states, and autonomous dynamics.

Assume that the target equilibrium is a sustainable coevolution state $\tilde{\mathbf{P}}$. In order to prove the accessibility of $\tilde{\mathbf{P}}$ from $\mathbf{P}_0$ by explicitly constructing the geo-cybernetic corridor $\mathfrak{C}(\mathbf{P}_0 \curvearrowright \tilde{\mathbf{P}})$ or some *sub-corridor* $\bar{\mathfrak{C}}(\mathbf{P}_0 \curvearrowright \tilde{\mathbf{P}}) \subset \mathfrak{C}(\mathbf{P}_0 \curvearrowright \tilde{\mathbf{P}})$, the *stepping-stone approach* may be particularly useful:

Draw an appropriate line (not necessarily a straight one) in coevolution space, which connects $\mathbf{P}_0$ to $\tilde{\mathbf{P}}$, and choose a sequence $\{\mathbf{P}_l\}; l = 1, \ldots, L$, of roughly equidistant intermediate points. Then try to verify the accessibility relations

$$\mathbf{P}_l \curvearrowright \mathbf{P}_{l+1} \tag{196}$$

one by one, starting from the target state, i.e. beginning with

$$\mathbf{P}_L \curvearrowright \tilde{\mathbf{P}} \ . \tag{197}$$

Note that $\mathbf{P}_0 \curvearrowright \tilde{\mathbf{P}}$ holds if all the intermediate relations are fulfilled, as the finite sum of finite passage times is again finite. If certain accessibility relations do not hold, however, try to devise appropriate intermediate detours – quite similar to the way we identify a stepping-stone passage across a shallow brook. Such a procedure also resembles the analytic continuation of functions in the complex plane $\mathbb{C}$, but we have to emphasize that the latter task will be much better defined and easier to achieve, in general.

Within the geo-cybernetic context the stepping-stone approach may become feasible with excessive computer assistance only, as the exploration of multiple combinatorial options tends to be extremely time-consuming.

4.5.3 Quality of Passage

If the desired generalized equilibrium $C[\mathbf{S}]$ is really accessible then, usually, there exist a variety of ways to approach it from $\mathbf{P}_0$. What are the criteria for selecting the transitory management (episode) $\tilde{\mathbf{M}} \in \mathfrak{M}(\mathbf{P}_0 \curvearrowright C[\mathbf{S}])$? In the following, we briefly discuss several plausible modes of determining the concrete avenue to stability.

Example 1 ("Gentleness"):
The most conspicuous feature of any access to $C[\mathbf{S}]$ is the *geometry* of the connecting coevolution trajectory segment $\boldsymbol{\Pi}_{T(\tilde{\mathbf{M}})}(\mathbf{P}_0, 0 \mid \tilde{\mathbf{M}})$ with respect to the target structure. The generalized equilibrium may be approached, e.g., *directly* or *smoothly* or *asymptotically*. This is illustrated in Fig. 30 under the assumption that $\mathbf{S}$ is a periodic orbit in coevolution space. If the Global Subject $\mathcal{S}$ actually picks SD paradigm $\mathcal{P}_4$, then one of the crucial decisions of this "Earth System pilot" will be the choice between bringing about stabilization in a *crash-halt* or a *soft-landing scenario.*

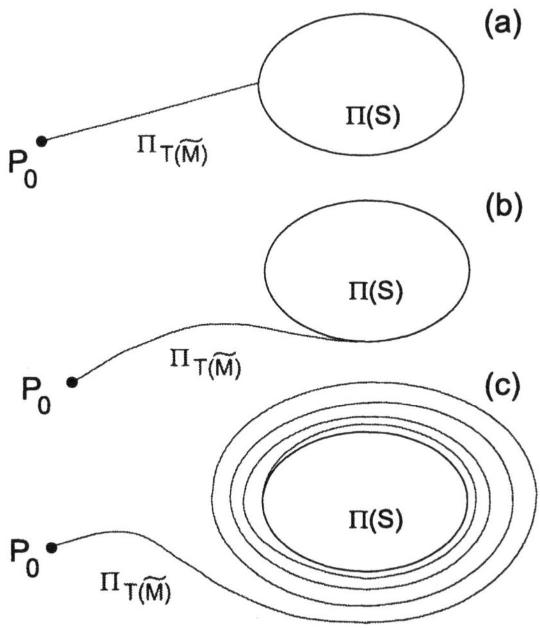

Figure 30. Various ways of approaching the periodic coevolution orbit $\mathbf{S}$, arranged in the order of increasing gentleness. **(a)** Geodesic passage, i.e. minimizing the length of the connecting trajectory segment. **(b)** Tangential passage, i.e. observing differentiability requirements. **(c)** Encircling passage, i.e. gradually relaxing to periodic motion.
Note that the passage nevertheless has to comply with the condition $T(\tilde{\mathbf{M}}) < \infty$.

Example 2 ("Duration"):
An obvious and particularly popular criterion for transitory management is the time needed for entering the target equilibrium. One could, for instance, search for the *minimum passage-time management* $\tilde{\mathbf{M}}^* \in \mathfrak{M}(\mathbf{P}_0 \curvearrowright C[\mathbf{S}])$, such that

$$T(\tilde{\mathbf{M}}^*) \leq T(\tilde{\mathbf{M}}) \quad \text{for any} \quad \tilde{\mathbf{M}} \in \tilde{\mathfrak{M}}(\mathbf{P}_0 \curvearrowright C[\mathbf{S}]) \ . \tag{198}$$

Such a time-optimal approach, however, may involve sub-optimal or even unpleasant features regarding other criteria like gentleness.

Rather than minimizing the duration of passage to equilibrium, the Global Subject may have to organize the transition to stable coevolution in a quite *limited stretch of time*. Such a pressure for action might arise from the foreseeable exhaustion of pertinent natural resources or, more probably, from the time restrictions on the political mandate of national and global decision makers: the public may either lose patience and ask for more radical action or install a restorational regime, if the transition process takes too long.

If the time limit is set to be $\hat{T}$, say, then the main geo-cybernetic task is to identify the *sub-passage*

$$Z_{\leq \hat{T}}(\mathbf{P}_0 \curvearrowright C[\mathbf{S}]) \subset Z(\mathbf{P}_0 \curvearrowright C[\mathbf{S}]) \ , \tag{199}$$

which contains the coevolution path segments that are run through in no more than $\hat{T}$ time units. This sub-passage is therefore generated by management episodes

$$\tilde{\mathbf{M}} \in \tilde{\mathfrak{M}}_{\leq \hat{T}}(\mathbf{P}_0 \curvearrowright C[\mathbf{S}]) \subset \tilde{\mathfrak{M}}(\mathbf{P}_0 \curvearrowright C[\mathbf{S}]) \ , \tag{200}$$

that satisfy

$$T(\tilde{\mathbf{M}}) \leq \hat{T} \ . \tag{201}$$

The actual management sequence employed may be singled out from the elements of $\tilde{\mathfrak{M}}_{\leq \hat{T}}(\mathbf{P}_0 \curvearrowright C[\mathbf{S}])$ eventually by convenience considerations.

Example 3 ("Costs"):
Another eligible criterion for qualifying the management sequences contained in $\tilde{\mathfrak{M}}(\mathbf{P}_0 \curvearrowright C[\mathbf{S}])$ is the costliness of passage to equilibrium. As a pre-requisite for such an evaluation we need a functional

$$G[\mathbf{P}(\mathbf{P}_0, 0; t \mid \tilde{\mathbf{M}})] \equiv G(t \mid \tilde{\mathbf{M}}) \tag{202}$$

that determines the *incremental economic costs* along the coevolution path segment $\mathbf{P}(\mathbf{P}_0, 0; t \mid \tilde{\mathbf{M}})$. To be more precise, $G(t \mid \tilde{\mathbf{M}})$ measures the current costs per time unit in percentage of gross global product, which arise from employing the geo-cybernetic management $\tilde{\mathbf{M}}$ instead of doing nothing to steer the Earth System ("Business as Usual").

The overall incremental costs of the transitory management $\tilde{\mathbf{M}}$ are then given by

$$G(\tilde{\mathbf{M}}) := \int_0^{T(\tilde{\mathbf{M}})} G(t \mid \tilde{\mathbf{M}}) \, dt \ . \tag{203}$$

One could now look for the management sequence $\tilde{\mathbf{M}}^*$ that *minimizes the total costs* of passage to equilibrium, i.e.

$$G(\tilde{\mathbf{M}}^*) \leq G(\tilde{\mathbf{M}}) \quad \text{for any} \quad \tilde{\mathbf{M}} \in \tilde{\mathfrak{M}}(\mathbf{P}_0 \curvearrowright C[\mathbf{S}]) \ . \tag{204}$$

From the point of view of economics, however, it makes more sense to keep the temporal (for instance, annual) costs of geo-cybernetic management below a certain *threshold* $\hat{G} \in \mathbb{R}$. This defines again a sub-set

$$\tilde{\mathfrak{M}}_{\leq \hat{G}}(\mathbf{P}_0 \curvearrowright C[\mathbf{S}]) \subset \tilde{\mathfrak{M}}(\mathbf{P}_0 \curvearrowright C[\mathbf{S}]) \tag{205}$$

of passing management sequences which satisfy

$$G(t \mid \tilde{\mathbf{M}}) \leq \hat{G} \quad \text{for} \quad t \in [0, T(\tilde{\mathbf{M}})] \ . \tag{206}$$

The signatory states of the Framework Convention on Climate Change (*UN*, 1995 [235]) are clearly confronted with problems of that kind: What amount of "mitigation costs" for stabilizing the atmospheric CO_2-concentration, or even the climate, is still tolerable for the various players involved? However, note that all their problems would vanish into the air if the set

$$\tilde{\mathfrak{M}}_{\leq 0}(\mathbf{P}_0 \curvearrowright C[\mathbf{S}]) \tag{207}$$

turned out to be non-void: this is the set of *"no-regret options"* for passage to equilibrium.

Let us finally point out that, instead of the purely economic costs, also the social, ethical, or ecological costs of transitory management may (or must) be considered. An integrated evaluation can be based again on the "Lagrangian" $L(\mathbf{P})$.

4.6 Complex Paradigms

Now that we have finished our tour d'horizon of the *pure SD paradigms* listed in Tab. 1, we might ask ourselves which of these paradigms will be chosen by the Global Subject $\mathcal{S}$ as the relevant guiding light for geo-cybernetic management. We will not, however, try to give an answer that obviously has to transpire from a democratic election process rather than from a scientific selection procedure.

We shall instead point out that in future reality (or real future, if you like) such an election process will probably not generate a clear-cut decision for one single paradigm. So $\mathcal{S}$ may have to match almost incommensurate SD strategies, and we will briefly investigate in the rest of this section the formal aspects arising from the task of satisfying *coincident* pure paradigms.

Let us begin with the observation that a multi-objective mandate can reduce or enhance the decision dilemmata which tend to plague the actors in most control problems. Specifically within our geo-cybernetic context, we have to remember that some of the pure paradigms may provide us with a precise recipe for determining *the* "correct" management sequence $\mathbf{M}^*$ (Optimization, Equitization), while the other paradigms leave us with considerable ambiguity on that score. So, given a desired coincidence of several pure paradigms, the worst case arises if the unique management sequences selected according to the respective rules of the individual paradigms are not identical. We may qualify such a situation as the *competition of pure paradigms*. The best case, on the other hand, arises if the individual paradigms define a sequence of nested sub-sets of $\mathfrak{M}$ where the innermost set might embrace one single element only, namely $\mathbf{M}^*$. We may qualify this situation as the *cooperation of pure paradigms*.

In practice, however, neither the competition nor the cooperation of pure paradigms will occur "by accident" and the Global Subject will be confronted with a quite mixed case. Such cases can be successfully dealt with by defining *complex SD paradigms*, where the individual components are *ranked* or *compared* in a clear-cut and consistent manner. An appropriate methodology has been developed in the last decades by *multicriterion decision theory* (see, e.g., *Bell* et al., 1977 [17]). In the following we will briefly discuss the most important ways of constructing complex paradigms and some of the problems associated with these constructions.

Whether we deal with just one SD paradigm or several of them simultaneously, the ultimate goal of all our efforts is to identify the "best management $\mathbf{M}^*$ for running planet Earth". The distinct pure paradigms provide us with specific information (or even algorithms) for selecting $\mathbf{M}^*$, and practically all this information may be compressed into *quality functionals* $Q_{\mathcal{P}_l}[\mathbf{P}(\mathbf{M})]$, where the index $l = 1, \ldots, L$ counts the eligible fundamental strategies (see Sect. 3). Our analysis of complex paradigms will be based on these functionals, which we assume to be scalar entities for the sake of simplicity. For any $\mathcal{P}_l$ there exists a non-void set $\mathfrak{M}^*_{(l)} \subset \mathfrak{M}$ of available management sequences such that

$$Q_{\mathcal{P}_l}[\mathbf{P}(\mathbf{M})] \leq Q_{\mathcal{P}_l}[\mathbf{P}(\mathbf{M}^*_{(l)})] \tag{208}$$

for all $\mathbf{M}^*_{(l)} \in \mathfrak{M}^*_{(l)}$ and arbitrary $\mathbf{M} \in \mathfrak{M}$.

Example 1 ("Pareto Efficiency"):
Let us assume that the Global Subject has to comply with all L SD paradigms simultaneously. This will not pose any problem if

$$\mathfrak{M}^{**} := \mathfrak{M}^*_{(1)} \cap \mathfrak{M}^*_{(2)} \cap \ldots \cap \mathfrak{M}^*_{(L)} \neq \emptyset \quad , \tag{209}$$

i.e., if there is at least one management sequence $\mathbf{M}^{**}$ that maximizes all quality functionals at the same time. A pictorial representation of such a situation is given in Fig. 31.

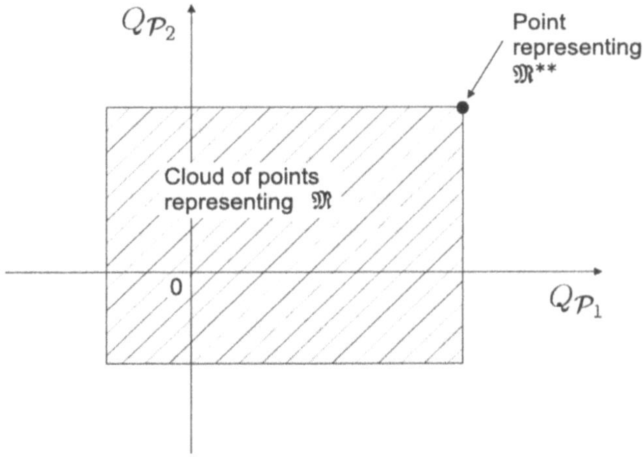

Figure 31. Assuming $L = 2$, the "option space" $\mathfrak{M}$ is projected onto the "quality plane" $\mathbb{R}^2$ through the map $q(\mathbf{M}) = (Q_{\mathcal{P}_1}[\mathbf{P}(\mathbf{M})], Q_{\mathcal{P}_2}[\mathbf{P}(\mathbf{M})])$. Note that this map is generally not injective, so many different management sequences may lurk behind one single dot in $\mathbb{R}^2$. The elements in $\mathfrak{M}^{**}$ are optimal with respect to $\mathcal{P}_1$ *and* $\mathcal{P}_2$ and therefore generate the *upper-right corner point* of the $\mathfrak{M}$-cloud, which has to be *locally rectangular.*

Eq. 209 applies, for instance, when the coincident paradigms $\mathcal{P}_1, \ldots, \mathcal{P}_L$ are *indifferent* with respect to each other, so any management sequence $\mathbf{M} \in \mathfrak{M}$ can be split up in the following way:

$$\mathbf{M} = (\mathbf{M}_1, \ldots, \mathbf{M}_L) \quad , \tag{210}$$

where the *management component* $\mathbf{M}_l$ acts exclusively on processes pertinent to $\mathcal{P}_l$, and in no way affects the compliance with all the other pure paradigms. If, for example, the economic growth mechanisms around the globe could be completely decoupled from the state of the natural ecosphere, then Optimization and Pessimization strategies might be pursued in parallel.

As a consequence of the general *separability* of $\mathbf{M}$, the quality functionals would be of the type

$$Q_{\mathcal{P}_l}[\mathbf{P}(\mathbf{M})] = Q_{\mathcal{P}_l}[\mathbf{P}(\mathbf{M}_1, \ldots, \mathbf{M}_L)] \equiv Q_{\mathcal{P}_l}(\mathbf{M}_l) \quad . \tag{211}$$

Thus

$$\begin{aligned} Q_{\mathcal{P}_l}[\mathbf{P}(\mathbf{M}_1, \ldots, \mathbf{M}_l, \ldots, \mathbf{M}_L)] &= Q_{\mathcal{P}_l}(\mathbf{M}_l) \\ &= Q_{\mathcal{P}_l}[\mathbf{P}(\overline{\mathbf{M}}_1, \ldots, \mathbf{M}_l, \ldots, \overline{\mathbf{M}}_L)] \end{aligned} \tag{212}$$

for any choice of the L-1-tupel of management components

$$(\overline{\mathbf{M}}_1, \ldots, \overline{\mathbf{M}}_{l-1}, \overline{\mathbf{M}}_{l+1}, \ldots, \overline{\mathbf{M}}_L) \quad . \tag{213}$$

Eq. 212 directly implies that the $\mathfrak{M}$-cloud in quality space is *globally rectangular*, as exemplified in Fig. 31. In particular, $\mathfrak{M}^{**} \neq \emptyset$, and this sub-set of

$\mathfrak{M}$ consists of management sequences of the form $(\mathbf{M}_1^*, \ldots, \mathbf{M}_L^*)$, where each component $\mathbf{M}_l^*$ maximizes the respective quality functional $Q_{\mathcal{P}_l}$.

In general, however, Eq. 209 will *not* be satisfied and we have to deal with really conflicting multiple objectives. This situation is typically reflected by the shape of the $\mathfrak{M}$-cloud in L-dimensional quality space. If we generalize the formalism sketched in the context of Fig. 31, then this "cloud" is given by the set

$$Q(\mathfrak{M}) := \{q(\mathbf{M}) \mid \mathbf{M} \in \mathfrak{M}\} \subset \mathbb{R}^L \quad , \tag{214}$$

where

$$q(\mathbf{M}) := (Q_{\mathcal{P}_1}[\mathbf{P}(\mathbf{M})], \ldots, Q_{\mathcal{P}_L}[\mathbf{P}(\mathbf{M})]) \quad . \tag{215}$$

As a rule, the boundary of $Q(\mathfrak{M})$ is *curvilinear*, as illustrated in Fig. 32.

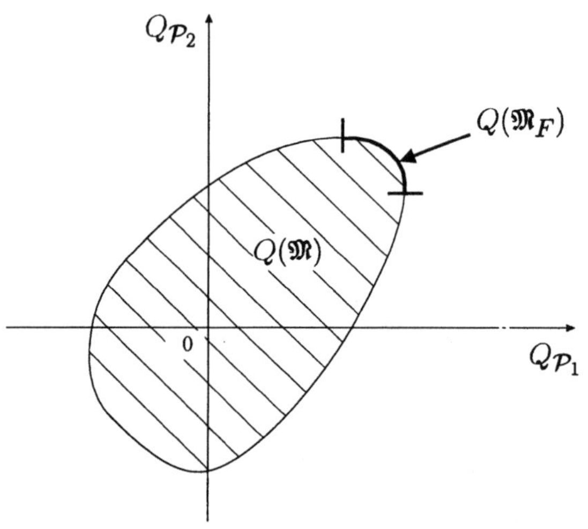

Figure 32. Typical shape of the $\mathfrak{M}$-cloud, $Q(\mathfrak{M})$, in the two-dimensional setting familiar from Fig. 31. Instead of an optimal corner-point, $Q(\mathfrak{M}^{**})$, we now observe an extended *Pareto frontier*, $Q(\mathfrak{M}_F)$. Moving along this boundary segment of $Q(\mathfrak{M})$ consistently improves the quality with respect to the one paradigm, yet worsens the quality with respect to the other. Moving away from $Q(\mathfrak{M}_F)$ to other parts of $Q(\mathfrak{M})$, however, necessarily worsens the quality with respect to *both* paradigms.

Under these generic circumstances a *complex paradigm* is defined by the obligation to ensure at least *Pareto efficiency*, i.e., choose management sequences only that generate points on the Pareto frontier. The corresponding sub-set $\mathfrak{M}_F \subset \mathfrak{M}$ is determined in the general case by the following requirement:

$\mathbf{M} \in \mathfrak{M}_F \Leftrightarrow$ There is no $\overline{\mathbf{M}} \in \mathfrak{M}$ such that

$$\begin{aligned} Q_{\mathcal{P}_l}[\mathbf{P}(\overline{\mathbf{M}})] > Q_{\mathcal{P}_l}[\mathbf{P}(\mathbf{M})] \text{ for some } l \in \{1, \ldots, L\} \text{ and} \\ Q_{\mathcal{P}_{l'}}[\mathbf{P}(\overline{\mathbf{M}})] \geq Q_{\mathcal{P}_{l'}}[\mathbf{P}(\mathbf{M})] \text{ for } l' \neq l \ . \end{aligned} \tag{216}$$

Note that Pareto efficiency is a rather weak selection criterion, so we may be stuck with a considerable amount of geo-cybernetic ambiguity if we built our complex paradigm upon this criterion.

Example 2 ("Hierarchical Order"):
The integration of coincident pure paradigms within one complex paradigm is rather straightforward when the $\mathcal{P}_l$ involved can be *ranked* in an unambiguous way. Such a ranking may directly result from the preferences of $\mathcal{S}$, i.e. from the "democratically averaged" volitions of all actors in the global E & D process. The so-established hierarchy of pure paradigms defines a *"super-paradigm"* for management selection, which operates like this:

Let Λ be a permutation of the set $\{1, 2, \ldots, L\}$ such that $\mathcal{P}_l$ is on level $\lambda = \Lambda(l)$ of the paradigm hierarchy. So, if we introduce the symbol ${}^{\lambda}\mathcal{P}$ to indicate the pure paradigm of λth priority, we have the relation

$$\mathcal{P}_l = {}^{\Lambda(l)}\mathcal{P} \quad . \tag{217}$$

Conversely

$${}^{\lambda}\mathcal{P} = \mathcal{P}_{\Lambda^{-1}(\lambda)} \quad , \tag{218}$$

where Λ^{-1} is the permutation inverting Λ.

Let us further introduce sub-sets ${}^{(\lambda)}\mathfrak{M}^*$ of $\mathfrak{M}$ in the following inductive way:

$${}^{(1)}\mathfrak{M}^* := \mathfrak{M}^*_{\Lambda^{-1}(1)} \quad . \tag{219}$$

For $\lambda = 1, \ldots, L-1$ we define ${}^{(\lambda+1)}\mathfrak{M}^*$ to be the quality-maximizing sub-set of ${}^{(\lambda)}\mathfrak{M}^*$ with respect to paradigm ${}^{(\lambda+1)}\mathcal{P}$, i.e.,

$${}^{(\lambda+1)}\mathfrak{M}^* \subset {}^{(\lambda)}\mathfrak{M}^* \tag{220}$$

and

$$Q_{{}^{\lambda+1}\mathcal{P}}\left[\mathbf{P}({}^{(\lambda)}\mathbf{M}^*)\right] \leq Q_{{}^{\lambda+1}\mathcal{P}}\left[\mathbf{P}({}^{(\lambda+1)}\mathbf{M}^*)\right] \tag{221}$$

for all ${}^{(\lambda+1)}\mathbf{M}^* \in {}^{(\lambda+1)}\mathfrak{M}^*$ and arbitrary ${}^{(\lambda)}\mathbf{M}^* \in {}^{(\lambda)}\mathfrak{M}^*$.

The operational scheme for the super-paradigm in question, namely the ordered set $({}^{1}\mathcal{P}, {}^{2}\mathcal{P}, \ldots, {}^{L}\mathcal{P})$, is now readily formulated in terms of a selection algorithm. In the first step, determine the set ${}^{(1)}\mathfrak{M}^*$, which is optimal with respect to the top-priority pure paradigm. If ${}^{(1)}\mathfrak{M}^*$ contains more than one element, identify the sub-set ${}^{(2)}\mathfrak{M}^*$, which is *conditionally* optimal with respect to the second-most important pure paradigm, and so on. The end product of this algorithm is the set ${}^{(L)}\mathfrak{M}^* \subset \mathfrak{M}$, which is supposed to account for all possible ecological, economic, social and cultural preferences and values. Any remaining management ambiguity has to be removed ad hoc.

Let us conclude here with a specific example of a potential super-paradigm for managing humanity's interference with the Earth's climate. With respect to this crucial aspect of geo-cybernetics the following hierarchy of pure paradigms seems appropriate:

$$\begin{aligned}&({}^1\mathcal{P},{}^2\mathcal{P},{}^3\mathcal{P},{}^4\mathcal{P}) = (\mathcal{P}_2,\mathcal{P}_4,\mathcal{P}_3,\mathcal{P}_1) \equiv \\ &\equiv (\text{Pessimization}, \text{Stabilization}, \text{Equitization}, \text{Optimization}) \ .\end{aligned} \tag{222}$$

Thus the grand strategy is to avoid catastrophe domains at any rate, to achieve climate stabilization far away from the disaster zones, to bring about stability in an equitable way in consideration of the needs of the developing countries and, finally, to carry out the climate management in the most cost-effective way. We will come back to this super-paradigm at the very end of our essay.

Example 3 ("Conversion"):
The basic assumption here is that all possible pure paradigms may matter, but that their respective qualities can be converted into one single "currency" – US dollars, letters of indulgence, neg-entropy, or what have you. In other words, we are able to define a *generalized quality functional*

$$Q[\mathbf{P}(\mathbf{M})] := \sum_{l=1}^{L} \alpha_l Q_{\mathcal{P}_l}[\mathbf{P}(\mathbf{M})] \ , \tag{223}$$

which measures the overall "value" of a given coevolution $\mathbf{P}(\mathbf{M})$ in appropriate common units. The fixed *weights* $\alpha_l; l = 1, \ldots, L$, are specific conversion factors that reflect the comparative importance of the distinct pure paradigms for the global community. Usually we have

$$\alpha_l \geq 0 \quad \text{for all} \quad l \ , \tag{224}$$

and a normalization condition

$$\sum_{l=1}^{L} \alpha_l = 1 \tag{225}$$

may be satisfied without loss of generality.

The resulting complex paradigm operates in an obvious manner: simply try to identify the set $\mathfrak{M}^* \subset \mathfrak{M}$, the elements of which maximize the generalized coevolutionary quality, i.e.,

$$Q[\mathbf{P}(\mathbf{M})] \leq Q[\mathbf{P}(\mathbf{M}^*)] \tag{226}$$

for all $\mathbf{M}^* \in \mathfrak{M}^*$ and arbitrary $\mathbf{M} \in \mathfrak{M}$.

We emphasize that all sorts of cost-benefit analyses of geo-cybernetics measures (like reduction of greenhouse-gas emissions) rely, implicity but heavily, on the convertibility of distinct E & D qualities through the concept of "monetizing".

* * *

We close this section with several remarks.

(i) Total quality functionals for the comparison of pure paradigms do not have to be of the linear-combination type as expressed in Eq. 223.
A more general ansatz assumes that

$$Q[\mathbf{P}(\mathbf{M})] = H\left(Q_{\mathcal{P}_1}[\mathbf{P}(\mathbf{M})], \ldots, Q_{\mathcal{P}_L}[\mathbf{P}(\mathbf{M})]\right) \quad , \tag{227}$$

where the function H may be *non-linear* in some or all of its arguments.

(ii) Neither a complete hierarchy nor a full convertibility of coincident pure paradigms may transpire from a global volition survey. As a matter of fact, a *"caste system"* of paradigms is more likely to result from democratic election processes. This means that there is a macro-hierarchy of paradigms, where each caste may contain several convertible elements – but convertibility does not apply across the "class barriers". The operational scheme for the corresponding complex paradigm is easily constructed by amalgamating the procedures described in Examples 2 and 3, respectively.

(iii) Finally, we should point out that the relationship between pure and complex paradigms, as we have defined them, is quite different from the relationship between pure and *mixed* strategies in mathematical game theory (see, e.g., *Binmore*, 1990 [24]). Mixed strategies refer to the *aleatoric choice* of tactics from a set of pure game strategies according to a given probability distribution, while complex paradigms are basically deterministic concepts. The latter restriction may and has to be relaxed very often, however, for the practical purposes of geo-cybernetics. This relaxation affects primarily the implementation of management sequences as we will discuss in Sect. 6.

5 Integrated Modelling: Exploring Virtual Planetary Futures

The general structure of the task as defined by geo-cybernetics is the same as the problem structure encountered in *ordinary control theory* (see, e.g., *Bellmann*, 1957 [18]; *Pontryagin*, 1962 [185]; *Lee* and *Markus*, 1967 [130]; *Dantzig* and *Veinott*, 1968 [52]). There is archaeological evidence that, e.g., the control of irrigation systems in Mesopotamia was a well-developed art at least by the 20th century BC, but a serious scientific study of the field began only after World War II. Control theory can be viewed as the art of regulating the environment, in the physical, biological, or even social sense, and geo-cybernetics "simply" extends this art to the global scale.

The selection of overall objectives, or paradigms, is undoubtedly a crucial element of any control exercise – we have extensively discussed this aspect regarding global environmental management in the last section. Yet other elements matter as well, for instance, the various types of possible readjustments of an initially chosen control strategy. We may analyse the *problem anatomy* of geo-cybernetics by comparing it to the task of navigating a ship in foreign-going trade. We will encounter many analogies, but there are fundamental differences as well: managing the coevolution is certainly a unique challenge and asks also for techniques that cannot be provided by traditional control theory and empirics.

The following table lists and confronts the main control aspects of navigation at sea and geo-cybernetics, respectively.

Table 2: Elements of Control

Number of Element	**Foreign-Going Trade**	**Geo-Cybernetics**
1	*Selection of targets and boundary conditions* by orderer (ship-owner, charterer etc.). Example: Ship cargo X to harbour Y before deadline Z at minimal costs and risks	*Selection of pure or complex paradigm* according to cosmos of individual and collective preferences as reflected particularly by the ethical-voluntative component $\mathcal{V}$ of the Global Subject $\mathcal{S}$
2	*Installation of crew*, mainly composed of *executive staff* (captain, officers, etc.) and *technical staff* (navigators, machine operators, etc.)	*Election or emergence* of the *executive component* $\mathcal{M}$ of $\mathcal{S}$, i.e. political institutions and representatives in charge of international E & D management. $\mathcal{M}$ is assisted by a crowd of established scientific and technical advisors, constituting the *component* $\mathcal{B}$ ("Global Brain") of the Global Subject
3	*Synopsis of available relevant information*, e.g., nautical charts, data on climate and ocean currents, information about accessible harbours and political conditions in stop-over countries, files on cargo and crew. Based on that information, *composition of a navigation model* for planning games	*Survey of all available information* about – the Earth System – external driving forces – control options and instruments. Based on that information, *composition of (qualitative and quantitative) simulation models* for coevolution gaming exercises

continued on next page

continued from previous page

Number of Element	Foreign-Going Trade	Geo-Cybernetics
4	*Determination of course*, i.e. optimal cruising strategy within the framework of the map exercise, using elementary logic and geometry, nautical instruments, risk analysis, expert knowledge, etc.	*Determination of optimal management sequence* $\mathbf{M}^*$ within the framework of coevolution modelling, using mathematical calculus, numerical search algorithms, educated guesswork, etc., as employed by the representatives of $\mathcal{B}$
5	Navigation, i.e. *translation pre-calculated of cruising strategy into real motion* at sea through control devices. Steering is supported by *instruments* indicating vessel parameters and *relative* movement with respect to transport medium (e.g., speedometer)	Geo-cybernetic action, i.e. *implementation of* $\mathbf{M}^*$ through the organs of $\mathcal{M}$. Coevolution management is supported by sets of *indicators* monitoring, e.g., actor compliance with geo-cybernetic agreements and *relative* movement in coevolution space
6	*Location*, i.e. checking of absolute geographical position for course validation by means of fixed sky- or landmarks	*Determination of absolute position* in coevolution space, in particular with respect to catastrophe domains and political target states. Example: Inspection of actual CO_2-concentration in atmosphere as a complement to continuous emission monitoring
7	*Course correction* if absolute information reveals deviation from *pre-calculated optimal cruising strategy*	*Readjustment of E&D measures* if absolute information reveals insufficiency of envisaged steps to keep coevolution on *desired track*
8	*Local evading manoeuvres* in response to short-term information about *unpredictable events* (iceberg, storm, etc.)	*Episodical small-scale modification* of coevolution strategy due to *stochastic phenomena* (volcanic eruption, political crisis, etc.)

continued on next page

continued from previous page

Number of Element	Foreign-Going Trade	Geo-Cybernetics
9	*Major revision of cruising strategy* due to additional *information of great (i.e. large-scale and/or long-term) moment* Example: Shutdown of Suez Canal as a consequence of war	*Revision of entire management strategy* $\mathbf{M}^*$ due to – new scientific insights into the basic functioning of the Earth System – new technological break-throughs extending the options of coevolution management, i.e. additional *information of global importance*
10	*Re-evaluation of targets and boundary conditions* as a consequence of overall development of facts and interests	*Substitution of initial coevolution paradigm* by a new one reflecting modified knowledge and attitude of Global Subject

Our table is obviously an extremely compressed account and comparison of two rather involved control problems, and a number of comments could be made. We will confine ourselves, however, to just two direct remarks.

(i) A fundamental difference between foreign-going trade and geo-cybernetics becomes evident when considering, for instance, Element 3: while sea passages to distant shores are, in principle, repeatable enterprises that may draw heavily on former experience, the man-made passage of the Earth System through coevolution space is a one-way trip of a unique vessel. Thus, any experience in steering this vessel has to be gathered along the way. This fact dramatically enhances the *importance of simulation modelling*, as we will discuss a bit later.

(ii) At least four different types of revisionary actions are involved in any complex control exercise (Elements 7 to 10). The order in which these readjusting steps and the other elements listed above have to be executed is not completely fixed. As a matter of fact, certain control measures like episodical course correction in response to stochastic events may be necessary over and over again. This general observation implies that the Global Subject has to perform its geo-cybernetic task with utmost *flexibility*. The latter is *even more important than precision*, as we will explain in Sect. 6 which deals with "Fuzzy Control".

In the rest of this section, we will focus on those two technical aspects that are most important and characteristic for the control object considered in geo-cybernetics, namely (full or partial) *"Earth System modelling"* and *"sustainability indication"*.

Let us first re-emphasize that our planet is a *historical unicum* and thus suitable for experimental investigation and testing only in a very restricted sense. This almost self-evident observation rules out the conventional heuristic approach as employed especially in the natural sciences. Physics, for instance, which is often considered to be the model discipline of the 20th century, basically deals with *ahistoric universals* and therefore, probably unconsciously, carries on the tradition of Abelard's mediaeval scholasticism (*Gilson*, 1955 [87]). According to *Popper* [186] the methodological credo of physics can be summarized as follows:

Based on evidence of any type, construct a hypothesis with respect to the behaviour of a given universal (e.g., a quark or a black hole). In the next step of the heuristic process, set up experiments with individual representatives of the universal considered. Reject the hypothesis — for all representatives everywhere and forever — if it is falsified by one single spatiotemporal observation ("event"). Retain the stock of yet unfalsified hypotheses as knowledge ready for recall. Note that, according to this recipe, our net wisdom can only grow if the production rate of hypotheses exceeds the falsification rate. Present-day science seems to satisfy the latter condition quite easily ...

Regardless of the quality (and falsifiability!) of Popper's epistemological hypothesis in view of the real research process, the methodo-logic just outlined does certainly *not* apply to Earth System Analysis. We are not willing or allowed to sacrifice the integrity of the one and only planetary specimen we have got for the sake of scientific progress. So most of those wonderful geo-experiments we might design are not at our disposal.

As a consequence, we are left with a fundamental epistemological and ethical dilemma. If anybody makes conjectures about, say, the impacts of CO_2-tripling in the atmosphere or of the shut-down of the Conveyor Belt or of the loss of 90% of the planet's biodiversity – do we *really* want to put these conjectures to the test? Yet the degree of truth of any such hypothesis is crucial as regards any mitigative or adaptive action of the Global Subject. Thus we can formulate the following "theorem" of Earth System Analysis:

> "Hypotheses about Global Change are the less falsifiable the more they are relevant to humanity!"
>
> (*Schellnhuber*, 1997 [205])

So, whenever the rejection or validation of a truly pertinent conjecture about the Earth System takes place, then this evaluation is brought about "by accident", i.e., as the result of economic, political, etc. action without scientific rhyme or reason. There is some hope, of course, that the careful inspection of accessible geological palaeo-archives (ice cores, lacustrine sediments, stromatolites and so on) will reveal the responses of the planetary ecosystem to certain major perturbations. However, note that this very system has evolved considerably since the fixation of the proxy data at hand (through plate-tectonic motion, for example), so the Earth System analyst is in a similar position as the historian who tries to make predictions about the future of a given society: the lessons of the past may provide useful hints, but they may also turn out to be completely worthless because some decisive control factor or boundary condition has changed.

There is one way out of this dilemma, however, namely *virtual* falsification or verification of Global Change hypotheses with the help of *artificial copies of the Earth System* or of crucial parts of the latter. These copies are (geo-)cyberspace representatives of *integrated simulation models* that can be perceived as *synthetic universals* substituting the individual real system in question. The so-created digital globes or regions provide us with testbeds for no-regret experimentation with major manipulations or perturbations of the ecosphere-anthroposphere complex: a runaway greenhouse event, e.g., in virtual computer reality may cost us one CPU year on the most advanced CRAY machine, but not our lives! In such a case, we simply restart the digital game and try to employ a more careful strategy. In this way we may eventually be able to explore the *plume of potential coevolutionary futures* as generated from the present state of the Earth System by the management options contained in $\mathfrak{M}$.

Note that simulation exercises have become a valuable training technique in many fields of complex decision-making, where reality gaming is ruled out or at least problematical for practical and principal reasons (see, for instance *Dörner*, 1996 [58]). In particular, the military-industrial sector has discovered the usefulness and convenience of cyberspace simulation, as exemplified by the recently booming activities in digital tank battles, air force encounters, or nuclear weapon explosions. The Pentagon even seems to be planning a complete digital inventory of the Earth's surface ("Project Number 2851") for all conceivable types of cyberspace warfare. Why shouldn't we make full use of the knowledge and methods involved for less destructive purposes like the preservation of our environment?

Before reviewing briefly the character and state-of-the-art of integrated modelling in Global Change research, let us emphasize two points.

First, a mysterious law of evolution seems to warrant a dynamic balance between humanity's means to destabilize its life support systems and its means to prevent such a destabilization. The ozone hole over Antarctica, for example, was detected with the help of sophisticated remote-sensing equipment (*Farman* et al., 1985 [67]) "just in time" to alarm the world public and to initiate counter-measures which seem sufficient to keep negative impacts at a tolerable level. Furthermore, there is considerable evidence that the breathtaking development in supercomputing will soon enable our climatologists to provide us with the necessary "on-line" information for appropriate climate protection policy. In the same way, the advent of integrated simulation modelling appears to be in good time for making the geo-cybernetic dream come true. So the global civilization seems to subsist in a perpetual state of "self-organized criticality" (*Bak* et al., 1987 [12]; *Bak* and *Chen*, 1991 [11]; *Drossel* and *Schwabl*, 1992 [61]; *Andrade*, 1995 [5]; *Bak*, 1997 [10]), where *fluctuation and regulation power nearly compensate* each other. However, note that even self-organized critical systems collapse when they go too far astray ...

Second, the power of simulation modelling of the environment and development process must not be overestimated. If used properly, integrated models prove to be excellent instruments for exploring the topology of the coevolution space, for detecting the presence or even the proximity of catastrophe domains, for analysing the sensitivity of pertinent Earth System parameters with respect to perturbations. Yet these models will seldom be able to make precise predictions about quantities of interest due to notorious *cognitive uncertainties*, which tend to grow non-linearly with the size of the forecast range. The situation becomes worse if we take into account "*voluntative noise*", i.e. the a-priori stochasticity inherent in human decision making. So whenever we employ simulation models of Global Change, we should be well aware of the fact that such models will never serve as "history machines". We will demonstrate in Sect. 6, however, how decision-makers (including $\mathcal{S}$) can profit tremendously even from imperfect models.

5.1 Integration and Integrity

Simulation modelling tends to be a "high-tech enterprise" with respect to the methodologies and the equipment involved. However, the modelling does not have to be a "high-soph(istication) venture" as well, if the system to be mimicked is just *complicated*, but not truly complex. The simulation of the collective dynamics of thousands of interacting molecules in a liquid is, for instance, a straightforward exercise when the equations of motion are well-defined and adequate hard- and software is available. Simulation modelling simply replaces unfeasible analytical prediction by brute-force numerical forecasting in such a case.

However, what are we going to do if the system considered exhibits *genuine complexity*, if it cannot be decomposed neatly into identical or similar elements of well-known properties, and if the interrelations between the constituents vary largely in character, range, strength, explicitness, crispness, etc.? Then the system in question is characterized by *fundamental inhomogeneity* that cannot be reduced or reproduced by concise algorithms. Note that the latter qualification excludes from the club even intricate objects like fractals as these may be constructed from a compact set of generating rules. Truly inhomogeneous systems cannot be mimicked by "ordinary" simulation modelling any more – we have to resort to *"integrated modelling"*.

Integrated modelling tries to catch the nature of the system without attempting to create a one-by-one cyberspace copy of it. In order to achieve this, almost incommensurable parts and pieces, widely differing spatiotemporal scales, outright reductionistic disciplinary descriptions, and clearly distinct actor levels have to be coupled. Even more important, the notoriously inhomogeneous information structure and precision with respect to the various system components and processes has to be harmonized intelligently by a spectrum of methods ranging from mathematical analysis to gaming exercises and educated guesswork (*Dowlatabadi* and *Morgan*, 1993 [60]). As a consequence, integrated modelling (*Nakićenović* et al., 1996 [160]) sometimes appears to be less a science than an art, which primarily draws on intuition and imagination. We prefer, however, to describe it as an ingenious way to *preserve the integrity of the object to be simulated without violating scientific integrity.* The latter imperative requires that any unavoidable assumption, approximation, simplification or omission is clearly identified and acknowledged in each step of the simulation enterprise. In particular, much care has to be devoted to data quality assessment.

Rather diverse issues in Earth System science may be tackled by integrated simulation modelling, for instance,

- the emergence of quaternary glaciations as a non-linear co-operative response of atmosphere, hydrosphere, biosphere and pedosphere to rather weak insolation fluctuation of the Milankovich type,
- the multidimensional costs and benefits of long-term climate change management, and
- the potential development of highly vulnerable regions (like the semi-arid Northeast of Brazil or the Maghreb) under the driving forces of Global Change.

But whatever topic is dealt with, two main principles have to be observed, namely

(i) *Relative Comprehensiveness*, and
(ii) *Weak Endogeneity.*

Here the qualifications "relative" and "weak", respectively, are crucial, as absolute comprehensiveness and full endogeneity are almost impossible to achieve in complex systems analysis.

The first principle refers to the fact that integrated modelling is primarily supposed to answer specific questions or even to solve certain problems; it is less concerned with creating novel micro-elements of disciplinary methodology. Therefore, *all* the system's components and processes *pertinent to the issue considered* have to be identified and taken into account, while the complementary elements should be ignored in an intelligent way.

The second principle refers to the notorious *closure problem* for the scientific description of an open or semi-open system, which is non-trivially coupled to an external world. As it would be neither feasible nor reasonable to model the outside universe in comparable detail to the system in question, *the import-export relations have to be taken care of* in an efficient and rather smart manner. Note that even "Earth Models" can only provide extremely selective descriptions of our planet, ignoring, e.g., the dynamics of the Earth's mantle, or the micro-organismic cosmos, or the ultimately complex sphere composed of billions of human souls. So there is always an inside and an outside world, separated by ill-defined borderlines (or, in general, border hyper-surfaces).

Evidently, these two principles of integrated simulation modelling are interrelated. Both imperatives may be complied with simultaneously by the following operational recipe:

(a) *Identify and account* for all those pertinent system components whose properties have to be evaluated in order to deal competently with the issue at stake;
(b) *Exogenize* all those elements of either the system itself or of the "outer universe" which are influenced by the pertinent system components, yet do not affect them, by truncation;
(c) *Exogenize* all those elements of either the system itself or of the "outer universe" which are not influenced by the pertinent system components, yet do affect them, by scenarios, look-up tables, proxy data or the like;
(d) *Neglect* all other elements, irrespective of their affiliation to the inner or the outer world.

A symbolic illustration of this strategy is given in Fig. 33.

For any given simulation task the general modelling design as epitomized by Fig. 33 has to be cast in a specific mould that accounts for the character and the peculiarities of the complex environmental system in question. Let us consider, for instance, the so-called "integrated assessment models" (*Kaya* et al., 1993 [119]; *Nakićenović* et al., 1994 [159]; *Nakićenović* et al., 1996 [160]; for a recent review see *Schellnhuber* and *Yohe*, 1997 [207]) which are supposed to serve as scientific detectors of optimal climate protection strategies. In this case, there is no need of geographical exogenization as the object at stake is the entire globe. By way of contrast, a number of relevant processes like demographic development and technological progress cannot be completely endogenized in these models because such processes are strongly affected by factors well outside the climate-related sub-cosmos of reality. Appropriate global scenarios, calibrated by empirical and theoretical wisdom about local societal dynamics and cultural specifica, will have to do the closure job here.

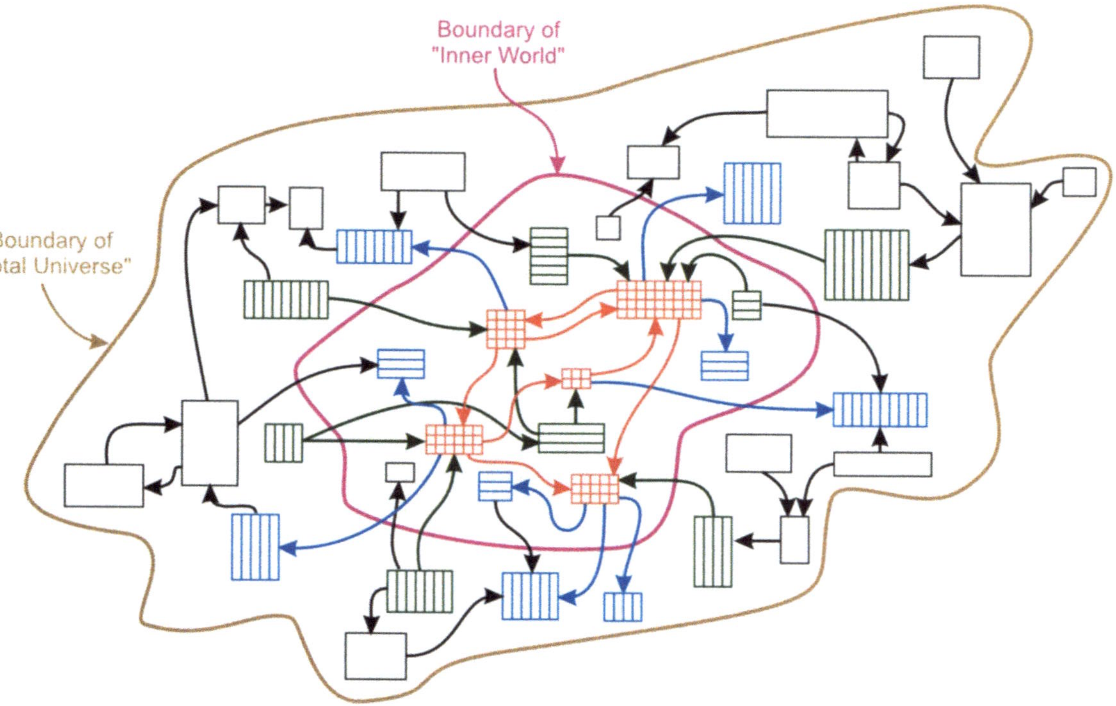

Figure 33. Wiring diagram and cut-off strategy for a fictitious system to be simulated by integrated modelling. The "inner world" (as delimited by the *pink membrane*) embraces all pertinent elements (*full red hatching*) as well as additional components irrelevant to the issue under investigation. Among the latter, those components directly influencing the pertinent system elements are indicated by *horizontal green hatching*, and those components directly influenced by the relevant elements by *horizontal blue hatching*, respectively. The remaining inner components are characterized by a *black contour*. The same indicative logic applies to the components of the "outer world" (as delimited by the *brown contour*) that encompasses the core domain of interest. Note, however, that the *hatching* of outer elements is vertical. According to the modelling strategy described above, all the *red arrows* must be taken care of by dynamic coupling, all the *green arrows* should be taken care of by appropriate scenarios or input from independent external simulation, and the remaining *blue* and *black arrows* may be neglected.

If we wish to construct, on the other hand, a regional integrated model, then a "nesting approach" will be eligible: within the limited focal area as many elements of the overall inventory as possible should be endogenized through coupled simulation. There is, however, an "outside world" (in the true sense of the word) whose influences can be represented by the outputs from coarse-grained global models for the evolution of, say, climate, economy and life-style.

Apart from their degree of integration according to the general principles just discussed, simulation models of complex E & D objects may differ strongly with respect to *realism and sophistication*. The latter qualifications have to correspond to basic determinants or limiting factors such as purpose, ambition or resources (regarding time, data, computer equipment, etc.). By and large, we may distinguish here four main types of models.

A. Tutorial models.
These models mainly serve as "cyberspace toys" that help us, on the one hand, to *train our skills* for the rough climate of the "serious" simulation world and, on the other hand, to *identify fundamental effects and phenomena* characteristic of the system under investigation. Tutorial models are, therefore, rather barbaric and eclectic in nature; they tend to heavily reduce or mutilate the original entity and to keep just the crudest outlines of the remainder! If executed properly, however, such atrocious simplifications may be rewarded not just by savings in time and effort but also by a sharp exposition of the intrinsic mechanisms and major interrelations that shape the system's structure and behaviour. Figuratively speaking, this is *surgery using Occam's razor*, aiming to lay open dominant feed-back loops or catastrophic reaction paths in the entrails of the specimen. As a matter of fact, modelling through symbolic dynamics (*Robinson*, 1994 [196]; *Lind* and *Marcus*, 1995 [136]) often suffices to reveal characteristic properties of the system in question.

If we employ yet another metaphor and compare simulation modelling with the art of portrayal, then tutorial models correspond to the first and the last human documents of *abstract* painting as represented by the unknown genius of Lascaux and the well-known eccentric Picasso.

For illustrations of the tutorial approach we again refer to geophysiological models like the famous Daisy-World toy (*Watson* and *Lovelock*, 1983 [245]), which was devised to support the GAIA hypothesis. The rather simple nonlinear mathematics of the Daisy-World parable actually encapsulate an extremely robust principle of homeorhetic self-control. As already indicated in Sect. 2, appropriate extensions of the original Daisy-World model reveal generic relations between structure and stability of complex eco-systems (*Svirezhev* and *von Bloh*, 1996 [226]; *von Bloh* et al., 1997 [242]).

B. Conceptual Models.
The models in this class generally attempt to catch the spirit of the original *in its entity*, but not necessarily to reproduce the proportions of the real

counterpart. In other words, a conceptual model will reflect correctly the *topology* – the composition and connectivity – of the system to be simulated while the *metric* – the actual numbers along the dimensional axes involved – is of inferior interest.

So, in the best case, the modelling activity creates a *caricature* which is even more characteristic and recognizable than the original itself. The caricature does not have to preserve the dimensionality of the system: under certain conditions even zero-dimensional simply-wired box models may successfully simulate the *qualitative behaviour* of the real object. The model is supposed to generate, in particular, the right phase diagram, i.e. the true mosaic of (dynamic) equilibrium states in parameter space up to some homeomorphic distortion.

The construction of conceptual simulation models is an extremely difficult task. The risk of conjuring up a "Dalí", i.e. a beautiful but surrealistic misconception of the original, is rather high, especially when fundamental processes are not well understood. Some of the current efforts are still rooted in the world models of first generation (*Meadows* et al., 1974 [149]). An important example for a carefully designed integrated global model of the conceptual type is TARGETS (*Rotmans* and *de Vries*, 1997 [200]), which is described in some detail also in this book (*Rotmans*, 1998 [199]). Another fraction of conceptual models consists mainly of stepwise extensions of climate models (*Gallee* et al., 1991 [77]; *Gallee* et al., 1992 [76]; *Stocker* et al., 1992 [223]; *Xiao* et al., 1997 [260]). Yet another illustration is provided by the "Potsdam Earth System Models" (POEM) for the entire ecosphere as composed of the interacting components atmosphere, hydrosphere, biosphere and pedosphere. The POEM hierarchy embraces models of widely differing sophistication and dimensionality, where the latter ranges from 0 to $2\frac{1}{2}$ (*Claussen* and *Ganopolski*, 1997 [44]; *Ganopolski* et al., 1997a [78]; *Ganopolski* et al., 1997b [79]).

C. Analogical Models.
According to a former statement, integrated simulation modelling is not supposed to create simply down-scaled copies of the original that take into account virtually all facets of reality. By way of contrast, pure geophysical simulation ventures like coupled ocean-atmosphere circulation models certainly go for that goal. The so-called analogical integrated models for the Earth System or large segments of it try at least partially to live up to the standards set by geophysical modelling: These models have to neglect, omit or simplify a multitude of traits of the original for the sake of feasibility, but they nevertheless attempt to mimic a number of pertinent (natural and socio-economic) dimensions *true to scale*. This means, in particular, that analogical modelling tries to preserve some spatiotemporal metric through *geographical explicitness*. The resulting 3-dimensional construction is then animated by

a portfolio of dynamic prescriptions, including partial differential equations and discrete decision rules of the cellular-automaton type.

If we employ our painting metaphor once again, then analogical modelling corresponds to *naturalistic* portrayal as has been mastered perfectly by, e.g., Dürer, Velázquez or Leibl. Among the most advanced and powerful models of the "isomorphic" type, the IMAGE family (*Alcamo* et al., 1994 [3]) plays a leading role.

D. Hybrid Models.
Many integrated (simulation or assessment) models are concocted from rather disparate ingredients. In particular, some of the intellectual devices introduced for conducting cost-benefit analyses of anthropogenic climate change are composed in a rather ill-defined way of tutorial, conceptual and analogical modules. This is partly due to uneven availability of the necessary components from the various disciplines involved, but also to the continuing dominance of economists in this adolescent field. It is, of course, perfectly legitimate to focus on the things one understands best.

Some of these difficulties may be surmounted by modelling teams, which are assembled *ab initio* according to transdisciplinary criteria and which try to harmonize the inhomogeneous cognitive pool by projecting the latter on a uniform level of sophistication. The resulting *reduced-form* or *effective* integrated models are thus tuned to the weakest scientific module involved and do not run the risk of producing numbers of unwarranted precision. All "integrationists" should be warned of apodictive use of exact figures by the famous joke about the museum over-attendant who told the visitors that the dinosaur on exhibit was 90,000,006 years old. Upon questioning, the man explained that he had been told the dinosaur was 90 million years old when he was hired six years before (*Paulos*, 1995 [172]).

Another way of handling strong disparaties and uncertainties in integrated modelling is *"fuzzification"* (see Sect. 6 for detailed references). Much of the analysis of the complex system considered may consist of qualitative reasoning based on expert knowledge, educated guesswork, comparative case studies, and scientific intuition. In spite of this, it may be useful and reasonable to present the outcome of this reasoning in a geographically explicit form, for instance by a set of regional maps. There is nothing wrong with such a procedure, if the inherent vagueness of the results is properly expressed. Fuzzy set theory provides an elegant technique to achieve that. The strategy to be employed in such a case may be illustrated by the so-called "Syndromes Concept", a hybrid-type activity dedicated to the integrated assessment of Global Change (*Schellnhuber* et al., 1997 [206];

WBGU, 1997a [84]; *WBGU*, 1997b [85]). This activity starts out at the top phenomenological level by identifying the generic functional patterns of problematic civilization-nature interactions (the "syndromes"). The syndromes are finally made visible as partially overlapping, patchy structures representing the "clinical pictures" of the global environment. The most advanced analysis so far has been performed for the so-called Sahel syndrome, which represents the cause-effect complex governing the overexploitation of marginal land by displaced people and/or underprivileged segments of the population (*Cassel-Gintz* et al., 1997 [39]).

* * *

Now that we have briefly reviewed the various types of integrated models up and running, we may ask ourselves whether there is a "*natural evolution*" within the scientific world towards a "superior model species". The question is particularly valid with respect to global integrated modelling that is supposed to ultimately mimic the dynamic behaviour of the entire Earth System. By intuition, we may postulate that one should start out with the most simple variety and end up with the most complex one. We are, therefore, tempted to presume that the evolution arrow clearly points from tutorial to analogical integrated models.

Unfortunately (or fortunately?), the answer is not that straightforward: tutorial models may be highly sophisticated, e.g., from a mathematical point of view, while analogical models might constitute just a dull, uninspired aping of what is taken for "reality". As a matter of fact, the degree of abstraction, approximation and generalization which is respectively considered necessary or tolerable, largely depends on the ***disciplinary bias*** of the individual scientist confronted with the integration challenge. Physicists, for example, tend to put their trust almost naively in universal spatiotemporal principles reigning nature *and* society; thus they strive for utmost simplicity as represented by zero-dimensional models cast in differential equation form. By way of contrast, psychologists insist on describing *each single* individual or ethnic group or culture in its full complexity, which embraces not just the physiological facts but also a universe of "inner dimensions". When it comes to the question of geographic resolution of an integrated simulation model, we may therefore formulate – tongue-in-cheek – the following "scale theorem":

> "Natural scientists prefer a scale smaller than $1 : 10^6$;
> geographers prefer the scale $1 : 1$;
> social scientists prefer a scale larger than $10^6 : 1$."

This jocular and outrageous statement should not be taken seriously, yet it contains some grain of truth ...

We should emphasize, anyway, that a marked "biodiversity" in integrated modelling will subsist in the medium term and will provide us with a fertile heuristic pool for all sorts of Earth System analyses. In spite of this, a certain average trend towards hybrid or "collage" models can be observed.

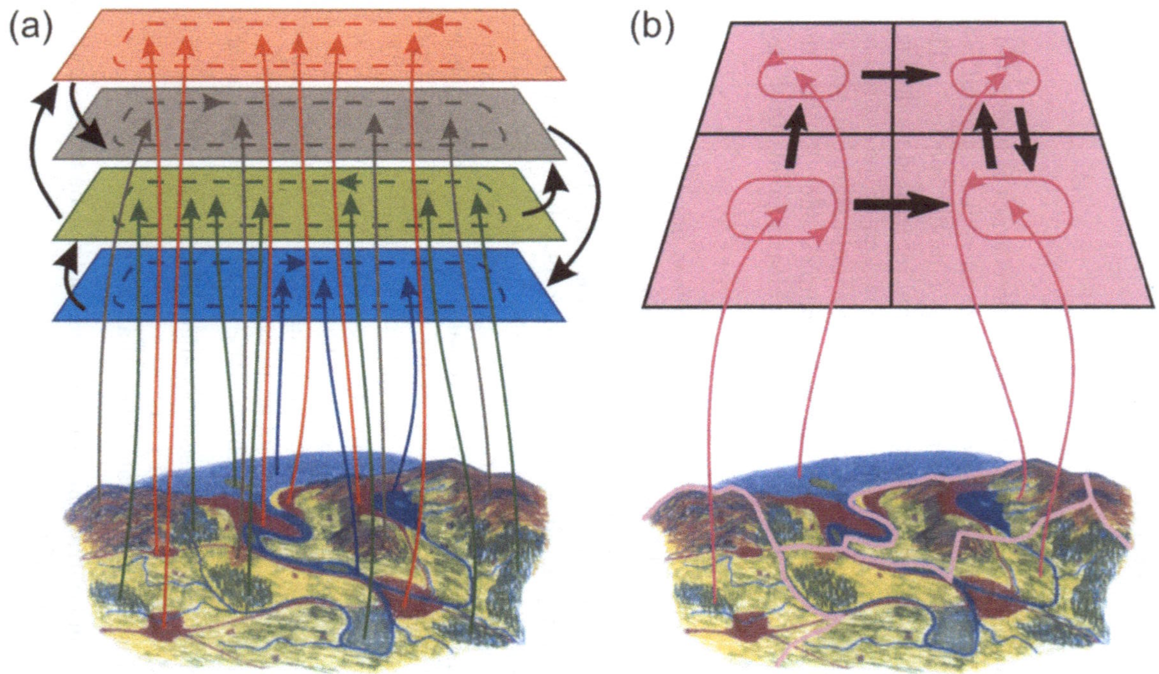

Figure 34. Alternative lumping strategies for constructing zero-dimensional global integrated models.
(a) Aggregation of sectoral elements from all over the world and dynamic coupling of the resulting sub-sphere modules.
(b) Aggregation of regional multi-sectoral inventories and dynamic coupling of the resulting simulation cells.

As far as the *collage technique* is concerned, fundamentally different philosophies can be identified, especially when it comes to global simulation. If such a model is assembled from a plethora of components in a cut-and-glue manner, then the integration of these modules consists of basically two steps:

(i) *Aggregation* by compressing and summarizing fine-grained information;
(ii) *Wiring* by constructing appropriate dynamic interfaces between the lumps emerging from (i).

Let us actually focus on the first step, which is directly related to "integration" in the mathematical sense of the word, i.e. in the sense of an operation which collapses a whole range of data onto a resulting single entity. We recognize that at least two alternative strategies for this procedure are available in the context of Earth System Analysis. The *traditional strategy* is to consider the disciplinary sub-spheres of the planet, like the atmosphere, the biosphere or the world economy, separately and to integrate out most geographical details within each of these global sectors. The resulting fields of suitable dimensionality are finally connected in a way that can animate the substitute of the real system. Most of the existing "World Models", like the ones in the famous Forrester-Meadows family (*Forrester*, 1971 [73]; *Meadows* et al., 1974 [149]; *Meadows* et al., 1992 [148]), are constructed along these lines.

By way of contrast, a completely different "Frankensteinian" assembling technique might be adopted. This *alternative strategy* seeks for a decomposition of the Earth's surface into an ensemble of sufficiently well-defined natural or administrative geographical units ("geotopes") like river catchments, culture domains or countries. By integrating out most of the ecological and socio-economic variables representing the total inventory of a given individual cell, the corresponding geotope module is generated. Finally, the resulting modules have to be wired with a special emphasis on the lateral import-export relations regarding water, energy, substances, species, goods, etc.

Fig. 34 sketches and compares the two fundamental integration techniques just described.

A crucial question is, of course, whether the end product, i.e. the integrated model, depends on the mode of assembling, or – speaking in mathematical terms – whether the *sectoral* and the *geographical* "integration" may be freely *interchanged*. As the operations involved are certainly of the nonlinear type, a lot of care has to be invested here.

We should also point out that the second composition strategy is *inherently parallel* and should therefore lend itself to computational simulation with distributed resources. We may even dream of a "world model" in the literal sense, which is run on a communicating network where the nodes are formed by selected computers based at the respective geotopes.

5.2 Playing the Game

Let us assume that we have created a sound and operable integrated model of the interacting complex composed of the ecosphere $\mathcal{N}$ and the anthroposphere $\mathcal{A}$. This is a necessary condition for geo-cybernetic map exercises, yet not a sufficient one. According to the control systematics described at the beginning of this section, three *main ingredients* have to be added:

(i) The certain as well as the possible *external driving forces and perturbations* of natural origin. Most of these factors can be provided by astrophysical and geophysical analysis.

(ii) The *pool of control options* $\mathfrak{M}$ at the disposal of the executive component $\mathcal{M}$ of the Global Subject. The portfolio of E & D measures includes legal, political, economic, educational and technological elements. This pool has to be explored, modelled and simulated at least as carefully as the $\mathcal{N}$-$\mathcal{A}$ "hardware".

(iii) The *spectrum of methods capable of determining optimal solutions* $\{\mathbf{M}^*\}$ of the geo-cybernetic exercise for any given SD paradigm $\mathcal{P}$. Among these methods, which have to be employed by the "intellectual" component $\mathcal{B}$ of $\mathcal{S}$, we can identify several distinct types, namely
   - mathematical analysis in the tradition of classical control and optimization theory;
   - numerical search algorithms of deterministic and/or stochastic character (Newtonian techniques, dynamic programming, Monte Carlo methods, evolutionary procedures, etc.);
   - educated computer-assisted guesswork, based on advanced expert and data management systems.

Then, finally, the E & D planning game has to be played with all these beautiful though sometimes elusive toys ... The players as recruited from the "Global Brain" $\mathcal{B}$ are supposed to dice for the hypothetical optimal control strategy $\mathbf{M}^*$ under the given boundary conditions. However, note that the geo-cybernetic map exercise has a remarkably dynamic and even *self-referential* character (as observed already in Sect. 3): instead of just playing the game with fixed rules, we may modify these rules; we may watch ourselves playing, modifying, and responding to our own actions; and *we may model all this including the modelling itself and the implications thereof*!

Some of the so-called "Integrated Assessment Models" (IAMs) for climate change management, which we have briefly mentioned above, try to take into account at least the most elementary aspects of this truly intricate situation. Such models may be qualified as "Evolutionary IAMs" (*Schellnhuber* and *Yohe*, 1997 [207]) because they allow for readjustable strategies, i.e. *mid-course corrections due to endogenous (and exogenous) learning* (see, e.g.,

Kelly and *Kolstadt*, 1996 [120]; *Dowlatabadi* and *Morgan*, 1993 [60]; *Lempert* et al., 1994 [133]).

However, that is only a first step towards genuine self-referentiality: Beyond the digestion of accidental cognitive shocks from an independent knowledge-producing machinery responsible for "the scientific progress", we may *integrate the latter process* into our map exercise. As a consequence, appropriate measures for improving the boundary conditions for knowledge production about $\mathcal{N}$, $\mathcal{A}$, and $\mathfrak{M}$ itself have to be considered a part of an optimal control strategy, as well as the flexible and versatile design of elements of the management pool, in order to make maximum capital out of potential new insights. This means, in particular, that the rigid reservoir $\mathfrak{M}$ has to be replaced by an evolutionary one.

We may even go one step further by allowing also for the evolution of the SD paradigm in the course of the geo-cybernetic process, and by *trying to anticipate such an evolution* within the framework of the simulation game! One has to be extremely careful, however, not to run this way into logical paradoxes or vicious circles. In order to avoid some of the intellectual pitfalls lurking behind self-referentiality and to bring the entire reasoning down to Earth (!), we can make use of a "human interface" between the virtual and the real world.

The interface is provided by exemplary *policy exercises* (*Tòth*, 1986 [228]; *Tòth*, 1988 [229]; *Tòth*, 1989 [230]) for E & D decision-makers, who are playing the geo-cybernetic game interactively on the model surface, and who are assisted and observed by the "pilots" from the scientific expert community. The test persons are to be recruited from politics, from economy or from the reservoir of public pressure groups and movements. The exercises may be perceived as "study courses for global environmental management based on simulated control", and these courses should reveal crucial elements of the intrinsic dynamics of Earth System steering. Note, in particular, that policy exercises of that kind properly reflect the *reality of geo-cybernetics* – however great our ignorance about the precise functioning of the $\mathcal{N}$-$\mathcal{A}$ complex may be. For the test persons *are* the "global players", the very actors that will shape the future of this planet to a sufficient extent. It is they, after all, who have to interpret the vague SD paradigms ventilated by society, who have to adopt and correct the coevolution path under considerable cognitive and voluntative uncertainty, and who have to learn from wrong advice and decisions. *So watching these players is a bit like watching the manufacturing of the environmental future* (or the future environment, if you like).

5.3 Orientators and Indicators: From Virtual to Real Reality

Once the ideal coevolution-management strategy $\mathbf{M}^*$ has been determined within the framework of our geo-cybernetic map exercise through some combination of the techniques described above, this brainchild has to be *converted into an adequate sequence of real control actions.* These actions are supposed to bring the $\mathcal{N}$-$\mathcal{A}$ tanker on to the right course with respect to the preferred SD paradigm and, most importantly, to keep it there. Such a task can only be completed if a perpetual stream of information links the virtual to the real reality and helps to bridge the gap between theoretical and practical operation. The geo-cyberneticist needs, in particular, *on-line data* describing the current position of the system in coevolution space as well as the present status of the propulsive equipment and the control devices. The corresponding stages of the navigation process have been identified and described as the Elements 5 to 9 of the "control table" presented at the beginning of this section (Tab. 2).

However, note that a virtual "Earth manager" in the planning game is capable of digesting instantaneously the full bulk of data provided by the digital sources, and that this electronic agent might even be able to evaluate the information prospectively for any desired time horizon. By way of contrast, the real geo-cyberneticist generally only disposes of rather patchy and fragmentary data, which is nevertheless far beyond his/her receptive and ruminative capacities. As a consequence, the human agents have to be provided with "the essential information" only, preferably *compressed into concise orientators and indicators* that decisively facilitate the various control decisions to be made. As a matter of fact, we will subsequently subsume any type of decision-supporting criteria under the common term "indicators", as the word "orientator" has recently acquired a rather specific scientific meaning (*Bossel*, 1996a [29]; *Bossel*, 1996b [30]).

Given a complex environmental system like a landscape or the total ecosphere, appropriate indicators are supposed to keep us informed on

- the system's *status* (instantaneous-local indicators),
- the system's *dynamics* (differential-infinitesimal indicators), and
- the system's overall *evolution* (integrated-global indicators).

Under more technical aspects, geo-cybernetic indicators may be classified according to their degree of compositional or operational complexity. We distinguish between

- *simple* indicators (e.g., the CO_2-concentration in the atmosphere as measured in ppmv),
- *compound* indicators (e.g., the "Human Development Index" (*UNDP*, 1990 [239])), and

– *systemic* indicators (e.g., the number and strength of positive feedback loops promoting soil degradation by marginal agriculture).

As reflected by the suggested terminology, systemic indicators are devised for detecting and evaluating generic systems properties like complexity, versatility, stability, resilience, brittleness, connectivity, autonomy, criticality, and so on.

Qualities of the type just listed have primarily been discussed and explored by ecosystems theory, but even there the task of constructing adequate indicators has by no means been completed. Concise criteria to indicate, for instance, the "pristinity" or, even more difficult, the "robustness" of a biotope are not yet available.

A crucial question is whether indicators should be endowed with *normative functions*, which allow value judgements to be read off directly. Within the context of the canonical approach to control theory, any conceivable indicator is meant to support the implementation of a *pre-determined* ideal strategy and to be instrumental in correcting unintentional deviations (due to human or technical failure, for instance) from the desired course. In other words, all sorts of valuation acts are performed already before or during the planning game, and the indicators are purely analytical tools for realizing the geo-cybernetic ideal.

However, what are we going to do if both the long-term coevolution desideratum $\mathbf{P}^*(t)$ and the generating management sequence $\mathbf{M}^*(t)$ are not definable – because of fundamental ethical and political disagreement among the decision makers and/or serious cognitive deficiencies that cannot be fixed by the scientific community? Under such circumstances the geo-cyberneticist has to resort to the "ultimato ratio", namely to steer the system *at short sight* by using certain indicators as the primary information basis for any navigational decision without reference to some teleological optimum.

It is evident that indicators of the type needed here have to act as *normative on-line monitors*, which continuously signal the quality of the system's state and the temporal modifications thereof. Instantaneous normative indicators may evaluate, e.g., whether a tentative change of state would represent an immediate improvement for the ecosphere, for humanity as a whole or for selected indigenous cultures. To give more specific examples, such an indicator might detect the proximity of locally critical manifolds in real coevolution space, or identify the direction of maximum relative increase in global or regional prosperity.

As a matter of fact, the present sustainability debate is about to marginalize the importance of integrated long-term concepts and strategies for perpetual coevolutionary progress, and to back an opportunistic ad-hoc management instead. This trend, however, is not so much the result of judicious reasoning as of the persisting lack of consensus regarding geo-cybernetic paradigms and paths. As a consequence, ***so-called sustainability indicators are often taken for the entity they are supposed to indicate*** – we are on the right track if monitor X, which has been devised according to the instructions of, say, the Commission on Sustainable Development (CSD), exhibits a green light! Such an attitude is reminiscent of the rather naive identification of guaranteed socio-cultural progress with certain minimum growth rates of gross national product (say, 3 % p.a.).

From the point of view of Earth System Analysis, simplistic approaches based on local sustainability indication are rather problematic. It is a fact, well-established by both practical experience and scientific reasoning, that small-scale and short-term optimal response may eventually head for disaster ***if the response is not guided by a holistic strategy*** (see also the following section). This has to be borne in mind when assessing the usefulness of the plethora of products at present produced by an entire SD indicator industry (see the aforementioned references, particularly those on pp. 47, 49 and 54). The reader who is not familiar with the habits of this sector should be warned of perceiving each and every specimen presented as "the real thing", which convicts nature or humanity of any possible aberration with factual authority. Actually many indicators promoted in the field of Global Change and Sustainable Development are more or less sophisticated ***hybrids*** of real variables (***"observables"***) and simulated activities (***"simulables"***). Meteorological research, for instance, frequently makes use of "computer-ameliorated observations" where patchy data gathered from the available monitoring stations are complemented or "improved" by simulation modelling. This is illustrated by the semi-artificial statistics that have been employed recently to furnish proof to the anthropogenic enhancement of the natural greenhouse effect (*Hegerl* et al., 1997 [102]; *Hasselmann*, 1997 [99]; *Kaufmann* and *Stern*, 1997 [118]; *Hasselmann*, 1997 [98]).

* * *

We conclude this section with the statement that it is not always (in fact, rather seldom) reasonable to stick to a pre-determined ideal course: due to newly emerging information, not the implementation of the optimal strategy may have to be adjusted but the strategy itself! This implies various types of intentional course corrections, namely evading manoeuvres, revision of cruising design, or re-evaluation of targets as summarized in the control table (Tab. 2). The subtle interplay between planning and perpetual readjustment in order to produce a winning strategy will be discussed in some detail in the following section.

6 Fuzzy Control: Soft Decision Making under Uncertainty *or* A Tale of Two Demons

Suppose that the "Global Subject" had made up its/her/his "Global Mind" and definitely selected a pure or a complex SD paradigm to be realized. However, is it actually possible to control the E & D process?

We might, of course, immediately dismiss this question by qualifying it as either unacceptable or irrelevant: for all present efforts related to global E & D policy certainly rely on the fundamental hypothesis that coevolution *can* be influenced, intentionally and significantly, for the better! Losing that *faith* would leave us with such unpleasant alternatives as total resignation or paralysis. In addition, note that even a deliberate resolution on the part of humanity to act *not at all* as a global entity would be thwarted, anyway, by the real existence of countless visible and invisible threads already interlinking almost all the people on this planet.

The sheer existence, though, of some global E & D policy does not warrant success, so the crucial question regarding the *feasibility of global environmental management* remains. The Earth System, as defined in Sect. 3 (see Eq. 16), is probably among the most complex dynamic entities in the universe. It is composed of myriads of constituents that mostly interact non-linearly at all conceivable spatiotemporal scales. The ozone-hole lesson teaches us that this system *can* be destabilized by human interference, but what about command in the sense of positive perturbation? The laws of thermodynamics, stating that entropy-enhancing "destruction" is much easier to achieve than entropy-reducing "creation", do not provide much encouragement here. Shouldn't we conclude that the whole notion of geo-cybernetics is, at best, foolishness, and, at worst, perilous demagogy?

Instead of rushing to such conclusions, let us elaborate a bit on the problem of controlling complex systems like "spaceship Earth". Evidently, the two main aspects to be considered are

(i) the *effectiveness* of available steering options, and
(ii) the *knowledge* (or, even more important, the ignorance) about the functioning and evolution of the system in question.

The first aspect, concerning the manoeuvrability of truly big and complicated vessels in high-dimensional phase space, will be dealt with below. Let us beforehand turn to the second aspect, which is closely related to the daunting complexity of the system to be controlled.

Our *systemic wisdom* is notoriously *uncertain*, *fragmentary* or *fuzzy* with respect to the

- components and compositions,
- characters and dynamics of relevant processes,
- initial, current and boundary conditions, and
- external perturbations

involved. The total ignorance may be functionally decomposed into *cognitive* and *voluntative* elements, where the first class can be further sub-divided into *removable* and (for all practical purposes) *irremediable* deficiencies. This taxonomy needs some specification:

1. The ensemble of removable cognitive uncertainties defines the scientific domain, where intensive research may finally generate *precise information.*

2. The ensemble of practically irremediable cognitive uncertainties embraces all those elements and processes of the complex system in question, which are rigorously determined – in principle – by the laws of nature yet *cannot be specified crisply* at a given point in time.
   It seems to be impossible, for instance, to monitor in full detail the prevailing stress and strain fields in the Earth's crust, although we are dealing here with a well-defined problem of "19th century" continuum mechanics. As a consequence, deterministic geophysical processes enter our perception as aleatoric tectonic phenomena culminating in the insidiousness of earthquakes or volcanic eruptions. In much the same way, extrinsic perturbations of our ecosphere, like fluctuations in solar activity, meteorite bombardment or passage of cosmic dust plumes, appear to be purely stochastic in nature.

3. Fundamentally irremediable voluntative uncertainties "simply" result from the *"freedom of will" of billions of individuals*, which constitute the driving-force cosmos of the anthroposphere. Thus any predictions about the further development of the Earth System are cursed by a self-referential vagueness of almost quantum-mechanical character. As a matter of fact, some eminent physicists are at present striving to explain the incalculability of human decision-making from the theory of quantum fluctuations (see, e.g., *Penrose*, 1974 [178]) .... In spite of this *fundamental indeterminacy at the individual actor level*, we are capable of formulating and validating useful statistical laws concerning the behaviour of large samples of co-operating and/or competing human beings (see, e.g., *Schweitzer*, 1997 [213]). Thus "socio-thermodynamics" may come into being as a novel scientific field in the approaching decades.

For the purpose of our discourse, we assume that the respective SD paradigm selected by the Global Subject is consistent with the tohubohu of innumerable micro-level voluntative processes, yet contains a holistic element that transcends the summation of atomic will vectors. The emergence of global decision-making from the individual acts of volition of micro-, meso- and macro-actors (ranging from single consumers to entire nation states) is a most subtle topic, though, which cannot be dealt with in any detail here. We will slightly touch upon it only in the concluding section.

6.1 The Teleological Dream: Laplace's Demon and Contemporary Company

Having specified the major sources of uncertainty regarding the Earth System, *what can we do about them*? While our ignorance of the first kind may be considerably reduced by intensified research (focusing, for example, on the dominant atmosphere-hydrosphere-biosphere interactions or the crucial relations between social dynamics and environmental resource use), there is no decent way to eliminate our ignorance of the third kind. Even the brutal suppression of individual "degrees of freedom" by totalitarian regimes ultimately fails, as evidenced by history. Voluntative indeterminacy may be, at best, "tamed" or "civilized" by democratic processes resulting in medium-term political mandates for social executives.

So it is the *second type* of uncertainty which really gives us a headache. If even the biogeophysical part of the Earth System, the ecosphere, obstinately retained its character of a huge roulette wheel generating random events – how could we even dare to hope for prediction, let alone control of coevolution? *Do we have to resort to metaphysics for salvation*?

A fictional solution to this dilemma is in fact provided by the recent debate among physicists concerning the post-Newtonian mechanistic paradigm. Imagine a godlike creature, who knows at a given point in time the *simultaneous positions and momenta of all particles in the universe.* By integrating the well-defined Hamiltonian equations of motion, our creature would be able, in principle, to predict the further development of the universe along with its inventory. This supernatural being is a brain-child of the French polymath Pierre Simon Laplace and therefore goes by the name of *"Laplace's Demon"*.

We will not deal here with the predicaments inflicted on the poor ancien-régime demon by the quantum mechanics revolution. Even if we restrict ourselves to the realm of classical mechanics as reigning the macroscopic world, we have to recognize that modern physics sets a considerably higher standard for the miraculous creature: the demon actually lacks any long-term predictive power unless it knows all initial conditions for all particles with *infinite precision*! The latter statement is a consequence of the deterministic chaos that prevails in many non-linear dynamic systems possessing more than one degree of freedom. The non-linearities involved namely induce the existence of so-called *hyperbolic points* in the system's phase space causing "sensitive dependence on initial conditions". Hyperbolic points can be viewed as dynamic crossroads, where alternative passages lead to globally different system evolutions. (For popular introductions into the theory and phenomenology of deterministic chaos see, e.g., *Stewart*, 1989 [222] or *Ruelle*, 1991 [201].

Deterministic chaos can already be observed (and probably most impressively) in rather simple systems which by no means deserve the label "complex". Since the days of Henri Poincaré, whose monumental essay ignited the "non-linear revolution" (*Poincaré*, 1899 [184]), the *three-body problem* (see Fig. 35) has served as a paradigm and an icon of chaotic behaviour.

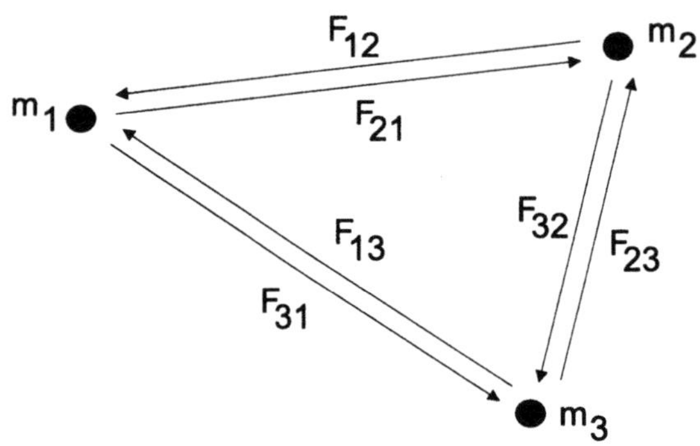

Figure 35. Simplest version of a celestial three-body system – point masses interact through standard gravitational forces.

The forces indicated in Fig. 35 obey Newton's law, i.e.,

$$\mathbf{F}_{ij} = G m_i m_j \frac{(\mathbf{r}_i - \mathbf{r}_j)}{\| \mathbf{r}_i - \mathbf{r}_j \|^3} \quad ; \quad i,j = 1,2,3 \quad , \tag{228}$$

where m_i, r_i denote the mass and the position vector, respectively, of the i-th body, and G is the gravitational constant.

A rigorous proof that the three-body dynamics is generically *non-integrable* ("chaotic") has been achieved only recently (*Xia*, 1994 [259]). This rather technical statement has a quite dramatic corollary for the solar system: the long-term stability of this "astronomical clockwork" turns out to be a mere illusion, although excessive computer simulations indicate that it will not decay in the next ten million years ... Other numerical experiments reveal the non-stability of the asteroid belt (see Sect. 1), the chaotic dance of Hyperion (one of Saturn's satellites) and even the irregularity of Pluto's trajectory (*Wisdom* et al., 1984 [251]; *Wisdom*, 1985 [250]; *Sussman* and *Wisdom*, 1988 [225]; for a review, see also *Peterson* (1993) [179]).

Given the fact that even the motion of simple celestial systems defies any concise analytic description, it seems that only a demon which is by far superior to Laplace's creature might be able to predict the fate of a truly complex system like planet Earth. Does this definitely mean *"Lasciate ogni speranza, voi que entrata!"* for *human* geo-cyberneticists? The answer is *"No"*.

We have to realize that the natural sciences – and particularly modern theoretical physics – apotheosize the mundane cosmos of complex systems into a theatre world where only supernatural beings might survive the chaotic rat-race. However, we do not need demons in order to do very well in our non-linear everyday world! Let us support this contention with a few examples:

Example 1 (Swinging Child):
Without having the foggiest idea of Hamiltonian mechanics, not to mention parametric resonance or KAM theory (*Arnold*, 1978 [7]), almost any child is able to "operate" a garden swing regularly with any desired amplitude. This can be observed in spite of the fact that even a considerably simplified version of the physical child-swing system (the mathematical double pendulum, see Fig. 36 and Sect. 2) exhibits deterministic chaos of the finest quality (*Richter* and *Scholz*, 1985 [195]).

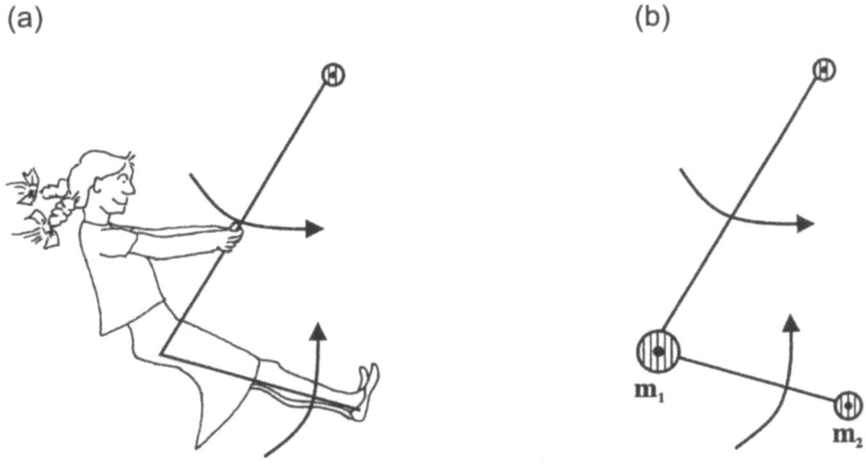

Figure 36. The double pendulum subject to a homogeneous gravitational field **(b)** represents a physical caricature of the swinging girl **(a)**. On closer inspection, the two "dynamic systems" prove to be rather different, of course.

Example 2 (People in a Hotel):
Imagine a huge crowd, say the fans of a famous rock band, waiting for admittance to a hotel where their idols are giving an autograph session. Once the entrance has been opened, a long multiple flight of revolving doors has to be passed through in order to push forward to the waiting stars. This rather typical situation of present-day life is sketched on the left-hand side of Fig. 37.

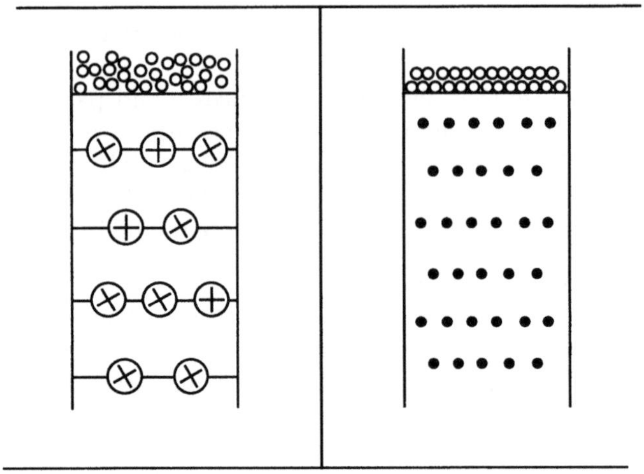

Figure 37. Comparison of human throng passage through flight of revolving doors (*left*) with physical passage of marbles through a nail-board (*right*).

In spite of the intricacy and explosiveness of the described situation, which may horrify not only the road managers of our rock band, the crowd will almost certainly manage to get along in an acceptably smooth way: only few individuals will collide, and probably no one will be squeezed in a revolving door and seriously hurt. (The validity of this assessment strongly depends, of course, on age, gender and mental state of the people involved.) This means that *a random sample of human beings solves, as a matter of routine, a most sophisticated dynamic problem which would even frustrate a Laplacian demon of the advanced non-linear type* ("Laplace-Poincaré demon"). The tremendous cognitive challenges involved here may be best illustrated by conceiving, in a first approximation, the members of the crowd as will-less particles, whose dynamics is completely determined by the basic laws of mechanics in conjunction with the specific set of initial and boundary conditions. The so-defined situation can be epitomized by the physical problem of marbles trickling down a tilted nail-board (see the right-hand side of Fig. 37): once the barrier on top of the board has been removed, the marbles will perform a gravitation-driven *chaotic* downward motion through a cascade of collisions with each other and the fixed nail-heads. Hereby the nails act like "hyperbolic points in position space", which may be passed by in two fundamentally different ways.

Note that the marbles problem is, in turn, a particularly complicated variant of the so-called *billiard systems*, the investigation of which has given rise to an entire branch of modern non-linear dynamics (see, e.g. *Tabor*, 1989 [227]). For the simplest types of billiards it can be proven rigorously, in fact, that these systems are under the rule of clear-cut deterministic chaos (see, for instance, *Sinai*, 1976 [217]). Thus for all practical purposes, it is *unfeasible* to predict precisely the motion of the marbles in a nail-board and to determine in advance, for example, the one which arrives first at the bottom line.

However, note also that the passage of the throng through the flight of revolving doors is infinitely more complicated than the marbles problem and is actually *unpredictable in principle.* This is due to the fact that the throng problem is dominated by a whole universe of *voluntative degrees of freedoms* that complement the physical ones. In addition, the former degrees of freedoms, i.e., the spontaneous locomotive decisions of the various individuals in the crowd are incalculable in nature – they cannot be monitored directly or even determined a priori by fellow fanatics or an external observer!

The reader might object here that voluntative indeterminacy generating cognitive uncertainty can be reduced significantly by *convention*: before the entrance door flies open, the waiting crowd could agree upon precise routing-schemes for each individual in order to keep subsequent chaos at bay. Yet even if the people involved were capable of working out such an agreement and were willing to comply with the accord, all pre-determined co-operative strategies would be wrecked in no time at all by the various *inherent hyperbolicities.* In summary, there seems to remain no scientifically justified way to survive the revolving-door race, but the trick is done each day with no or few casualties.

Example 3 (Traversing the Piazza San Marco):
Tourists who wish to cross the overcrowded St. Mark's Square in a certain direction on a sunny August afternoon are confronted with a manoeuvring problem not dissimilar to the revolving-doors predicament just discussed. Again, experience tells us that the seemingly unfeasible control task can be solved with a tolerable amount of effort and trouble. It is evident, however, that any conceivable type of *fixed initial instruction* of the actor in the sense of a locomotive ("ballistic") programme would completely fail and ultimately result in a fist fight with other tourists run down by our single-minded sleep-walker.

The cognitive and voluntative uncertainties involved here are generally the same as in the hotel episode. In addition, the Venetian holiday example provides us with a crucial new aspect of control, namely *global orientation*: even if the tourist succeeds in avoiding unpleasant encounters with other people, he/she may lose his/her general bearings in the jostling crowd. This can be avoided, however, by *sporadic re-orientation with the help of prominent landmarks* that repeatedly come into sight (say, the red-brick Campanile).

* * *

The purpose of our three examples was to demonstrate that individual or collective actors are very capable of *managing even the most complicated situations and of controlling the most complex dynamic systems in everyday life.* When analysing these miraculous achievements, it is necessary to distinguish between *passive performance* (like muddling through intricate terrain) and *active steering* (like intentionally influencing capricious processes).

However, the secret of success remains the same under all circumstances: passive or active control becomes incredibly effective through *perpetual revision and up-dating of approximative management strategies!* The strategies have to be tentative due to the *temporal vagueness and incompleteness* of the available local as well as global information about the system in question. The above-formulated intuitive-empirical principle for surviving in a ruleless and irregular world will therefore be referred to in the following as *"Fuzzy Control"*. We owe this nomenclature as well as pioneering insights into the overall issue to the work of *Zadeh* (1965) [261]. In the last two decades the general field of "fuzzy analysis" has become a flourishing branch of science and, in particular, engineering. Both the recent development of this field and the more formal description of the fuzzy control principle will be dealt with further below.

First we shall try to understand in some mathematical-quantitative way the basic philosophy of *iterated approximative control*, which proves to be so successful in the everyday-life situations described above. This analysis will provide useful hints for future transformation of the common-sense principle "Fuzzy Control" into a well-defined *scientific strategy for steering complex systems.* We will restrict ourselves, however, to illustrative examples of passive performance in very simple, and often even deterministic, situations. For the standard motivation and interpretation of fuzzy systems analysis, we refer the reader to the "classic" literature (see, e.g., *Zimmermann*, 1991 [262]; *Kandel*, 1986 [112]).

6.2 Newton's Root-Finding Method – a Paradigm for Fuzzy Control

Let us perform the following *Gedankenexperiment*: A plane is forced to land on a make-shift runway during a foggy night. The aircraft will be assisted by an optical localizer beam only. So the pilot tries to reduce the plane's velocity *in accordance with the distance-dependent brightness* of the ground-based light source. After considerable simplification, this problem may be formulated as a one-dimensional root-finding task:

We assume the vehicle's position to be described by the variable $x \geq 0$ and the beam source to be located at $\hat{x} = 0$. Let the perceptible brightness, $H(x)$, be a function of the (qualitative) shape depicted in Fig. 38, and let $H_0 \equiv H(0)$ denote the absolute brightness of the source.

Thus $H(x)$ is characterized by

- non-vanishing derivative for all values of x and particularly at the origin
- concave shape for small values of x
- approximately linear shape at intermediate distances from the source

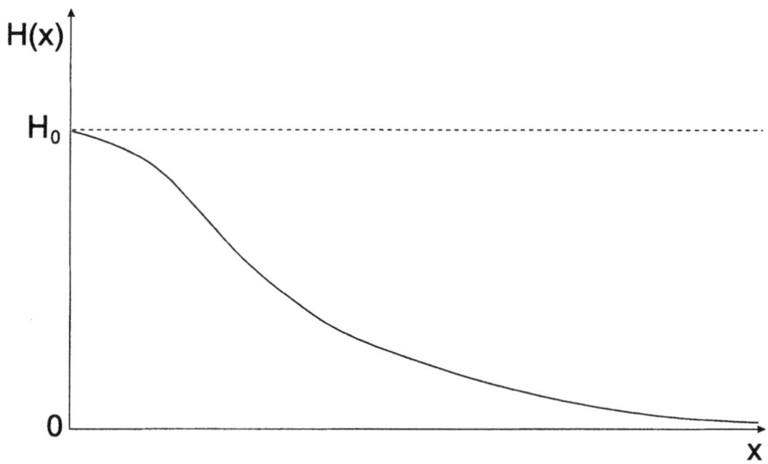

Figure 38. Perceptible brightness of localizer beam as a function of distance from its source, which is located at the origin.

– convex shape for large values of x, and
– asymptotic decay towards zero.

The so-defined overall behaviour may serve as a not-too-unrealistic description of the combined effect of various physical processes involved (like absorption, scattering, defocusing, etc.). The crucial hypothesis is, anyway, that we are dealing with a *non-linear function.*

Introducing

$$f(x) := H(x) - H_0 \tag{229}$$

we have $f(0) = 0$, i.e., the target location $\hat{x} = 0$ is the unique root of the new function. Thus the pilot, who cannot determine the aircraft's absolute position x, is able to safely land the plane by "nullifying" the proxy distance measure f. In order to achieve this, only *local* brightness information and the calibration value H_0 are needed.

However, how to organize the actual root-finding process in an effective and robust way? From the plethora of available recipes we pick just two procedures, which represent *diametrically opposed search strategies* and, therefore, *fundamentally different control philosophies.*

A. Conventional Perturbation Theory:
Let us assume that the plane has already reached a position x_0 in the vicinity of the airport, so $f(x_0)$ is rather small. Let

$$\delta := \hat{x} - x_0 \tag{230}$$

denote the initial (directed) distance from the root, i.e., from the runway. Let us further assume that the local brightness can be measured sufficiently precisely for determining – with the help of standard aircraft instruments like the speedometer – not only the function value $f(x_0) \equiv f^{(0)}(x_0)$, but also *all the derivatives* $f^{(n)}(x_0)$. Then the root may be found by exploiting the *Taylor expansion* of f around x_0.

Note that

$$0 = f(\hat{x}) = f(x_0 + \delta) = \sum_{n=0}^{\infty} \frac{1}{n!} f^{(n)}(x_0)\delta^n. \tag{231}$$

Due to the fact that $f^{(1)}(x_0) \neq 0$ for all $x \geq 0$, we may slightly rearrange this equation to obtain

$$\delta + \sum_{n=2} \left[\frac{1}{n!} \frac{f^{(n)}(x_0)}{f^{(1)}(x_0)} \right] \delta^n = -\frac{f(x_0)}{f^{(1)}(x_0)} =: \varepsilon. \tag{232}$$

The quantity ε, which decays to zero at the origin, may be interpreted here as a *"smallness parameter"* for the perturbation analysis. Eq. 232 defines a power series in the variable δ, which can be formally inverted. To that end we make the ansatz

$$\delta(\varepsilon) = b_1\varepsilon + b_2\varepsilon^2 + b_3\varepsilon^3 + \dots , \tag{233}$$

where the quantities $b_l, l \in \mathbb{N}$, are real coefficients yet to be determined, and insert (233) in (232). For the resulting power series in ε to vanish identically, all the respective aggregated coefficients have to be zero. As a consequence, we obtain the following relations:

$$\begin{aligned} b_1 &= 1 , \\ b_2 &= -\frac{1}{2}\frac{f^{(2)}(x_0)}{f^{(1)}(x_0)} , \\ b_3 &= \frac{1}{2}\left[\frac{f^{(2)}(x_0)}{f^{(1)}(x_0)}\right]^2 - \frac{1}{6}\frac{f^{(3)}(x_0)}{f^{(1)}(x_0)} , \\ b_4 &= -\frac{5}{8}\left[\frac{f^{(2)}(x_0)}{f^{(1)}(x_0)}\right]^3 + \frac{5}{12}\frac{f^{(2)}(x_0)f^{(3)}(x_0)}{[f^{(1)}(x_0)]^2} - \frac{1}{24}\frac{f^{(4)}(x_0)}{f^{(1)}(x_0)} , \\ &\text{etc.} \end{aligned} \tag{234}$$

Our pilot is now able to construct from the local optical information a *sequence* $\{x_n\}$ *of approximations* of the target position $\hat{x}$ *of monotonically increasing quality.*

This sequence is defined as

$$x_n := x_0 + \sum_{l=1}^{n} b_l\varepsilon^l \quad , \; n \in \mathbb{N} \quad , \tag{235}$$

and evidently converges to the desired root in an *exponential* way: Let

$$\varepsilon_{n+1} := x_{n+1} - x_n = b_{n+1}\varepsilon^{n+1} \quad , \tag{236}$$

implying, in particular,

$$\varepsilon_1 = x_1 - x_0 = \varepsilon \equiv -\frac{f(x_0)}{f^{(1)}(x_0)} \quad . \tag{237}$$

Thus the difference between successive approximations of $\hat{x}$ is of the order

$$\varepsilon^{n+1} = \varepsilon \cdot e^{\ln \varepsilon \cdot n} \quad . \tag{238}$$

Note that the set $\{x_n\}$ does *not* represent a sequence of real stop-over points during the actual landing process, but a sequence of mathematical approximations in virtual navigation space. Within the framework of the perturbation approach, the pilot tries to compute *once and for all* via Eq. 235 an approximation of the highest possible order for the remaining travelling section $\Delta x := \hat{x} - x_0$. This approximation will form a fixed basis for the design of the landing manoeuvre. In other words, the chosen strategy relies on a single and rigid set of initial instructions of *utmost precision*. Here we strike a huge snag: Eq. 234 tells us that the computation of higher-order coefficients needed for satisfactory assessment of Δx becomes extremely involved and may be spoiled completely by minute errors infesting the bulk of derivative information! Thus the seemingly clever perturbation approach illustrates once more the predicament of teleological control strategies, even when the situation is by no means intricate.

B. Newton's Method:
Fortunately, there are much smarter ways of root-finding, so our aircraft does not have to be navigated by some sort of demon. All these recipes take advantage of the principle of *iterated use of weak or even imprecise local information*, and therefore belong to the set of *plesiological* strategies (where "plesios" is the Greek word for "near"). A famous example for strategies of this type, which are at the same time highly effective, robust and simple, is provided by Newton's root-finding method. It is worthwhile to analyse the latter method in some detail within the context of our Gedankenexperiment, as this analysis exposes, in a most transparent formal way, crucial elements of the rather mundane secret of controlling complex systems.

We now assume that the pilot is only able to determine at each position x the absolute value $f(x)$ and the first derivative $f^{(1)}(x)$ of the brightness measure f. This piece of local information allows, after all, for a *linear approximation* of the distance between the initial position $\bar{x}_0 (= x_0)$ and the target position $\hat{x}$. Thus the first estimate $\bar{x}_1$ of the runway's location is given by the simple formula (see Fig. 39)

$$\bar{x}_1 = \bar{x}_0 - \frac{f(\bar{x}_0)}{f^{(1)}(\bar{x}_0)} \quad . \tag{239}$$

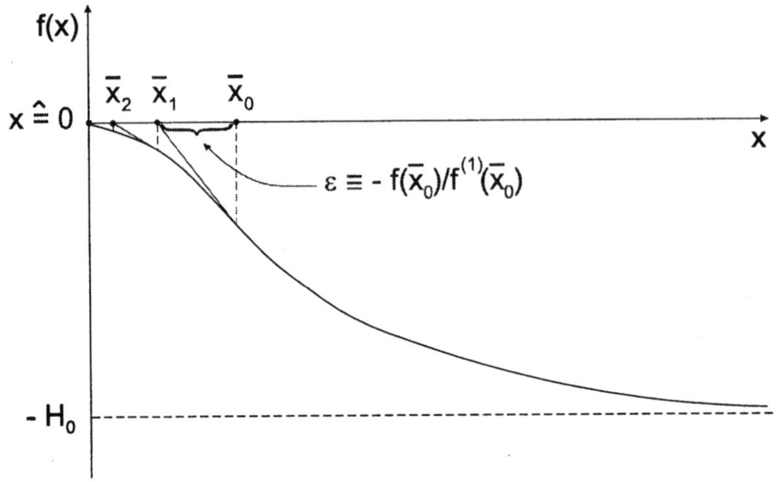

Figure 39. Approach to root of brightness measure $f(x)$ through iterated linear approximation based on local first derivative.

Now the critical point is that our pilot will orientate himself/herself by the "inaugural" estimate $\bar{x}$, yet will *not commit* himself/herself to this figure! For in the further course of the landing manoeuvre it may well turn out that the first distance approximation is quite a bit off target (as illustrated by Fig. 39). That is no tragedy, however, as the linear estimation procedure can be easily repeated – *but now from the new basis* $\bar{x}_1$.

Thus the second estimate for the target position is

$$\bar{x}_2 = \bar{x}_1 - \frac{f(\bar{x}_1)}{f^{(1)}(\bar{x}_1)} , \tag{240}$$

and the general algorithm for continuing the approximation procedure is described by *Newton's famous formula*

$$\bar{x}_{n+1} = \bar{x}_n - \frac{f(\bar{x}_n)}{f^{(1)}(\bar{x}_n)} \quad ; n = 0, 1, 2, \ldots . \tag{241}$$

The so-generated iterative approach to the root of a function is indicated for the brightness measure in Fig. 39. Note, however, that Newton's method only works if the first derivatives $f^{(1)}(\bar{x}_n)$ do not vanish.

Let us contrast the efficiency of this relatively primitive procedure with the convergence behaviour of the above-mentioned relatively complicated perturbation approach. This is achieved in a most lucid way by employing again the "perturbation (or smallness) parameter"

$$\varepsilon \equiv -\frac{f(\bar{x}_0)}{f^{(1)}(\bar{x}_0)} , \tag{242}$$

which gains a direct meaning within the framework of Newton's method (see again, Fig. 39). Introducing

$$\bar{\varepsilon}_{n+1} := \bar{x}_{n+1} - \bar{x}_n, \tag{243}$$

where $\{\bar{x}_n\}$ is the *Newtonian sequence of approximations* of the target position $\hat{x}$, we have, in particular, $\bar{\varepsilon}_1 = \varepsilon$. Now the decisive step is to relate generally the quantity $\bar{\varepsilon}_{n+1}$ to its algorithmic predecessor $\bar{\varepsilon}_n$. Making ample use of Taylor expansion, we obtain

$$\begin{aligned}\bar{\varepsilon}_{n+1} &= -\frac{f(\bar{x}_n)}{f^{(1)}(\bar{x}_n)} = -\frac{f(\bar{x}_{n-1}+\bar{\varepsilon}_n)}{f^{(1)}(\bar{x}_{n-1}+\bar{\varepsilon}_n)} \\ &= -\frac{f(\bar{x}_{n-1}) + f^{(1)}(\bar{x}_{n-1})\bar{\varepsilon}_n + \frac{1}{2}f^{(2)}(\bar{x}_{n-1})\bar{\varepsilon}_n^2 + \mathcal{O}(\bar{\varepsilon}_n^3)}{f^{(1)}(\bar{x}_{n-1}) + f^{(2)}(\bar{x}_{n-1})\bar{\varepsilon}_n + \mathcal{O}(\bar{\varepsilon}_n^2)} \quad ,\end{aligned} \tag{244}$$

where the "order of magnitude" symbol $\mathcal{O}(.)$ indicates powers of the smallness parameter and its iterates here.

However, note that

$$f(\bar{x}_{n-1}) + f^{(1)}(\bar{x}_{n-1})\bar{\varepsilon}_n = 0 \quad , \tag{245}$$

according to Eqs. 241 and 243. If we account for this fact in the nominator of Eq. 244 and expand the denominator of the same expression via the general formula for the function $\frac{1}{1+\xi}$, then we end up with the following result:

$$\bar{\varepsilon}_{n+1} = \left[-\frac{1}{2}\frac{f^{(2)}(\bar{x}_{n-1})}{f^{(1)}(\bar{x}_{n-1})}\right]\bar{\varepsilon}_n^2 + \mathcal{O}(\bar{\varepsilon}_n^3). \tag{246}$$

By repeating the whole calculation for $\bar{\varepsilon}_n$, we obtain

$$\bar{\varepsilon}_{n+1} = \left[\frac{1}{4}\frac{f^{(2)}(\bar{x}_{n-1})}{f^{(1)}(\bar{x}_{n-1})}\frac{f^{(2)}(\bar{x}_{n-2})}{f^{(1)}(\bar{x}_{n-2})}\right]\bar{\varepsilon}_{n-1}^4 + \mathcal{O}(\bar{\varepsilon}_{n-1}^6). \tag{247}$$

Concluding by induction, we ultimately find that the difference $\bar{\varepsilon}_{n+1} \equiv \bar{x}_{n+1} - \bar{x}_n$ is of the order of

$$\varepsilon^{2^n} = (e^{\ln\varepsilon})^{e^{(\ln 2)n}} = e^{(\ln\varepsilon)e^{(\ln 2)n}}. \tag{248}$$

This means that the Newtonian root-finding process converges *super-exponentially* fast!

* * *

Let us now draw the *conclusions* from our comparative analysis: we can certainly state that Newton's approach is vastly superior to conventional perturbation theory. This implies for our landing example that the pilot reaches the goal rapidly and safely, if he/she employs the *current (relatively imprecise) approximation as the point of departure for the succeeding estimation,* which is again relatively imprecise in nature. He/she will be far worse off by trying to work out a sophisticated perturbational determination of the target position in *one single step*.

We hasten to emphasize here that Newton's method typifies much more than an overall plain "wait and see" strategy. As a matter of fact, the Newtonian approach becomes so successful through a trick of ingenious simplicity, namely by making the desired root $\hat{x}$ of an arbitrary function $f(x)$ the *super-attractive fixed point of the associated Newton map* $N(x)$. To be specific, let us assume again, without loss of generality, that $\hat{x} = 0$, i.e. $f(0) = 0$, and $f^{(1)}(0) \neq 0$. Furthermore, we define the Newton map $N(x)$ by

$$N(x) := x - \frac{f(x)}{f^{(1)}(x)} \quad , \tag{249}$$

and we consider a small deviation $\xi := x - \hat{x}$ from the root, i.e., $|\xi| \ll 1$.

An excellent representation of $N(\xi)$ is therefore provided by the Taylor expansion

$$N(\xi) = N(0) + N^{(1)}(0)\,\xi + \frac{1}{2}N^{(2)}(0)\,\xi^2 + \mathcal{O}(\xi^3) \tag{250}$$

around the reference point $\hat{x} = 0$. However, straightforward calculation yields

$$N(0) = N^{(1)}(0) = 0 \quad , \tag{251}$$

so the root $\hat{x}$ of $f(x)$ is in fact a *fixed point of the map* $N(x)$ and the associated *Lyapunov exponent* (*Lyapunov*, 1949 [142]; for a modern review see, e.g. *Schuster*, 1988 [212])) vanishes. The latter fact qualifies the fixed point $\hat{x} = 0$ as *super-attractive* – in the terminology of non-linear dynamics – based on the following observation:

$$\begin{aligned} N(\xi) &= \frac{1}{2}N^{(2)}(0)\,\xi^2 + \mathcal{O}(\xi^3) \\ &= \frac{1}{2}\frac{f^{(2)}(0)}{f^{(1)}(0)}\xi^2 + \mathcal{O}(\xi^3) \quad . \end{aligned} \tag{252}$$

Repeated application of the Newton map N yields, therefore,

$$N^{(n)}(\xi) \equiv \underbrace{N \circ N \circ \ldots \circ N}_{n \ times}(\xi) = \left[\frac{1}{2}\frac{f^{(2)}(0)}{f^{(1)}(0)}\right]^n \xi^{2^n} + \mathcal{O}(\xi^{3 \cdot 2^{n-1}}) \quad . \tag{253}$$

Thus the initial deviation ξ is *contracted in a super-exponential way* under iterated mapping!

It should be mentioned here that the "jewel in the crown" of modern dynamic systems theory, the celebrated Kolmogorov-Arnold-Moser (KAM) Theorem proving the existence of ordered quasi-periodic motion under genuine non-linear circumstances, primarily rests upon *super-convergent analysis*. This analysis, in turn, is based on a sophisticated Newton's method for operator-valued maps. Thus most fundamental insights into the nature of deterministic chaos in conservative non-linear systems are owed to the Newtonian strategy of *perpetual readjustment of the approximation basis*. This statement includes the recent investigation of the famous 3-body problem discussed above. For a comprehensive discourse of the general topic we refer the reader to Berry's brilliant review (*Berry*, 1978 [23]), which formed the basis for our convergence assessments, and to the monographs by *Arnold* and *Avez* [8] and by *Lichtenberg* and *Liebermann* [135].

* * *

By now, we have illustrated via Newton's method the general principle of control by readjusted approximation within a purely *deterministic* context only – thus that control strategy appears less "fuzzy" than "lazy". Simple deterministic conditions prevail, in particular, whenever the crucial system variables can be described by analytic functions (in space and/or time). Then we may be able to *predict*, for instance, the complete future evolution of some observable F solely from the knowledge of the behaviour of this entity *in an arbitrary small section of its history*:

Let, e.g., $F(t)$ be analytic on the entire time axis t and assume that the tiny interval $(-\tau, \tau)$, where $0 < \tau \ll 1$, has been monitored carefully. Then all the derivatives $F^{(n)}(0), n \in \mathbb{N}$, can be computed at the instant $t = 0$ and the corresponding Taylor series, which converges for all $t > 0$, yields the complete "future" of the observable F. Even under the restriction that only a finite number of higher derivatives were available, we would still succeed in determining at least a polynomial evolution corridor via Lagrange's remainder formula.

The interesting aspect of such an observation is the insight that, in principle, forecasting may well do *without* any knowledge of the system's equations of motions – if only complete local information is available (but see the abovementioned "curse of perfection" haunting Laplace's demon). Unfortunately, almost all relevant systems to be controlled are not of the simple, deterministic type: due to the multitude and non-analyticity of the pertinent functions involved, a naive global prognostic is not feasible for them. So, as a rule, any successful control strategy has to take into account and to master the handicap of fragmentary knowledge as well as apparent and/or genuine stochasticity. It is precisely under those circumstances, however, where the full power of perpetual readjustment based on updated local and global information, i.e. Fuzzy Control, unfolds. Before casting our still casual description of the latter principle into the mould of an authoritative definition, let us return once more to our aeronautical master example for demonstrating how to cope with uncertainty.

Here the unavoidable cognitive deficits can be represented concisely through the, possibly progressive, *imprecision in the determination of higher derivatives* of $H(x)$ or $f(x)$, respectively. The main sources of noise and/or error will be meteorological perturbations, aerosols and competing optical phenomena. Let us assume, therefore, that the actually locally measured derivatives scatter around the true values $f^{(n)}(x)$ according to *Gaussian statistics.* Introducing the *random derivative of* n^{th} *order* of f at the point x, $D_n(x)$, we thus presume that the probability for observing the value $D_n(x)$ is given by the formula

$$p(D_n(x)) = \frac{1}{(2\pi)^{1/2}\sigma_n} \exp\{\frac{1}{2}\frac{[D_n(x) - f^{(n)}(x)]^2}{\sigma_n^2}\} \quad , \quad n \in \mathbb{N} \quad . \tag{254}$$

Let σ_n denote the *standard deviation,* which will generally grow with the order n of the derivative, i.e.,

$$\sigma_n = \omega(n) \quad , \tag{255}$$

with

$$\omega(n) = \frac{1}{10}n^2 \quad , \tag{256}$$

for instance.

We have studied the influence of this stochastic complication on the landing (or root-finding) manoeuvre in some detail; the result can be summarized as follows: *Newton's method is largely resistant* to those stochastic distortions of the local information and finds its way unwaveringly to the target. By way of contrast, *conventional perturbation theory fails completely* as this technique does not take advantage of the option to correct faulty data in the course of rapprochement. We will illustrate this general finding by a very simple, yet non-trivial calculation. For that purpose, we consider instead of the still-too-complicated brightness measure $f(x)$ the mundane function

$$g(x) = \tan(\frac{\pi}{4} + x) - 1. \tag{257}$$

Evidently, $g(0) = 0$, i.e., $\hat{x} = 0$ and $g^{(1)}(0) \neq 0$.

We now compare the numerical performance of four distinct root-finding scenarios for this function, namely

(i) the *deterministic Newton's method,*
(ii) *Newton's method under uncertainty,* i.e. fuzzy derivative information,
(iii) the *deterministic perturbation method,* and
(iv) *the perturbation method under uncertainty.*

In each case we start our root-finding procedure from the point $x_0 = 0.2$ and generate a sequence $\{x_n\}$ of approximations of the root $\hat{x} = 0$. For the two noisy scenarios, the probability distribution of Eq. 254 is employed, but for the standard deviation we use the flat progression

$$\sigma_n = \frac{1}{10} \quad \forall \; n \in \mathbb{N} \tag{258}$$

instead of (256). Tab. 3 contrasts the results, which are represented in the stochastic cases by *typical* sequences.

Table 3. Comparison of four basic root-finding scenarios for the test function $g(x)$ and the common point of departure $x_0 = 0.2$.

	Newton's Method		Perturbation Method	
	Deterministic	Noisy	Deterministic	Noisy
n	x_n	x_n	x_n	x_n
0	0.2	0.2	0.2	0.2
1	$0.4476033\ 10^{-1}$	$0.5162266\ 10^{-1}$	$0.4476033\ 10^{-1}$	$0.4671267\ 10^{-1}$
2	$0.2061910\ 10^{-2}$	$0.4757890\ 10^{-2}$	$0.8406512\ 10^{-2}$	$0.1126749\ 10^{-1}$
3	$0.4257312\ 10^{-5}$	$0.9068513\ 10^{-5}$	$0.1140281\ 10^{-2}$	$0.4271971\ 10^{-2}$
4	$0.1812476\ 10^{-10}$	$0.6112070\ 10^{-6}$	$0.2275428\ 10^{-4}$	$0.3209610\ 10^{-2}$
5	$-0.5382865\ 10^{-21}$	$0.2854914\ 10^{-7}$	$-0.5181396\ 10^{-4}$	$0.3139614\ 10^{-2}$
6	$-0.1079167\ 10^{-42}$	$0.8589053\ 10^{-9}$	$-0.2210679\ 10^{-4}$	$0.3167149\ 10^{-2}$
7		$0.4926537\ 10^{-10}$	$-0.5976345\ 10^{-5}$	$0.3181912\ 10^{-2}$
8		$0.1745666\ 10^{-11}$	$-0.1124674\ 10^{-5}$	

We clearly recognize that Newton's method is significantly slowed down by noise, but this type of Fuzzy Control still achieves its goal *exponentially fast*! The perturbation method, which already performs rather poorly under deterministic circumstances, rapidly ***gets stuck in the noise*** and leads us nowhere. These observations can be further corroborated by comparing the respective ***ensemble behaviour*** of multiple runs of Newton's method and the perturbation method, respectively, under uncertainty (see Fig. 40). The superiority of the fuzzy-control strategy is most impressively demonstrated by the fact that after 9 iterations, 100% of the Newtonian sequences have entered the 10^{-10}-neighbourhood of $\hat{x} = 0$, while 0% of the perturbation sequences have reached that domain!

6.3 Coping with Uncertainty: Grand Entrance of Maxwell's Demon

In much the same way as the fictitious pilot in our Gedankenexperiment, the heroes of everyday life are used to solving complex tasks with the help of fragmentary or noisy information: the secret of their success consists to a large extent in perpetual updating of decisions and behaviour in view of additional fuzzy local (in space and time) data. The child that flies a kite responds intuitively to the stochastic fluctuations in the prevailing wind conditions, and the pedestrian who fights his way through a crowd accounts successfully even

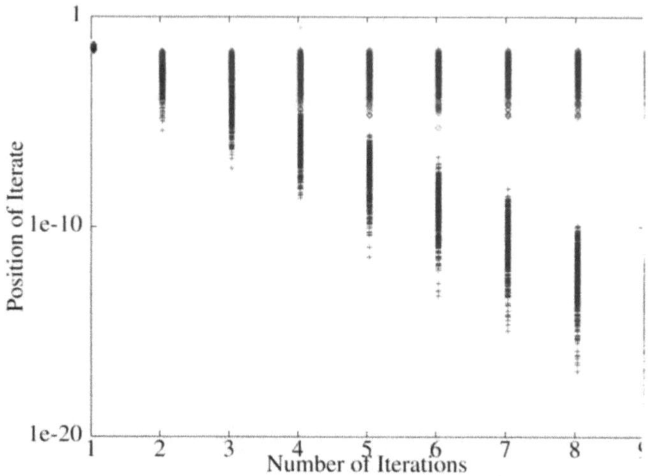

Figure 40. Statistical comparison of root-finding performance of Newton's method and perturbation method under noise for the function $g(x)$. The *blue crosses* depict the various Newtonian sequences generated by stochasticity, while the *green diamonds* indicate the perturbation sequences. Note the logarithmic representation of the position axis.

for the fundamentally unforeseeable changes in direction of his fellow sufferers. By way of contrast, the insistence on a fixed predetermined sequence of actions will rapidly and inevitably generate embarrassing misadventures in such situations – regardless of the sophistication of the initial programming. We have ventilated these ideas quite a bit in the preceding sections.

However, note that successful iterative control strategies are generally not constructed as clearly as Newton's procedure, which manages to eliminate *exactly* the linear deviation from the target. Even "super-Newtonian methods" may be devised (e.g. by exploiting parabolic local approximations) that converge considerably faster than the familiar algorithm even in multi-dimensional settings. But for almost all purposes, *non-smart readjusting strategies* will do the job sufficiently well – if a *minimum responsiveness* to fresh information is warranted!

The latter principle and its miraculous performance under appropriate conditions may be best symbolized by yet another creature from the pandemonium of theoretical physics, namely *"Maxwell's Demon"*. This fabulous djinni was imagined by J.C. Maxwell in 1871, to illustrate the possibility of violating the Second Law of thermodynamics. The demon is capable of detecting and reacting to the motions of individual molecules. By operating, e.g., a microscopic "door" connecting two vessels filled with gases at the same temperature (see Fig. 41), the creature can generate excess energy from molecular chaos in order to perform *macroscopic work.* "All" the demon has to do is to allow fast-moving particles to pass from room A to room B and to allow slow-moving molecules to migrate only from B to A.

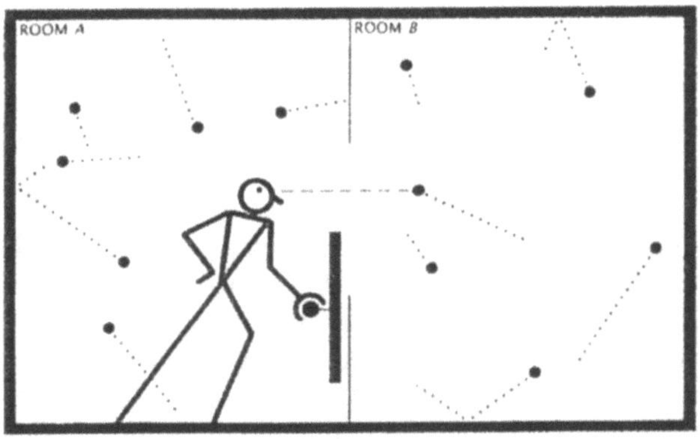

Figure 41. Maxwell's Demon, described in 1871 by James Clerk Maxwell, seems able to violate the second law of thermodynamics. The demon controls a sliding door that blocks a hole in a wall between rooms containing gas at equal temperatures and pressures. It observes molecules approaching the hole and opens and closes the door to allow fast-moving molecules to pass from room A to room B but not vice versa. Slow-moving molecules, conversely, are allowed to pass only from B to A. As the demon sorts, B heats up and A cools. According to the second law, a certain amount of work is required to create a temperature difference, but the work of sliding the door can be made negligibly small (taken from *Bennett* [19], courtesy of Jerome Kuhl).

Information-Theoretical Excursion:
Laplace's Demon "only" needs to know the present condition of all molecules in the universe, as determined by their positions and momenta, in order to be able to provide the condition of those molecules at any later point in time. The creature by no means intervenes in the unfolding of cosmic history here. Its activity is pure prediction, its demonic talent consists exclusively in being able to comprehend all particle conditions simultaneously - an achievement that clearly transcends human perceptive capacities.

Maxwell's Demon, on the other hand, is a *creative* djinni, who can bring about order. The reason for this lies in the fact that it can handle "information" in a superior way. The demon is

- not provided with the present, but with target conditions (i.e., the desired order);
- able to recognize molecular conditions autonomously (i.e., without being told);

- not simply a passive observer, but intervenes actively;
- acting continuously not just once;
- able to tune its intervention in view of the currently molecular conditions, namely by comparing them with the target ones.

The very last capability realizes a *negative feedback*, as introduced by N. Wiener as the fundamental phenomenon of his *cybernetics*. In this sense, the demon performs as a *steersman* ($\kappa\upsilon\beta\epsilon\rho\nu\eta\tau\eta\zeta$) who can control the Brownian motion of the gas particles. Thus the djinni should more precisely be referred to as the "*Maxwell-Wiener Demon*". ■

Whatever name we assign to it, Maxwell's demon has been a *tantalizing riddle* for theoretical physicists for more than a century now. Although several exorcising attempts have been made (for example, by *Brillouin* (hjs:Brillouin.1950) [?], the troublesome case is not settled yet (see *Bennett*, 1987 [19], for a highly readable review of this topic).

We are not concerned here with the mutual consistency of physical axioms, but rather with the control paradigm represented by Maxwell's djinni. This *"plesi-visionary" creature* constitutes the perfect counterpart of the "tele-visionary" Laplace-Poincaré demon. The former owes its success not to the divine capability of exact global prognosis, but to its *animal obstinateness* in transforming strictly local (microscopic, to be precise) fuzzy knowledge into rough action. That is the proper way to defeat molecular stochasticity...

In spite of its stunning achievements, Maxwell's demon is not suitable to serve as the unique figurehead of complex-systems control. For the creature has to pay a price for its myopia implying the absence of *large-scale and long-term orientation* – and however vague this orientation may be, it often turns out to be indispensable. Without at least a *blurred panoramic view* of the total problem setting, the risk is quite high of passing through points of no return, of entering dead-end streets, of approaching abysses, etc. These are examples for wrong tracks that might be avoided with the help of an even quite imprecise *conception of the system as a whole.*

Because without some "holistic" orientation, no mountain range can be climbed over, no forest can be traversed and – in fact – no plane can be landed safely. Let us illustrate this general insight again within the context of our aircraft Gedankenexperiment. Our previous characterization of the pilot's task was certainly an over-simplistic one. For the aircraft commander has to distinguish by any means between the *"right"* and additional *"wrong" light sources* on the ground during the entire landing approach.

A formal caricature of this challenge is again provided by the task of finding the roots of a function – but this time of a function with *more than one root in its multi-dimensional domain.* For the sake of simplicity we restrict our demonstration to *complex* functions $f(z)$, defined in the entire plane $\mathbb{C}$. To be specific, we select as an example the innocent-looking cubic polynomial

$$f(z) = z^3 - 1, \tag{259}$$

which obviously has the roots

$$z^{(1)} = 1, z^{(2)} = e^{i\frac{2\pi}{3}}, z^{(3)} = e^{i\frac{4\pi}{3}} \tag{260}$$

situated on the unit circle.

We assume that $z^{(1)}$ corresponds to the target location, i.e., the position of the airport, while $z^{(2)}$ and $z^{(3)}$ are mock marks which bring disaster upon the unwary ones. Presupposing further that the value and the first derivative of f are available in each point of $\mathbf{C}$, we may again employ (the deterministic version of) Newton's method starting from an arbitrary initial position $z_0 \in \mathbf{C}$. This means that an iterative sequence $\{z_0, z_1, z_2, \dots\} \equiv \{z_n\}$ of vertices of the search motion is generated by the complex Newton map

$$N(z) := z - \frac{f(z)}{f^{(1)}(z)} = \frac{2z^3 + 1}{3z^2} \tag{261}$$

via the prescription

$$z_{n+1} = N(z_n); \quad n = 0, 1, 2, \dots \ . \tag{262}$$

We observe that

$$\frac{df}{dz}(z^{(i)}) \neq 0 \quad \text{and} \quad \frac{dN}{dz}(z^{(i)}) = 0 \ ; \quad i = 1, 2, 3, \tag{263}$$

so all the roots of f are in fact super-attractive fixed points of N.

Here we are specifically interested in determining the *basin of attraction* of the saving fixed point $z^{(1)}$, i.e. the set of all starting points z_0 that lead to the target through repeated application of the map (261). Let us formally define the latter set, denoted $A(z^{(1)})$, by

$$A(z^{(1)}) := \{z_0 \in \mathbf{C} |\ N^{(n)}(z_0) \to z^{(1)} \quad \text{as} \quad n \to \infty\} \ , \tag{264}$$

where again

$$N^{(n)} \equiv \underbrace{N \circ N \circ \cdots \circ N}_{n \text{ factors}} \ . \tag{265}$$

Let us point out that the analysis of the basins of attraction of iterated Newton maps derived from complex polynomials is a *legendary problem of pure mathematics*. The basic research on this topic is associated especially with the names of Cayley, Julia and Fatou (for more information, see *Peitgen* and *Richter* [177]).

One might presume that the basins of attraction of the three fixed points of $N(z)$, $A(z^{(1)})$, $A(z^{(2)})$, and $A(z^{(3)})$, constitute dull-looking simply-separated domains that cover $\mathbf{C}$ in a trivial way. Under such conditions it would be relatively easy to readjust an iterative sequence $\{z_n\}$, that happens to go astray, by mid-way corrections. However, a numerical investigation immediately convinces us that the above presumption is as wrong as can be: the three basins of attraction actually form an *infinitely-nested fractal structure* as depicted in Fig. 42.

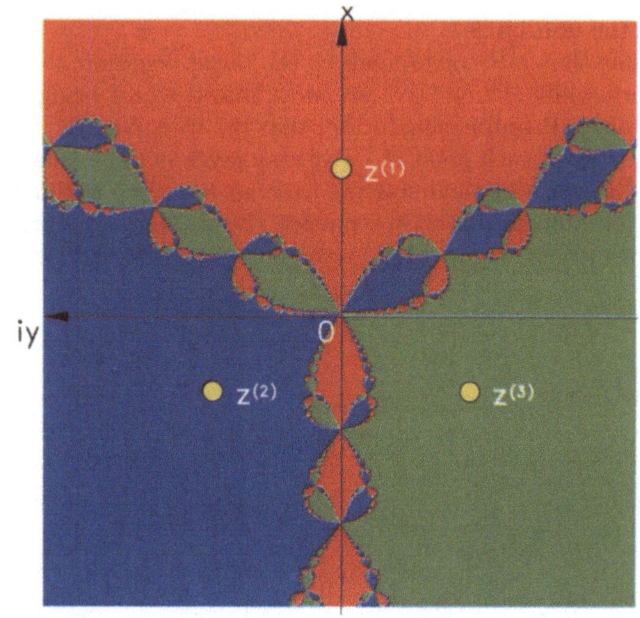

Figure 42. Basins of attraction of the fixed points of $N(z)$. $A(z^{(1)})$ is indicated in *red*, $A(z^{(2)})$ in *blue* and $A(z^{(3)})$ in *green*, respectively.

The basin structure shown in this figure is sometimes called the *Cayley fractal* and it belongs by now to the icons of "chaos science" – together with such famous comrades as the Mandelbrot set, the Hofstadter butterfly and the Lorenz attractor. The self-similar composition of Cayley's fractal may be derived and evaluated by analytic methods (*Nauenberg* and *Schellnhuber*, 1989 [162]).

From Fig. 42 it becomes clear that the Newtonian search motion for the root $z^{(1)}$ of $f(z)$ represents a highly non-trivial task, as $A(z^{(1)})$ is screened off by *fractal barriers*. Thus the design of appropriate mid-way corrections from positions that appear to be iterated to forbidden fixed points seems to be unfeasible. Things look black, in particular, when the starting point z_0 is situated somewhere in the infinitely intricate entrails of the pattern: the slightest imprecision in determining the initial position will imply a completely wrong prediction of the actual terminal reached through the Newtonian walk. So do we have to evoke a demon of the Laplace-Poincaré breed again?

No, we do not – but evidently Newton's myopic method has to be complemented by some element of *non-local information processing.* We will demonstrate in the following how this might be achieved. Our illustrative recipes have the advantage of being both easy to grasp and highly robust; they could be replaced, of course, by much more sophisticated procedures.

The first exemplary algorithm is sketched in Fig. 43.

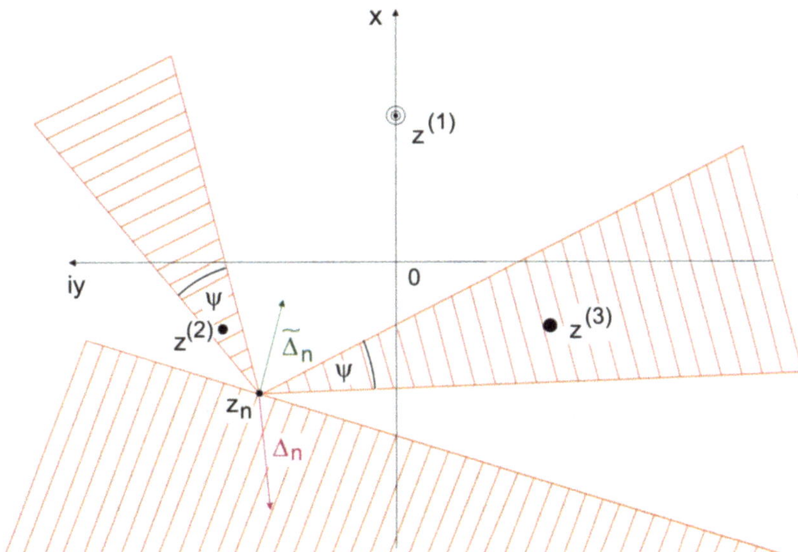

Figure 43. Modified Newtonian algorithm taking into account also (weak) global information about the location of the fixed points in the Cayley problem. The forbidden zones emanating from the intermediate position z_n (the far semi-plane of $z^{(1)}$ as well as Ψ-sectors with respect to $z^{(2)}$ and $z^{(3)}$) are *hatched in red.* Thus the "regular" Newton stride Δ_n is prohibited here; it is replaced by an aleatoric step $\tilde{\Delta}_n$ into the accessible domain.

The basic assumption here is that some rather imprecise information concerning the *directions* towards the distinct fixed points of $N(z)$ is available at each intermediate position visited "on the way home" – in addition to the knowledge of the Newtonian map $N(z)$ itself. As a consequence, it can be determined whether the next Newton step realizing the displacement

$$\Delta_n := z_{n+1} - z_n = -\frac{f(z_n)}{f^{(1)}(z_n)} = N(z_n) - z_n \tag{266}$$

leads *"forward"* (in the widest sense of the word) with respect to the target $z^{(1)}$ and, simultaneously, turns *"away"* (in a narrower sense) from the *fatal attraction points* $z^{(2)}$ *and* $z^{(3)}$. To be specific, we presuppose that it is possible to find out whether the angle defined by Δ_n at z_n lies within the far 180°-sector with respect to $z^{(1)}$ and/or within one/both of the Ψ-sectors with respect to $z^{(2)}$ and $z^{(3)}$, where Ψ is some angular uncertainty between 0 and $\frac{\pi}{2}$. If this is actually the case in one or the other way, then the regular (deterministic) Newton displacement Δ_n is discarded and replaced by an "allowed" step $\tilde{\Delta}_n$ selected by proper random choice. The resulting new position in $\mathbb{C}$ is subsequently identified with the $n + 1$-th vertex of the walk, i.e.,

$$\tilde{z}_{n+1} := z_n + \tilde{\Delta}_n \quad , \quad \tilde{z}_{n+1} \to z_{n+1} \quad . \tag{267}$$

The amazing observation is that such an unsophisticated strategy really works: the so-constructed *hybrid algorithm* triumphs over fractality and *inevitably finds its way* through the self-similar barriers towards the target point $z^{(1)}$! We will not support this statement by numerical studies here; the performance of hybrid strategies will be demonstrated computationally instead for a slightly advanced procedure below.

Let us emphasize that the algorithm just described may be rather slow under certain circumstances and, in fact, only operates in a satisfactory way when

(i) the angle Ψ is chosen properly,
(ii) the length of regular Newtonian steps $\{\Delta_n\}$ is artificially bounded from above (especially important in the neighbourhood of the origin), and
(iii) the length of aleatoric steps $\{\tilde{\Delta}_n\}$ is moderately bounded from below (especially important in the neighbourhood of the false targets $z^{(2)}$ and $z^{(3)}$).

The overall strategy proposed here may be interpreted within the framework of our aircraft allegory in the following way. The pilot has at his/her disposal two qualitatively different means of orientation. The first one comes from crisp knowledge regarding the local ("microscopic") behaviour of the total brightness measure and allows the plane to be landed safely once the vicinity of the airport has been reached. The second one comes from vague information regarding some additional indicators (like the specific colour or polarization of light rays) that allows the approximate global ("macroscopic") directions towards the right and wrong targets, respectively, to be identified. The pilot adopts a hybrid strategy by implementing a precise linear approximation of the current distance from the runway – if the fuzzy complementary directional information is compatible with that step. If this is not the case, then an aleatoric displacement commensurate with global orientation is performed. This all boils down to the common practice of marking airport elements with headlights of unambiguous colours ... .

Returning to the pure mathematics of the Cayley problem, let us propose and test a *more effective algorithm* for approaching the "correct" final point

$z^{(1)}$ of the Newton map $N(z)$. That approach is considerably facilitated if we permit (restricted) displacements also *in the direction of* the "bad" points $z^{(2)}$ *and* $z^{(3)}$. An approximate realization of such an improved strategy is sketched in Fig. 44.

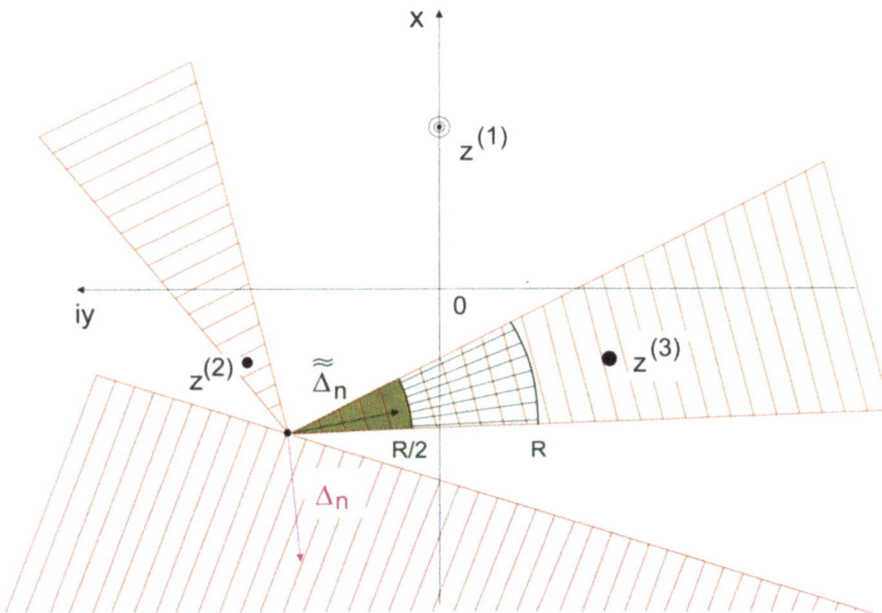

Figure 44. Improved hybrid algorithm taking into account also (intermediate) mesoscopic (or regional) information about the location of the wrong targets. Segments of radius $R/2$ of the Ψ-cones surrounding $z^{(2)}$ and $z^{(3)}$, respectively, are re-opened for deterministic or aleatoric movement (like the displacement $\tilde{\tilde{\Delta}}_n$) if the distance from z_n to the wrong fixed point considered is larger than R.

This second search procedure is similar to the first one, except for a crucial additional assumption: there exists now the possibility to determine whether the undesired roots of $f(z)$ are at a *minimum distance* $R > 0$ from the current position z_n. If the latter is actually the case for one (or both) of the undesired attraction points for Newton's method, then an appropriate tip of the forbidden Ψ-cone(s) associated with the wrong point(s) is declared accessible for deterministic or aleatoric displacements from z_n. For the sake of "safety", i.e. error tolerance, the re-opened sector is restricted to the radius $R/2$.

The new strategy amalgamates three distinct types of information, namely

(a) *precise microscopic knowledge* about $N(z)$ and its derivative at the current position,

(b) *intermediate mesoscopic knowledge* about a minimum distance R from the undesired roots of $f(z)$, and
(c) the *vaguest possible macroscopic knowledge* about the angular sector to be searched for the desired root of $f(z)$.

In practice, the performance of this algorithm depends heavily on the tuning of the adjustable parameters Ψ and R. We have carried out this optimization exercise that leads to a hybrid search strategy, which does the job of root-finding in a multifractal world in a most reliable and effective way. Our findings are illustrated in Figs. 45 and 46; in both cases the choice $\Psi = \frac{\pi}{4}$ and $R = 0.3$ is made. In Fig. 46 the overall performance of the

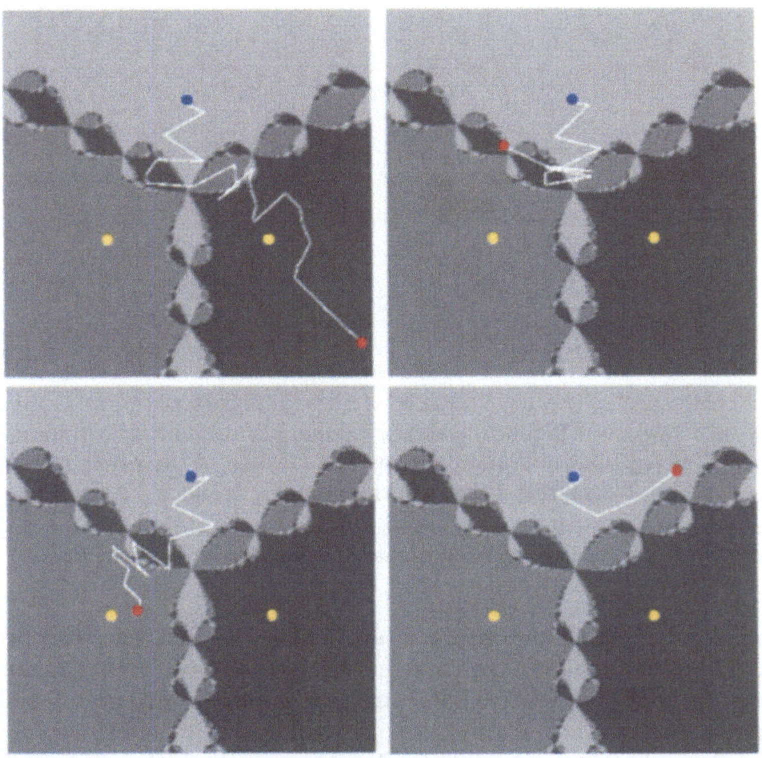

Figure 45. Typical trajectories of root-finding walks across the fractal Cayley structure (*grey background shading*) for improved hybrid search strategy. The different starting points z_0 are indicated in *red*, the undesired roots $z^{(2)}$ and $z^{(3)}$ in *yellow*, and the target point $z^{(1)}$ is marked in *blue*.

second algorithm is demonstrated. For that purpose we consider a cloud of 600×600 starting points z_0 which are homogeneously distributed over the complex square

Figure 46. Stroboscopic coverage of the evolution of a quadratic starting set under improved hybrid iteration. The initial cloud embraces 600 × 600 points, and only 16 steps from a sequence of 44 iterations are presented for the sake of brevity. The target point $z^{(1)}$ is again marked in *blue*.

$$\{x + iy \mid -2 \leq x, y \leq 2\} \ . \tag{268}$$

The initial dust is successively iterated by the hybrid method towards the target point $z^{(1)}$ – much in the same way as a liquid is drawn down a sink.

From the two figures it becomes evident that some rather primitive modifications of the original Newtonian method *guarantee one-hundred percent success*, i.e., the rapid infallible convergence to the pre-selected root of $f(z)$ from any possible starting point. This is true in spite of the intimidating complexity of the mathematical landscape to be wandered through!

We may translate also this somewhat advanced hybrid algorithm into the context of our aircraft Gedankenexperiment. The pilot now makes simultaneous use of local, regional and global information. The *regional information* may be provided, e.g. by searchlights or some primitive type of radar equipment of range R for scanning the hazardous spatial sectors. If the mock targets are beyond the range of the detection gadget, then a (deterministic or aleatoric) step of maximum length $R/2$ towards the correct target seems to be a no-regret option.

We have to emphasize here that we can do very well *without precision* in our model world: even if the information concerning the derivative of the brightness measure f regarding the hazardous sectors associated with the mock targets, or concerning the finite searchlight cones, are *moderately noisy*, then the procedure proposed here still finds its way!

6.4 Fuzzy Control and Geo-Cybernetics

The lessons from our journey through allegoric domains can now be formulated. This will help us to face the real world again and its notorious problem of judicious decision-making under uncertainty. Summarizing as well as generalizing, we may say that Fuzzy Control is the (everyday) art of solving complex (or just complicated) tasks *by soft and imprecise measures* in a satisfactory (or even perfect) way. To be more specific, *the fundamental operational principle of Fuzzy Control* reads as follows:

Based on uncertain and/or fragmentary information, adopt a rough long-term and/or large-scale strategy, which has to be continuously readjusted in an approximate fashion according to all sorts of generally imprecise additional data.

This principle of *soft decision-making under uncertainty* has three main ingredients:

1. *Existence of Leeway,*
2. *Moderate Responsiveness, and*
3. *Rough Panoramic View.*

Without *Ingredient 1*, any element of "fuzziness" would be absolutely forbidden – because only perfect action based on exact data could then produce the *unique solution* of the control problem in question. The expression "leeway" must not be restricted here to the option of deviating moderately from an ideal spatiotemporal course without wreaking havoc. As a matter of fact, and typically, the topological conditions of the pertinent high- or even ∞-dimensional control spaces have to be explored and scrutinized. Only if the (possibly precise) target structure corresponds to *open sets*, i.e. *solution corridors*, in the space of available control strategies, the imperfect steering can be employed successfully. In the case of crisp once-and-for-all decisions – like filling in a lotto coupon – the fuzzy-control principle is absolutely useless.

Without *Ingredient 2*, the influxes of fresh and/or complementary information in the course of the steering process would lead to nothing – perpetual readjustment according to improved data requires an adequate ability to react to new circumstances. "Adequateness" may be defined here via critical ratios between the *"velocities" of system variability*, *information updating* and *measure implementation*, respectively. The latter element is evidently crucial; it may be most negatively influenced by rigid rules and inflexible organs or institutions, for example.

Finally, without *Ingredient 3*, all softly-iterated control strategies would be necessarily myopic and dyspnoeic, in perpetual danger of entering potentially *disastrous cul-de-sacs.* By way of contrast, Fuzzy Control converts even most sparse and vague large-scale information into hints for *global orientation*, which may be revised, however, in the end stage of the control process – just like the local course. To illustrate this point, we consider a medical student who becomes increasingly concerned about her professional perspectives. The reason is that the "labour market for physicians" continues to look bleak for the n-th consecutive year. Although the latter observation provides only a most sweeping and unstable basis for long-term job re-orientation, the student may finally make up her mind to go for a position in the research department of a big pharmacy combine. And she may be right ...

In wrapping up the main insights from this section's discourse, let us return for the last time to our allegoric world of demons: Fuzzy Control requires a *humanized cross-breed of the teleological-contemplative Laplace-Poincaré demon and its plesiological-reactive Maxwellian comrade.* The hybrid creature is humanized insofar as the supernatural capabilities of the two original djinnis may be safely replaced by rather *amateurish skills* of the respective type. Even dull and slow individuals survive in an incredibly intricate and dynamic world – if they are not stubborn! The famous Russian mathematician and physicist A.N. Kolmogorov summarized the praise of *"correct imprecision"* in the phrase:

"It is better to be right than to be rigorous!"

* * *

The fuzzy-control principle in the form derived and summarized above has *by no means been sufficiently explored* yet. This statement holds with respect to its theoretical foundation, as well as with respect to its applicability to a wide variety of complex steering tasks. In recent years, however, the overall issue or certain aspects of it have become important research topics. Within this large field of scientific activities two specific lines of investigation can be identified:

The *first front* develops in the *tradition of* Zadeh and approaches the subject via the thematic sequence "fuzzy sets" – "fuzzy systems" – "fuzzy control" (see, for instance, *Kandel* and *Langholz*, 1993 [113]). The *second front* has emerged from the *modern theory of non-linear dynamics* and focuses on mastering deterministic-chaotic systems with the help of rather sophisticated methods (see, e.g., *Ott* et al., 1990 [169]; *Romeiras* et al., 1992 [197]); a remarkable paper on the virtual taming of El Niño's irregular appearance has been published quite recently (*Tziperman* et al., 1997 [234])). There have been actually efforts that try to combine those two lines of investigation for the analysis of relatively simple tasks like the manipulation of a bouncing ball on an oscillating table (see, e.g. *Karr* and *Gentry*, 1993 [115]).

In spite of all these impressive accomplishments, even more remains to be done in a field which has still not attained the attention it deserves in view of its *fundamental and practical importance.* The persisting gaps in our understanding are particularly large and deplorable regarding the task of active Fuzzy Control of *genuinely complex systems.* The well-known and well-explained showcases of deterministic chaos (like the mathematical double pendulum discussed earlier in this text) are almost exclusively very simply systems that consist of few components. One might conclude, therefore, that the control of many-component systems with non-linear interactions between their constituents was an impossible mission ...

Yet it would be wrong to draw such a sweeping and apodictic conclusion. First of all, *different types of complexity* have to be distinguished when we consider the rich cosmos of non-simple systems. The elaboration of a consistent (and concise) taxonomy for that cosmos is a gigantic task, however, and may always remain a scientific dream (see in this context, for instance, the efforts of *Haken*, 1983 [95]; *Holling*, 1987 [105], or *Chaitin*, 1987 [41], who inspect the complexity issue from rather different perspectives). However, let us advance three arguments that speak for the feasibility of active control of many truly complex non-linear systems – if the principle of iterated fuzzy adjustment is observed:

(i) The *non-linearity* of the systems in question, that is, the disproportionate (and sometimes even discontinuous) response to internal and/or external driving forces, can render such systems even *better to control than linear ones*! A necessary prerequisite for exploiting that characteristic is sufficient knowledge about the strategic variables and the sensitive (or critical) sectors of the dynamic complex considered. Then even *"inverse butterfly effects"* may be accomplished, i.e. the almost complete suppression of massive perturbations by non-linear attenuation mechanisms.

(ii) Complex systems are generally *highly structured* (for example, in a hierarchical way) and *intrinsically involved* by a multitude of interactions between their components – the latter property really inducing a definition. Now the number of interrelations, bonds and mutual restrictions grows, as a rule, over-proportionally with the system's size, so the *relative magnitude of degrees of freedom tends to decline.* It has to be emphasized here that this subtle and gradual reduction of transformability and variability is not exclusively due to trivial enhancement of connectedness, but also to *dynamic co-operative effects.* Haken's "Slaving Principle" (*Haken*, 1975 [94]) and Bak's "Theory of Self-Organized Criticality" as mentioned above (*Bak* et al., 1987 [12]) shed some light on the resulting stability and even resilience of many complex systems.
As a matter of fact, the trend towards *decreasing flightiness with increasing complexity* is massively supported by empirical observations: as mentioned before, it is rather difficult to discover concrete examples for deterministic chaos in large natural systems (like the atmospheric machinery). One pertinent reason for that failure is the fact that these multicomponent systems generate their own quasi-stochastical noise which subdues and dissolves the sharp characteristics of strange attractors, fractal structures or similar specimens from the chaos zoo. Otherwise it would be completely hopeless, say, to go for climate predictions extending over several centuries ahead of us.

(iii) Finally, and especially when dealing with *environmental systems* that deserve our primary interest, we must not neglect the point that complex systems are often the *selective product* of remarkably *lengthy and intricate evolutionary processes.* The relatively high robustness of such systems is therefore by no means surprising.
Take for example the human body with its incredible adaptive and regenerative abilities – without these abilities our much-admired medical treatment would probably have no chance whatever of influencing the physical conditions all that positively, without really understanding most

of the physiological, biochemical and molecular mechanisms involved! A similar remark applies to the *Earth System* – independently of the specific interpretation of the ecosphere as a "super-organism" or just as a huge "biogeophysicochemical machine".

In summary, truly complex systems are much less inclined to the caprices and escapades of moderately simple ones, and they may often be influenced or even controlled *"the soft way"*. Note that it is virtually impossible to "hold together a swarm of fleas" (according to a familiar German saying), while a flock of sheep may be efficiently controlled by an indefatigable German shepherd, and a German crowd can usually be directed by a few loudspeaker announcements (the latter fact being a mixed blessing). Thus the steering of complex systems succeeds in particular when the internal regulative mechanisms are exploited in the sense of stimulated *fuzzy self-control.*

It has to be emphasized, however, that the controllability of complex systems dramatically subsides when these systems become *critical*, for instance, by approaching phase transition lines due to intrinsic dynamics or external driving forces. In such situations, self-amplifying processes might be ignited that lead to inevitable *transformations of structure and operating mode* of the system in question. Using again the human body as an illustration, we may think of crises as uncontrollable (and therefore deadly) auto-immune reactions or sudden circulatory collapses. The planetary ecosystem is no less susceptible to critical developments (see, e.g., the glaciation events triggered by minor fluctuations in insolation) – so it is imperative for humankind to *identify potential roads to criticality and to bypass them at a respectful distance* ...

* * *

This has been a long and winding, yet by no means exhaustive, excursion into the field of passive and active control of simple as well as complex systems. This field trip is no such thing as a proof of the feasibility of geo-cybernetics, but it should raise at least some hope in the realization of one or other variant of fuzzy global E & D management. This hope may only come true, however, if the general prerequisites for Fuzzy Control (see pp. 176 – 178) are specified in the geo-cybernetic context as follows:

1. Observance of the *precautionary principle*, i.e., securing of error-tolerant operating by judiciously staying away from the currently presumed civilizatory and ecological load limits (*"Leeway"*).

2. Creation and cultivation of *flexible instruments* for global E & D management, i.e., international consultation and decision mechanisms, institutions and infrastructures that allow for perpetual readjustment in accordance with cognitive and voluntative developments (*"Responsive-*

ness"). The present "geo-cybernetic" means, including the system of existing global conventions (*WBGU*, 1996 [83]), seem much too brittle from the point of view of systems analysis or operations research.

3. Incessant and *intensive exploration of virtual coevolution futures* in computer-animated model earth systems for identifying potential dead-end streets and catastrophe domains (*"Panoramic View"*).

These ingredients are especially instrumental in the very test case for fuzzy E & D management, namely *global climate protection policy*. For the latter task unites all main aspects of soft decision-making under uncertainty – the necessity to accept temporal "ecological overburdens" without losing all safety margins; the ability of short-term revisions of long-term greenhouse-gas emissions quota; the intelligent handling of the blurry crystal balls provided by climate and climate impact models, etc.

The interested members of the scientific community have recently begun to focus on the control-theoretical elements of the climate protection issue. For a recent review see, e.g., the book by *Nakićenović* et al. (1996) [160] and in particular the contributions by *Manne* (1996) [145], and *Kelly* and *Kolstadt* (1996) [120]. A topical taxonomy of the pertinent variety of *integrated assessment models* has recently been provided by *Schellnhuber* and *Yohe* [207], where the next-generation models are divided into three classes, namely *evolutionary models*, *guardrail models*, and *combinations of the two pure breeds* (see also the paper by *Hasselmann* and *Hasselmann* (1998) [100] in this book). All these new approaches are implicitly or explicitly suited for taking into account uncertainty and implementing the fuzzy-control principle. This is illustrated for the evolutionary class, e.g., by the papers of *Dowlatabadi* (1996) [59] and *Lempert* et al. (1994) [133], as mentioned already above.

The guardrail models, on the other hand, try to define "tolerable windows" (*WBGU*, 1996 [83]; *Tòth* et al., 1997 [232]; *Petschel-Held* and *Schellnhuber*, 1997 [180]) or "safe corridors" (*Alcamo* and *Kreileman*, 1996 [4]) for climate evolution, and to determine by inverse methods the admissible socio-economic dynamics. Such an approach is *domain-based* instead of trajectory-based and, therefore, easily fuzzified.

At the very end of this section, we would like to re-emphasize the essence of our argumentation in just a few sentences:

The future of the Earth System cannot be predicted – due to irreducible cognitive and voluntative uncertainties. Yet the global E & D process may be shaped – to a certain degree and in an iterative way. Allons corriger le futur!

7 Epilogue

Those who read these lines belong to one of two distinct groups:

The first (and probably dominant) breed embraces the people interested in the overall ESA subject, who therefore absorb the Prologue, "skim and skip" their way through the five main sections, and jump on to the final passages with a certain expectation for conclusions. However, how to conclude an emergent panorama?

The second breed represents those hardy readers who bravely struggle, along with the author, on the tortuous intellectual road to the very end. Some members of this group may actually deplore the *brevity* of my essay, which touches many subtle points quite superficially and downright ignores other important aspects. What I have produced is, in fact, rather a "Picture Book of Earth System Analysis and Sustainable Development" than a comprehensive manual of E & D management. I am convinced, however, that the "Anschauung" – a German notion that amalgamates vision and contemplation – is the most powerful tool for grasping truly complex correlations.

The general opinion about this piece of work will be shaped, anyway, by the real sovereign of popular as well as scientific literature, namely the non-reader ...

Instead of presenting crisp conclusions, I would like to re-emphasize in this brief final section the paramount importance of *wise and equitable* use of our ever-growing knowledge about the Earth System. The fundamental predicament of global modern society is that, at the same time, *too little and too much information is available.* We have discussed the various aspects of cognitive deficiency in some detail in the previous chapters, so let us just recall here that even our purely geophysical wisdom is a rather bizarre melange of "high-know" and sheer guesswork. However, note that geo-cybernetic success does not necessarily depend upon a perfect information basis: From a pragmatic point of view, we need not understand the things we *cannot* control, and – as pointed out in Sect. 6 – we need not understand *precisely* the things we actually can control!

The knowledge we do have, on the other hand, is more than sufficient to give rise to a plethora of serious problems, that may eventually shatter the ethical fundament of our civilization. With the increasing indicative and predictive power of the "Global Brain", we are rapidly approaching the second expulsion from the paradise of ignorance: Imagine that we could tell "wrong from right" in the sense of being able to foresee the entire cascade of long-term consequences of any possible interference with the regional or global environment. If we were ready to take full responsibility of some trigger action in the presence, we would therefore have to anticipate the infinite ensemble of so-generated potential E & D futures as spanned by the diverging tree of alternative decisions to be made in the course of time ahead. It is impossible to make *the* correct choice under these circumstances, because minuscule differences in the initial action might raise hell instead of establishing heaven on Earth; and who would dare to rate the well-being of the, say, 21st generation from now higher than the containment of environmental damages to the 35th generation, for instance?

If we think through this argumentation, then we may arrive at the conclusion that "Sustainable Development" is, above all, an *intellectual problem.* People have been exploiting non-renewable resources like the silver mines of Goslar or Potosí, or have been overexploiting renewable resources like the Mediterranean forests throughout history. They have done this, however, in a "state of innocence" due to the lack of reliable forecasts about the reserves and consequences passed onto their descendants. We are not enjoying this comfortable state anymore, as science and technology confront us relentlessly with the possible implications of any single step we take. So whether we like it or not, *we are morally obliged to behave in a sustainable way* and condemned to make nasty decisions like weighing the life of thousands of coastal settlers threatened by sea-level rise against the prosperity of industries elsewhere on the globe.

Some people will therefore long for the times of splendid ignorance, when they entered a supermarket without worrying for the subsistence of the ozone layer in the stratosphere. For them, and actually for all of us, the *certitude* that our knowledge about the global environment will always contain elements of *uncertainty* may provide some consolation. For this uncertainty preserves at least a pale reflection of arbitrariness, alternatively called freedom of will ...

Yet the problem of prospective decision making, fully conscious of the far-reaching consequences involved, remains. However, let us give rein for a moment: The Global Subject will have to take advantage of modern instruments for moulding and expressing its volition anyway. Why not turn to *"cyberspace democracy"*, where the actors and stake-holders in the co-evolution process grind out crucial decisions via interactive planning games based on Earth System modelling? This would ensure a broad distribution of responsibility through equitable access to knowledge, if not wisdom.

Whatever geo-cybernetics will look like in the next millennium, it should contain one element of rationality which the contemporal E & D debate often lacks. I mean the willingness and the ability to assess problems by *the right order of magnitude and importance.* The present generation is, for instance, ready to take incredibly high social and ecological risks by carelessly promoting the replacement of cultures by lifestyle – a development that will ultimately eradicate hundreds of deeply-rooted regional civilizations. By way of contrast, billions of public US-Dollar, D-Mark or Yen are annually spent in the industrialized countries to fend off even the most marginal or elusive environmental menaces for the well-being of human individuals or pet species.

My favourite anecdote illuminating the reigning irrationality and inconsistency is the touching story about the Prince of Wales, who recently announced to switch from lead to bismuth ammunition for wildfowl hunting. The motivation behind this imperial act of environmental consciousness is to prevent the unintentional poisoning by scattered buckshot of those animals that escaped the gun. It seems that the Global Subject is still a long way from home ...

References

1. R. Abraham and J. E. Marsden. *Foundations of mechanics.* Benjamin/Cummings, Reading, Mass., 1977.
2. S. G. Akl. *Parallel Computation: Models and Methods.* Prentice Hall, Upper Saddle River, 1997.
3. J. Alcamo, editor. *IMAGE 2.0: Integrated Modeling of Global Climate Change.* Kluwer, Dordrecht, 1994. Reprinted from Journal of Water, Air and Soil Pollution, Vol. 76 Nos. 1-2, 1994.
4. J. Alcamo and G., J. J. Kreileman. Emission scenarios and global climate protection. *Global Environmental Change*, 6(4):305 pp., 1996.
5. J. S. Andrade. Self-organized criticality in the El Niño Southern Oscillation. *Physica A*, 215(3):331 pp., 1995.
6. V. I. Arnold. Instability of dynamical systems with several degrees of freedom. *Soviet Mathematics – Doklady*, 5:581 pp., 1964.
7. V. I. Arnold. *Mathematical Methods of Classical Mechanics.* Springer, New York, 1978.
8. V. I. Arnold and A. Avez. *Ergodic Problems of Classical Mechanics.* Benjamin, New York, 1968.
9. P. Baccini and P. H. Brunner. *Metabolism of the Anthroposphere.* Springer, Berlin, 1991.
10. P. Bak. *The Science of Self-Organized Criticality.* Oxford University, Oxford, 1997.
11. P. Bak and K. Chen. Self-organized criticality. *Scientific American*, 264(1):46 pp., 1991.
12. P. Bak, C. Tang, and K. Wiesenfeld. Self-organized criticality. *Physical Review Letters*, 59(4):381 pp., 1987.
13. W. L. Baker. Longterm response of disturbance landscapes to human intervention and global change. *Landscape Ecology*, 10(3):143 pp., 1995.

14. E. B. Barbier. *Economics, Natural-Resource Scarcity and Development.* Earthscan, London, 1989.
15. C. J. Barrow. *Developing the Environment: Problems and Management.* Longman Scientific & Technical, Essex, 1995.
16. R. J. Baxter. *Exactly Solved Models in Statistical Mechanics.* Academic, London, 1990.
17. D. E. Bell, R. L. Keeney, and H. Raiffa, editors. *Conflicting Objectives in Decisions.* Wiley & Sons, London, 1977.
18. R. E. Bellmann. *Dynamic Programming.* Princeton University, Princeton, 1957.
19. C. H. Bennett. Demons, engines and the second law. *Scientific American*, 257(5):108 pp., 1987.
20. P. Bergé. Chaos and unusual attractors. *Physikalische Blätter*, 46(7):209 pp., 1990.
21. P. Bergé, Y. Pomeau, and C. Vidal. *Order within Chaos.* Wiley & Sons, New York, 1984.
22. H. O. Bergesen and G. Parmann, editors. *Green Globe Yearbook 1996.* Oxford University, Oxford, 1996.
23. M. V. Berry. Regular and irregular motion. In S. Jorna, editor, *Topics in nonlinear dynamics. A tribute to Sir Edward Bullard*, number 46 in AIP conference proceedings, La Jolla, CA, 1978. American Institute of Physics.
24. K. Binmore. *Essays on the Foundations of Game Theory.* Basil Blackwell, Cambridge, 1990.
25. C. Blackburn, editor. *Summary, Conclusions, and Recommendations*, Global Change and the Human Prospect: Issues in Population, Science, Technology, and Equity, Research Triangle Park, NC, 1992. Sigma Xi, The Scientific Research Society.
26. P. M. Blaikie and H. Brookfield. *Land Degradation and Society.* Methuen, London, 1987.
27. D. Bohm. *Quantum Theory.* Prentice Hall, New York, 1951.
28. H.-R. Bork. *Bodenerosion und Umwelt.* Landschaftsgenese und Landschaftsökologie, No. 13. Technische Universität Braunschweig, Braunschweig, 1988.
29. H. Bossel. Deriving indicators of sustainable development. *Environmental Modeling & Assessment*, 1(4):193 pp., 1996.
30. H. Bossel. Ecosystems and society: implications for sustainable development. *World Futures*, 47:143 pp., 1996.
31. H. Breitmeier. *Ozonschicht und Klima auf der globalen Agenda.* Number 17 in Tübinger Arbeitspapiere zur internationalen Politik und Friedensforschung. Arbeitsgruppe Friedensforschung, Tübingen, 1992.
32. W. S. Broecker. Chaotic climate. *Scientific American*, 273(5):62 pp., 1995.
33. L. R. Brown, C. Flavin, and S. Postel. *Saving the Planet.* Norton, New York, 1991.
34. N. Brown. *The Strategic Revolution: Thoughts for the Twenty-First Century.* Brassey's, London, 1992.
35. M. J. Budyko and G. S. Golitsyn. *Climatic Catastrophes.* Springer, New York, 1988.
36. M. Buitenkamp, H. Venner, and T. Wams. *Sustainable Netherlands.* Friends of the Earth Netherlands, Amsterdam, 1992.
37. H. B. Callen. *Thermodynamics and an Introduction to Thermostatistics.* Wiley, New York, 1985.
38. M. Carley and I. Christie. *Managing Sustainable Development.* University of Minnesota Press, Minneapolis, 1993.

39. M. A. Cassel-Gintz, M. K. B. Lüdeke, G. Petschel-Held, F. Reusswig, M. Plöchel, G. Lammel, and H. J. Schellnhuber. Fuzzy logic based global assessment of the marginality of agricultural land use. *Climate Research*, 8(2):135 pp., 1997.
40. L. Cesari. *Optimization Theory and Applications*. Springer, New York, 1983.
41. G. J. Chaitin. *Algorithmic Information Theory*. Cambridge University, Cambridge, 1987.
42. W. C. Clark. Managing planet earth. *Scientific American*, 261(3):46 pp., 1989.
43. W. C. Clark and R. E. Munn, editors. *Sustainable Development of the Biosphere*. Cambridge University, Cambridge, 1986.
44. M. Claussen and A. Ganopolski. *Klimasystemmodelle*. Umwelt- und Klimabeeinflussung durch den Menschen. VDI Verlag, Düsseldorf, 1997.
45. W. R. Cline, editor. *The Economics of Global Warming*. Institute for International Economics, Washington DC, 1992.
46. C. W. Cobb and J. B. Cobb. *The Green National Product - A Proposed Index of Sustainable Economic Welfare*. University Press of America, Lanham, 1994.
47. R. Costanza, R. d'Arge, R. de Groot, S. Farber, M. Grasso, B. Hannon, K. Limburg, S. Naeem, R. O'Neill, J. Paruelo, R. Raskin, P. Sutton, and M. van den Belt. The value of the world's ecosystem services and natural capital. *Nature*, 387(6630):253 pp., 1997.
48. W. Cronon. *Uncommon Ground. Towards Reinventing Nature*. Norton, New York, 1995.
49. H. E. Daly. Toward some operational principles of sustainable development. *Ecological Economics*, 2:1 pp., 1990.
50. H. E. Daly and J. B. Cobb. *For the Common Good*. Beacon Press, Boston, 1994.
51. H. E. Daly and K. N. Townsend, editors. *Valuing the Earth. Economics, Ecology, Ethics*. MIT Press, Cambridge, 1993.
52. G. B. Dantzig and A. F. Veinott Jr. *Mathematics of the Decision Sciences*. American Mathematical Society, Providence, 1968.
53. J. Darmstädter. Global Development and the Environment - Perspectives on Sustainability. Washington D.C., 1992.
54. P. C. W. Davies. *The Ghost in the Atom: A Discussion of the Mysteries of Quantum Physics*. Cambridge University, Cambridge, 1986.
55. J. de Rosnay. *L'homme symbiotique: regards sur le troisiéme millénaire*. Seuil, Paris, 1995.
56. E. A. Desloge. *Classical Mechanics*. Wiley, New York, 1982.
57. B. d'Espagnat. Quantum theory and reality (reprint). *Scientific American*, 241(5):158 pp., 1979.
58. D. Dörner. *The Logic of Failure: Why things go wrong and what we can do to make them right*. Metropolitan Books, New York, 1996.
59. H. Dowlatabadi. Adaptive management of climate change mitigation: a strategy for coping with uncertainty. Discussion paper, unpublished, Center for Integrated Study of the Human Dimensions of Global Change, Carnegie Mellon University, 1996.
60. H. Dowlatabadi and G. M. Morgan. Integrated assessment of climate change. *Science*, 259:1813–1814, 1993.
61. B. Drossel and F. Schwabl. Self-organized criticality in a forest-fire model. *Physica A*, 191(1-4):47 pp., 1992.
62. W. Ebeling, H. Engel, and H. Herzel. *Selbstorganisation in der Zeit*. Akademie-Verlag, Berlin, 1990.

63. M. Efinger and H. Breitmeier. *Zur Theorie und Praxis der Verifikation einer globalen Klimakonvention.* Number 3 in Berichte des Forschungszentrums Jülich. Arbeitsgruppe Friedensforschung, Jülich, 1992.
64. Environmental Impact Assessment Review (EIAR). Sustainable development. *Environmental Impact Assessment Review*, 12(1-2), 1992.
65. K. Falconer. *Fractal Geometry. Mathematical Foundations and Applications.* Wiley & Sons, Chichester, 1990.
66. S. Fankhauser. *Valuing Climate Change.* Earthscan, London, 1995.
67. J. C. Farman, B. G. Gardiner, and J. D. Shanklin. Large losses of total ozone in Antarctica reveal seasonal CO_x/NO_x interaction. *Nature*, 315(6016):207 pp., 1985.
68. Federal Planning Office of Belgium (FPOB). *Report of the workshop on indicators of sustainable development for decision-making*, Ghent, 1995. Submitted to the UN Commission on Sustainable Development.
69. J. Feichter, U. Lohmann, and I. Schult. The atmospheric sulfur cycle in ECHAM-4 and its impacts on the shortwave radiation. *Climate Dynamics*, 13:235 pp., 1997.
70. R. P. Feynman and A. R. Hibbs. *Quantum Mechanics and Path Integrals.* McGraw-Hill, New York, 1965.
71. W. H. Fleming and R. W. Rishel. *Deterministic and Stochastic Optimal Control.* Springer, New York, 1975.
72. Food and Agriculture Organization of the United Nations (FAO). *Lessons from the green revolution*, 1996. http://www.FAO.org/wfs/final/e/volume2/t06sum-e.htm.
73. J. W. Forrester. *World dynamics.* Wright-Allen, Cambridge, 1971.
74. Friends of the Earth Europe (FEE). *Towards Sustainable Europe.* Brussels, 1995.
75. B. Fritsch, S. Schmidheiny, and W. Seifritz. *Towards an Ecologically Sustainable Growth Society. Physical Foundations, Economic Transitions, and Political Constraints.* Springer, Berlin, 1994.
76. H. Gallee, J. P. van Ypersele, T. Fichefet, Marsiat C., C. H. Tricot, and A. Berger. Simulation of the last glacial cycle by a coupled sectorially averaged climate-ice sheet model. II. Response to insolation and CO_2 variations. *Journal of Geophysical Research*, 97:15713 pp., 1992.
77. H. Gallee, J. P. van Ypersele, T. Fichefet, C. H. Tricot, and A. Berger. Simulation of the last glacial cycle by a coupled sectorially averaged climate-ice sheet model. I. The climate model. *Journal of Geophysical Research*, 96:13139 pp., 1991.
78. A. Ganopolski, V. Brovkin, M. Claussen, and V. Petoukhov. A study of biogeophysical feedbacks with a coupled climate-biosphere model CLIMBER-2. *Annales Geophysicae*, 15(II):352 pp., 1997.
79. A. Ganopolski, S. Rahmstorf, V. Petoukhov, and M. Claussen. Simulation of modern and glacial climates with a coupled climate model. *Nature*, 1997. Accepted.
80. A. Ganopolski and H. J. Schellnhuber. Influence of salinity on the structure of the thermohaline ocean circulation. Unpublished, 1997.
81. German Advisory Council on Global Change (WBGU), Annual Reports 1993-1997, English and German editions.
82. German Advisory Council on Global Change (WBGU). *World in Transition: The Threat to Soils. Annual Report 1994*, Economica, Bonn, 1995.
83. German Advisory Council on Global Change (WBGU). *World in Transition: Ways Towards Global Environmental Solutions. Annual Report 1995*, Springer, Berlin, 1996.

84. German Advisory Council on Global Change (WBGU). *World in Transition: The Research Challenge. Annual Report 1996*, Berlin, 1997. Springer.
85. German Advisory Council on Global Change (WBGU). *World in Transition: Ways Towards Sustainable Management of Freshwater Resources. Annual Report 1997*, Springer, Berlin, 1997. Forthcoming.
86. R. Gersonde, F. T. Kyte, U. Bleil, B. Diekmann, J. A. Flores, K. Gohl, G. Grahl, R. Hagen, G. Kuhn, F. J. Sierro, D. Völker, A. Abelmann, and J. A. Bostwick. Geological record and reconstruction of the late Pliocene impact of the Eltanin asteroid in the Southern Ocean. *Nature*, 390(6658):357 pp., 1997.
87. E. H. Gilson. *History of Christian Philosophy in the Middle Ages*. Sheed and Ward, London, 1955.
88. H. Goldstein. *Classical Mechanics*. Addison-Wesley, Reading, 1982.
89. R. Goodland, editor. *Environmentally Sustainable Economic Development: Building on Brundtland*. World Bank, Washington, 1991.
90. A. Gore. *Earth in the Balance - Ecology and the Human Spirit*. Houghton Mifflin Company, Boston, 1992.
91. J. R. Gribbin. *In Search of Schrödinger's Cat*. Black Swan Books, London, 1991.
92. R. B. Griffiths. Correlations in separated quantum systems: a consistent history analysis of the EPR problem. *American Journal of Physics*, 55(1):11 pp., 1989.
93. A. Grübler and A. McDonald. The drive to cleaner energy. *Options*, Winter 1995:8 pp., 1995.
94. H. Haken. Cooperative phenomena in systems far from thermal equilibrium and in non-physical systems. *Reviews of Modern Physics*, 47(1):67 pp., 1975.
95. H. Haken. *Advanced Synergetics*. Springer, Berlin, 1983.
96. P. R. Halmos. *Measure Theory*. Springer, Berlin, 1974.
97. G. A. Harrison and D. Jeffries. Human biology in urban environments. *Nature and Resources*, 12(1):2 pp., 1976.
98. K. Hasselmann. Climate change - are we seeing global warming? *Science*, 276(5314):914 pp., 1997.
99. K. Hasselmann. Multi-pattern fingerprint method for detection and attribution of climate change. *Climate Dynamics*, 13(9):601 pp., 1997.
100. S. Hasselmann and S. Hasselmann. Multi-actor optimization of greenhouse gas emission paths using coupled integral climate response and economic models. In *Earth System Analysis: Integrating Science for Sustainability*. Springer, Berlin, 1998. Contribution in this proceedings.
101. HDP, editor. *Global Change, Local Challenge*, volume 1 and 2 of *HDP Third Scientific Symposium 20-22 September 1995*, Geneva, 1996. Human Dimensions of Global Environmental Change Programme. Report 8.
102. G. C. Hegerl, K. Hasselmann, U. Cubasch, J. F. B. Mitchell, E. Roeckner, R. Voss, and J. Waszkewitz. Multi-fingerprint detection and attribution analysis of greenhouse gas, greenhouse gas-plus-aerosol and solar forced climate change. *Climate Dynamics*, 13(9):613 pp., 1997.
103. F. Hole. Environmental instabilities and urban origins. In G. Stein, editor, *Chiefdoms and Early States in the Near East: The Organizational Dynamics of Complexity*, number 18 in Monographs in World Archaeology. Prehistory, Madison, 1994.
104. F. Hole. Evidence for mid-Holocene environmental change in the western Habur drainage, Northeastern Syria. In *Third Millennium BC Abrupt Climate Change and Old World Social Collapse*, New Haven, 1994. Yale University. NATO Advanced Research Workshop, September 1994.

105. C. S. Holling. Simplifying the complex: the paradigm of ecological function and structure. *Journal of Operational Research*, 30:139 pp., 1987.
106. R. M. Hough, I. Gilmour, C. T. Pillinger, J. W. Arden, K. W. R. Gilkes, J. Yuan, and H. J. Milledge. Diamond and silicon carbide in impact melt rock from the Ries impact crater. *Nature*, 378:41 pp., 1995.
107. Intergovernmental Panel on Climate Change (IPCC). *Climate Change 1995. Working Group I: The Science of Climate Change. Working Group II: Impacts, Adaptations and Mitigation of Climate Change. Working Group III: Economic and Social Dimensions of Climate Change*, Cambridge, 1996. University Press.
108. The International Geosphere-Biosphere Programme IGBP. *Global Change. Report 28, IGBP in Action: Work Plan 1994-1998*, Stockholm, 1994.
109. M. D. Intriligator. *Mathematical Optimization and Economic Theory.* Prentice-Hall, Englewood Cliffs, 1971.
110. U. Irmer. Personal communication, Federal Environmental Agency of Germany (UBA).
111. G. Julia. Sur l'iteration des fonctions rationelles. *Journal Mathematique Pure et Appl.*, 8:47 pp., 1918.
112. A. Kandel. *Fuzzy Mathematical Techniques with Applications.* Addison-Wesley, Reading, 1986.
113. A. Kandel and G. Langholz, editors. *Fuzzy Control Systems.* CRC Press, Boca Raton, 1993.
114. I. Kant. *Schriften zur Metaphysik und Logik.* Suhrkamp, Frankfurt a. M., 1991. Reprint.
115. C. L. Karr and E. J. Gentry. Control of a chaotic system using fuzzy logic. In A. Kandel and G. Langholz, editors, *Fuzzy Control Systems.* CRC Press, Boca Raton, 1993.
116. H. G. Kastenholz, K. H. Erdmann, and M. Wolff, editors. *Nachhaltige Entwicklung - Zukunftschancen für Mensch und Umwelt.* Springer, Berlin, 1996.
117. J. F. Kasting, O. B. Toon, and J. B. Pollack. How climate evolved on the terrestrial planets. *Scientific American*, 258(2):90 pp., 1988.
118. R. K. Kaufmann and D. I. Stern. Evidence for human influence on climate from hemispheric temperature relations. *Nature*, 388(6637):39 pp., 1997.
119. Y. Kaya, N. Nakićenović, W. D. Nordhaus, and F. L. Tòth, editors. *Costs, Impacts, and Benefits of CO_2 Mitigation*, Laxenburg, 1993. IIASA.
120. D. L. Kelly and C. D. Kolstadt. The climate change footprint: Will we see it before it is upon us? In N. Nakićenović, W. D. Nordhaus, R. Richels, and F. L. Tòth, editors, *Integrating Science, Economics, and Policy*, Laxenburg, 1996. IIASA.
121. G. Kirsch. *Neue Politische Ökonomie.* Werner, Düsseldorf, 1993.
122. T. R. Knutson and S. Manabe. Impact of increased CO_2 on simulated ENSO-like phenomena. *Geophysical Research Letters*, 8(21):2295 pp., 1994.
123. E. Kreyszig. *Introductory Functional Analysis with Applications.* Wiley, New York, 1978.
124. E. Kreyszig. *Statistische Methoden und ihre Anwendungen.* Vandenhoek & Ruprecht, Göttingen, 1991.
125. W. E. Krumbein and H. J. Schellnhuber. Geophysiology of carbonates as a function of bioplanets. In V. Ittekkot, S. Kempe, W. Michaelis, and A. Spitzy, editors, *Facets of Modern Biogeochemistry, Festschrift for E. T. Degens.* Springer, Berlin, 1990.
126. F. T. Kyte, L. Zhou, and J. T. Wasson. New evidence on the size and possible effects of a latePliocene oceanic asteroid impact. *Science*, 241:63 pp., 1988.

127. D. P. Landau. Theory of phase transitions beyond mean-field theory. In K. Binder and G. Cicotti, editors, *Euroconference on Computer Simulation in Condensed Matter Physics Chemistry*, Bologna, Italy, 1996. Italian Phys. Soc.
128. K. Lanius. *Die Erde im Wandel.* Spektrum, Heidelberg, 1995.
129. M. Latif, T. P. Barnett, M. A. Cane, M. Fluegel, N. E. Graham, H. von Storch, J.-S. Xu, and S. E. Zebiak. A review of ENSO prediction studies. *Climate Dynamics*, 9:167 pp., 1994.
130. E. B. Lee and L. Markus. *Foundations of Optimal Control Theory.* Wiley & Sons, New York, 1967.
131. S. M. Lélé. Sustainable development - a critical review. *World Development*, 19(607), 1991.
132. S. Lem. *Solaris.* Faber and Faber, London, 1991.
133. R. J. M. Lempert, M. E. Schlesinger, and J. K. Hammitt. The impact of potential abrupt climate changes on near-term policy choices. *Climatic Change*, 26:351 pp., 1994.
134. J. S. Lewis. *Rain of Iron and Ice: The Very Real Threat of Comet and Asteroid Bombardment.* Addison-Wesley, Reading, 1995.
135. A. J. Lichtenberg and M. A. Liebermann. *Regular and Stochastic Motion*, volume 38 of *Applied Mathematical Sciences.* Springer, New York, 1983.
136. D. Lind and B. Marcus. *An Introduction to Symbolic Dynamics and Coding.* Cambridge University, Cambridge, 1995.
137. E. N. Lorenz. Deterministic non-periodic flow. *Journal of the Atmospheric Sciences*, 20:130 pp., 1963.
138. R. Loske. *Zukunftsfähiges Deutschland: ein Beitrag zu einer global nachhaltigen Entwicklung.* Birkhäuser, Basel, 1996. Wuppertal Institut für Klima, Umwelt, Energie.
139. R. Loske and W. Sachs. *Industrial Societies & Sustainability: Proposals for Managing the Transition.* Humanities, UK, 1997.
140. J. E. Lovelock. *Gaia, The Practical Science of Planetary Medicine.* Gaia Books, London, 1991.
141. L. J. Lundgren, editor. *Views of Nature.* Swedish Council for Planning and Coordination of Research, Stockholm, 1993.
142. A. M. Lyapunov. Problème général de stabilité du mouvement. *Annals of Mathematics Studies*, 17, 1949. French translation of the original work of 1892.
143. S. Manabe and R. J. Stouffer. Century-scale effects of increased atmospheric CO_2 on the ocean-atmosphere system. *Nature*, 364:215 pp., 1993.
144. B. B. Mandelbrot. *The Fractal Geometry of Nature.* Freeman, New York, 1991.
145. A. S. Manne. Hedging strategies for global carbon dioxide abatement: a summary of poll results, EMF 14 subgroup - analysis for decisions under uncertainty. In N. Nakićenović, W. D. Nordhaus, R. Richels, and F. L. Tòth, editors, *Integrating Science, Economics, and Policy*, Laxenburg, 1996. IIASA.
146. R. M. May and G. Sugihara. Applications of fractals in ecology. *Trends in Ecology & Evolution*, 5(3):79 pp., 1990.
147. D. McKenzie-Mohr and M. Marien. *Visions of Sustainability.* Future. Butterworth-Heinemann, Oxford, 1994.
148. D. H. Meadows, D. L. Meadows, and J. Randers. *Beyond the Limits.* Chelsea Green Publishing Co., Post Mills, 1992.
149. D. L. Meadows, W. W. Behrens III, D. H. Meadows, R. F. Naill, J. Randers, and E. K. O. Zahn. *Dynamics of Growth in a Finite World.* Wright-Allen, Cambridge, MA, 1974.

150. G. A. Meehl, G. W. Branstator, and W. M. Washington. Tropical Pacific interannual variability and CO_2 climate change. *Journal of Climate*, 6(1):42 pp., 1993.
151. N. D. Mermin. Is the moon there when nobody looks? Reality and the quantum theory. *Physics Today*, 38(4):38 pp., 1985.
152. M. Milankovitch. Mathematische Klimalehre und astronomische Theorie der Klimaschwankungen. In W. Köppen and R. Geiger, editors, *Handbuch der Klimatologie.* Gebrüder Borntraeger, Berlin, 1930.
153. B. Moldan and S. Billharz, editors. *Sustainability-Indicators: Report of the Project on Indicators for Sustainable Development.* SCOPE 58. John Wiley, Chichester, 1997.
154. M. Montanari. *Der Hunger und der Überfluß: Kulturgeschichte der Ernährung in Europa.* Beck, München, 1995.
155. P. D. Moore, W. G. Chaloner, and P. A. Stott. *Global Environmental Change.* Blackwell Science, Oxford, 1996.
156. A. Mosier, W. J. Parton, D. Valentine, D. S. Ojima, D. S. Schimel, and O. Heinemeyer. CH_4 and N_2O fluxes in the Colorado shortgrass steppe: 2. Longterm impact of land use change. *Global Biogeochemical Cycles*, 11(1):29 pp., 1997.
157. R. A. Muller and G. J. MacDonald. Glacial cycles and astronomical forcing. *Science*, 277:215 pp., 1997.
158. R. E. Munn, J. W. M. La Rivière, and N. van Lookeren Campagne, editors. *Policy Making in an Era of Global Environmental Change.* Kluwer, Dordrecht, 1996.
159. N. Nakićenović, W. D. Nordhaus, R. Richels, and F. L. Tòth, editors. *Integrative Assessment of Mitigation, Impacts, and Adaptation to Climate Change*, Laxenburg, 1994. IIASA.
160. N. Nakićenović, W. D. Nordhaus, R. Richels, and F. L. Tòth, editors. *Integrating Science, Economics, and Policy*, Laxenburg, 1996. IIASA.
161. NASA. *Earth System Science.* NASA, Earth System Sciences Committee, Washington, 1988.
162. M. Nauenberg and H. J. Schellnhuber. Analytical evaluation of the multifractal properties of a Newtonian Julia set. *Physical Review Letters*, 62(16):1807 pp., 1989.
163. J. W. Negele. The mean-field theory of nuclear structure and dynamics. *Reviews of Modern Physics*, 54(4):913 pp., 1982.
164. W. D. Nordhaus. An optimal transition path for controlling greenhouse gases. *Science*, 258:1315 pp., 1992.
165. W. D. Nordhaus and J. Tobin. Is growth obsolete? In *Economic Growth*, number 46 in General Series, 96E, National Bureau of Economic Research, 1972. Columbia University.
166. J. B. Opschoor. *Environment, Economy and Sustainable Development.* Wolters-Noordhoff, Rotterdam, 1992.
167. J. B. Opschoor and L. Reijnders. Towards sustainable development indicators. In O. Kuik and H. Verbruggen, editors, *In Search of Indicators of Sustainable Development.* Kluwer Academic Publishers, Dordrecht, 1991.
168. J. B. Opschoor and J. v. d. Straaten. Sustainable development: an institutional approach. *Ecological Economics*, 7:203 pp., 1993.
169. E. Ott, C. Grebogi, and J. A. Yorke. Controlling chaos. *Physical Review Letters*, 64(11):1196 pp., 1990.
170. M. Papageorgiou. *Optimierung: statische, dynamische, stochastische Verfahren für die Anwendung.* Oldenbourg, München, 1991.
171. W. Pareto. *Traité de sociologie générale.* Payot, Lausanne, 1917.

172. J. A. Paulos. *A Mathematician Reads the Newspaper.* Penguin Mathematics. Penguin, London, 1995.
173. D. Pearce. *Economic Values and the Natural World.* Earthscan, London, 1993.
174. D. Pearce and R. K. Turner. *Economics of Natural Resources and the Environment.* Harvester Wheatsheaf, New York, 1990.
175. D. W. Pearce. *Blueprint 2: Greening the World Economy.* Earthscan, London, 1991.
176. D. W. Pearce, A. Markandya, and E. Barbier. *Blueprint for a Green Economy.* Earthscan, London, 1989.
177. H. O. Peitgen and P. H. Richter. *The Beauty of Fractals.* Springer, Berlin, 1988.
178. R. Penrose. *The Emperor's New Mind: Concerning Computers, Minds, and the Laws of Physics.* Oxford, Oxford University, 1989.
179. I. Peterson. *Newton's Clock: Chaos in the Solar System.* W. H. Freeman, New York, 1993.
180. G. Petschel-Held and H. J. Schellnhuber. The tolerable windows approach to climate control. In F. L. Tòth, editor, *Cost-benefit analyses of climate change: the broader perspective.* Birkhäuser, Basel, 1997.
181. S. G. H. Philander. *El Niño, La Niña and the Southern Oscillation.* Academic, New York, 1989.
182. K. T. Pickering and L. A. Owen. *An Introduction to Global Environmental Issues.* Routledge, London, 1994.
183. D. Pitt and P. Samson, editors. *Beyond the Biosphere: The Noosphere and Global Problems.* Routledge, London, 1997.
184. H. Poincaré. *Les methodés nouvelles de la mécanique céleste.* Gauthier-Villars, Paris, 1899.
185. L. S. Pontryagin. *The Mathematical Theory of Optimal Processes.* Interscience, New York, 1962.
186. K. R. Popper. *The Logic of Scientific Discovery.* Routledge, London, 1992. Reprint.
187. I. C. Prentice, M. T. Sykes, and W. Cramer. A simulation model for the transient effects of climate change on forest landscapes. *Ecological Modelling,* 65(1-2):51 pp., 1993.
188. S. Rahmstorf. Bifurcations of the Atlantic thermohaline circulation in response to changes in the hydrological cycle. *Nature,* 378:145 pp., 1995.
189. Rat von Sachverständigen für Umweltfragen (RSU). *Umweltgutachten 1994. Für eine dauerhaft-umweltgerechte Entwicklung,* Statistisches Bundesamt, Wiesbaden, 1994.
190. J. Rawls. *A Theory of Justice.* Harvard University, Cambridge MA, 1971.
191. S. Rayner. *Integrated Climate Change Assessment: Fairness and Equity Issues. An Overview.* Battelle Northwest National Laboratory, Washington DC, 1997.
192. M. Redclift. Reflections on the "Sustainable Development" debate. *International Journal of Sustainable Development and World Ecology,* 1(1):3 pp., 1994.
193. W. E. Rees. *Defining "Sustainable Development".* University of British Columbia, Vancouver, 1989.
194. P. ReVelle and C. ReVelle, editors. *The Global Environment. Securing a Sustainable Future.* Jones and Bartlett, Boston, 1992.
195. P. H. Richter and H.-J. Scholz. *Das ebene Doppelpendel.* IWF Göttingen, 1985.

196. C. Robinson. *Dynamical Systems: Stability, Symbolic Dynamics, and Chaos.* CRC Press, Boca Raton, 1994.
197. F. J. Romeiras, C. Grebogi, E. Ott, and W. P. Dayawansa. Controlling chaotic dynamical systems. *Physica D*, 58(1-4):165 pp., 1992.
198. N. Rosen. *The dilemma of Einstein, Podolsky and Rosen, 60 years later.* Institute of Physics Pub., Bristol, 1996.
199. J. Rotmans. Global Change and Sustainable Development: Towards an Integrated Conceptual Model. In H.-J. Schellnhuber and V. Wenzel, editors, *Earth System Analysis: Integrating Science for Sustainability*, Berlin, 1998. Springer. Contribution in this Proceedings.
200. J. Rotmans and H. J. M. de Vries. *Perspectives on Global Change: The TARGETS Approach.* Cambridge University, Cambridge, 1997.
201. D. Ruelle. *Chance and Chaos.* Princeton University, New Jersey, 1991.
202. C. Safina. The world's imperiled fish. *Scientific American*, 273(5):46 pp., 1995.
203. P. A. Samuelson and W. D. Nordhaus. *Economics.* McGraw-Hill, New York, 1995.
204. P. Saunders. *Introduction to Catastrophe Theory.* Cambridge University, Cambridge, 1980.
205. H. J. Schellnhuber. Global environmental change: the physics and the metaphysics. *Global Environmental Change*, 1997. Submitted.
206. H. J. Schellnhuber, M. A. Cassel-Gintz, J. Kropp, G. Lammel, W. Lass, R. Lienenkamp, C. Loose, M. K. B. Lüdeke, O. Moldenhauer, G. Petschel-Held, M. Plöchel, and F. Reusswig. Syndromes of global change. *GAIA*, 6(1):19 pp., 1997.
207. H. J. Schellnhuber and G. W. Yohe. Comprehending the economic and social dimensions of climate change by integrated assessment. Proceedings of the WCRP-Conference, Geneva, 1997. Forthcoming.
208. F. Schmidt-Bleek. *Wieviel Umwelt braucht der Mensch? MIPS, das Maß für ökologisches Wirtschaften.* Birkhäuser, Berlin, 1994.
209. H. Schmitz. *Anaximander und die Anfänge der griechischen Philosophie.* Bouvier, Bonn, 1988.
210. E. Schrödinger. *Naturwissenschaften* 23:807 pp., 1935, translated in J. A. Wheeler and W. H. Zurek, editors, *Quantum Theory and Measurement*, Princeton University, Princeton, NJ, 1983.
211. I. Schult, J. Feichter, and W. F. Cooke. The effect of black carbon and sulfate aerosols on the global radiation budget. *Journal of Geophysical Research*, 1997. Accepted.
212. H. G. Schuster. *Deterministic chaos: an introduction.* VCH, Weinheim, 1988.
213. F. Schweitzer, editor. *Self-Organization of Complex Structures: From Individual to Collective Dynamics.* Gordon & Breach, London, 1997.
214. Scientific Committee on Problems of the Environment (SCOPE), Series No. 1-58, 1971-1997.
215. U. E. Simonis. *Beyond Growth. Elements of Sustainable Development.* Edition Sigma, Berlin, 1990.
216. U. E. Simonis. *Weltumweltpolitik.* Wissenschaftszentrum für Sozialforschung, Berlin, 1996.
217. Y. G. Sinai. *Introduction to Ergodic Theory.* Math. Notes 18. Princeton University, Princeton, 1976.
218. R. M. Solow. *An Almost Practical Step to Sustainability.* Resources for the Future, Washington, 1992.

219. J. H. Spangenberg. *Ein zukunftsfähiges Europa. Towards sustainable Europe.* Wuppertal Papers, Bd. 42. Wuppertal Institut für Klima - Umwelt - Energie, Wuppertal, 1995.
220. D. Sprinz and U. Luterbacher. International relations and global climate change. PIK Report 21, Potsdam Institute for Climate Impact Research (PIK), Potsdam, 1996.
221. N. Stehr and H. von Storch. Das soziale Konstrukt des Klimas, 1997. http://w3g.gkss.de/G/Mitarbeiter/storch/vdi.html.
222. I. Stewart. *Does God play Dice? The Mathematics of Chaos.* Basil Blackwell, Oxford, 1989.
223. T. F. Stocker, D. G. Wright, and L. A. Mysak. A zonally averaged coupled ocean-atmosphere model for paleoclimate studies. *Journal of Climate*, 5:773 pp., 1992.
224. M. B. Stoff, J. F. Fanton, and R. H. Williams, editors. *The Manhattan Project.* Temple University, Philadelphia, 1991.
225. G. J. Sussman and J. Wisdom. Numerical evidence that the motion of Pluto is chaotic. *Science*, 241(4864):433 pp., 1988.
226. Y. M. Svirezhev and W. von Bloh. A minimal model of interaction between climate and vegetation: qualitative approach. *Ecological Modelling*, 92:89 pp., 1996.
227. M. Tabor. *Chaos and Integrability in Nonlinear Dynamics.* John Wiley & Sons, New York, 1989.
228. F. L. Tòth. Practicing the future: implementing "the policy exercise concept". Working Paper 86-23, IIASA, Laxenburg, 1986.
229. F. L. Tòth. Practicing the future. Part 2: Lessons from the first experiments with policy exercises. Working Paper 88-12, IIASA, Laxenburg, 1988.
230. F. L. Tòth. Policy exercises. Research Report 89-2, IIASA, Laxenburg, 1989.
231. F. L. Tòth, editor. *Fairness Concerns in Climate Change*, Proceedings of the International Workshop held at PIK, Potsdam, 1997. Forthcoming.
232. F. L. Tòth, T. Bruckner, H. Füssel, M. Leimbach, G. Petschel-Held, and H. J. Schellnhuber. The tolerable windows approach to integrated assessment. In *Asia-Pacific Workshop on Integrated Assessment*, Tokyo, 1997. IPCC. Forthcoming.
233. R. P. Turco. *Earth under Siege: From Air Pollution to Global Change.* Oxford University, Oxford, 1996.
234. E. Tziperman, H. Scher, S. E. Zebiak, and M. A. Cane. Controlling spatiotemporal chaos in a realistic El Niño prediction model. *Physical Review Letters*, 79:1034 pp., 1997.
235. *United Nations Climate Change Bulletin*, Geneva, 1995. Interims Secretariat for the UN Climate Change Convention. Issue 7.
236. UNESCO. *Man Belongs to the Earth: International Co-operation in Environmental Research. UNESCO's Man and the Biosphere Programme.* UNESCO, Paris, 1988.
237. United Nations. *Indicators of Sustainable Development: Framework and Methodologies*, United Nations, New York, 1996.
238. United Nations Conference on Environment and Development (UNCED), Rio de Janeiro. *Earth Summit '92*, London, 1992. The Regency.
239. United Nations Development Programme (UNDP). *Human Development Report 1990*, Oxford University Press, New York, 1990.
240. U.S. National Committee (USNC). *The United States Man and the Biosphere Program*, U.S. MAB Secretariat, Washington, 1996.

241. D. G. Victor. The Montreal Protocol's non-compliance procedure: lessons for making other international environmental regimes more effective. In W. Lang, editor, *The Ozone Treaties and Their Influence on the Building of Environmental Regimes*. Federal Ministry of Foreign Affairs, Vienna, 1996.
242. W. von Bloh, A. Block, and H. J. Schellnhuber. Self-stabilisation of the biosphere under global change: a tutorial geophysiological approach. *Tellus*, 49B:249 pp., 1997.
243. J. von Neumann. *Theory of Games and Economic Behaviour*. Princeton University, Princeton, 1944.
244. L. Walras. *Elements d'économie politique pure ou theorie de la richesse sociale*. Lausanne, 1874.
245. A. J. Watson and J. E. Lovelock. Biological homeostasis of the global environment: the parable of daisy world. *Tellus*, 35B:286 pp., 1983.
246. World Bank (WB), editor. *Monitoring Environmental Progress*, The World Bank, Washington, 1995.
247. WELT NEWS. Malaysia, 1996. Special supplement to the newspaper Frankfurter Allgemeine Zeitung, 23.12.1996, p.2, p.14.
248. R. Weterings and J. B. Opschoor. *The Ecocapacity as a Challenge to Technological Development*. Advisory Council for Research on Nature and Environment, Rijswijk, 1992.
249. P. Williamson and P. S. Liss. Understanding the earth system. In R. E. Munn, J. W. M. La Rivière, and N. van Lookeren Campagne, editors, *Policy Making in an Era of Global Environmental Change*. Kluwer, Dordrecht, 1996.
250. J. Wisdom. Meteorites may follow a chaotic route to Earth. *Nature*, 315(6022):731 pp., 1985.
251. J. Wisdom, S. J. Peale, and F. Mignard. The chaotic rotation of Hyperion. *Icarus*, 58(2):137 pp., 1984.
252. J. Wittig. Experimentelles zur Rayleigh-Bénard-Konvektion. Skript zum Ferienkurs der KFA Jülich, 28.2-12.3.1983, "Nichtlineare Dynamik in kondensierter Materie".
253. World Climate Research Programme (WCRP). *World Climate Research Programme*, 1996. http://www.wmo.ch/web/wcrp/wcrp-home.html.
254. World Commission on Environment and Development (WCED). *Our Common Future (The Brundtland Commission)*, Oxford University, Oxford, 1989.
255. World Resources Institute (WRI). World Resources, Annual Reports 1986-1997.
256. World Resources Institute (WRI). *World Resources 1992-93*, Oxford, 1992. Oxford University Press.
257. Worldwatch Institute (WI). *State of the world: Worldwatch Institute Reports on Progress Towards a Sustainable Society*. Annual Reports 1984-1997.
258. Worldwatch Institute (WI). *State of the world 1994. A Worldwatch Institute Report on Progress Towards a Sustainable Society*, Norton, New York, 1994.
259. Z. Xia. Arnold diffusion in the elliptic restricted three body problem. *Journal of Dynamics and Differential Equations*, 5:219 pp., 1993.
260. X. Xiao, D. W. Kicklighter, J. M. Melillo, A. D. McGuire, P. H. Stone, and A. P. Sokolov. Linking a global terrestrial biogeochemical model and a 2-dimensional climate model: implications for the global carbon budget. *Tellus*, 49B(1):18 pp., 1997.
261. L. A. Zadeh. Fuzzy sets. *Information and Control*, 8(3):338 pp., 1965.
262. H.-J. Zimmermann. *Fuzzy Set Theory and its Applications*. Kluwer, Dordrecht, 1991.
263. X. Zolotas. *Economic Growth and Declining Social Welfare*. New York University, New York, 1981.

Commentary: Earth System Analysis – Explorations in a Research Frontier

Peter Nijkamp

1 Orientation

Earth system analysis is one of the promising new research areas in the field of global change, with a particular focus on climate change. There is an urgent need to improve and to enhance our insights in the complex web of the rise in atmospheric concentrations of greenhouse gases (e.g., CO_2, CH_4), climatic change, sea level rise, changes in the ozone layer and the behaviour of mankind. Since the publication of the Brundtland Report this issue has provoked world-wide attention. Furthermore, at the United Nations Conference on Environment and Development (UNCED) in Rio (1992) global change was even the main item on the agenda. This concern is once more warranted, as the background of this phenomenon is closely connected with anthropogenic causes. Therefore, research at the interface of natural sciences and social sciences has to be welcomed. The Potsdam Institute for Climate Research (PIK) has to be complemented for the initiative to bring the issue of climate change from an interdisciplinary perspective to the forefront. In addition, the well thought and challenging introductory discourse by Schellnhuber on "Earth System Analysis: The Scope of the Challenge" sets the tone for a new research endeavour in a field fraught with many difficulties and uncertainties.

Global change – and also climate change – research distinguishes itself in various respects from other research issues. It is full of uncertainties and risks; it is transnational in nature; it has long range impacts covering multiple generations; it concerns many people on our planet; it is related to human behaviour (e.g., consumption, mobility, technology, demography etc.) but it has also its indigenous development; it is connected with almost all economic activities, directly or indirectly; and it is very difficult to develop preventive policies.

Since the Rio Conference many initiatives have been developed to cope with the negative consequences of climate change, but fundamental scientific knowledge is still missing to identify clear policy strategies. The activity plan of PIK looks very promising in this perspective. The Netherlands has in the past years also developed a national research programme on global change. And it is interesting to see that there is quite some congruence between the Dutch plans and the PIK mission statement. The objectives of the Dutch research programme are:

- to reduce uncertainties in causes, processes, effects and measures
- to integrate research results in Dutch global change and climate policy
- to define and quantify environmental quality standards to enhance insight into risk (e.g., via risk assessment and perception analysis)
- to seek international cooperation in climate research
- to improve the awareness of the climate problem among the Dutch population.

There is much scope for new research collaboration, and it may be hoped that European countries will join forces to study the backgrounds, implications and policy responses to climate change. PIK is certainly able to play a crucial role in such a research undertaking.

2 Substance

The discourse by Schellnhuber is fascinating and innovative in that it combines knowledge on natural earth system analysis with socio-political issues and formal modelling. From the wealth of his thoughts I will address here three important issues:

- the complexity of the climate change problem, not because of physical reasons but because of the *human factor.*
- the difficulties involved in global engineering of our planet, not because of lack of natural science insights but because of the *social dilemma* (or social trap).
- the lack of strategic thinking, not as a result of lack of formal modelling but as a result of the need for complex *contingency strategies.*

I will now succinctly discuss these three themes.

2.1 The human factor

The human factor refers to anthropogenic causes and bottlenecks in the global change problem. In recent years, numerous attempts have been made in the natural sciences to come to grips with the complex relationships and multiple causalities involved in global environmental change (*Clark* and *Munn*, 1986 [6]; *Böttcher*, 1993 [3]). In most cases however, the research efforts have not yet been convincingly successful, as various outcomes of modelling experiments were inconclusive, statistical tests were not satisfactory in light of the relatively short observation period of phenomena, and many disturbances and unexplained variations in facts were left as a consequence of the inability to model *"the human factor"* (Clark, 1990 [5]; *Jacobson*, 1992 [10]; *Kates* et al., 1990 [12]; *Miller* and Jacobson, 1992 [13]). It is noteworthy that in a recent publication of the Dutch Ministry of Education and Science the scope of "global environmental change" research is explicitly widened towards a more comprehensive social science orientation, as it is claimed that research in this field would comprise: processes and relationships in the natural systems (sea, land, atmosphere), impacts of anthropogenic destruction of these systems (e.g., climate change), implications of the human factor (causes, effects, control mechanisms), and monitoring of phenomena and observation of facts [14].

The latter multidisciplinary approach to global environmental change with a clear social science component is also found back in a publication of *Jacobson* and *Price* [11] on "A Framework for Research on the Human Dimensions of Global Environmental Change":

> "Research on this topic must consider human activities both as they contribute to and as they are affected by global environmental change. The human activities that interact with the Earth's natural systems are driven by three fundamental factors: the number of human beings and their distribution around the globe; human needs and desires, as conditioned by psychological, cultural, economic, and historical factors, which provide individuals and societies with motivations to act; and the cultural, social, economic, and political structures and institutions and norms and laws that shape and mediate their behaviour" (p.13).

The above observations are confirmed by a publication of the United Nations Environment Programme (UNEP [24]), which calls explicit attention to the causes and consequences of world-wide environmental changes, amongst others in the field of agriculture, fishery, industry, energy use, tourism, population, resources, settlement patterns, human health, and safety and peace. As a result of such a social science based view on global environmental change, research in this area should in particular focus on appropriate response and management strategies, such as increasing environmental awareness, changes in environmental perceptions and attitudes, (local, regional and national) policy responses, and international control strategies.

It is also evident from the above observations that the range of a social science orientation toward global environmental change is broad and comprises many disciplines: economics, geography, demography, planning, political science, management, psychology, sociology, history, law, philosophy, theology and ethics.

Given the global interactions of human behaviour including the concern about future generations (as advocated inter alia in the Brundtland Report), questions of social ethics in global environmental change are of paramount importance, but they provoke at the same time intriguing research questions for social scientists.

2.2 The social dilemma

Problems of global environmental change are not isolated from the way individuals, groups and governments take decisions in our society. A number of fundamental causes of these problems lie in the organisation and functioning of decision-making systems in this society.

In a fully operating market system, where all actors would incur all full costs of activities undertaken by them (including indirect consequences), the "hidden hand" (à la Adam Smith) would ensure the attainment of a balanced production and consumption system, where all external costs and benefits would be internalized via an appropriately functioning price system. In the absence of a market for (global) environmental goods and in the absence of a democratic system where also the interests of future generations or distant nations would be fully incorporated, a distorted outcome is likely to occur. Thus if the "hidden hand" fails, a biased result will emerge, where the actual decisions and behaviour are no longer driven by undistorted signals based on "real" prices for scarce commodities.

The above remarks can be clearly illustrated by an anecdotic story from Adam *Smith* in his "Theory of Moral Sentiments" [22]. Suppose a "man of humanity" who lives in Europe hears the sad news of a natural disaster in the Far East which has caused the death of thousands of people. This "man of humanity" would perhaps reflect for a while on questions of life and death, and he would certainly consider the consequences of this catastrophe for his business, but then he would continue with his normal work as though nothing had happened. If however, the same man would hear from his medical doctor that he might have to miss one of his fingers as a consequence of a small accident, he would have sleepless nights and might probably be mentally unable to continue his daily business life. Does that mean that the loss of a finger has a higher value than the tragic fate of thousands of human beings?

Apparently, there is no decision-making mechanism which forces us to incorporate in our behaviour socio-economic externalities to such an extent that the interests of other parties are fully respected and reflected. This dilemma is clearly illustrated by the economic historian *Heilbroner* [9] who in his book "An Inquiry into the Human Prospect" raises the provocative question: "What has posterity ever done for me?" The above remarks imply that new institutional and decision-making frameworks supporting the design and implementation of environmental benign strategies, insights and new attitudes, based on a deeper awareness of cultural and ethical dimensions of sustainability, have to be developed.

An additional severe complication is that global environmental change is not just a marginal distortion in an otherwise perfectly operating market system. The size and intensity of human activities leading to devastating impacts on the earth's physical, chemical and biological systems have dramatically increased in the past century, with an exponential growth rate in the past decades. By the turn of the century it seems plausible to assume that all areas on our planet are under severe environmental threat. This holds for both the geosphere and the biosphere. Despite many uncertainties, it is generally taken for granted that global environmental change is – via the intermediate greenhouse gases such as CO_2 and CFC– directly and indirectly connected with the functioning of our industrial and economic system, notably the use of energy and raw materials, production, consumption, demographic growth, and land use (*Arrow* and *Fisher*, 1974 [1]; *Opschoor*, 1987 [19]; *Zoeteman*, 1987 [25]; *Brouwer* and *Nijkamp*, 1989 [4]).

The essence of social science research questions in the area of global environmental change is the so-called "social dilemma". Awareness of the social costs of global change does not automatically generate a behavioural change compatible with the ecological paradigm, even though the consequences may be severe and may be felt by all individuals on earth. The main question is: which are appropriate incentives to induce a needed change and who is responsible for such incentives?

Various economists have tried to cope with the social dilemma by looking for correction mechanisms which would fully incorporate all (marginal) social costs of (global) environmental decay. The founding father of modern welfare theory, *Pigou* [21], has introduced the construct of the government as the "trustee for unborn generations". By introducing a proper tax system the market solution might again lead to a social optimum. This was necessary, as the recognition of the finiteness of life and the individual (rational but myopic) behaviour based on self interest would otherwise never allow for an incorporation of interests of other parties in the economic behaviour of actors.

In later debates on the trustee principle, it was recognized that for multi-generational trade-off problems this assertion would need some modification (*Solow*, 1974 [23]). By introducing the notion of a set of distinct generations to whom individuals belong in each period of their life, it seems plausible to assume that individuals will die (and hence base their decisions on short-run interests), but that generations will survive. By using then an *overlapping multi-generation* model it might – in theory – still be possible that the one generation cares about the other one, provided satisfactory and effective measures for coping with intergenerational externalities are developed and implemented. Thus if next individuals would respect their specific generational interest, a balanced or sustainable solution might still be possible. However, the validity of this multi-generation model has never been demonstrated, and it may thus be extremely risky to resort to this behavioural model for global sustainable development. On the other hand, rejection of this model as a realistic option would bring society back into the realm of the "trustee" principle with emphasis on governance by public decision-makers. However, this has the obvious disadvantage of the implicit but unproven hypothesis that the government is legitimized by individual myopic citizens to take a non-myopic view on sustainability issues. Nevertheless, the "trustee"-principle seems to be one of the few possibilities to solve the social dilemma caused by a discrepancy between egocentric behaviour in case of normal market transactions by individuals and altruistic behaviour in case of socio-political trade-off problems at a collective level. Thus the "trustee"-principle bridges this gap by taking for granted that selfish individuals without sufficient respect (or need) for a sustainable future elect a democratic government in the expectation that the government will take care of the failures in their own market behaviour.

It should be added that others claim that government interventions aiming to correct *market failures* may also embody *new significant failures*, as the government does often not have full information on complicated feedback behaviours of individual actors. Furthermore, it seems plausible that the government may have other objectives than the fulfilment of *collective* interest in society, so that the "trustee"-principle does not necessarily comply with actual governmental behaviour.

It turns out that global environmental change problems generate a complex array of social science questions. If there is no global environmental market, and if there is no institutional mechanism for imposing a system of Pigovian taxes on a world-wide scale, then the only economic possibility left is to define precisely property rights (in space and time) on the management and use of scarce environmental resources, like the ozone layer, tropical forests, oceans etc. (*Pearce* and Turner, 1990 [20]). Discussions at international fora such as the Rio conference have taught however, that such international property conditions are hard to specify, since there is no international environmental law. As a consequence, we see a tendency toward voluntary behavioural adjustments via protocols and international agreements (such as the Montreal protocol on a gradual ban on CFCs). Such recent developments suggest clearly that the problem of global environmental change is not so much caused by the "tragedy of the commons", but rather by the absence of well defined property, entry and management rules (*Hardin*, 1968 [8]).

Consequently, international coordination and management of global environmental goods (such as oceans, tropical forests, the stratosphere etc.) is a *sine qua non* for sustainable global development. Unfortunately, the field of international environmental law and global political-environmental science is as yet still in its infancy, so that the uncertainties on global environmental resource management are still formidable. Transaction costs theory, option theory and dynamic game theory might perhaps be meaningful theoretical tools, but the pace to actual sustainable environmental development may perhaps be unacceptably slow (*Giaoutzi* and Nijkamp, 1993 [7]; *Beltratti*, 1993 [2]).

In light of the great many obstacles in ensuring a sustainable development, many people resort to another *"deus ex machina"*, viz. *technological progress*. It is often believed that new technologies may alleviate many of the current global environmental problems. Competitive behaviour and user signals would be sufficient to cope with long-term environmental problems.

Despite the great appeal of the new technology option, there is historically no convincing evidence that the technology option will meet these aspirations. On the contrary, many current environmental problems are just caused by modern technologies. Thus it seems as though a risk assessment of new technologies does not necessarily provide sufficient guarantees.

Consequently, in reality we see a mix of various options, the trustee option, the regulatory option, and the technology option, but the optimal shares of these options are still largely unknown. Policy agenda formation on global environmental issues is hence still underdeveloped.

2.3 Contingency strategies

It is clear that research on such long-term changes needs a considerable research effort in the field of climatology and related disciplines. Significant progress has indeed recently been made in understanding such long-term processes. However, at the same time it is recognized that thus far the social sciences have not yet built up any research tradition in the area of global environmental change. Yet it should be noted that – as far as global environmental change is caused by anthropogenic factors – it is of extreme importance to investigate their driving forces as well as stimuli to change these forces. Even if the assumption of anthropogenic climate (and global environmental) change would not be fully accepted by the field, it is evident that social science research would be needed on ways of coping with the negative impacts of such global changes on the man-made and natural systems of our earth. There is thus a need for an intensified action for social science research on global environmental change.

A prominent focal point in global environmental change research would have to be uncertainty and risk assessment. It is often taken for granted that uncertainty is lower in natural science phenomena governed by fixed laws and regularities. This is reflected in advances made in the area of biogeochemical cycles, climate systems, stratospheric ozone destruction, acidification, changes in biodiversity and so forth. However, in recent years it is increasingly recognized that also in these fields there is an increasing uncertainty on causalities, time horizons and non-linear feedback effects to such an extent that natural sciences are sometimes facing the same degree of uncertainty as the humanities. This also explains the current popularity of *chaos* principles in scientific research. Incorporation of the "human factor" in global environmental change research increases even further the basic uncertainty in global change research, especially as far as its predictive capability is concerned.

In view of lack of operational insight into complex global change phenomena, many researchers have taken resort to scenario experiments in order to test on the computer the validity and sensitivity of various model assumptions which are not based on "hard facts". Such scenarios may generate useful learning insights into complex dynamic systems, but they can at best – due to lack of empirical verification – be seen as *flight simulators:* they serve to make decision-makers aware of sensitivities, critical points, flexibility in behaviour, systems constraints and basic uncertainties, but they cannot be regarded as econometrically or statistically validated models. This is clearly a weak point in the current modelling experiments at the interface of natural and social sciences.

In dealing with basic uncertainties, policy-makers have to face two types of *response failures.* First, an *overreaction to a temporary* trend which later on proves to be less harmful is extremely costly; this type of question plays a role in for instance, policy-making regarding sea level rise (*Nijkamp*, 1992 [15]). Secondly, *inert policy responses* to an ambiguous trend may lead to very serious consequences and disasters which may also incorporate excessive social costs (e.g., a volcanic eruption). Thus the balance between preventive and curative measures may be difficult to find. In current discussions on global change policies very often the principle of "no regret" strategies is propagated, which means that – despite many uncertainties – it may be necessary to develop alert response strategies, in particular if it can be shown that the implementation of such policies may be beneficial from other viewpoints than those considered at the outset in the basic uncertainty trade-off. For instance, even though the long-range global implications of CO_2 rise are not fully understood, it may be meaningful to economize on energy and resource use for direct financial benefits (*Nordhaus*, 1987 [16]; *Nordhaus*, 1991 [17]; *Nordhaus*, 1993 [18]).

It is evident that much research on contingency strategies is needed on the question *how to target problems that threaten us most.* For the time being, this question might perhaps be answered by applying a cost-effectiveness analysis. Such analysis would require an assessment of risks involved in the existence of each of the global change components, an assessment of all costs involved in coping with these problems and a specification of target levels for an ecologically sustainable economy. However, in view of the great many uncertainties characterizing global change processes such a cost-effectiveness is very hard to carry out, so that new research directions in this field will necessarily include subjective expert and ad hoc political opinions.

It is clear that the field of global environmental change research is fraught with many difficulties. This is also reflected in the existence of various international research and policy organizations, such as the Intergovernmental Panel on Climate Change (IPCC), the International Group of Funding Agencies for Global Change (IGFA), the World Climate Research Programme (WCRP), the International Geosphere-Biosphere Programme (IGBP), and the Human Dimensions of Global Environmental Change of the International Social Science Council (ISSC). All these organisations have worked on the formulation and implementation of research programmes in this field, sometimes even in a fragmented way. In most proposals however, the need for contingency studies is apparent. If we look at the environment and the climate as a "collective" good, then it is necessary to consider also ownership rules to be imposed as restrictive measures by public bodies. Such measures will demand a high degree of social acceptance and evaluation studies, while also "free rider" behaviour would be a central component. An additional problem may be formed by the fact that unforeseen changes in the external environment may also generate new attitudes, intentions and behaviour, which might sometimes be at odds with an environmentally benign policy. In this context strategic contingency studies may be a very helpful tool in social science research.

3 Concluding remarks

In general, the range of global change problems is so wide and diverse, that a uniform research strategy is not feasible for the time being. Nevertheless, it would be interesting to start with a set of carefully selected pilot and experimental studies in which knowledge from different research centres could be pooled. Interesting candidate areas might be delta-areas (in relation to sea level rise) or health effects (in relation to climate change). Such experimental studies could easily be undertaken as a collaborative effort of research institutes where already a strong tradition in the area of social science analysis of environmental problems exists.

It is increasingly recognized that environmental degradation and poverty cannot be sufficiently addressed by simple recovery measures; it requires proactive strategies of both governments and the business sector. It is sometimes tempting to resort to the development of new technology. However, it is recognized now that technological progress as such cannot be expected to generate all medicines against environmental decay (leave aside the point that technology may also be the source of much evil). Clearly, a number of fundamental causes of these problems lies in the organization and function of society itself, so that a thorough analysis of decision-making mechanisms is a *sine qua non*. In this framework, sufficient attention to decision risks in connection with basic uncertainties inherent in social science research in global environmental change is necessary. The initiatives taken by PIK seem to be very promising in this respect.

References

1. K. J. Arrow and A. C. Fisher. Environmental preservation, uncertainty, and irreversibility. *Quarterly Journal of Economics*, 88(3):312–319, 1974.
2. A. Beltratti. Climate change and option values. Technical report, Enrico Mattei Foundation, Milano, 1993.
3. F. Böttcher. Science and fiction of the greenhouse effect and carbon dioxide. *Change*, 13:3–6, April 1993.

4. F. M. Brouwer and P. Nijkamp. Modelling interactions between economic development and environmental change: A policy life-cycle interpretation. Dept. of Economics, Free University, Amsterdam, 1989. Research Memorandum.
5. W. C. Clark. *The Human Ecology of Global Change*. J.F. Kennedy School of Government, Harvard University, Cambridge, 1990.
6. W. C. Clark and R. E. Munn, editors. *Sustainable Development of the Biosphere*. Cambridge University, Cambridge, 1986.
7. M. Giaoutzi and P. Nijkamp. *Decision Support Models for Regional Sustainable Development*. Avebury, Aldershot, 1993.
8. J. Hardin. The tragedy of the commons. *Sciences*, 13(162):1243–1248, 1968.
9. R. L. Heilbroner. *An Inquiry into the Human Prospect*. Norton, New York, 1974.
10. H. K. Jacobson. Human dimensions of global environmental change. *Earthquest*, pages 1–2, 1992.
11. H. K. Jacobson and M. Price. *A Framework for Research on the Human Dimensions of Global Environmental Change*. International Social Science Council, Human Dimensions of Global Environmental Change, Paris, 1990.
12. R. W. Kates, W. C. Clark, V. Norberg-Bohm, and B. L. Turner III. *Human Sources of Global Change*. J. F. Kennedy School of Government, Harvard University, Cambridge, 1990.
13. R. B. Miller and H. K. Jacobson. Research on the human components of global change. *Global Environmental Change*, pages 170–182, 1992.
14. Ministerie van Onderwijs en Wetenschappen. *Wetenschapsbudget 1993*, Zoetermeer, 1993.
15. P. Nijkamp. Climate change, sea level rise and Dutch defense strategies. *Project Appraisal*, 6(3):143–148, 1992.
16. B. G. Nordhaus. Exploitation, conservation and preservation approaches to time preference. In G. Kirsch, P. Nijkamp, and K. Zimmermann, editors, *The Formulation of Time Preference in a Multidisciplinary Perspective*, pages 167–188. Avebury, Aldershot, 1987.
17. W. D. Nordhaus. To slow or not to slow: The economics of the greenhouse effect. *The Economic Journal*, 101:920–937, July 1991.
18. W. D. Nordhaus. Rolling the 'DICE': An optimal transition path for controlling greenhouse gases. *Resource and Energy Economics*, 15:27–50, 1993.
19. J. B. Opschoor. *Duurzaamheid en Verandering*. Inaugural address, Free University, Amsterdam, 1987.
20. D. Pearce and R. K. Turner. *Economics of Natural Resources and the Environment*. Harvester Wheatsheaf, New York, 1990.
21. A. C. Pigou. *The Economics of Welfare*. MacMillan, London, 1924.
22. A. Smith. *The Theory of Moral Sentiments*. A. Millar, London, 1759.
23. R. Solow. Intergenerational equity and exhaustible resources. In *Review of Economic Studies*, volume 41, pages 29–45, 1974. Symposium.
24. UNEP. *The World Environment 1972-1992*. Chapman & Hall, London, 1992.
25. B. C. J. Zoeteman. Soil pollution, vulnerability of soil and groundwater to pollutants. In W. van Duyvenboden and H. G. Waegeningh, editors, *Proceedings and Information*, volume 38, pages 17–27. TNO Committee on Hydrological Research, 1987.

Commentary: Surprises in the Climate Change Course

G. S. Golitsyn

1 Introduction

Climate was always changing, but the rates and amplitudes of the changes now seem to be somewhat larger than those, say, extracted from tree rings or paleorecords of various kinds. The last IPCC-5 report for the first time said that there is a discernible signal of the human influence on the climate changes during the last three decades. In the 20th century we have observed surprisingly large regional climatic changes of considerable duration such as Arctic warming in 1920-30, Sahel drought, and drastic changes in the level of the Caspian Sea: a dramatic fall in the 1930s and the continuous rise since 1978. These surprises cause large socio-economic problems and in the cases of the Sahel and the Caspian Sea require great geopolitical considerations because they involve many countries. We believe that the climate models should strive for prediction of such surprising changes and thus to serve their societal purpose. Here we briefly describe the last problem, that of the Caspian Sea and will try to outline a perspective of how the goal might be achieved.

2 The Caspian Sea case

The Caspian Sea is the world's greatest closed water reservoir occupying now just below 400 000 km^2. This is about the area of Germany and the Netherlands combined. Its main feeder is Volga river which supplies about 80% of the run-off to the Sea. The area of the water-shed of all rivers streaming to the Sea is about 2 000 000 km^2, or about 1.5% of the global land surface. The water balance of the Sea involves besides run-off the precipitation and evaporation from its water surface minus the outflow to the bay of Kara-Bogaz-Gol on its eastern part. The bay is connected with the Sea by a narrow passage and is below its level. The rate of the flow depends on the level of the Sea. Detailed data on the Sea level exist since 1837 and adequate meteorological observations since the mid-1920s.

Golitsyn and *Panin* [2] using the data were able to estimate evaporation from the Sea by standard methods and knowing from observations the other terms of the water balance equation including the outflow to Kara-Bogaz-Gol calculated the level. Annual mean values of the calculated level for 1930-1986 when compared with the observed levels revealed a good correlation ($r = 0.8$).

The decadal accumulated changes in the level were quite good: for 1930-1940 the observed fall was 170 cm and the calculated one 175 cm; for 1978-1986 the observed rise was 117 cm and the calculated one 103 cm. So one can be comfortable that major, if not all, parts of the level changes are due to climatological reasons (some contributions might be due to regional tectonic motions).

During the 19th and early 20th century the Caspian level fluctuated between −25 m and −26 m (minus means below the ocean level), then there was a sharp fall in the 1930s and then there was a slow fluctuating decline by another 1.3 m with a minimal level of −29 m in 1977. Since then it rose continuously until 1996. In 1995 the level was −26.6 m but for the first time after 18 years at the end of 1996 it was 15 – 20 cm below the level of the end of 1995.

In the 1970s the USSR Ministry of Water Management started to develop a project of diversion of some northern rivers' flow into Volga "to save the Caspian Sea". It was the greatest environmental battle in the Soviet Union. As a part of the operation of "saving the Sea" the Bay of Kara-Bogaz-Gol was dammed in 1980 in a fast almost secret manner without any open discussions. With the beginning of "perestroyka" and bad signs in the country economy the diversion was postponed indefinitely in 1986.

During the period after World War II when the Sea was receding people followed it constructing buildings, port facilities, roads, etc. All this is now either under water or greatly damaged by sea storms. In May 1994 UNEP organized in Moscow a workshop connected to the rise. The total damage in the five countries around the Sea – Iran, Turkmenistan, Kazakhstan, Russian Federation and Aserbaijan – was estimated then between 30 and 50 billion US$, about 50 thousand people have been relocated, and dams many tens of kilometers to protect oil excavation sites were constructed even during the Soviet time. During the next three years the damage only increased. Local authorities are crying for help and for the forecast of the future levels. The situation does show a great economic value of the regional climatic forecast. At the present a very crude forecast is based on the water balance equation and the assumption of preserving its components averaged for the last two decades, and accounting for the morphometry of the coastal area is that the increased evaporation from an enlarged water mirror should stop the level at about −25 m, i.e. another meter or so rise might occur during the next few decades (*Kazansky*, 1994 [3]).

In April 1992 Turkmenistan destroyed the dam and now the water is flowing again from the Caspian Sea to Kara-Bogaz-Gol. This now is taking 5 – 7 cm per year of the level rise. This should be compared to the mean 15cm per year of the 1978-1995 rise.

The situation is aggravated by the fact that there are now five independent states on the Sea shores. There is no legal status of the Sea, whether it is a lake or sea which implies quite different status for economic zones etc. The

Sea and its shores are rich with oil and gas; most of it is still unexploited so the future activities require again a forecast for the level.

3 Modelling activities

I would like to report how we are trying to approach the problem of the level forecast with the climate models. As a first step we have analyzed the results of the Atmospheric Model Intercomparison Project, AMIP. In this exercise about 30 climate models are involved. They use observed monthly mean ocean surface temperature for the decade 1979-1988 and reproduce the rest of the climate. In March 1994 the AMIP diagnostics subproject N 24 was approved named "The Caspian Sea Climate". *Golitsyn* et al. [1] reported in short the results of analyses of 15 model outputs for the value of P-E (precipitation minus evaporation) for the Volga watershed which was taken as the run-off to the Sea. P-E was also calculated for the Sea itself. The results have been compared with the actual course of the Sea level. Several models with good spatial resolution produced the right sign for the level change and some models, like MPI-Hamburg, MPO-Sankt-Petersburg, and few others gave the value of the rise not far from the observed one. More detailed consideration of 21 models can be found in *Meleshko* [4].

These results show that the climate of this part of the world is governed mostly by Sea Surface Temperature (SST). Of course, better parametrizations for the hydrological cycle are needed in the models. In the next step we plan to analyze the results of climate model projections for the next century with gradually increasing concentration of the greenhouse gases. In this way requiring some years of the work with several models we hope to obtain an ensemble averaged outlook for the future levels of the Sea.

4 Conclusions

We believe that the Caspian Sea rise is a very severe case of the regional climate change. Comprehensive climate models should strive to reproduce such changes involving large areas of our planet's surface. It is also a startling example of a case study for impacts on society of the sea level rise. The world ocean may rise by 30 – 80 cm by the middle of the 21st century, but here we have almost a 2.5 m rise in 18 years with all kinds of impacts on society and ecosystems.

References

1. G. S. Golitsyn and V. P. Meleshko et al. GCM simulation of water balance over Caspian Sea and its watershed. In *Proc. of First Intern. AMIP Scientific Conf.*, pages 113–118, Monterey, 1995. WCRP-92, WMO TD-732.
2. G. S. Golitsyn and G. N. Panin. Modern changes of the Caspian Sea level and its causes. *Soviet Meteorology and Hydrology*, (1):57–64, 1989.
3. A. B. Kazansky. Possible approach to the Caspian Sea level rise forecast. *Trans. (Doklady) Russ. AC. Sci./Earth Sci. Sect.*, 338(4):531–533, 1994.
4. V. P. Meleshko and G. S. Golitsyn et al. Calculation of the water balance components on the Caspian Sea watershed with ensemble of general atmospheric circulation models. *Izv.- Atmos. Oceanic Physics*, 1997. Accepted for publication.

Commentary: Sustainable Development – Teleology and Ambiguity

Volker Wenzel

Logic is the Moral of Thinking.
Moral is the Logic of Acting.

Jean Piaget

1 Introductory remarks

A particular merit of H.J. Schellnhuber's discourse is the development of a proto-analytic formalism and its introduction to the systematic foundation of earth system analysis as geocybernetics (*Schellnhuber*, 1998 [8]). An essential part of this is Sect. 4, in which the role to be played by the expression *sustainable development* within the new scientific discipline is specified more precisely.

The expression itself is subjected to a linguistic analysis and is introduced as a "lightly masked synonym for continuable progress". I believe, however, that progress represents a subjectively coloured interpretation of the results of development, which should be drawn into the calculus only later. Then the semantic analysis leads to a result differing in nuance, that applies particularly in the valuation of the role of the paradigm of standardization.

2 Semantic consideration

The term *development* depicts dynamic processes which may respectively be characterized through their initial and their final conditions. It is important that these two conditions differ from each other. If the beginning and the end are the same, time has indeed passed, but no development in the conditions of interest has taken place.

The scope of meaning takes in such heavy expressions as *genesis* and *evolution*, but also such everyday terms as *unwrapping* (which even has the same linguistic root as *developing* in German), and further *unfolding* or perhaps even *displaying*. In daily use of speech, something new, which previously did not really exist as such, is always coming out of something already on hand. The thus-described semantics of a word, which should not be confused with the ontology of the process described, stand at any rate for its reception as a purposeful, or in other words *teleologically* determined process. Something unfolds, which however can neither be defined nor created by a critical observer. The less this observer knows about the inherent dynamics of the process, the more surprising its result is likely to be to him.

So much for the thesis, the exclusively objective aspect or – closer to the vocabulary of earth system analysis – the natural dimension. The antithesis now comes into play within this given natural connection with its attribute *sustainable*. This is linked with the critical observer – with Schellnhuber's Global Subject, who, over and above his power to observe, is in a position also to influence processes. If these processes however are to remain in essence original, natural ones, then this influence is limited to creating the conditions which cause these unfolding processes to be broken off or impeded or, conversely, which cause blocks to be removed so that unfolding processes can run their course which would not otherwise have been able to do so. In particular, the Global Subject - under consideration of a more complex causal relationship – can allow conditions to unfold which guarantee that certain other unfolding processes can never again take place.

In order to be able to make precise assertions at all, *sustainable development (SD)* should thus be defined as an entirety (E) of (possibly cyclic) unfolding processes, which should be self-sustaining.

The dilemma of the Global Subject lies in his ambivalence as regards this entirety, in other words, his inability to define it clearly and to guarantee with certainty its existence through his dealings (or lack of dealings) with it. There are two fundamental reasons for this which will be elucidated in the following:

– the subjectivity of ideas about values and
– a "fatal" logical incompleteness of any potential method for description.

3 The role of values

The precise definition of E requires the choice and setting of priorities which must be based upon moral values, which unfortunately cannot be ascribed to a single or several maxims like a "categorical imperative". Moreover, the Global Subject here disintegrates into many individual subjects or groups of subjects, each with their own incomplete state of knowledge, contradictory value concepts, specific imbalances or even irrationality in judgement.

Only the pragmatic requirement of finding a *consensus* can counter this situation, which at a first glance seems apparently hopeless. Historically developed consent structures in cultural circles on various levels and in more-or-less spontaneously-formed interest groups do exist. However, over and above this there is a need for this to be directed, a need for stimulation and organization.

Just here lies the significance of Schellnhuber's essay, in which he seeks for structures which are on the one hand an expression and a result of historical consent-finding processes, and at the same time the instruments which purposively drive these processes. If these instruments can furthermore be mathematically formalized or adapted in conjunction with already existing methodology, then we are already on relatively firm ground in laying a scientific foundation for SD.

It is exactly in this sense that he speaks of teleological paradigms and introduces four basic principles for SD – *optimization, pessimization, equitization* and *stabilization* – which may be operationalized and further studied by means of mathematical theorems. On the other hand it is also immediately clear that this does not exhaust the whole field of problems. Precisely for that reason, *standardization* is given a role as a further paradigm. Essentially, however, we cannot speak of this paradigm in the same sense as the previous four, *teleological* paradigms. This commentary aims above all to illuminate more clearly both this boundary line between the teleological and the remaining paradigms and the substance and range of standardization.

Firstly a little more attention should be paid to the dimension which has developed historically of the four teleological paradigms. In a language suited to cultural theory, the term "teleological paradigm" should better be given as "aspiration". In contemporary cultural theory there is a nomenclature which distinguishes cultures by their respective world view (*Thompson*, 1990 [9]) which corresponds clearly to dominant aspirations. Prototypes of the most important cultures characterized in this way are the *individualistic*, the *fatalistic*, the *hierarchical* and the *egalitarian*. Thus, by the way, a link is made to J. Rotman's chapter in this book, in which to define scenarios for the study of SD with the aid of the TARGETS model, these aspirations are operationalized and allocated to particular regions (*Rotmans*, 1998 [7]).

It is surely not unjustifiably arbitrary if in an initial approximation we make the following classification:

World view	$\Longrightarrow$	**dominant aspiration**
individualistic	$\Longrightarrow$	optimization
fatalistic	$\Longrightarrow$	pessimization
egalitarian	$\Longrightarrow$	equitization
hierarchical	$\Longrightarrow$	stabilization.

4 Incompleteness

A sharper logical analysis reveals however a deeper reason for the difficulties with SD, which affects also the subjectiveness of values and the need to find a consensus.

Assertion 1:
No selection procedure can be found which is able to exhaust our entirety E completely.

For a proof, half the evidence has already been provided in the semantic analysis performed at the beginning. Concerning the rest one has only to notice that conditions and processes which lead to cascades of subsidiary conditions and processes are not simply analogous but are the logical equivalent of a predicate logic calculus system in which predicates are developed(!) or proved to be true out of premises through chains of logical conclusions, which in their turn form the premises for further chains of conclusions. For sufficiently complex systems of this sort – and the arithmetic of the figures is already complex enough – the principle of Gödel, with its following basic conclusion, applies.

> **Essence of Gödel's Theorem**: However comprehensive a subsystem (system of axioms) A of an entirety of predicates E is chosen, other predicates would be found belonging to the difference set E minus A, which cannot be derived from the premises belonging to the chosen subsystem A (*Gödel*, 1931 [4]).

For our pragmatically directed endeavours this means that the conditions for the processes that are indispensable, that are recognised as needing to be maintained, must be included in A step by step. These are stipulations for which in their entirety the paradigm of standardization is responsible. Individual settings are to be sure trivial acts, but the whole, indispensable paradigm nonetheless proves to be much more deeply founded.

5 Standardization paradigm

Scientific methods can also support the practical introduction of standards to a much greater extent than at first seemed possible. Standards only make sense when they have a knowledge basis and when this basis corresponds with the latest scientific findings; arbitrary setting of standards is seldom sufficient.

The most important scientific goal for this purpose consists of locating threshold values – a generalized form of natural constants, where quantity is changed into quality. Of course, computer-aided mathematical models are extremely helpful in these studies. A well known example is the carrying capacity of habitats, which in the equations (models) of mathematical ecology has a fixed size, in other words, has thus a sufficient scientific basis. An important discipline in the support of standardization lies before us in the broad field of dose-effect relationships, which are primarily researched by pharmaco-kinetics and toxico-kinetics. The methods for the study of stress-strain relationships are transferred directly to the scale level and the agencies of earth system analysis, where instead of MAC-*values* (maximal allowable concentrations), *critical loads* and *critical levels* are determined.

Since threshold values, although they may be determined with sufficient precision, cannot themselves be set as standards for reasons of practicality or cost, further risk analyses are carried out and other statistical methods applied whose aim is the valuation of tiered aspiration levels.

Standardization thus does not bear the character of elementary aspirations, but that of the residual quantity or complement of those. It consists of innumerable partial reachings of consensus, and sometimes of dictates or at least bridges to communication. It can only practically be carried through as an inductive-approximative process, although we know that the residual quantity is inexhaustible in this and any other way. Thus it is also clear that while *ambiguity* can with some effort be contained, it cannot as a matter of principle be removed. It remains virulent even in degrees, since assent and consensus can be withdrawn and auxiliary scientific disciplines can cease to have validity, be it through falsification (*Popper*, 1966 [6]) or through a new generation of scientists coming to the fore (*Kuhn*, 1967 [5]).

6 Concluding remarks

Due to this logical structure, the paradigm is predestined to have an especially close connection with the most up-to-date "number crunchers", i.e. the most powerful computers in terms of speed and memory. If, say, results such as the recognition that the *continuum hypothesis* is independent of the axiomatic system of arithmetic (*Cantor*, 1878 [2]; *Cohen*, 1964 [3]) are compared with the solution of the *four-colour problem* (*Appel* and *Haken*, 1986 [1]), it becomes clear that, aided by these computers, scientific progress today is no longer achieved through the depth of the ideas alone but ever more frequently through the extent of the calculations to be organized. This organization itself represents the content of a demanding scientific discipline with rather far-reaching methods. The accompanying phenomena which result from this and which are relevant to earth system analysis are an intensive observation and continually progressing measurement of the earth system, ever more exact mathematical models representing the structured accumulation of knowledge on the objects and subsystems of these, use of these procedural *vicarious* objects in numerical experiments or scenario analyses in grand style, the continuing posing and checking of hypotheses and conjectures, etc.

All this can and will be put to the service of the paradigm of standardization. And the assistance offered by the new possibilities for managing quantity is extremely effective. This should not seduce us into relying on the words of the poet "Wer vieles bringt, wird manchem etwas bringen" (Give much, and you'll have given to many minds). Because of the logical inexhaustibility, a considerable complement will always remain which is still not regulated by standards, for which moral values, readiness to consent and responsibility will play a so much greater role.

References

1. K. Appel and W. Haken. The four-colour proof suffices. *Mathem. Intell.*, 8(1):10–20, 1986.
2. G. Cantor. Ein Beitrag zur Mannigfaltigkeitslehre. *Journal für die reine und angewandte Mathematik*, 84:142–258, 1878.
3. P. J. Cohen. The independence of the continuum hypothesis. *Proceedings of the National Academy of Sciences*, 51, 1964. 105-110.
4. K. Gödel. Über formal unentscheidbare Sätze der Principia Mathematica und verwandter Systeme I. *Monatshefte für Mathematik und Physik*, 38:173–198, 1931.
5. T. S. Kuhn. *Struktur wissenschaftlicher Revolutionen.* Number 25 in stw. Suhrkamp Taschenbuch Wissenschaft, Frankfurt/Main, 1967.
6. R. K. Popper. *Logik der Forschung.* Mohr, Tübingen, 1966.
7. J. Rotmans. Global Change and Sustainable Development: Towards an Integrated Conceptual Model. In H.-J. Schellnhuber and V. Wenzel, editors, *Earth System Analysis: Integrating Science for Sustainability*, Berlin, 1998. Springer. Contribution in this Proceedings.
8. H.-J. Schellnhuber. Discourse: Earth System Analysis – The Scope of the Challenge. In H.-J. Schellnhuber and V. Wenzel, editors, *Earth System Analysis: Integrating Science for Sustainability*, Berlin, 1998. Springer. Contribution in this Proceedings.
9. M. Thompson, R. Ellis, and A. Wildavsky. *Cultural Theory. Political Cultures Series.* Westview Press, Boulder, 1990.

Part 2

Natural Dimensions

Chapter 1: Ecology and the Earth System

Iain Colin Prentice

Abstract Climate research today is ultimately driven by the need for a better understanding of the impacts of human activities on ecological, hydrological and socioeconomic systems. Theory and model development and testing are required in order to understand climate itself, and its impacts. The impacts disciplines however must progress rapidly if they are to meet the challenge. Ecosystem impacts models in particular need to be global in scope, in order to be physically consistent in their modelling of different regions, and if they are to predict feedbacks. Mounting evidence attests to the importance of feedbacks from ecosystems to climate, through biogeophysical interactions like surface albedo and conductance changes, and changes in biogeochemical cycles mediated by fluxes of CO_2 and other trace gases between ecosystems and the atmosphere. A major shift of focus in ecology from the micro- to the macro-scale is underway, spurred by the need for improved understanding of these linkages. Powerful tools are already at hand, including the routine use of numerical models at various levels of complexity, and the recognition of "fuzzy laws" that serve as globally applicable, robust elements of ecosystem models. Models are also moving rapidly towards integration across subsystems with different space and time scales. For example, biogeographic modelling has progressed from static formulations in terms of ecophysiological constraints to more explicitly mechanistic formulations including energy and water exchanges, nutrient cycling, resource competition, adaptive acclimation and allocation strategies, and vegetation dynamics. Increasingly complex models require ingenuity and rigour in testing. Again powerful tools exist, ranging from isotopic measurements on individual leaves, through CO_2 flux measurements on ecosystems, to remote atmospheric measurements of seasonal cycles, interannual variations and long-term trends in O_2, CO_2 and isotopic indicators of biospheric activity. Further challenges and possibilities for model validation are provided by palaeodata, if properly interpreted and analysed. Attempts to simulate equilibrium climate and biome distributions for past times, constrained by data that record shifting regional patterns in vegetation and hydrology, have produced surprising results that firmly establish the need to include feedbacks from climate-induced vegetation changes in simulations of future climates. The dynamics of the Earth system, including atmospheric trace gas composition, ocean circulation and ice-sheet dynamics during the past 150 000 years, present still greater challenges: we eventually must model the coupled Earth system, including non-linear interactions that may well generate complex transient behaviour in the future just as they have in the past.

1 Introduction

Climate impacts – effects of changes in the atmospheric environment on ecological, hydrological and socioeconomic systems – are central in the analysis of global change. These impacts generally will affect the human population to a far greater extent than climate change itself. Considerable effort is being expended trying to predict the consequences of different energy use and emissions scenarios on climate. This effort takes for granted that we can deduce the impacts of predicted climate changes on the systems that sustain human life, health and welfare. However, the problems of predicting impacts are at least as great as the problems of predicting climate changes, and the "impacts" disciplines generally have further to go in terms of theory and model development.

My specific concern here is with the impacts of climate change on terrestrial ecosystems. Terrestrial ecosystems may be natural, semi-natural (e.g. managed but naturally regenerating forests and rangelands) or artificial (e.g. planted forests and field crops). My research focuses on natural and semi-natural ecosystems (which even today occupy the major part of the global land surface) but has implications for all terrestrial ecosystems, which must operate within the same constraints of climatic extremes, resource availability and energetic efficiency. I wish to stress the need for fundamental ecosystem research of a theoretical and analytical nature, oriented towards the problem of understanding and predicting how ecosystems at large respond to climate change. Such research has to be global in scope, and interdisciplinary to an unprecedented degree.

An excellent illustration of the need for problem-oriented fundamental research on ecosystems is provided by the Framework Convention on Climate Change, which requires nations to take action on greenhouse gas emissions in order to avoid "dangerous" consequences for ecosystems. Yet no rational judgment on what might be dangerous is in fact possible, unless we have suitable quantitative models to predict the nature and magnitude of changes in ecosystem structure and function. Such models must satisfy tests of predictive power that go far beyond what one can ask of the "precise and general but unrealistic" (*sensu Levins*, 1966 [53]) models that dominate the literature of theoretical ecology. Ecosystem models for global change research must also satisfy criteria of physical consistency that are neglected in many of the less general, but more realistic, impact-oriented ecological models that abound in the literature of applied ecology, forestry and hydrology. Furthermore, the models we need will be genuinely global in scope, whereas most current impact models apply only to one region or one type of ecosystem: a restriction that seriously limits their credibility and usefulness.

2 Why global models?

There are several reasons to insist on the development of global models (as opposed to multiple, region- or biome-specific models) for ecosystems. One is almost trivial, i.e. that global changes potentially affect all ecosystems; any comprehensive assessment of impacts must therefore rely on global models. (If not, then at the very least a global suite of models is required. For consistency these must have a common structure, i.e. they must be formally identifiable special cases of a global model. Another reason is that climate change forces transitions between types of ecosystems that cannot easily be accommodated except within a model structure that embraces all ecosystems. Finally and decisively, a global treatment of the biosphere is unavoidable if we are to understand the interactions of the physicochemical and biotic parts of the earth system, since the atmosphere acts as a single entity dynamically coupled across the globe.

That biosphere-atmosphere interactions exist, and need to be considered in any attempts to understand long-term changes in the physics and chemistry of the atmosphere and oceans, is not only a tenet of the GAIA hypothesis. (Most writings on GAIA deal with mechanisms that operate on long geological time scales, commensurate with macroevolution and plate tectonics. We are not concerned with these longer time scales here. The GAIA hypothesis also emphasizes self-regulation, i.e. negative feedbacks. We are concerned with all feedbacks whether positive or negative.) Biospheric effects on the atmosphere (feedbacks) can be categorized as biogeophysical and biogeochemical. There is strong evidence for both these kinds of feedback (*Schimel* et al., 1991 [90]; *Melillo* et al., 1996 [58]), operating on annual to orbital time scales (*Milankovitch*, 1938 [60]).

Biogeophysical feedbacks occur because ecosystems are the main conduit for transfer of energy and water between the soil and the air above, and because key parameters governing these exchanges (such as surface albedo and canopy conductance), and their seasonal variations, vary with ecosystem structure. The likely importance of such feedbacks is indicated most persuasively by atmospheric general circulation model (AGCM) experiments in which the effects of external forcing (for example, orbital changes during the Holocene) have been shown to be amplified by large-scale changes in ecosystem structure that have been reconstructed from palaeoenvironmental proxy records (e.g. monsoon enhancement in the northern subtropics, *Street-Perrott* et al., 1990 [102], *Kutzbach* et al., 1996 [49]; warming in high latitudes, *Foley* et al., 1994 [26], *TEMPO*, 1996 [48]).

Biogeochemical feedbacks occur because terrestrial ecosystems annually exchange a substantial fraction of the total atmospheric inventory of CO_2 (and are the major source for other radiatively active trace gases such as CH_4 and N_2O) and because the fluxes of all of these chemical species between the atmosphere and biosphere are intimately tied to ecosystem function. Analyses of the global carbon budget based on different sets of assumptions have consistently shown an approximate balance overall, from the mid-1970s onwards, between CO_2 annually released to the atmosphere from deforestation and biomass burning, and carbon that must (by elimination of alternatives) be taken up by terrestrial ecosystems (*Taylor* and *Lloyd*, 1992 [103], *Houghton*, 1993 [40], *Moore* and *Braswell*, 1994 [62], *Friedlingstein* et al., 1995 [29], *Melillo* et al., 1996 [58]). Recent analyses based on a comparison of changes in atmospheric CO_2 and O_2 concentrations have confirmed this interpretation (*Battle* et al., 1996 [5]; *Keeling* et al., 1996 [43]). Sensitivity experiments with existing ecosystem models have further shown that climate-induced changes in ecosystem structure and function could considerably modify this source/sink balance, either in the direction of CO_2 uptake or release depending on the climate-change scenario and on the time scale considered (*King* and *Neilson*, 1992 [46]; *Smith* et al., 1992 [97]; *Smith* and *Shugart*, 1993 [98]; *Prentice* and *Sykes*, 1995 [76]; *Melillo* et al., 1996 [58]; *Kindermann* et al., 1996 [45]).

Ecosystem changes would also be expected to bring about major changes in the atmospheric burden of other radiatively active trace gases. For example, ice-core analyses have shown that atmospheric CH_4 changed very rapidly along with climate during the final stages of the last glaciation, and also varied significantly during the Holocene. These changes can plausibly be explained by climate-induced changes in the microbiological activity of the tropical wetland CH_4 source (*Chapellaz* et al., 1993 [12], *Street-Perrott*, 1994 [101]), and (during the later Holocene) by the buildup of northern wetlands (*Blunier* et al., 1995 [7]).

Feedbacks are logically distinct from impacts in the sense that many impacts of climate change on processes at the Earth's surface may not cause any discernible feedback to climate. However, a far-sighted approach to modelling the impacts of climate change on ecosystems much take account of the fact that one key application of ecosystem "impact models" in the future will be as components of integrated models of the physics, chemistry and ecology of the Earth's atmosphere, ocean and land surface.

3 Can ecology meet the challenge?

Applied ecologists are used to living with high expectations from society. They are already expected to have or to be actively developing a predictive understanding of highly complex issues, such as how to sustain yields of fisheries and forests, or how to conserve rare species in a fluctuating environment and in the face of landscape fragmentation. Such issues have been handled by the development of conceptual and numerical models that typically presume a stationary climate. Now ecologists are challenged to go even further, to predict how the state and function of the biosphere (including sustainable yields and biodiversity, as well as the physical and chemical characteristics that determine biosphere-climate feedbacks) may change in response to a time-varying atmospheric environment, and to do so in the context of the whole biosphere, not just on a local or regional basis.

Fortunately, many ecologists have recognized the intellectual challenge posed by the global change problematic, and indeed many have come to regard global change as the most important new direction for ecology. This is abundantly clear from the ambitious goals set by the Global Change and

Terrestrial Ecosystems (GCTE) core project of the International Geosphere-Biosphere Programme (IGBP) (*Steffen* et al., 1992 [100]), and from the extent to which global change problems now engage the attention of leading ecologists (see e.g. the discussion by *Allen* and *Hoekstra*, 1992 [1]). I emphasize "now", because until recently, the central approach in ecological research was extremely fine-scale in its choice of processes and patterns to study. Even now, there are ecologists who happily disregard the behaviour of ecological systems on spatial scales greater than a field plot or time scales longer than a research grant! Thus, the recent shift of focus towards global-scale issues reverses a strong reductionist trend.

The microscale approach, with its associated neglect of interactions between organisms and their physical environment, has been encouraged by a syndrome that seems to be far more widespread than "physics envy". The chief symptom is the delusion that good science must be based on direct experiment. I call this syndrome "experimental angst". If any envy is involved, it must be envy of the large research grants attracted by molecular biologists, with whom ecologists (at least in English-speaking countries) have been obliged to compete. Whatever the explanation, this belief has formerly been so pervasive in ecology that quantitative theory, indirect observations, and all but the simplest explanatory models were treated with scepticism by many ecologists. Crude and inaccurate models such as the *Holdridge* [39] life-zone system have persisted, bizarrely, because of a sort of "despairing empiricism" – large-scale systems have been perceived as unanalysable because of their complexity, so that simplistic empirical relationships are all that is left.

I believe that ecology is poised to meet the challenge of global change, but it is important to recognize the obstacles that still have to be faced. These obstacles are connected partly with common prejudices against interdisciplinary research (*Schneider*, 1992 [91]); and partly with the conventional position of ecology as a branch of the life sciences, where the obvious success of experimental approaches in other fields may have contributed to ecologists' angst. Ecology however is an earth science as much as a life science, and its successful pursuit requires interdisciplinary research spanning substantial areas of both. This perspective is at last becoming accepted, and underpins the rapid development of global ecology today.

4 The tools of global ecological modelling

Global ecology possesses several powerful tools that enable the building of large-scale process-based models. I emphasize process-based modelling, because purely empirical or statistical models based on contemporary observations cannot be expected to perform well outside the range of contemporary conditions, and do not advance understanding.

1. Explicit consideration of scale (spatial, temporal, and levels of organization). All scientists who make models of large physical systems are familiar with the simplifications that must be made due to the impossibility of combining full physical representations of subsystems with grossly incommensurate scales, either in space or in time. The application of this principle to ecology has been pointed out e.g. by *Levin* [52], *Allen* and *Starr* [2] and *Prentice* [70]. Explicit "scaling up" has been recognized as one of the keys to success in large-scale modelling (*Rosswall* et al., 1988 [83]). The key point that narrow reductionism misses is that by appropriate choices of scaling, a particular subset of processes can be effectively isolated from those at both coarser and finer scales, allowing a tractable model to be developed.
2. Use of mathematical and numerical models as normal tools of analysis, as in other theory-rich disciplines. An emphasis on modelling is essential because experimental approaches are impossible at the large spatial and temporal scales of interest. Explicit models are also required for scaling up findings from those scales at which experiments are possible (e.g. photosynthesis of leaves, nutrient uptake by plants, competitive relations within plant communities) to those larger scales at which experiments are not possible (see e.g. *Woodward*, 1990 [114]).
3. Identification of fuzzy laws governing ecosystem processes. A fuzzy law is one that holds approximately, most of the time. There is still debate about whether ecology, or indeed biology, can have laws (other than the underlying laws of chemistry and physics). This debate seems academic. In practice, ecological models cannot be constructed without invoking processes that are contingent on past evolutionary history. For example, carbon flux models routinely invoke the biochemical characteristics of the enzymes and light-harvesting structures involved in photosynthesis (*Farquhar*, 1989 [23]). These have remained essentially constant through higher plant evolution yet (a) it is not clear that the particular system that evolved is the only one that could have done so, and (b) variations in e.g. leaf structure mean that realized values of fundamental quantities (such as the quantum yield of photosynthesis in the absence of molecular oxygen) cannot be treated as true physical constants. Other good examples of fuzzy laws include the 3/2 thinning exponent governing biomass-density relationships in plant communities (*White*, 1981 [110], *Westoby*, 1984 [109], *Osawa* and *Sugita*, 1989 [67]), the negative exponential relationship between thermal time-to-budburst and the length of the antecedent chilling period in temperate woody plants (*Murray* et al., 1989 [63]), and the constant-stress relationship determining the shapes of tree trunks (*Cannell* and *Dewar*, 1994 [11]). The fact that any one of these "laws" would be demolished by applying a strict falsification criterion merely demonstrates the inappropriateness of that criterion in practice.

5 Patterns in the biosphere

Global regularities in vegetation structure reflect climatic patterns and transcend the differences in evolutionary history among biogeographic provinces. These patterns have been known since pre-Darwinian times. More recent research by the German school of plant ecophysiology, above all, led to major advances in understanding the causes of these regularities (see e.g. *Walter*, 1979 [106]; *Schulze*, 1982 [92]). However, the incorporation of this understanding into explicit predictive models has to wait for the development of fast computers, climate station networks, and the recent revival of interest in global ecological patterns (*Box*, 1981 [10]; *Woodward*, 1987 [113]; *Prentice* et al., 1992 [71]; *Neilson*, 1995 [65]; *Neilson* and *Marks*, 1995 [66]; *Haxeltine* et al., 1996 [35]; *Haxeltine* and *Prentice*, 1996 [33]).

The BIOME model of *Prentice* et al. [71] is a well-known example of a predictive model for global patterns in vegetation, combining elements of the *Woodward* [113] and *Box* [10] models. Biomes are constructed as combinations of plant functional types (PFTs, *Steffen* et al., 1992 [100]). The number of plant types recognized is the minimum needed to distinguish a given set of biomes. The basic classification of vegetation types into biomes and plants into PFTs is uncontroversial and represents the lowest common denominator of various classifications. Limiting values to each PFT for just three principal bioclimatic variables, mean temperature of the coldest month (Tc), growing degree days (GDD) over a base temperature of 5 °C, and a moisture index (the ratio of actual to equilibrium evapotranspiration), computed using a simple one-layer soil hydrology. Minimum values of Tc were inferred from data summarizing experiments on cold tolerance, which have shown a correspondence between cold tolerance and PFT (*Woodward*, 1987 [113]). The other limiting values were fitted to actual limits as inferred from vegetation maps. Limiting values were not assigned in the absence of a known or plausible underlying process. Competition between PFTs is only invoked in the trivial sense of a dominance relation, such that (for example) trees > shrubs > grasses and tropical trees > temperate trees where ">" means "always outcompete".

Application of this model to the global scale shows that even such a simple formulation as BIOME has a surprisingly good predictive power. Three dimensions of bioclimate can generate realistic nuances, such as the different zonation on tropical and temperate mountains, or the disjunct distribution of cold deciduous forests (lacking either boreal conifers or "temperate" deciduous genera such as *Quercus*) in hypercontinental, hyperoceanic, and semi-arid climates. The zonation difference is due to the way in which the annual range of temperature affects the relation between Tc and GDD, allowing broadleaved evergreens to ascend to treeline in the less seasonal environments of the tropics (*Prentice* and *Jolly*, [73]). The disjunct distribution of cold deciduous forests is due to the wide tolerance of boreal summergreen trees (relative to boreal evergreen conifers) with respect to moisture and chilling deficits as well as extreme low temperatures.

BIOME however has limitations, and sensitivity tests have shown that some of these are clearly due to problems of the model structure, not just problems with values of coefficients (limiting values for PFTs, or parameters in the hydrology submodel). The most obvious limitation of BIOME is its inability to distinguish savannas from woodlands. Tropical savannas and dry forests are lumped because they cannot be distinguished by critical values of any of the bioclimatic variables calculated by the model. Simulated temperate evergreen and deciduous forests extend too far into natural savannas or grasslands, e.g. in the Pampas of South America and the "prairie peninsula" of North America. In these summer-rainfall dominated areas, grasses compete effectively with trees. The competitive advantage of grasses in such environments is enhanced by clay soils, because of the slow downward percolation of water in clay, and reduced on sandy soils (*Walter*, 1979 [106], *Walker* and *Noy-Meir*, 1982 [105]). This is opposite to the texture effect predicted by BIOME or any other model relying on a single "bucket" representation of soil water dynamics. A process description of tree-grass competition incorporating such effects had to wait for the development of more explicitly process-based models.

6 Primary production and optimality

Patterns in global net primary production (NPP) were first charted as a result of the International Biological Programme (IBP), and were initially summarized as empirical global relationships with mean annual precipitation and temperature (*Lieth*, 1975 [54]). Such relationships suffer from the usual limitations of purely statistical models. As in the case of biome models, physiological process-based NPP models were slow to appear during the years following IBP, but there has been an explosion of activity during the last few years (see e.g. *Running* and *Coughlan*, 1988 [86]; *Bonan*, 1991 [8]; *Rastetter* et al., 1991 [81]; *Raich* et al., 1992 [80]; *McGuire* et al., 1992 [57]; *Potter* et al., 1993 [69]; *Kindermann* et al., 1993 [44]; *Kindermann* et al., 1996 [45]; *Foley*, 1994, 1995 [26, 27]; *Ruimy* et al., 1994 [85]; *Warnant* et al., 1994 [107]; *Woodward* et al., 1994 [115]; *Ruimy* et al., 1996 [84]). Today there are at least 30 such models, and a major intercomparison has been conducted at PIK (*Fischer*, 1997 [25]) to explore the degree of consensus, or otherwise, between their results.

Models that simulate NPP have become known as "terrestrial biogeochemical models" (TBMs) because the process representation requires the simulation not only of carbon fluxes but also water (and often nitrogen) fluxes and the coupling between these fluxes. Global assessments of CO_2 and climate change effects on NPP have been presented (*Melillo* et al., 1993 [59]), and others will surely follow.

Some TBMs are "diagnostic" models, which calculate global NPP on a monthly or annual basis using remotely sensed vegetation indices (related to foliage cover) as a key driving variable (e.g. *Ruimy* et al., 1994, 1996 [85, 84], *Potter* et al., 1993 [69], *Knorr* and *Heimann*, 1995 [47]). Such models will presumably evolve into "prognostic" formulations in which spatial and temporal foliage cover variations are predicted rather than prescribed. Then the remotely sensed data can be reserved for validation, and future scenarios can be explored. Already some diagnostic models can run in prognostic mode and vice versa. A further distinction among TBMs is between models where a set of parameters is estimated separately (mostly from IBP or more recent field studies) for each biome, and models in which only a small set of known, intrinsic characteristics (such as life form and phenological type) are prescribed for each PFT. The evolution will presumably be from the former, data-demanding type to the latter, more theoretically-based type in which the field measurements can be used for validation rather than calibration.

The one major obstacle to these developments is that in order for a model to do without both remotely-sensed inputs and biome-specific parameters, the model itself has to predict biome distributions: i.e. it has to be more than just a biogeochemistry model. The standard approach in assessments of climate change impacts on ecosystem function (including NPP circumvents this obstacle by means of a unidirectional coupling: AGCM scenarios drive changes in biome distributions, which in turn modify the response of TBMs to the change in atmopheric composition and climate (VEMAP Members, 1995 [104]). A better approach however is for the biome and biogeochemical modelling components to be interactively coupled, so that biome distributions both influence and are influenced by the biogeochemical fluxes. This type of coupling has been implemented in the BIOME2 model of *Haxeltine* et al. [35] and the BIOME3 model of *Haxeltine* and *Prentice* [33]. Some principles and findings that have emerged during the development of these models are summarized below, grouped under two key hypotheses.

1. A generally applicable relationship between absorbed photosynthetically active radiation (PAR) and gross and net carbon assimilation by foliage can be derived theoretically, assuming an optimal distribution of photosynthetic capacity through vegetation canopies (*Farquhar*, 1989 [23], *Field*, 1991 [24], *Sellers* et al., 1992 [93], *Haxeltine* and *Prentice*, 1997 [34]). This approach provides a theoretical underpinning for the empirical linear relationships typically used in diagnostic TBMs. It is an example of explicit scaling up from the leaf to the canopy scale that results in a major simplification of the mathematics required to simulate biosphere carbon uptake.
2. Various formulations of the relationship between sustainable leaf area and environment (*Specht*, 1972 [99], *Eagleson* and *Segarra*, 1985 [18], *Woodward*, 1987 [113], *Neilson*, 1995 [65], *Arris* and *Eagleson*, 1994 [3]) collapse to the single hypothesis that natural vegetation optimizes total NPP, subject to constraints of water, carbon and nitrogen availability (*Haxeltine* et al., 1996 [33]). The water availability constraint dominates

in semi-arid to arid environments in the tropics while the carbon and nitrogen constraints dominate in high latitudes. This approach can be used to simulate global patterns in foliage cover, yielding results that compare well with remotely sensed estimates. The same principle can be carried further, to simulate the competitive balance between different co-existing PFTs. The balance of trees and grasses can be simulated in this way provided that a two-layer soil hydrology model is employed.

These two hypotheses are of a different kind from the constraints underlying the original BIOME model. Those constraints correspond to evolutionarily conservative properties of plants, conferring tolerance or requirements for environments within a certain range. A classic example is the lowest temperature tolerated by the overwintering buds of many temperate summergreen trees, which approaches the anucleate freezing point of water (-40 °C). This is the limit to which liquid water can be supercooled in living cells. Plants that can withstand lower temperatures have to have an additional tolerance mechanism, such as extraorgan freezing in boreal conifers (*Sakai*, 1979 [89]). Another, more subtle example is the temperature of ca 22 °C above which C_4 photosynthesis has a higher quantum efficiency than C_3 plants. This limit is determined by the temperature-dependent kinetics of photosynthetic enzymes at present atmospheric O_2 and CO_2 concentrations (*Ehleringer* and *Björkman*, 1977 [19]; *Long* and *Hutchin*, 1991 [55]). It appears in BIOME as the mean warmest-month temperature below which "warm" (primarily C_4) grasses and shrubs are not a major component of the vegetation. The change to be expected in this limit as a function of changing CO_2 levels can be calculated on the assumption that ratios of internal to ambient CO_2 are conserved (*Ehleringer* and *Monson*, 1993 [20]).

Optimality hypotheses in contrast rely on the adaptability of ecosystem properties. Adaptation occurs through a variety of processes – physiological mechanisms within individual plants, microevolution within populations, adjustments of species composition within communities. (The fact that these processes have very different time constants can be ignored so long as the focus is on simulating equilibrium conditions.) Optimality hypotheses in models are a short-cut, whose plausibility must be assessed from independent evidence. Hypothesis (1) above is in fact strongly supported by long-standing observations on sun and shade plants and sun- and shade-adaptation within the canopy of individual plants, and by studies showing near-optimal distributions of photosynthetic capacity and nitrogen content within canopies (e.g. *Hirose* and *Werger*, 1987 [38], *Evans*, 1989 [21]). Its prediction of a linear production-PAR relationship also has strong, independent empirical support as reviewed e.g. by *Russell* et al. [88] and *Potter* et al. [69].

Hypothesis (2) is more speculative. It seems plausible that PFTs, or combinations of PFTs, with low productivity will tend to be ousted in competition with PFTs with high productivity. The scale at which the hypothesis is applied is that of the grid cell of global ecosystem models, i.e. in the order of 0.5° × 0.5°. Grid cells are large enough to include a mosaic of patches in different stages of succession (i.e. different times since "disturbance" due to natural or human agencies). NPP in many natural ecosystems undergoing succession increases to a plateau, followed eventually by a decline (*Peet*, 1992 [68]). Because disturbance probabilities generally increase with time (*Clark*, 1989 [14]), patches in the phase of declining NPP will soon be returned to the phase of increasing NPP. Such a mechanism may keep NPP in natural ecosystems within its optimal (plateau) range. These speculative arguments support a fuzzy version of hypothesis (2). More persuasively, the hypothesis when implemented in models seems to be able to account for key aspects of global vegetation patterns including foliage cover and duration and the distribution of savannas of different kinds.

7 Equilibrium and transient responses

Models incorporating a combination of evolutionary constraints and ecological optimization hypotheses can be considered analogous to equilibrium models in economics, with similar uses and limitations. The time constants of successional processes however can be in the order of hundreds of years, so any consideration of global change during the next centuries has to take account of transient effects arising through lags in the adaptation of vegetation composition to climate change (*Prentice* et al., 1992 [74]).

Simulations with biome models, representing equilibrium conditions, cannot be taken literally as predictions of future vegetation states. Simulations with biogeographic models can however be construed as predictions of potential vegetation in a general sense. For example, changes in the simulated distributions of different types of tree (based on constraints on where they can grow and regenerate) have direct implications for the kind of forestry that can be successful, even if natural transitions between forest types are expected to take centuries. Similarly, changes between biomes in semi-arid regions have direct implications for the type of land-use that can be supported without irrigation. Simulated competitive balances apply to the potential natural vegetation, and are more readily modified by human activities.

GCTE's most ambitious project, the development of a dynamic global vegetation model (DGVM), calls for mechanistic simulation of succession and disturbance processes and the way in which these processes would be affected by climate change and land use (*Steffen* et al., 1992 [100]). Transient simulation with a DGVM could be used to investigate the problem of how the terrestrial biospheric carbon sink may evolve as atmospheric composition and climate continue to change according to various scenarios. The development of models like BIOME3 has provided many of the building blocks necessary to construct a DGVM. The optimality hypothesis (2) is replaced in DGVMs by an explicit simulation of dynamic equilibrium between ecosystem carbon uptake and losses including respiration, mortality and disturbance. The equilibrium condition can then be relaxed and the model run in transient mode. This principle underlies the current development of DGVMs (e.g. *Foley* et al., 1996 [28]). The main obstacle to this development until recently has been the problem of devising a representation of vegetation dynamics that is appropriate for the global scale, while capturing the essential dynamic properties that can be simulated by region- or biome-specific ecological succession models (*Shugart*, 1984 [94];*Prentice* et al., 1989 [78]; *Shugart* and *Prentice*, 1992 [95]; *Prentice* et al., 1992 [74]). Current approaches to this problem range in complexity from "big leaf" formulations with just two canopy layers, to multiple replicates of simplified "gap models" based on simulating the interactions of individual plants. At the time of writing there are four or five functioning DGVMs, each with a distinct parametrization for vegetation dynamics, and a study comparing their behaviour under present and simulated future conditions is already underway at PIK.

8 The validation problem

Designing strong tests of terrestrial biosphere models is a challenge to ecologists' ingenuity. Some tests are reasonably obvious, calling on conventional ecological data sets. For example, TBMs can be tested at specific points against the available field measurements of NPP (*Melillo* et al., 1993 [59]), and against soil carbon measurements on some spatially averaged basis (*Foley*, 1995 [27]). An interdisciplinary perspective on the validation problem however allows more (and more powerful) tests, for example:

- Spectral reflectance data can be transformed to estimates of foliage cover (e.g. *Choudhury*, 1987 [13]), including its seasonal and interannual variation, on a geographic grid. These estimates provide a direct, spatially extensive test of simulated leaf area index or foliage projective cover and their phenological responses (*Sitch* et al., 1997 [96])
- CO_2-flux measurement series over land (e.g. *Fan* et al., 1990 [22], *Wofsy* et al., 1993 [112], *Grace* et al., 1996 [32]) will test key components of carbon flux models (e.g. dark respiration, quantum yield, photosynthetic capacity, stomatal closure responses) more thoroughly and precisely than aggregate annual NPP measurements.
- Zonally averaged net CO_2 fluxes should be consistent with seasonal and interannual variations of CO_2 at oceanic measurement stations. This comparison can be made rigorously by treating the terrestrial fluxes as input to an atmospheric transport model (*Fung* et al., 1987 [30]; *Heimann* and *Keeling*, 1989 [37]; *Hunt* et al., 1996 [41]; *Heimann* et al., 1996 [36]; *Sitch* et al., 1997 [96])
- The theory of isotope discrimination in photosynthesis (*Farquhar* et al., 1989 [23]), combined with simulated CO_2 and water fluxes and predictions

of the spatial and seasonal distribution of C_3 and C_4 photosynthesis, should analogously allow prognostic simulation of the observed zonal pattern and seasonal and interannual variations of ^{13}C in atmospheric CO_2.

These data sets all test relatively fast processes of the type simulated by TBMs. The possibilities for validating simulated biome distributions with contemporary observations are more limited. The best that can be hoped for is a biome map based on remote sensing (*Running* et al., 1994 [87]), which will presumably be more reliable than the existing maps. Still this biome map will represent just one state of the biosphere, deduced from one or a few years' observations. To test biome simulations under a changed climate it is necessary to look backwards, to earlier times during or before the Holocene when global climate boundary conditions and vegetation patterns were substantially different from today.

9 The uses of palaeodata

The broad outlines of climate change since the last glacial maximum (LGM), ca. 21 000 yr before present (18 000 yr B.P. in "^{14}C years"), have been established by AGCM simulations and comparisons with ^{14}C-dated palaeoenvironmental proxy data recording past vegetation types, lake levels and other indicators (COHMAP Members, 1988 [15]; *Wright* et al., 1993 [116]). Spatially extensive data of this kind are useful in testing equilibrium simulations. The equilibrium assumption is fully appropriate when comparing, for example, biosphere patterns at 6000 yr B.P. or the last glacial maximum (LGM), ca 21 000 *yr* B.P., with the present. Both times represent periods of relative stability under global boundary conditions unlike either present or future (high-CO_2) conditions.

One AGCM simulation forced by the known LGM boundary conditions of low CO_2, extensive continental ice sheets and generally low sea-surface temperatures was used to generate a simulated biome distribution (*Lautenschlager* and *Herterich*, 1990 [51]; *Prentice* et al., 1993 [77]). The simulation satisfied a weak quantitative test: it implied a reduction in total terrestrial carbon storage similar to that inferred from the slightly negative ^{13}C of the mean LGM ocean. Spatial vegetation patterns inferred from pollen data should provide a stronger test. The problem is that existing vegetation maps of the LGM have been constructed subjectively, used different and heterogeneous data sources, and massively disagree with one another. In addition, one much-cited reconstruction implies a far larger reduction in carbon storage than the carbon isotope data allow, even after several possible mechanisms affecting the isotopically based estimate have been taken into consideration (see e.g. *Bird* et al., 1994 [6]; *Crowley*, 1995 [17]).

A similar problem of data availability exists for 6000 *yr* B.P., although the data are more extensive. 6000 *yr* B.P. is the focus of a model experiment proposed as part of the current IGBP Global Analysis, Interpretation and Modelling (GAIM) action plan (GAIM Task Force and IGBP-DIS Strategic Task Force, 1994 [42]). The standard unidirectional coupling procedure is to be contrasted with coupled AGCM-biome model simulations, both with 6000 yr B.P. orbital forcing. The analogy often made between 6000 yr B.P. and future climates is false (*Webb* and *Wigley*, 1985 [108]; *Gallimore* and *Kutzbach*, 1989 [31]; *Mitchell*, 1990 [61]; *Crowley*, 1993 [16]; *Rind*, 1993 [82]); nevertheless 6000 yr B.P. is a convenient period for modelling, because it can be treated as a pure orbital forcing experiment (*Kutzbach* and *Guetter*, 1986 [50]). The Palaeoclimate Model Intercomparison Project (PMIP), sponsored by the World Climate Research Programme (WCRP) as well as by IGBP, has adopted both 6000 yr B.P. and the LGM as key times for climate simulation.

Preliminary results suggest that the unidirectional coupling approach underestimates the extent of warming in the high latitudes and monsoon expansion in the tropics at 6000 yr B.P. and that the shortfall can be at least partly remedied by the use of bidirectional coupling between atmospheric and biospheric models. But the data-model comparison needs to be carried out on a rigorous basis and at a global scale. IGBP is accordingly sponsoring the "BIOME 6000" data synthesis project, under the combined aegis of GAIM, the IGBP Data and Information System (IGBP-DIS) and the GCTE and Past Global Changes (PAGES) core projects (*Prentice* and *Webb*, 1994 [79]). The aim is to construct global data sets for 6000 yr and LGM using the best available pollen and plant macrofossil data, using a method inspired by fuzzy logic (*Prentice* et al., 1996 [72]) to infer biomes. The project explicitly focuses on the plant functional type as the entity allowing translation of data recording regional floras into information directly comparable with biome model output. The development of the 6000 yr B.P. global data set is well under way and is already being used as a benchmark with which to evaluate palaeoclimate model simulations within PMIP and other international projects.

10 Conclusions

One could argue that making models of the terrestrial biosphere is too easy, if the validation effort is confined to reproducing today's distribution of biomes and the few available NPP measurements. The physics-based validation approaches mentioned above can provide more stringent tests of TBMs, and will probably become routine in future. For biome models or coupled climate-biosphere models, however, palaeodata are indispensable. The problems of data access (there is no one central pollen data base, many data are not in any data base, publications are scattered, and regional expert knowledge is needed to interpret the data) are being addressed through co-operation, to the benefit of palaeoecology and global change research in general.

Data from the past also provide new challenges. A focus on relatively stable and well-understood periods like 6000 yr B.P. and the LGM is a way to get the most out of spatially extensive data, given the limitations of dating precision and the still very limited understanding of the causes of climate change on shorter time scales. The multidisciplinary proxy record of the last 21 000 years however has produced many surprises (*Lorius* and *Oeschger*, 1994 [56]). We do not yet have well-established explanations for phenomena such as the postglacial CO_2 increase (*Barnola* et al., 1987 [4]), the glacial Dansgaard-Oeschger events in the North Atlantic region and the longer cycles in which they are embedded (*Bond* et al., 1993 [9]), or the apparent variations of CO_2 during the early Holocene (*Neftel* et al., 1988 [64], *White* et al., 1994 [111]). It seems inescapable that these phenomena represent the outcome of non-linear interactions among ice sheets, ocean, atmosphere and biosphere. To model such interactions is not going to be easy – but it will be necessary, if we are to claim understanding of the carbon cycle and its behaviour during times of rapid global change.

Discussions of environmental change are often polarized by two, more-or-less equally anthropocentric viewpoints. Technological optimism posits a vast human capacity to remedy adverse biospheric effects. Environmental pessimism posits an equally vast human capacity to destroy the biosphere's function. The extraordinary magnitude and rapidity of changes that actually occurred in the ocean-atmosphere system during the last deglaciation, with drastic consequences for the land biota, should be a corrective to blind faith in the possibilities for technological remediation of anthropogenic climate change effects. Yet the biosphere and enormous numbers of species have evidently survived these remarkable climatic events, just as they survived multiple major oscillations in climate and atmospheric chemistry during over a million years of Quaternary glacial-interglacial cycles. Presumably the biosphere possesses resilience in the face of climate change, but the mechanisms that confer resilience may when tested prove to have costly or disruptive side-effects for human society.

Unfortunately no general rule can be adduced as to whether the biosphere's response to a given perturbation will be dominated by positive or negative feedbacks, or whether adverse effects on ecological systems will be exacerbated or alleviated by climate-biosphere interactions (see e.g. the discussion by *Prentice* and *Sarnthein*, 1993 [75]). There is thus no substitute for careful scientific analysis, including the development of models that can trade off negative and positive feedbacks in a quantitative manner. Data from the past provide one key source of data for testing such models. However, we also have a great deal of empirical information about the past which (at least) tells us something about what can happen. It tells us that the earth system is both more interesting, and harder to predict, than we would sometimes like to believe.

11 Acknowledgments

This article is based on a talk given at PIK as part of the symposium "Earth System Analysis: Integrating Science for Sustainability". I thank Professor H.-J. Schellnhuber for inviting me to participate in this symposium. The research summarized derives from the Global Systems Ecological Modelling Study at Lund University, an IGBP-GCTE core research activity that has been supported by contracts and grants from a variety of agencies including the Swedish Natural Science Research Council (NFR), the Swedish Environmental Protection Agency (Naturvårdsverket), the European Union Environment programmes, the US Electrical Power Research Institute (EPRI), the US Environmental Protection Agency (EPA) and the US National Science Foundation (NSF).

References

1. T. F .H. Allen and T. W. Hoekstra. *Toward a Unified Ecology.* Columbia University, New York, 1993.
2. T. F. H. Allen and T. B. Starr. *Hierarchy: Perspectives for Ecological Complexity.* University of Chicago, Chicago, 1982.
3. L. L. Arris and P. S. Eagleson. A water use model for locating the boreal/deciduous ecotone in eastern North America. *Water Resources Research,* 30:1–9, 1994.
4. J. M. Barnola, D. Raynaud, Y. S. Korotkevich, and C. Lorius. Vostok ice core provides 160,000-year record of atmospheric CO_2. *Nature,* 329:408–414, 1987.
5. M. Battle, M. Bender, T. Sowers, P. P. Tand, J. H. Butler, J. W. Elkins, J. T. Ellis, T. Conway, N. Zhang, and A. D. Clarke. Atmospheric gas concentrations over the past century measured in air from firn at the South Pole. *Nature,* 383:231–234, 1996.
6. M. I. Bird, J. Lloyd, and G. D. Farquhar. Terrestrial carbon storage at the LGM. *Nature,* 371:585, 1994.
7. T. Blunier, J. Chapellaz, J. Schwander, B. Stauffer, and D. Raynaud. Variation in atmospheric methane concentration during the Holocene epoch. *Nature,* 374:46–49, 1995.
8. G. B. Bonan. Atmosphere-biosphere exchange of carbon dioxide in boreal forests. *Journal of Geophysical Research,* 96:7301–7312, 1991.
9. G. Bond, W. Broecker, S. Johnsen, J. Mauns, L. Labeyrie, J. Jouzel, and G. Bonani. Correlations between climate records from north Atlantic sediments and Greenland ice. *Nature,* 365:143–147, 1993.
10. E. O. Box. *Macroclimate and Plant Forms: An Introduction to Predictive Modelling in Phytogeography.* Junk, The Hague, 1981.
11. M. G. R. Cannell and R. C. Dewar. Carbon allocation in trees: a review of concepts for modelling. *Advances in Ecological Research,* 25:60–104, 1994.
12. J. Chapellaz, T. Blunier, D. Raynaud, J. M. Barnola, J. Schwander, and B. Stauffer. Synchronous changes in atmospheric methane and Greenland climate between 40 and 8 kyr BP. *Nature,* 366:443–445, 1993.

13. B. J. Choudhury. Relationships between vegetation indices, radiation absorption, and net photosynthesis evaluated by a sensitivity analysis. *Remote Sensing of Environment*, 22:209–233, 1987.
14. J. S. Clark. Ecological disturbance as a renewal process: theory and application to fire history. *Oikos*, 56:17–30, 1989.
15. COHMAP. Climatic changes of the last 18,000 years: observations and model simulations. *Science*, 241:1043–1052, 1988.
16. T. J. Crowley. Use and misuse of the geologic "analogs" concept. In J. Eddy and H. Oeschger, editors, *Global Changes in the Perspective of the Past*, pages 17–28. Wiley, Chichester, 1993.
17. T. J. Crowley. Ice age terrestrial carbon changes revisited. *Global Biogeochemical Cycles*, 9:377–390, 1995.
18. P. S. Eagleson and R. I. Segarra. Water-limited equilibrium of savanna vegetation systems. *Water Resources Research*, 21:1483–1493, 1985.
19. J. Ehleringer and O. Björkman. Quantum yields for CO_2 uptake in C_3 and C_4 plants. *Plant Physiology*, 59:86–90, 1977.
20. J. R. Ehleringer and R. K. Monson. Evolutionary and ecological aspects of photosynthetic pathway variation. *Annual Review of Ecology and Systematics*, 24:411–439, 1993.
21. J. R. Evans. Partitioning of nitrogen between and within leaves grown under different irradiances. *Oecologia*, 31:255–267, 1989.
22. S.-M. Fan, S. C. Wofsy, P. S. Bawkin, and D. J. Jacob. Atmosphere-biosphere exchange of CO_2 and O_2 in the Central Amazon Forest. *Journal of Geophysical Research*, 95:16851–16864, 1990.
23. G. D. Farquhar. Models of integrated photosynthesis of leaves. *Philosophical Transactions of the Royal Society of London*, B323:357–367, 1989.
24. C. B. Field. Ecological scaling of carbon gain to stress and resource availability. In H. A. Mooney et al., editor, *Integrated Responses of Plants to Stress*, pages 35–65. Academic, San Diego, 1991.
25. A. Fischer. Seasonal features of global NPP models for the terrestrial biosphere. In B. Huntley, W. Cramer, A. V. Morgan, I. C. Prentice, and J. R. M. Allen, editors, *Past and Future Rapid Environmental Changes: The Spatial and Evolutionary Responses of Terrestrial Biota*, number 47 in ASI Series, pages 469–483. NATO, 1997.
26. J. A. Foley. Net primary production of the terrestrial biosphere: the application of a global model. *Journal of Geophysical Research*, 99:20773–20783, 1994.
27. J. A. Foley. An equilibrium model of the terrestrial carbon budget. *Tellus*, 47B:310–319, 1995.
28. J. A. Foley, I. C. Prentice, N. Ramankutty, S. Levis, D. Pollard, S. Sitch, and A. Haxeltine. An integrated biosphere model of land surface processes, terrestrial carbon balance, and vegetation dynamics. *Global Biogeochemical Cycles*, 10:603–628, 1996.
29. P. Friedlingstein, I. Fung, E. Holland, J. John, G. Brasseur, D. Erickson, and D. Schimel. On the contribution of CO_2 fertilization to the missing sink. *Global Biogeochemical Cycles*, 9:541–556, 1995.
30. I. Y. Fung, C. J. Tucker, and I. C. Prentice. Application of advanced very high resolution radiometer vegetation index to study of atmosphere-biosphere exchange of CO_2. *Journal of Geophysical Research*, 92:2999–3015, 1987.
31. R. G. Gallimore and J. E. Kutzbach. Effects of soil moisture on the sensitivity of a climate model to earth orbital forcing at 9,000 y BP. *Climatic Change*, 14:175–205, 1989.

32. J. Grace, J. Lloyd, J. McIntyre, A. C. Miranda, P. Meir, H. S. Miranda, J. Moncreiff, J. Massheder, I. Wright, and J. H. C. Gash. Fluxes of carbon dioxide and water vapour over an undisturbed tropical forest in southwest Amazonia. *Global Change Biology*, 1:1–13, 1995.
33. A. Haxeltine and I. C. Prentice. BIOME3: an equilibrium biosphere model based on ecophysiological constraints, resource availability and competition among plant functional types. *Global Biogeochemical Cycles*, 10:693–709, 1996.
34. A. Haxeltine and I. C. Prentice. A general model for the light use efficiency of primary production. *Functional Ecology*, 10:551–561, 1997.
35. A. Haxeltine, I. C. Prentice, and I. D. Creswell. A coupled carbon and water flux model to predict vegetation structure. *Journal of Vegetation Science*, 7:651–666, 1996.
36. M. Heimann, G. Esser, A. Haxeltine, J. Kaduk, D. W. Kicklighter, W. Knorr, G. H. Kohlmaier, A. D. McGuire, J. Melillo, B. Moore, R. D. Otto, I. C. Prentice, W. Sauf, A. Schloss, S. Sitch, U. Wittenberg, and G. Würth. Evaluation of terrestrial carbon cycle models through simulations of the seasonal cycle of atmospheric CO_2: First results of a model intercomparison study. submitted, 1996.
37. M. Heimann and C. D. Keeling. A three dimensional model of atmospheric CO_2 trnasport based on observed winds. II. Model description. In D. H. Peterson, editor, *Aspects of Climate Variablity in the Pacific and Western America*, Geophysical Monograph Series 55, pages 240–260, Washington, 1989. American Geophysical Union.
38. T. Hirose and M. J. A. Werger. Maximizing daily canopy photosynthesis with respect to the leaf nitrogen allocation pattern in the canopy. *Oecologia*, 7:520–526, 1987.
39. L. R. Holdridge. Determination of world formations from simple climatic data. *Science*, 105:367–368, 1947.
40. R. A. Houghton. Changes in terrestrial carbon over the last 135 years. In M. Heimann, editor, *The Global Carbon Cycle*, pages 139–158. Springer, Berlin, 1993.
41. E. R. Hunt Jr., S. C. Piper, R. Nemani, C. D. Keeling, R. D. Otto, and S. W. Running. Global net carbon exchange and intra-annual atmospheric CO_2 concentration predicted by an ecosystem process model and a three-dimensional atmospheric transport model. *Global Biogeochemical Cycles*, 10:431–456, 1996.
42. International Geosphere-Biosphere Programm. *Global Change, Report 30, IGBP Global Modelling and Data Activities 1994-1998*, Stockholm, 1994.
43. R. F. Keeling, S. C. Piper, and M. Heimann. Global and hemispheric CO_2 sinks deduced from changes in atmospheric O_2 concentrations. *Nature*, 381:218–221, 1996.
44. J. Kindermann, M. K. B. Lüdeke, F.-W. Badeck, R. D. Otto, A. Klaudius, Ch. Häger, G. Würth, T. Lang, S. Dönges, S. Habermehl, and G. H. Kohlmaier. Structure of a global and seasonal carbon exchange model for the terrestrial biosphere. The Frankfurt Biosphere Model (FBM). *Water, Air, and Soil Pollution*, 70:675–984, 1993.
45. J. Kindermann, G. Würth, G. H. Kohlmaier, and F. W. Badek. Interannual variation of carbon exchange fluxes in terrestrial ecosystems. *Global Biogeochemical Cycles*, 10:737–756, 1996.
46. G. A. King and R. P. Neilson. The transient response of vegetation to climate change: A potential source of CO_2 to the atmosphere. *Water, Air and Soil Pollution*, 65:365–383, 1992.

47. W. Knorr and M. Heimann. Impact of drought stress and other factors on seasonal land biosphere CO_2 exchange studied through an atmospheric tracer transport model. *Tellus*, 47B:471–489, 1995.
48. J. E. Kutzbach, P. J. Bartlein, J. A. Foley, S. P. Harrison, S. W. Hostetler, Z. Liu, I. C. Prentice, and T. Webb III. The potential role of vegetation feedback in the climate sensitivity of high-latitude regions: a case study at 6000 years B.P. *Global Biogeochemical Cycles*, 10:727–736, 1996.
49. J. E. Kutzbach, G. Bonan, J. Foley, and S. P. Harrison. Vegetation and soil feedbacks on the response of the African monsoon to orbital forcing in the early to middle Holocene. *Nature*, 384:623–626, 1996.
50. J. E. Kutzbach and P. J. Guetter. The influence of changing orbital parameters and surface boundary conditions on climate simulations for the past 18 000 years. *Journal of the Atmospheric Sciences*, 43:1726–1759, 1986.
51. M. Lautenschlager and K. Herterich. Atmospheric response to ice age conditions - climatology near the earth's surface. *Journal of Geophysical Research*, 95(22):547–557, 1990.
52. S. A. Levin. *Spatial Pattern and the Structure of Ecological Communities*, volume 8 of *Lecture Notes in Mathematics and Life Sciences*. American Mathematical Society, Providence, 1976.
53. R. Levins. The strategy of model building in population biology. *American Scientist*, 54:421–431, 1966.
54. H. Lieth. Modelling the primary production of the world. In H. Lieth and R. H. Whitakker, editors, *Primary Production of the Biosphere*, pages 237–263. Springer, New York, 1975.
55. S. P. Long and P. R. Hutchin. Primary productivity in grasslands and coniferous forests with climate change: An overview. *Ecological Applications*, 12:139–156, 1991.
56. C. Lorius and H. Oeschger. Palaeo-perspectives: reducing uncertainties in global change? *Ambio*, 23:30–36, 1994.
57. A. D. McGuire, J. M. Melillo, L. A. Joyce, D. W. Kicklighter, A. L. Grace, B. Moore III, and C. J. Vörösmarty. Interactions between carbon and nitrogen dynamics in estimating net primary production for potential vegetation in North America. *Global Biogeochemical Cycles*, 6:101–124, 1992.
58. J. Melillo, I. C. Prentice, E.-D. Schulze, G. Farquhar, and O. Sala. Terrestrial biotic responses to environmental change and feedbacks to climate. In J. T. Houghton, L. G. Meira Filho, B. A. Callander, N. Harris, A. Kattenberg, and K. Maskell, editors, *Climate Change 1995: The Science of Climate Change*, pages 445–482. Cambridge University, Cambridge, 1996.
59. J. M. Melillo, A. D. McGuire, D. W. Kicklighter, B. Moore III, C. J. Vörösmarty, and A. L. Schloss. Global climate change and terrestrial net primary production. *Nature*, 363:234–240, 1993.
60. M. Milankovitch. Astronomische Mittel zur Erforschung der erdgeschichtlichen Klimate. *Handb. Geophys.*, 9:539–698, 1938.
61. J. F. B. Mitchell. Greenhouse warming: is the mid-Holocene a good analogue? *Journal of Climate*, 3:1177–1192, 1990.
62. B. Moore III and B. H. Braswell Jr. Planetary metabolism: understanding the carbon cycle. *Ambio*, 23:4–12, 1994.
63. M. B. Murray, M. G. R. Cannell, and R. I. Smith. Date of budburst of fifteen tree species in Britain following climatic warming. *Journal of Applied Ecology*, 26:693–700, 1989.
64. A. Neftel, H. Oeschger, T. Staffelbach, and B. Stauffer. CO_2 record in Byrd ice core 20,000 - 5,000 years BP. *Nature*, 331:609–61, 1988.

65. R. P. Neilson. A mapped atmosphere-plant-soil system for predicting continental scale vegetation distributions. *Ecological Applications*, 5:362–385, 1995.
66. R. P. Neilson and D. Marks. A global perspective of regional vegetation and hydrologic sensitivities from climate change. *Journal of Vegetation Science*, 5:715–730, 1995.
67. A. Osawa and S. Sugita. The self-thinning rule: another interpretation of Weller's results. *Ecology*, 70:279–283, 1989.
68. R. K. Peet. Community structure and ecosystem function. In D. G. Glenn-Lewin, R. K. Peet, and T. T. Veblen, editors, *Plant Succession*, pages 103–151. Chapman & Hall, London, 1992.
69. C. S. Potter, J. T. Randerson, C. B. Field, P. A. Matson, P. M. Vitousek, H. A. Mooney, and S. A. Klooster. Terrestrial ecosystem production: a process model based on global satellite and surface data. *Global Biogeochemical Cycles*, 7:811–842, 1993.
70. I. C. Prentice. Some concepts and objectives of forest dynamics research. In J. Fanta, editor, *Forest dynamics research in western and central Europe*, pages 32–41. PUDOC, Wageningen, 1986.
71. I. C. Prentice, W. Cramer, S. P. Harrison, R. Leemans, R. A. Monserud, and A. M. Solomon. A global biome model based on plant physiology and dominance, soil properties and climate. *Journal of Biogeography*, 19:117–134, 1992.
72. I. C. Prentice, J. Guiot, B. Huntley, D. Jolly, and R. Cheddadi. Reconstructing biomes from palaeoecological data: a general method and its application to European pollen data at 0 and 6 ka. Manuscript, 1996.
73. I. C. Prentice and D. Jolly. Biome modelling. In P. J. Bartlein, J. Guiot, S. P. Harrison, and I. C. Prentice, editors, *Forward and Inverse Modelling in Palaeoclimatology*, NATO ASI series, 1996. In press.
74. I. C. Prentice, R. A. Monserud, T. M. Smith, and W. R. Emanuel. Modelling global vegetation change. In A. M. Solomon and H. H. Shugart, editors, *Vegetation Dynamics and Global Change*, pages 235–250. Chapman and Hall, New York, 1992.
75. I. C. Prentice and M. Sarnthein. Self-regulatory processes in the biosphere in the face of climate change. In J. Eddy and H. Oeschger, editors, *Global Changes in the Perspective of the Past*, pages 29–38. Wiley, Chichester, 1993.
76. I. C. Prentice and M. T. Sykes. Vegetation geography and global carbon storage changes. In G. M. Woodwell and F. T. MacKenzie, editors, *Biotic Feedbacks in the Global Climate System: Will the Warming Feed the Warming?*, New York, 1995. Oxford University.
77. I. C. Prentice, M. T. Sykes, M. Lautenschlager, S. P. Harrison, O. Denissenko, and P. J. Bartlein. Modelling global vegetation patterns and terrestrial carbon storage at the last glacial maximum. *Global Ecology and Biogeography Letters*, 3:67–76, 1993.
78. I. C. Prentice, R. S. Webb, M. T. Ter-Mikhaelian, A. M. Solomon, T. M. Smith, S. E. Pitovranov, N. T. Nikolov, A. A. Minin, R. Leemans, S. Lavorel, M. D. Korzukhin, J. P. Hrabovsky, H. O. Helmisaari, S. P. Harrison, W. R. Emanuel, and G. B. Bonan. Developing a global vegetation dynamics model: results of a summer workshop. Research Report 89(7), International Institute of Applied Systems Analysis, Laxenburg, 1989.
79. I. C. Prentice and T. Webb III. Global palaeovegetation data for climate-biosphere model evaluation. Unpublished report, 1994.
80. J. W. Raich, E. B. Rastetter, J. M. Melillo, D. W. Kicklighter, P. A. Steudler, B. J. Peterson, A. L. Grace, B. Moore III, and C. J. Vörösmarty. Potential net

primary production in South America: application of a global model. *Ecological Applications*, 1:399–429, 1992.
81. E. B. Rastetter, M. G. Ryan, G. R. Shaver, J. M. Melillo, K. J. Nadelhoffer, J. E. Hobbie, and J. D. Aber. A general biogeochemical model describing the responses of the C and N cycles in terrestrial ecosystems to changes in CO_2, climate, and N deposition. *Tree Physiology*, 9:101–126, 1991.
82. D. Rind. How will future climate changes differ from those of the past? In J. Eddy and H. Oeschger, editors, *Global Changes in the Perspective of the Past*, pages 39–50. Wiley, Chichester, 1993.
83. T. Rosswall, R. G Woodmansee, and P. G. Risser. *Scales and Global Change*. SCOPE, Wiley, Chichester, 1988.
84. A. Ruimy, G. Dedieu, and B. Saugier. TURC: a diagnostic model of continental gross primary productivity and net primary productivity. *Global Biogeochemical Cycles*, 10:269–286, 1996.
85. A. Ruimy, B. Saugier, and G. Dedieu. Methodology for the estimation of terrestrial net primary production from remotely sensed data. *Journal of Geophysical Research*, 99:5623–5283, 1994.
86. S. W. Running and J. C. Coughlan. A general model of forest ecosystem processes for regional applications. I. Hydrological balance, canopy gas exchange and primary production processes. *Ecological Modelling*, 42:125–154, 1988.
87. S. W. Running, T. R. Loveland, and L. L. Pierce. A vegetation classification logic based on remote sensing for use in global biogeochemical models. *Ambio*, 23:77–81, 1994.
88. G. Russell, P. G. Jarvis, and J. L. Monteith. Absorption of radiation by canopies and stand growth. In G. Russell et al., editors, *Plant Canopies: Their Form and Function*, pages 21–39. Cambridge University, Cambridge, 1989.
89. A. Sakai. Freezing avoidance mechanism of primordial shoots of conifer buds. *Plant Cell Physiology*, 20:1381–1390, 1979.
90. D. S. Schimel, T. G. F. Kittel, and W. J. Parton. Terrestrial biogeochemical cycles: global interactions with the atmosphere and hydrology. *Tellus*, 43AB:188–203, 1991.
91. S. H. Schneider. The role of the university in interdisciplinary global change research: structural constraints and the potential for change. *Climatic Change*, 20:vii–x, 1992.
92. E.-D. Schulze. Plant life-forms and their carbon, water and nutrient relations. In *Encyclopaedia of Plant Physiology 12B*, pages 615–676. Springer, Berlin, 1982.
93. P. J. Sellers, G. A. Berry, G. J. Collatz, C. B. Field, and F. G. Hall. Canopy reflectance, photosynthesis and transpiration. III. A reanalysis using improved leaf models and a new canopy integration scheme. *Remote Sensing of Environment*, 42:187–216, 1992.
94. H. H. Shugart. *A Theory of Forest Dynamics*. Springer, New York, 1984.
95. H. H. Shugart and I. C. Prentice. The use of individual-based forest simulation models in global change research. In H. H. Shugart, editor, *A Systems Analysis of the Global Boreal Forest*, pages 313–333. Cambridge University, Cambridge, 1992.
96. S. Sitch, D. McGuire, and M. Heimann. Seasonal cycles of atmospheric CO_2: comparison of prognostic and diagnostic simulations. submitted, 1997.
97. T. M. Smith, R. Leemans, and H. H. Shugart. Sensitivity of terrestrial carbon storage to CO_2 induced climate change: comparison of four scenarios based on general circulation models. *Climatic Change*, 21:367–384, 1992.
98. T. M. Smith and H. H. Shugart. The transient response of terrestrial carbon storage to a perturbed climate. *Nature*, 361:523–526, 1993.

99. R. L. Specht. Water use by perennial evergreen plant communities in Australia and Papua New Guinea. *Australian Journal of Botany*, 20:273–299, 1972.
100. W. L. Steffen, B. H. Walker, J. S. Ingram, and G. W. Koch, editors. *Global Change and Terrestrial Ecosystems. The Operational Plan*, Global Change Report 21, Stockholm, 1992. The Royal Swedish Academy of Sciences. IGBP Scretariat.
101. F. A. Street-Perrott. Palaeo-perspectives: changes in terrestrial ecosystems. *Ambio*, 23:37–43, 1994.
102. F. A. Street-Perrott, J. F. B. Mitchell, D. S. Marchand, and J. S. Brunner. Milankovitch and albedo forcing of the tropical monsoons: a comparison of geological evidence and numerical simulations for 9,000 y BP. *Transactions of the Royal Society of Edinburgh*, 81:407–427, 1990.
103. J. A. Taylor and J. Lloyd. Sources and sinks of CO_2. *Australian Journal of Botany*, 40:407–418, 1992.
104. VEMAP. Comparing biogeography and biogeochemistry models in a continental-scale study of terrestrial ecosystem responses to climate change and CO_2 doubling. *Global Biogeochemical Cycles*, 9:407–437, 1995. Vegetation/ecosystem modeling and analysis project (VEMAP).
105. B. H. Walker and I. Noy-Meir. Aspects of stability and resilience of savanna ecosystems. In B. Huntley, editor, *Ecology of Tropical Savannas*, pages 577–590. Springer, Berlin, 1982.
106. H. Walter. *Vegetation of the Earth and Ecological Systems of the Geobiosphere*. Springer-Verlag, New York, 1979.
107. P. Warnant, L. François, D. Strivay, and J. C. Gérard. CARAIB: A global model of terrestrial biological productivity. *Global Biogeochemical Cycles*, 8(3):255–270, 1994.
108. T. Webb III and T. Wigley. What past climates can indicate about a warmer world. In M. MacCracken and F. Luther, editors, *The Potential Climatic Effects of Increasing Carbon Dioxide*, pages 239–257. US Department of Energy, Washington, 1985. DOE/ER-0237.
109. M. Westoby. The self-thinning rule. *Advances in Ecological Research*, 14:167–223, 1984.
110. J. White. The allometric interpretation of the self-thinning rule. *Journal of Theoretical Biology*, 89:475–500, 1981.
111. J. W. C. White, P. Ciais, R. A. Figge, R. Kenny, and V. Markgraf. A high-resolution record of atmospheric CO_2 content from carbon isotopes in peat. *Nature*, 367:153–156, 1994.
112. S. C. Wofsy, M. L. Goulden, J. W. Munger, S.-M. Fan, P. S. Bawkin, B. C. Daube, S. L. Bassow, and F. A. Bazzaz. Net exchange of CO_2 in a mid-latitude forest. *Science*, 260:1314–1317, 1993.
113. F. I. Woodward. *Climate and Plant Distribution*. Cambridge University, Cambridge, 1987.
114. F. I. Woodward. Global change: translating plant ecophysiological responses to ecosystems. *Trends in Ecology and Evolution*, 5:308–311, 1990.
115. F. I. Woodward, T. M. Smith, and W. R. Emanuel. A global land primary productivity and phytogeography model. *Global Biogeochemical Cycles*, 9:471–490, 1995.
116. H. E. Wright Jr., J. E. Kutzbach, T. Webb III, W. F. Ruddiman, F. A. Street-Perrott, and P. J. Bartlein. *Global Climates Since the Last Glacial Maximum*. University of Minnesota, Minneapolis, 1993.

Commentary on "Ecology and the Earth System"

Gundolf H. Kohlmaier

1 Introduction

Professor Colin Prentice said that biosphere modelling was easy. I think this is typical British understatement. I say that good biosphere modelling is difficult. In my summary I should like to comment on the general remarks of Professor Prentice with respect to biosphere modelling, on the global Biome Model by *Prentice* et al. [7], generally referred to as BIOME1, and on the results from the BIOME2 model, which is not yet published, so I have a little difficulty to review it. As an introduction to my comments I should like, however, to relate to the general boundary conditions for modelling the biosphere and the global carbon cycle.

2 Boundary conditions for modelling the biosphere and the global carbon cycle

First of all, indeed we have a problem with the biosphere and the global carbon cycle. The problem is illustrated in Table 1: The IPCC report [2] states that the fossil fuel emissions (including cement production) have been on the average of 5.5 Gt C/yr for the period between 1980 and 1989. We know from measurements that the atmospheric increase was about 3.3 Gt C/yr or 1.56 ppm/yr in this period. Models tell us that the oceans could take up approximately 2.0 ± 0.8 Gt C annually. We need models, because we cannot measure the small change in CO_2 in the oceans, partly because of the low concentration of carbon, partly due to the Revelle buffer factor, which makes a 1% change in the atmosphere equal to a 0.1% change in the oceans. Now, what is left for the biosphere, looks like a neat net balance of approximately zero (actually 0.2 Gt C/yr as shown in Balance 1 of Table 1). However, if we take into consideration CO_2 sources from tropical deforestation of about 16 million ha/yr we arrive at an additional emission of carbon of about 1.6 Gt C. This is really a lot, although the estimates have been coming down a little bit in the past years. Taking the above estimate into consideration we have a missing sink in the biosphere or other regions aside from the biosphere of about 1.8 Gt C. The state-of-art explanation for the additional CO_2 uptake of the biosphere is composed of two factors:

- Uptake by Northern Hemisphere forest regrowth and
- CO_2-fertilization, nitrogen fertilization and climatic effects, which enhance the net primary production, which is in disequilibrium with the corresponding heterotrophic respiration.

Table 1. The IPCC results for the source-sink problem of the atmosphere derived from the update of 1995, Period 1980 – 1989: 10-year average.

Method	Source/sink	GtC/yr	Percentage relative to fossil fuel emissions
Statistics	Fossil fuel emissions	5.5±0.5	100
Measurements	Atmospheric increase	2.3 : 0.2 (1.56 ppm V/yr)	60
Modeling	Ocean uptake	2.0±0.8	36
Balance 1	Net uptake into the terrestrial biosphere	0.2±0.8	4
FAO Forest Statistics	Tropical deforestation	1.6±1.0	29
Balance 2	Missing sink: gross uptake by the terrestrial biosphere	1.8±1.5	33

3 Feedbacks and time delays in dynamic systems

In his first part Professor Prentice has been speaking also about general principles and he mentioned the importance of feedbacks. There are indeed very important positive and negative feedbacks. We need to consider and we have to quantify them in order to get conclusive estimates. However, we also have to consider time delays in dynamic systems: Each forcing process is accompanied by a specific time lag behavior of the system. Further, that is sometimes the most surprising feature of the system and the most difficult to deal with.

4 Vegetation models and climate

I am now coming to the part 2 of Professor Prentice's talk and his Biome Model 1: Most of us know that here in Berlin was the birthplace and home of *Alexander von Humboldt* (1769-1859), the great naturalist and founder of modern earth sciences. It was he who recognized the close relationship between a particular climate and a corresponding vegetation zone in the early nineteenth century (1807) and many vegetation/climate maps were published on the basis of this connection. Of the follow up work I shall only mention the investigations of *Köppen* and *Geiger* [4], *Walter* [8] and *Holdridge* [1]. I think, Professor Prentice successfully has stepped into the footsteps of *Alexander von Humboldt* and his successors. He has been able to establish rules between climate and plant functional types which are then combined to biomes under the additional criteria of dominance classes [7]. This variable composition of biomes may be more apt to describe the plant succession following an expected climate change. One of the most important requirements in the establishment of the climate-biome relationship is the availability of a good climate map with high spatial resolution ($1/2^{\circ} \times 1/2^{\circ}$ latitude by longitude), which requires approximately 60 000 data input points, considering the land surface alone. I think it is important to mention that the simultaneous development of a high-resolution climate map within the International Instititute for Applied Systems Analysis (IIASA) by Dr. Rik Leemans, now at the Dutch Institute of Public Health and Environmental Protection in Bilthoven, Holland and by Professor Cramer, now here at the Potsdam Institute for Climate Impact Research, helped greatly this advancement in biosphere modelling (*Leemans* and *Cramer*, 1991 [5]). Actually it is the combination of three things, namely climate, vegetation, and last but not least soils, which not only control the nutritional status but also the water balance which enables us for the first time to make an honest biosphere model. There are, however, some problems with the model as Professor Prentice has been pointing out: These are equilibrium models. What we all really want to know is the system's transient behavior and because of the great time delays, not only in the vegetation, which might consist of trees with life times of perhaps up to several hundred years, but even more in the soils, which respond to changes on a time scale of maybe 500 years or more, it is a transient problem that we have to solve. So we actually need to couple the biome models to process models not only of the vegetation but also of the soil.

I have been asked when I talked about the dynamic carbon models just recently: Actually where are the animals in your models? Where are the elephants and the butterflies? Or where is the most consuming animal, man? It was Paul Ehrlich and Peter Vitousek who said that about 40% of the net primary production of the biosphere was either used or changed by man. So we have a big problem: We need to get from the potential state of the system described in the climate-biome relations to the real world. There we have to do a lot of things.

5 Dynamic vegetation models

In the third part of his lecture Professor Prentice was speaking of the coupling of his biome model to a process model for carbon exchange between the atmosphere and the vegetation and soils. As these results of the Biome Model 2 are not published yet it is perhaps interesting to report of other efforts of coupling biomes with photosynthesis-respiration models. Professor Cramer has been pointing out that there was a first successful coupling of the Biome Model 1 with the Frankfurt Biosphere Model (*Kindermann* et al., 1993 [3]), a process model for short and long term carbon exchange, with a spatial resolution of $1/2 \times 1/2$ degrees (formerly $1° \times 1°$) and a time resolution of 1 hour which allows to follow the plant metabolism during the diurnal cycle of the sun in order to describe the photosynthesis correctly (*Plöchl* and Cramer, 1995 [6]).

6 Outlook

If you perform these simulations for the whole world you need a lot of computer capacity, but it is still a very interesting thing. So as a whole I would like to say that through Colin Prentice and some of the other people mentioned here we are for the first time really gaining a footing in modelling the biosphere in a decent way. We have real models, they are still very simplified, of course, but we have something to work with. Thus it is in that context that I am very optimistic and that I want to congratulate Professor Prentice for his achievements.

References

1. L. R. Holdridge. *Life Zone Ecology.* Tropical Science Center, San José, 1967.
2. Intergovernmental Panel on Climate Change (IPCC). *Climate Change 1995. Working Group I: The Science of Climate Change. Working Group II: Impacts, Adaptations and Mitigation of Climate Change. Working Group III: Economic and Social Dimensions of Climate Change*, Cambridge, 1996. University Press.
3. J. Kindermann, M. K. B. Lüdeke, F.-W. Badeck, R. D. Otto, A. Klaudius, Ch. Häger, G. Würth, T. Lang, S. Dönges, S. Habermehl, and G. H. Kohlmaier. Structure of a global and seasonal carbon exchange model for the terrestrial biosphere. The Frankfurt Biosphere Model (FBM). *Water, Air, and Soil Pollution*, 70:675–984, 1993.
4. W. Köppen. Das geographische System der Klimate. In W. Köppen and R. Geiger, editors, *Handbuch der Klimatologie*, pages 1–46. Gebrüder Borntraeger, Berlin, 1936.
5. R. Leemans and W. P. Cramer. The IIASA database for mean monthly values of temperature, precipitation and cloudiness on a global terrestrial grid. Technical Report RR-91-18, International Institute for Applied Systems Analysis, Laxenburg, 1991.

6. M. Plöchl and W. Cramer. Possible impacts of global warming on tundra and boreal forest ecosystems: comparison of some biogeochemical models. *Journal of Biogeography*, 22:775–783, 1995.
7. I.C. Prentice, W. Cramer, S. P. Harrison, R. Leemans, R. A. Monserud, and A. M. Solomon. A global biome model based on plant physiology and dominance, soil properties and climate. *J. Biogeography*, 19:117–134, 1992.
8. H. Walter. *Vegetation of the Earth and Ecological Systems of the Geo-Biosphere.* Springer Verlag, Berlin, 1985.

Commentary: Eco-System Modelling and the Social Sciences

Urs Luterbacher and Ellen Wiegandt

1 Introductory remarks

Ever since man's appearance on earth and especially since the emergence of agriculture, much of nature must be seen as a human creation. While physical forces have always constrained activities, so too have people modified landscape and climate through culturally defined technologies, social institutions, and individual strategies. To examine the human dimensions of global environmental change or the social dimensions of resource use is thus to ask both how political, economic and cultural factors mediate natural processes and how forces of nature shape certain institutional arrangements. In this commentary we contribute some complementary ideas to those laid out by Colin Prentice.

2 Global Change: an age old process

Our focus is clearly on the present and the perception that wide-spread industrialization and massive population growth in the recent past have already induced or will bring about significant modification of the global climate and environment. These changes may be unprecedented in magnitude but humans have always had to adapt to environmental fluctuations. Indeed, the very origins of man are to be sought in the progressive adjustments to transformations of the environment. Moreover, from the beginning, man has profoundly influenced the evolution of his natural environment by having at first an important impact on other animal species through hunting practices and then eventually by modifying the vegetation and the balance between plant and animal species through cultivation and herding. The industrial revolution constitutes in this sense only an acceleration of an age-old process. The nature of these man-environment interactions is shaped both by the kinds of environmental changes taking place and by the types of social-behavioral systems and categories of social actors that are implicated. A long time perspective provides useful insights into the types of pressures that climate changes have imposed on very different cultures in different regions at different times.

3 Origins of agriculture and links between climate change and human activities

Archaeologists can document real impacts that past climate changes have had and thus provide empirical data to test theoretical formulations. Numerous examples of interesting historical-ecological research could be cited. Among the most relevant to current questions of food production and land use is that examining the origins of agriculture. Joy *McCorriston* and Frank *Hole* [5] (and many others) have done extensive work on this problem. They first defined different types of environmental stress and suggested that each has distinct effects on cultural systems. Fluctuations in rainfall or temperature, for example, are characterized by some uncertainty in timing but the system knows they can occur. It thus anticipates conditions, perhaps through storage technology, perhaps through other mechanisms. Severe, sudden, or unanticipated conditions that exceed the limits the system can absorb, may well lead to significant system change or even failure. On the other hand, progressive evolution over relatively long periods can have several outcomes. Systems can recognize variations but adjust to them as little as possible, thus creating the conditions for collapse if an abrupt change should occur. These cumulative shifts can also create thresholds where systems persist for some time and then are precipitously transformed. The origin of agriculture in the Middle East, according to *McCorriston* and *Hole*, was a slow, progressive adaptation to changes in the physical and biological environment that ultimately led to a radically different form of subsistence and associated social organization. They describe a climate history that led to enhanced seasonality through higher summer temperatures and lower winter ones. These were accompanied by reduced quantities of surface water. The hunting and gathering populations living in the region pulled back into wetter zones and exploited the changing vegetation. They became more sedentary and had to devise storage technologies because the greater seasonality meant foods were not available all year. This new form of subsistence called upon different types of social organization capable of overseeing storage and assuring distribution of harvests. There is thus evidence of emerging stratification. These cultural developments in turn affected the environment because sedentism led to increased use of local resources and tended to deplete them. The strategy adopted in the Middle East was to replant and this voluntary replacement was the origin of agriculture. Hole has also described how further climate changes in later periods led to further technological innovations like irrigation and to greater aggregation of population which eventually produced the first urban settlements (*Hole*, 1994 [4]). The in-depth study of long trajectories identifies crucial relationships and thus helps define a research agenda that can guide current efforts. Over the very long term it becomes possible to examine the links between climate changes and human activities and to attempt to define their causal relationships. The inventory of indigenous knowledge and technique is evidence for the extensive cultural diversity that has existed over time and across space. Much indigenous knowledge has now been lost or abandoned but, viewed as analogous to biodiversity, it may suggest alternative strategies for confronting contemporary uncertainties and risks.

4 Interdependencies between forms of social organization and the environment

This ecosystem approach in archaeology has clearly brought significant new understanding of the interconnections between culture and environment and can provide important insights into contemporary research on the human dimensions of global climate change. With its emphasis on whole populations and cultural systems, however, it has been less sensitive to the dynamics of social groups within systems and thus to the importance of coalitions, conflict, and negotiation in the evolution of particular systems and their interactions with the environment. Elizabeth *Brumfield* has made just these points in her call for an archaeology more attuned to human actors and their strategies within the constraints imposed by environment and culture [1]. She argues that this is not simply a function of available data and that artifact categories, settlement types and their organization, architecture, and the use of space can all provide information about the claim and counterclaims in ongoing social negotiations (Brumfield, 1992 [1, p. 559]). Other social sciences such as ethnology, sociology, and political science have been somewhat more attentive to the notion that cultural systems are the outcomes of complex individual and group strategies to meet goals. These various organizational forms have very different consequences for the environment. In some cases, strong local institutions encourage the conservation of water, forest, and land for the use of the whole population. In some other instances, the environment creates high-risk agriculture but the social system devises few risk-sharing mechanisms so individuals are more powerful but also inclined to further over-exploit the environment. The ultimate strategy often leads to the abandonment of agriculture and to departure from the region.

5 Common vs. private property rights

These different aspects of social organizations stress the fact that the management of the environment, like other assets, is achieved through specific property right structures. This particular aspect has been emphasized in Garrett *Hardin*'s work on *The Tragedy of the Commons* (1977 [3]). He tried to point out the inefficiency and environmental inadequacy of common property arrangements. Private property situations should lead to a more efficient and therefore environmentally more adequate use of resources. Subsequent research has shown, however, that Hardin's criticism of common property systems was misguided and that the deficiencies that he points out only characterize uncontrolled open access property regimes. Common property structures that limit the access to resources and control their utilization can be as effective as private property ones in managing the environment. Indeed, such common property regimes can even be more adequate than private ones if certain conditions that are necessary for the efficient working of private ownership are absent such as a limited variation in the yield of the resource, enforceability of property rights, low costs of surveillance and maintenance, and an adequate level of information about the legal system which is supposed to guarantee the property right. If these conditions are not present, the famous Coase theorem (*Coase*, 1960 [2]) on private property structures breaks down and other arrangements will lead to more efficient results. The rational choice analysis of Pareto optimal and suboptimal outcomes is useful here in determining how and under which conditions private or common property arrangements at the national and/or international level are most efficient. The efficiency or, respectively, inefficiency of such arrangements will have a significant impact on land use and thus on ecosystems and their evolution.

6 The resulting research agenda

The emphasis on understanding social institutions and their consequences for the environment and for the evolution of individual members of societies derives from the notion that these considerations are relevant to our capacity to analyze and react to global environmental change. Social institutions and property structures are important for agriculture because this has historically been the dominant mode of production. It still is for the majority of the world's population and vitally concerns the rest. Moreover, the relationships evoked are universal as they concern the organization of production, family strategies, ethnic group organization, and power. One of the major problems in assessing both the consequences for man and the man-induced causes of global environmental change is to know which groups will be most affected and how they will adapt and how their choices will further influence the environment. From what we can learn from the past we know that the particular characteristics of specific societies will make them as systems and their members as individuals more or less vulnerable to certain kinds of pressures. The responses will have significant impacts for other areas. In this context, we view an important research agenda as one that assesses the capacity of marginal cultivators, who number in the millions, to adjust to temperature shifts, desertification or coastal flooding. One of the characteristic responses, already noted in historical research, is for populations under stress to move. In identifying those cultures whose local institutions are insufficient to accommodate current populations above certain threshold changes, we may be able to predict likely migration patterns and thereby assess the cascading pressures this process will bring to other regions of the country concerned and areas much farther away. Local and regional studies provide the detail on coping mechanisms, demographic patterns, family strategies and cultural cleavages. To understand the wider process it is necessary to place this detail in a context of global climate change, regional and international trade, and national government and international incentives and regulations. This approach requires interdisciplinary cooperation and models that allow for the shift in levels of analysis from the local to the international.

References

1. E. Brumfield. Distinguished lecture in archaeology: Breaking and entering the ecosystem - gender, class, and faction steal the show. *American Anthropologist*, 94(3):551–567, 1992.
2. R. H. Coase. The problem of social cost. *The Journal of Law and Economics*, 3(1):1–44, 1960.
3. G. Hardin. The tragedy of the commons. In G. Hardin and J. Baden, editors, *Managing the Commons*. Freeman and Company, 1977. Reprinted from Science 162: 1243-1248, pp. 16-30.
4. F. Hole. Environmental shock and urban origins. In J. K. Mitchell, S. Rothman, and G. Stein, editors, *Chiefdoms and Early States in the Near East*. Prehistory Press, Madison, 1994.
5. J. McCorriston and F. Hole. The ecology of seasonal stress and the origins of agriculture in the near east. *American Anthropologist*, 93(1):46–69, 1991.

Chapter 2: Climate Change and Land Use: Global and Regional Analyses

M. L. Parry, C. Rosenzweig[1], *J. E. Hossell*[2], *P. Jones*[2], *T. Rehmam*[3], *R. B. Tranter*[3], *J. S. Marsh*[3], *and J. C. Taylor*[3]

[1] Columbia University/Goddard Institute for Space Studies, New York City, U.S.A.
[2] Department of Geography, University of Birmingham, U.K.
[3] Centre for Agricultural Strategy, University of Reading, U.K.

Abstract The potential effects of climate change on agricultural land use in any given region are likely to stem as much from the response of the world food system to global climate change as to specific changes in climate in that region. To achieve this level of analysis requires evaluation of inputs at the regional level but within a global price environment. This, in turn, requires the development of a hierarchy of land use/climate models. In this paper we employ results of a global analysis of the sensitivity of world food prices to climate change, using these as inputs to a regional land use allocation model, in order to evaluate the potential effects of climate change on regional patterns of land use.

There are two components to the global estimation: Firstly, the estimation of potential changes in national grain crop yields using crop models and a decision support system. Secondly the estimation of world food trade responses. The yield changes were used as inputs into a world food trade model, the Basic Linked System (BLS). Outputs from simulations by the BLS provided information on global food production potential, and food prices.

The second part of the paper illustrates how an analysis of these two global elements (crop yields and food prices) can be made in order to evaluate their integrated effect in land use. A case study is used to illustrate the process; in this instance, for England and Wales using a Land Use Allocation Model and manipulating it in the context of a) global price changes due to climate and b) changes in yield potential in England and Wales due to climate. The combination of price and yield potential is used to calculate the highest value use for each of 250 000 1-km^2 cells of land in England and Wales for a future economic environment in 2060 (without climate change) and then for that same environment with climate change. The difference provides an estimated effect of climate change.

1 Introduction

The potential effects of climate change on agricultural land use in any given region are likely to stem as much from the response of the world food system to global climate change as from specific changes in climate in that region. In general, we might expect that changes in global production due to climate, leading to differential changes in prices for different crops, would affect the competitive position of different crops and thus what crops would be grown in what proportions.

At the same time, it is reasonable to expect that changes in climate which, for a given region, are likely to lead to changes in the yields of different crops, would lead to changes in the competitive advantage of different regions as growing areas for food crops.

If we could estimate these two types of effects we could infer the different weight of effects on land use that might stem directly from regional-level changes in climate per se and those that may stem indirectly from global climate-induced changes in agricultural prices. To achieve this level of analysis requires evaluation of inputs at the regional level but within a global price environment and the development of a hierarchy of land use/climate models. In this paper we employ results of a global analysis of the sensitivity of world food prices to climate change, using these as inputs to a regional land use allocation model. Research on the former has been published in full elsewhere (*Rosenzweig* and *Parry*, 1994 [13]). The latter is the product of recent, previously unpublished, research relating to land use and climate change in Britain. A schema of the research approach is given in Fig. 1.

2 Effects on regional competitiveness and global prices

The first part of this paper summarises current knowledge concerning the potential effects of climate change on the production potential of global agriculture. This will lead to a discussion of effects on prices of agricultural products and the altered competitiveness (or comparative advantage) of different agricultural regions with respect to the world market.

There are two components to this global estimation:

- *The estimation of potential changes in crop yield.* Potential changes in national grain crop yields were estimated using crop models and a decision support system developed by the U.S. Agency for International Development's International Benchmark Sites Network for Agrotechnology Transfer (IBSNAT, 1989). The crops modelled were wheat, rice, maize, and soybeans which account for more than 85% of the world traded grains and legumes (for details, see *Rosenzweig* and *Parry*, 1994 [13]).

- *The estimation of world food trade responses* . The yield changes were used as inputs into a world food trade model, the Basic Linked System (BLS), (*Fischer* et al., 1988 [4]). Outputs from simulations by the BLS provided information on global food production potential, and food prices.

2.1 Modelling effects on crop yields

Crop models. The IBSNAT crop models were used to estimate how climate change and increasing levels of carbon dioxide could alter yields of the selected crops over the major production areas of the world and the major worldwide

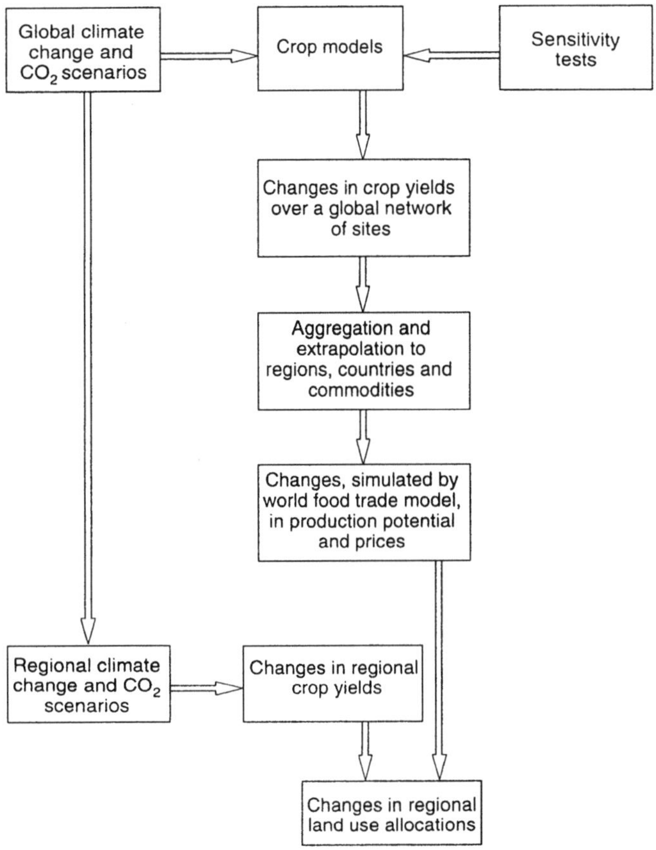

Figure 1. The research approach

vulnerable regions at low, mid, and high latitudes. These models employ simplified functions to predict the growth of crops as influenced by the major factors that affect yields, i.e., genetics, climate (daily solar radiation, maximum and minimum temperatures, and precipitation), soils, and management practices. Models used were for wheat (*Ritchie* and *Otter*, 1985 [11]; *Godwin* et al., 1989 [5]), maize (*Jones* and *Kiniry*, 1986 [8]; *Ritchie* et al., 1989 [12]), paddy and upland rice (*Godwin* et al., 1992 [6]), and soybean (*Jones* et al., 1989 [9]). Reasons for the selection of these models, and their calibration and validation, and their limitations, are given in *Rosenzweig* et al. [14].

The crop models account for the beneficial physiological effects of increased atmospheric CO_2 concentrations on crop growth and water use through the use of ratios between measured daily photosynthesis and evapotranspiration rates for a canopy exposed to high CO_2 values, based on published experimental results (*Allen* et al., 1987 [1]; *Cure* and *Acock*, 1986 [3]; and *Kimball*, 1983 [10]).

Crop modelling simulation experiments were performed at the 112 sites in 18 countries for the baseline climate (1951-80) and general circulation model (GCM) doubled CO_2 climate change scenarios with and without the physiological effects of CO_2.

Aggregation of site results. Estimates of altered yields were aggregated by weighting regional yield changes (based on current production) to estimate changes in national yields. This assumed the current mix of rainfed and irrigated production, and the current crop varieties, nitrogen management and soils. Changes in yields of crops that were not simulated were estimated based on similarities to modelled crops and growing conditions, and on a wide search of previous published and unpublished climate change impact studies.

The primary source of uncertainty in the estimates lies in the sparseness of the crop modelling sites which may not adequately represent the variability of agricultural regions within countries, the variability of agricultural systems within similar agro-ecological zones, or dissimilar agricultural regions. However, since the site results relate to regions that account for about 70% of world grain production, both the method adopted and the conclusions for the scale of the study adopted here are considered to be appropriate (*Rosenzweig* and *Parry*, 1994 [13]).

2.2 Modelling effects on food supply and prices, worldwide

The estimates of climate-induced changes in yields were used as inputs to a dynamic model of the world food system (the Basic Linked System, BLS) in order to assess possible impacts on future levels of food production, and food prices. (A full description of the BLS is given in *Fischer* et al., 1988 [4].)

The BLS is a general equilibrium model system which consists of 20 national models that cover about 80% of the world food system. The remaining 20% is covered by 14 regional models for the countries which have broadly similar attributes (e.g. African oil exporting countries, Latin American high income exporting countries, Asian low income countries etc.)(Fig. 2). In the model, countries are linked through trade, world market prices and financial flows. It is a recursively dynamic system: a first round of exports from all countries is calculated for an assumed set of world prices, and international market clearance is checked for each commodity. World prices are then revised, using an optimising algorithm and again transmitted to the national model. Next, these generate new domestic equilibria and adjust net exports. This process is repeated until the world markets are cleared in all commodities. At each stage of the reiteration domestic markets are in equilibrium. This process yields international prices as influenced by governmental and inter-governmental agreements in a given year. The system is solved in annual increments, simultaneously for all countries. Effects of changes in climate were introduced to the model as changes in the average national or regional yield per commodity as estimated above.

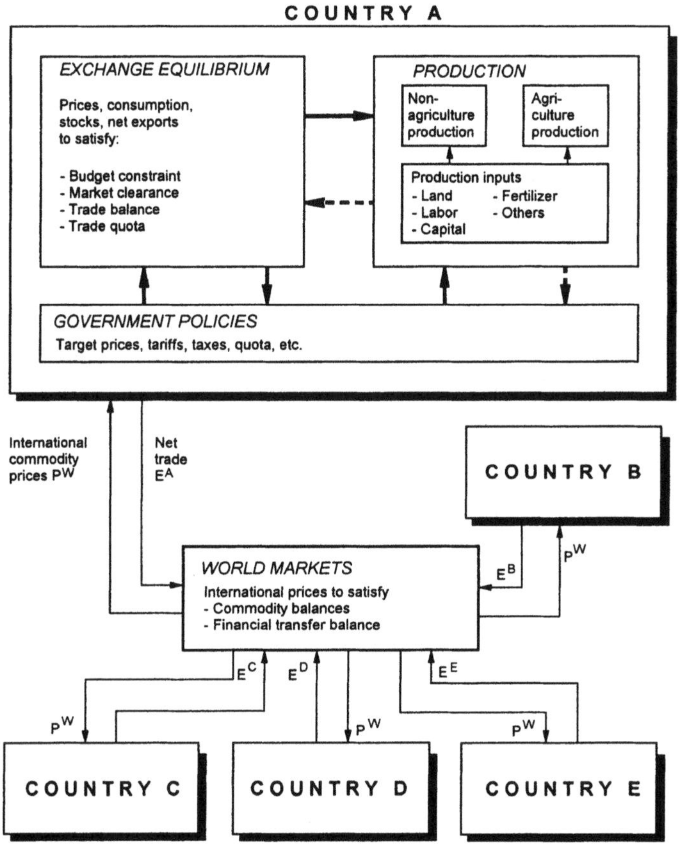

Figure 2. The Basic Linked System: Relationships between country components and world markets. Arrows to countries represent international commodity prices; arrows to world markets represent net trade

2.3 Assumptions about the future

The Reference scenario. This involved projection of the world agricultural system to the year 2060 assuming no climate change and some changes in the political or economic context of the world food trade. It assumes:

- United Nations (UN) medium population estimates (10.2 billion by 2060)(UN, 1989; International Bank for Reconstruction and Development (IBRD)/ World Bank, 1990)

- 50% trade liberalisation in agriculture introduced gradually by 2020
- Moderate economic growth (ranging from 3.0%/year in 1980-2000 to 1.1%/year in 2040-2060)
- Technology is projected to increase yields over time. Cereal yields of world total, developing countries and developed countries are assumed to increase annually by 0.7%, 0.9% and 0.6%, respectively
- No climate change.

Cereal prices are estimated at an index of 121 (1970 - £100) for the year 2060, reversing the trend of falling real cereal prices over the last 100 years (see Table 1). There are two phases of price development, these being strongly affected by the assumptions relating to tariff liberalisation.

Table 1. Index of world prices simulated by the Basic Linked System reference case (1970=100).

	1980	2000	2020	2060
Cereals	102	125	126	121
Other crops	110	118	110	94
All crops	108	120	115	102
Livestock	105	131	135	119
Agriculture	107	123	121	107

Scenarios of CO_2 and climate change. These are projections of the world food system including effects of climate on agricultural yields under three different climate scenarios. The scenarios for this study were created by changing observed data on current climate (1951-80) according to doubled CO_2 simulations of three general circulation models (GCMs). The GCMs used are those from the Goddard Institute for Space Studies (GISS), Geophysical Fluid Dynamics Laboratory (GFDL), and United Kingdom Meteorological Office (UKMO). The temperature changes of these GCM scenarios (4.0 - 5.2 °C) are nearer to the upper end of the range (1.5 - 4.5 °C) projected for doubled CO_2 warming by the IPCC (1990, 1992). Mean monthly changes in climate variables from the appropriate gridbox were applied to observed daily climate records to create climate change scenarios for each site.

CO_2 concentrations were estimated to be 555 ppm in 2060. The effective CO_2 doubling will occur around the year 2030, if current emission trends continue. The climate change caused by an effective doubling of CO_2 may be delayed by 30 to 40 years or longer, hence the projections for 2060 in this study.

Adaptation. Two adaptation levels to cope with potential effects of climate change on yield and agriculture were considered: those adaptations at the farm level that would not involve any major changes in agricultural practices; these included changes in planting date, in amounts of irrigation, and in choice of crop varieties that are currently available; and major changes in agricultural practices, e.g. large shifts of planting date, the availability of new cultivars, extensive expansion of irrigation and increased fertiliser application. These would be likely to involve policy changes at the national and international level and involve significant costs.

2.4 Altered potential cereal production and food prices

Fig. 3 shows estimated potential changes in world cereal production for the GISS, GFDL, and UKMO 2 × CO_2 climate change scenarios allowing for the direct effects of CO_2 on plant growth. These yield estimations are used as inputs to the dynamic simulations by the BLS to indicate how the world food system might respond to climate-induced supply shortfalls of cereals and consequently higher commodity prices through dynamic increases in production factors, cultivated land, labour and capital, and inputs such as fertiliser.

World cereal production is estimated to decrease between 1% and 7%, depending on the climate scenario. Under the UKMO scenario global production is estimated to decrease by more than 7%, while under the GISS scenario (which assumes lower temperature increases) cereal production is estimated to decrease by just over 1%. The largest negative changes occur in developing countries, which average −9 % to −11 %. By contrast, in developed countries production is estimated to increase under all but the UKMO scenario (+11% to -3%). Thus existing disparities in crop production between developed and developing countries are estimated to increase.

Effects on world prices are shown in Table 2. These are estimated to be 24% to 145% higher than those estimated for the reference case (i.e. the future in 2060 without climate change). Both minor and major levels of adaptation help restore world production levels (Fig. 3), but quite large price differences remain, particularly under the more extreme climatic scenarios (Fig. 4). These climate- affected prices can now be used as inputs to the model of regional land use.

2.5 Effects on regional-level land use

Changes of climate are likely to affect the use of agricultural land at a given place in two ways: Firstly, through changes in yield potential at that place and, secondly, through effects on the competitive position of that land *vis à vis* land elsewhere. The aim of the second part of this paper is to illustrate how an analysis of these two elements can be made in order to evaluate their integrated effect. A case study will be used to illustrate the process; in this instance, for England and Wales, using a Land Use Allocation Model and manipulating it in the context of a) global price changes due to climate and b) changes in yield potential in England and Wales due to climate. The combination of price and yield potential is used to calculate the highest value

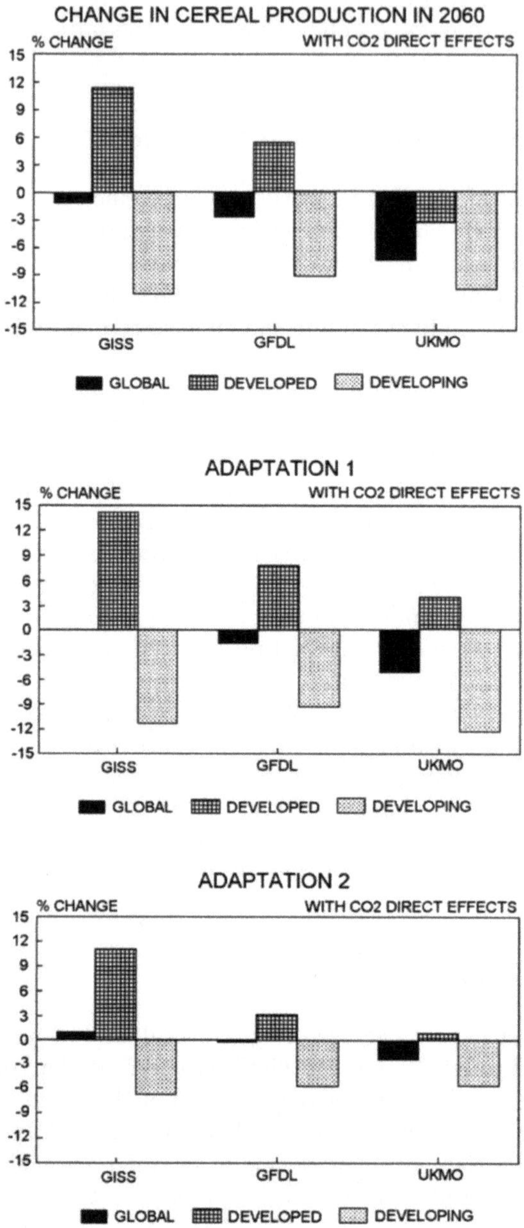

Figure 3. Change in world, aggregated developed country and developing country cereal production in 2060 projected by the BLS under climate change scenarios. Reference scenario 2060 assumes no climate change (world 3286 mmt, developed 1449 mmt, developing 1836 mmt).

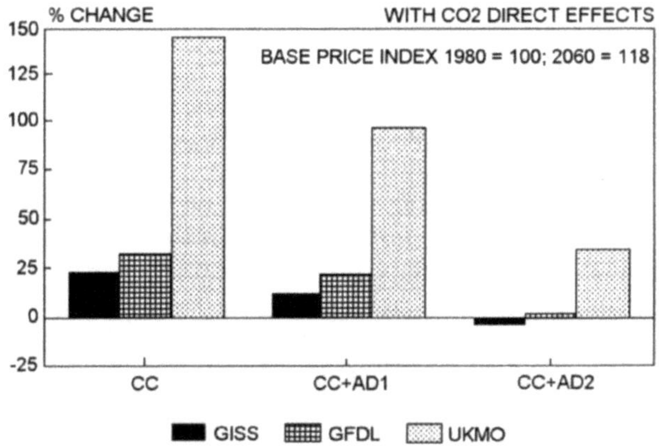

Figure 4. Change in cereal price index in 2060 calculated by the Basic Linked System under climate change scenarios. Reference scenario for 2060 assumes no climate change (price index is 18% above 1980 level).

Table 2. Change in cereal production, prices and number of people at risk of hunger in 2060 under GCM 2 × CO_2 climate change scenarios.

A. Change in cereal production.

Region	Reference scenario[1]	GISS %	GFDL %	UKMO %
Global	3286	-1.2	-2.8	-7.6
Developed	1449	11.3	5.2	-3.5
Developing[2]	1836	-11.0	-10.8	-10.8

B. Change in cereal price index (1970=100).

	Reference scenario[1]	GISS %	GFDL %	UKMO %
Cereal prices	121	24	33	145

C. Change in number of people at risk of hunger.

Region	Reference scenario[1]	GISS %	GFDL %	UKMO %
Developing[2]	641	10	17	58

[1]Reference scenario is for 2060 assumming no climate change
[2]Estimates for developing countries do not include China.

use for each of 250 000 1-km^2 cells of land in England and Wales. These cells form the basis of a land classification system which has been designed, in part, to examine the sensitivity of local land use allocation to policy, price and environmental variables.

2.6 The Land Use Classification System (LCS)

The LCS was devised as a means of providing an objective framework for ecological sampling. It divides Great Britain into 32 classes at a 1 km^2 resolution, with associated landscape, land use and land cover characteristics. The divisions are based upon 76 key attributes derived from map data e.g. topography, distance from south coast, slope, geology, distance from major road, climate etc. A Climate Land Classification (CLC) was then devised using the structure of the LCS drawing upon the detailed field survey data of that system. It assigns 1-km^2 cells of land into 32 Land Classes, with their associated land use and vegetation characteristics, according to climatic conditions. The CLC assigns squares to classes using 51 monthly climatic attributes. Discriminant function analysis was then run to see if the variables could predict the same Land Class pattern as the LCS. Comparison of Figs. 5a and 5b demonstrates the way the CLC closely mimics the LCS. The climatic classification can therefore be used as a surrogate of the land classification.

2.7 The Climate Land Use Allocation Model

The Climate Land Use Allocation Model (CLUAM) was derived from an earlier version, which had been originally designed and built to analyze the environmental and ecological consequences of changes in agricultural practices in response to adjustments in agricultural policies and market conditions (*Harvey* et al., 1992 [7]). In constructing the CLUAM, data on the various factors that were important in determining land use in the countryside were structured within the unified framework of a linear programming model that treats the agriculture of England and Wales as a single farm. On this England and Wales farm a variety of production activities are employed to produce various agricultural outputs using a range of inputs and resources and different types of land as captured in the LCS. The CLUAM allocates land of various types to these different land-using activities subject to three basic types of constraint:

- the availability of land of various types within each Land Class and the availability to shift this resource between different uses according to variation in soils, topology and climate;
- the total amount of production and quantity of input used, specifying the England and Wales demand for the products and supplies of inputs; and
- "policy" constraints which restrict production activities or impose specific land utilisation patterns in certain areas to conform to environmental or other objectives.

Figure 5. (**a**) The pattern of Land Classes in Great Britain as defined by the Land Classification System (LCS). (**b**) The pattern of Land Classes in Great Britain as defined by the Climate Land Classification.

Within these constraints, land utilisation is decided according to the economic margins that can be earned by the various activities. In the model the mix of land types in England and Wales is, therefore, represented by 15 Land Classes, some of which combine several LCS Classes. Within each Land Class, four types of land were identified: arable, ley, permanent pasture and rough grazing. Within each land type, land is further delimited by yield categories reflecting the range of production potentials possible for any crop or grass/herbage grown. There are three yield classes for arable land uses and five for each of the grassland types. In any Land Class there are up to 66 possible activities for the model to choose from. Thus with 15 Land Classes the model consists of over 990 activities.

The essential components are shown in Fig. 6. Ecological and environmental data expressed through the LCS and thus, indirectly, through the CLUAM. The "objective" that is optimized by the CLUAM is the gross margin and is defined as the difference between the output value of an enterprise and all those "variable" costs related directly to the scale of the enterprise, e.g., fertiliser use varies directly in proportion to the area of the crop grown.

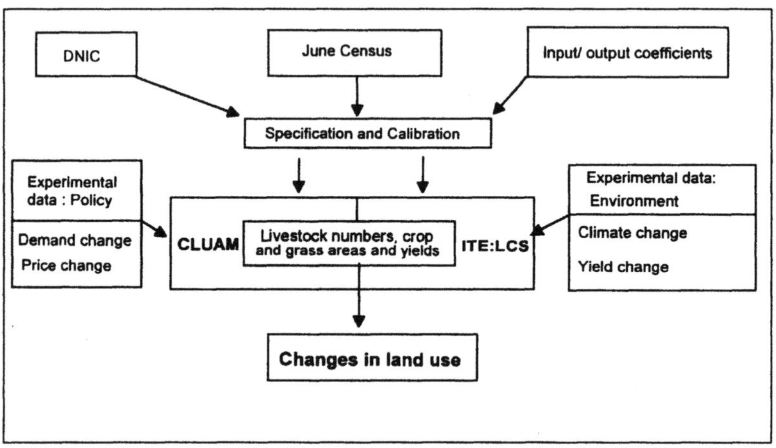

Figure 6. The structure of the CLUAM

Using real world field-level observation, a series of physical and financial parameters are calculated for all crop and livestock production and land use types. These parameters include physical yield, stocking rates, value of output, variable costs of production, levels of input use and gross margins; where appropriate their parameters can be measured in terms of both unit and area of production. In the base case areas are calculated of the various crops and grass/herbage types by yield level after allowing for the above two changes. When the model is run for both the altered climatic or economic conditions in 2060, the economic optimum mix of enterprises that is chosen by the model depends on the areally delineated yield potentials of the land available for the various enterprises, combined with the various price inputs and outputs and is subject to different levels of demand.

2.8 Effects of climate change on agricultural potential in England and Wales

Changes of climate may have two types of effect on agricultural potential in the UK:

a) The distribution of crop potential could be affected across the country by climate change. Local conditions would determine the degree of which yield changes reflect this response.
b) Crop yields could be directly affected by changes in ambient levels of the atmospheric carbon dioxide.

Estimates of the combined effects of a) and b) were calculated using results from crop model simulations described in Part 1 of this paper (Table 3) and the percentage changes applied to the yield classes in the CLUAM. For all crops the climate information was used to define a limit beyond which a given crop would not grow. Given the coarse resolution of climate data available and the large number of crops included in the model these limitations were defined in terms of simple thermal limits, with soil moisture conditions used to assess the ability to work the land sufficiently to both sow and harvest the crop. Similar assumptions have been used to assess the potential suitability of land in Britain and Europe for a number of crops under present and future climate conditions. (*Carter* et al., 1991 [2]; *Rowntree* et al., 1989 [15]). Table 4 provides the limitations used to define areas of crop growth. Production of a crop in an area was prohibited if either the thermal or moisture conditions severely limited growth. Moisture conditions were only limiting if the levels were high enough to prevent use of machinery during either the entire preparation/sowing period or the whole harvesting period.

2.9 Land use in England and Wales in a future without climate change

The projection of land use changes through the reference run to 2060 (REF-2060) allows the effects due to economic changes to be separated from the response due to climate perturbation alone. Table 5 shows the assumptions used by the BLS model which have been used to determine the 2060 conditions in the CLUAM. These take the form of proportionate changes in yields, prices and demand for products expected to take place between 1990 and 2060. Yield changes are fairly uniform over enterprise, ranging from increases of 21% to a maximum of 48% by 2060. Wheat and barley yields are projected to increase by 32% or 0.48 % annually. Prices for all enterprises are adjudged to fall in real terms, with wheat and barley falling by a little under 16% while

Table 3. Percentage yield changes in CLUAM, commodities under three general circulation model climate change scenarios for 2 × CO_2 conditions (*Rosenzweig* and *Parry*, 1994 [13]).

Commodity	GFDL GCM	GISS GCM	UKMO GCM
Barley	1%	1%	2%
Grain maize	9%	9%	9%
Grass	11%	20%	10%
Oats	1%	1%	2%
Oilseed rape	7%	15%	10%
Peas/beans	7%	15%	10%
Potatoes	7%	15%	10%
Sugar beet	1%	1%	2%
Sunflower	25%	25%	25%
Wheat	−15 %	8%	−9 %

other arable prices fall by as much as 30%. Demand seems fairly constant with the only significant change projected being an increase in coarse grains of 43%.

Figs. 7a and 7b show that the national area under cereals declines from 3.476 Mha to 2.069 Mha by 2060, a fall of some 40%, with the main production centring much more in SE England, the area most favourable to production. Elsewhere, however, large areas of the country, particularly in the north and Lincolnshire, fall out of cereal production altogether.

All land, under the CLUAM model baseline of 1984, was in agricultural production. However, the future economic conditions simulated for the 2060 scenario results in some areas of land not being economically viable for continued agricultural production. This land falls out of agricultural production and in the absence of alternative uses becomes "idle". The area of such "idle" land rises from a base position of zero to 2.299 Mha for former ley/arable land and 0.885 Mha for permanent pasture land (Figs. 8 and 9).

2.10 Effects of climate changes on land use in England and Wales

As with the 2060 reference run (REF-2060), the BLS provides a range of assumptions about the changes in yields, demand and prices under the climate change scenarios for which the CLUAM is run. In this paper we report only the estimations under GISS scenario (see Table 6). These coefficients are used by the model in exactly the same manner as for the reference run. In summary the following is evident:

Table 4. Climatic limits used to define areas suitable for crop production (AT means accumulated temperature)

Crop	Soil moisture deficit (mm)		Temperature limit	Rainfall limit (mm)
	Sowing	Harvest		
Barley	> 25 or < 0 between Sept and Feb	> 25 or < 0 between July and Sept	Mean annual temperature > 7.0 °C	
Grain maize	> 25 or < 0 between Oct and April	> 25 or < 0 between March and May	AT > 1200 °C base 8 °C	
Oilseed rape	> 25 of < 0 between Aug and Sept	> 25 or < 0 between Aug and Sept		
Peas/beans	> 20 or < 0 between Sept and May	> 20 or < 0 between June and Sept	AT > 1724 °C (base 0 °C)	
Potatoes	> 25 or < 0 between Sept and Feb	> 25 or < 0 in Sept		
Sugar beet	> 20 or < 0 between Sept and May	> 20 or < 0 between Sept and Dec		< 1000 mm
Sunflowers	> 20 or < 0 between Oct and April	> 20 or < 25 in Sept	AT > 1580 °C (base 6 °C)	
Wheat	> 25 or < 0 between Sept and Feb	> 25 or < 0 between July and Sept	Mean annual temperature > 7.5 °C	

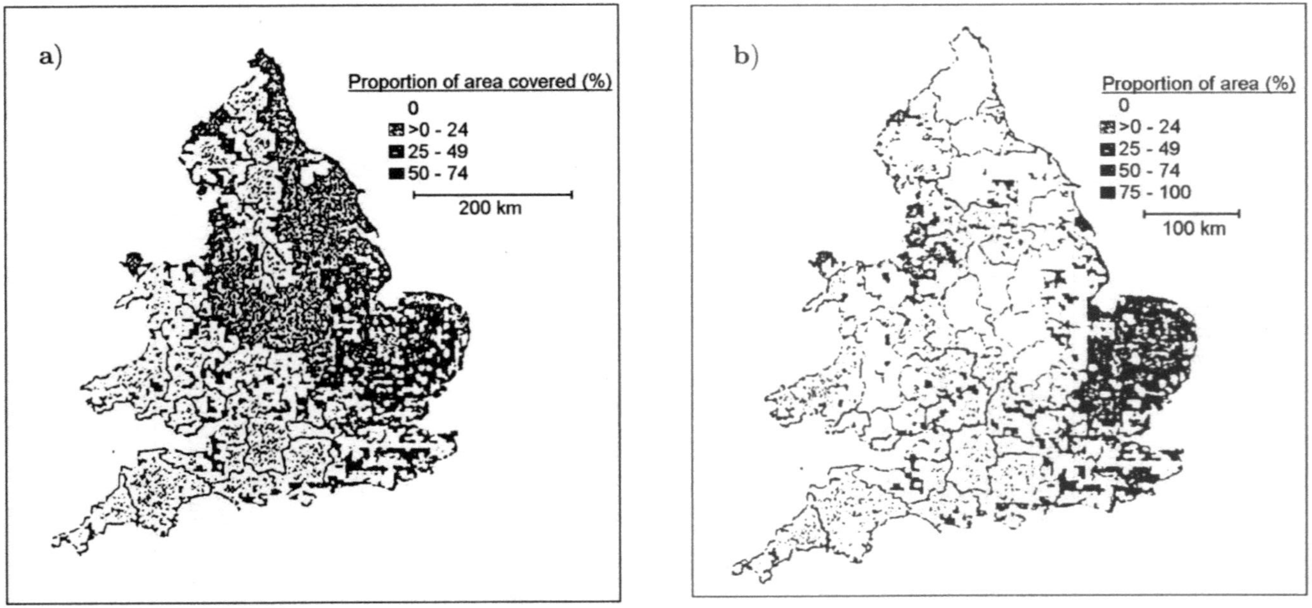

Figure 7. (**a**) The distribution of cereals in England and Wales under the mid-1980's optimal run, shown by the proportion of agricultural land covered. (**b**) The distribution of cereals in England and Wales under the Ref-2060, shown by the proportion of agricultural land covered.

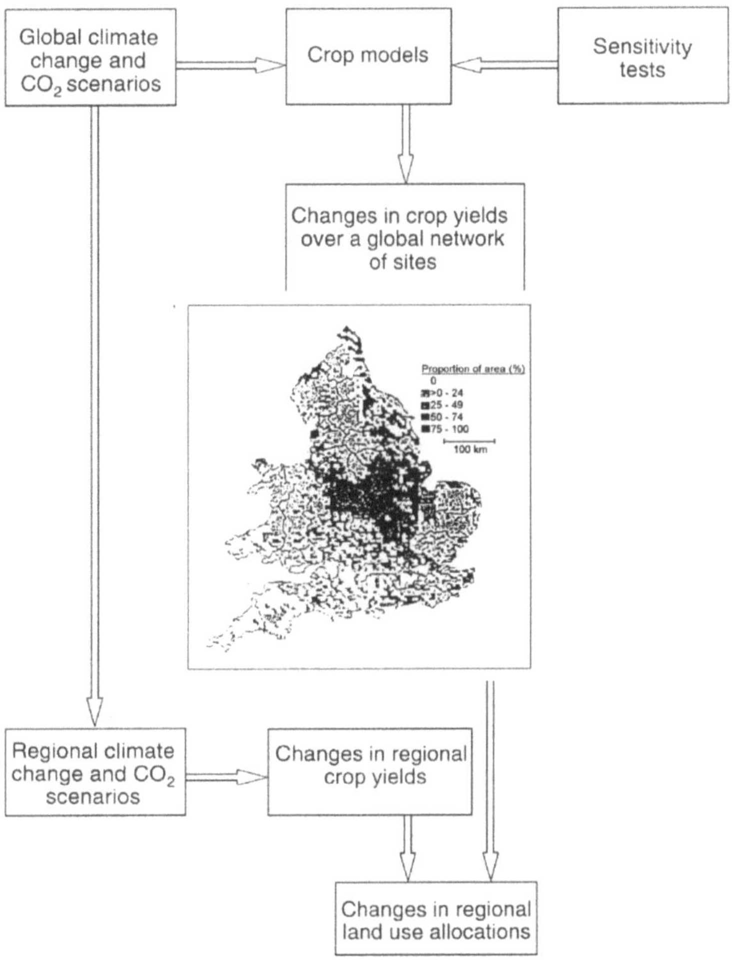

Figure 8. The distribution of "idle" former arable and ley land under the REF-2060 run, shown by the proportion of the agricultural land area covered.

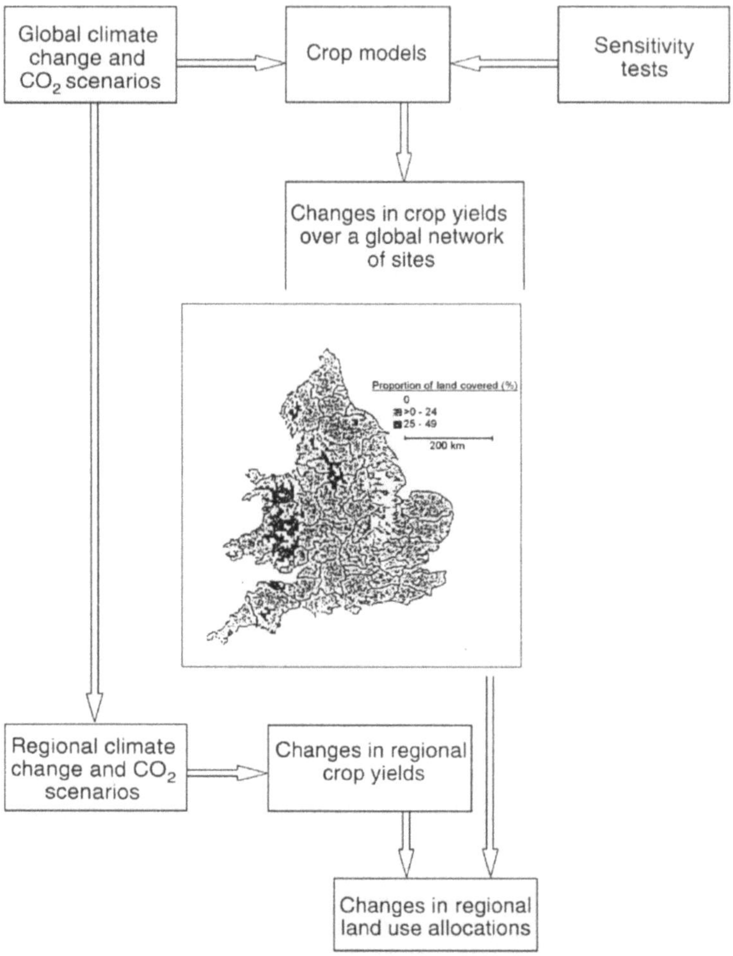

Figure 9. The distribution of "idle" former permanent pasture land under the REF-2060 run, shown by the proportion of the agricultural land area.

Table 5. The difference between BLS estimated yield and demand for CLUAM enterprises for mid-1980's Optimal and REF-2060. (Expressed as a proportion: Mid 1980's Optimal =1)

	Enterprise	Yield change	Price change	Demand change
	Wheat	1.32	0.84	0.99
	Barley	1.32	0.84	0.99
	Coarse grains	1.48	0.78	1.43
	Sugar beet	1.21	0.70	1.01
	Potatoes	1.21	0.70	1.01
	Peas and beans	1.21	0.70	1.01
	Oilseed rape	1.21	0.70	1.01
Dairy	(milk)	1.21	0.83	0.96
	(calf)	1.21	0.88	–
	(cull)	1.21	0.88	–
Beef	(meat)	1.21	0.88	0.93
Sheep	(lamb)	1.21	0.88	0.93
	(wool)	1.21	0.56	0.87
	(cull)	1.21	0.88	–

Yields

- Wheat yields changes are higher under the GISS scenarios than for REF-2060.
- Changes in barley yield are not different from those projected for REF-2060.
- Yield increases between 1990 and 2060 are estimated to be significantly higher for all remaining crop and livestock enterprised under GISS than under the REF-2060.

Prices

- Wheat prices show only moderate variation between the REF-2060 and GISS scenarios.
- Between 1990 and 2060 prices for crops such as sugar beet and potatoes are estimated to fall by about 30% in real terms, under the REF-2060 and similarly under the GISS scenario.

Demand

- Cereals demand is estimated to remain fairly constant between 1990 and REF-2060, and is not greatly affected under the GISS scenario.
- Beef and sheep meat demand is estimated to fall by roughly 7% in the REF-2060 and this is maintained under the GISS scenario.

While there is very little difference in the national cereals area between the REF-2060 and the GISS 2 × CO_2 scenario run, there occurs a major shift of the area northwards (Fig. 10). The explanation for this is the projected reduction of soil moisture in traditional production area in S.E. England. Cereal production is essentially seeking out those land classes where moisture

Table 6. BLS estimated yield, price and demand changes for CLUAM enterprises between 1990 and 2060 GISS scenario. (Expressed as a proportion: 1990=1)

	Enterprise	Yield change	Price change	Demand change
	Wheat	1.43	0.87	1.00
	Barley	1.33	0.87	1.00
	Coarse grains	1.49	0.91	1.38
	Sugar beet	1.22	0.69	1.03
	Potatoes	1.39	0.69	1.03
	Peas and beans	1.39	0.69	1.03
	Oilseed rape	1.39	0.69	1.03
Dairy	(milk)	1.45	0.86	0.96
	(calf)	1.45	0.90	–
	(cull)	1.45	0.90	–
Beef	(meat)	1.45	0.90	0.93
Sheep	(lamb)	1.45	0.90	0.87
	(wool)	1.45	0.59	0.87
	(cull)	1.45	0.90	–

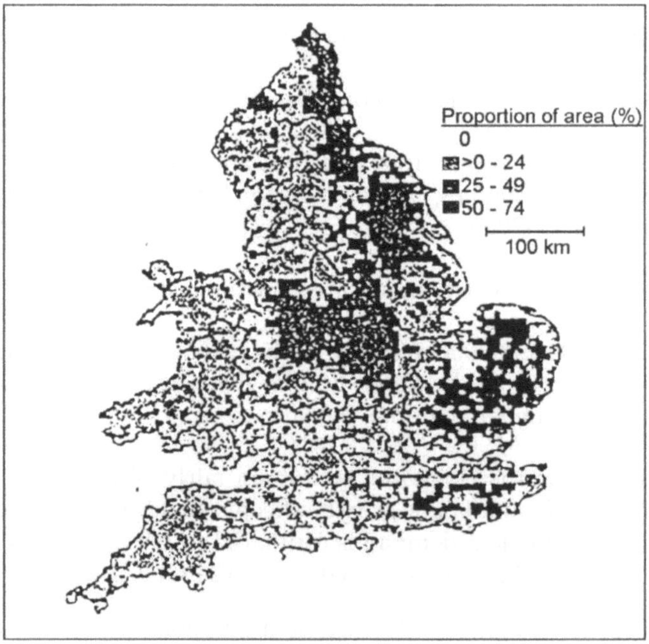

Figure 10. The distribution of cereals under the GISS 2 × CO_2 GCM run, shown as the proportion of the agricultural land under this enterprise

and temperature are at their optimum levels for output. These tend to be further north and further west under the GISS scenario than under current climate.

3 Conclusion

The purpose of this paper has not been simply to report estimates of land use change for specified changes of climate. Its objective has been to describe a means by which this can be done. Regional land-use models can be developed which capture the broadscale changes in the global agricultural environment that may occur as a result of climate change. These models can then be used to estimate the finer scale land-use response. As in any exercise such as this the sensitivity of the models and the use of them in combination needs to be explored in relation to the whole array of assumptions both economic and physical. There has not been the space to explore these tests fully in this paper; and not until they have been fully examined can the effects of climate change be properly evaluated. However, the means exist by which this can be done.

References

1. L. H. Allen Jr., K. J. Boote, J. W. Jones, P. H. Jones, R. R. Valle, B. Acockand, H. H. Rogers, and R. C. Dahlman. Response of vegetation to rising carbon dioxide: Photosynthesis, biomass and seed yield of soybean. *Global Biogeochemical Cycles*, 1:1–14, 1987.
2. T. R. Carter, M. L. Parry, and J. H. Porter. Climatic change and future agroclimatic potential in Europe. *International Journal of Climateology*, 11:251–69, 1991.
3. J. D. Cure and B. Acock. Crop responses to carbon dioxide doubling: A literature survey. *Ag. and For. Meteor.*, 38:127–145, 1986.
4. G. K. Fischer, M. A. Frohberg, Keyzer, and K. S. Parikh. *Linked National Models: A Tool for International Food Policy Analysis.* Kluwer, Dordrecht, 1988.
5. D. Godwin, J. T. Ritchie, U. Singh, and L. Hunt. *A User's Guide to CERES-Wheat - V2.10.* International Fertiliser Development Center, Muscle Shoals, 1989.
6. D. Godwin, U. Singh, J. T. Ritchie, and E. C. Alocilja. *A User's Guide to CERES-Rice.* International Fertiliser Development Center, Muscle Shoals, 1992.
7. D. R. Harvey, T. Rehman, P. Jones, and M. Upton, editors. *The centre for agricultural strategy land use model*, Aberdeen, 1992. Agricultural Economics Society Annual Conference.
8. C. A. Jones and J. R. Kiniry. *CERES-Maize: A Simulation Model of Maize Growth and Development.* Texas A&M Press, College Station, 1986. 194 pp.

9. J. W. Jones, K. J. Boote, G. Hoogenboom, S. S. Jagtap, and G. G. Wilkerson, editors. *SOYGRO V5.42: Soybean Crop Growth Simulation Model. Users' Guide*, Gainesville, 1989. Department of Agricultural Engineering and Department of Agronomy. University of Florida.
10. B. A. Kimball. Carbon dioxide and agricultural yield. An assemblage and analysis of 430 prior observations. *Agronomy Journal*, 75:779–788, 1983.
11. J. T. Ritchie and S. Otter. Description and performance of CERES-Wheat: A user-oriented wheat yield model. In W. O. Willis, editor, *ARS Wheat Yield Project*, pages 159–175. Department of Agriculture, Agricultural Research Service ARS-38, Washington, 1985.
12. J. T Ritchie, U. Singh, D. Godwin, and L. Hunt. *A User's Guide to CERES-Maize - V2.10.* International Fertiliser Development Center, Muscle Shoals, 1989.
13. C. Rosenzweig and M. L. Parry. Potential impact of climate change on world food supply. *Nature*, 367:133–138, 1994.
14. C. Rosenzweig, M. L. Parry, G. Fischer, and K. Frohberg. Climate change and world food supply. Research Report 3, pp. 28, Environmental Change Unit, Oxford, 1993.
15. P. R. Rowntree, B. A. Callendar, and J. Cochrane. Modelling climate change and some potential effects on agriculture in the UK. *Journal of the Royal Society of England*, 149:120–26, 1989.

Commentary: On the Inconsistency at the Interface of Climate Impact Studies and Global Climate Simulations

Martin Claussen

1 Introductory remarks

In their paper, *Parry* et al. [5] analyse the potential effects of climate change on agricultural land use owing to the response of the world food system to global climate change as well as to the specific changes in climate in various regions of the world. As input for their crop yield model Parry et al. use observed climate data and results of climate simulations performed with three different general circulation models (GCMs) of the atmosphere. The models provide the information on potential climate change owing to an increase in greenhouse gases which is expected to happen within the next century.

In my comment, I will not criticize Parry et al.'s discussion on their crop yield model; instead, I would like to draw the reader's attention to a problem often encountered in climate impact studies in general – not only in the paper of Parry et al. – that is the inconsistency at the interface of climate impact studies and global climate simulations. This inconsistency is related to the difference in spatial scales at which, on the one hand, information is needed for climate impact assessment and at which, on the other hand, climate models can provide reliable information. Secondly, the results of a climate simulation are often taken as a projection of climate change. However, it is rarely recognized that a single climate simulation is just one (numerical) realization of climate out of infinitely many other, equally likely realizations. Owing to internal variability of climate, these realizations more or less differ.

2 The problem of spatial scales

Climate impact studies often use output of GCMs which is implicitly considered as a continuous field. GCM modellers sometimes provoke this impression by presenting their results as interpolated fields in a high-resolution, fine-scale geographic map. Consequently, the gridding is thought to be just a convenient way to store the GCM output economically. The information resolved by the grid is considered reliable, and the subgrid-scale information may be recovered from the gridded data simply by spatial interpolation.

This view is quite problematic (*v. Storch*, 1995 [7]). Because of numerical reasons, the GCM output is reliable for scales considerably larger than the grid; features smaller than two times (some would argue four or even nine times) the grid-box size of a GCM cannot be seen by the numerics.

As a consequence, to give consistent, reliable information, climate impact studies could be done in a grid which is at least as coarse as the GCM's grid from which the forcing data are taken, or alternatively, the coarse scale information of the GCM has to be brought down to the scale needed for the climate impact study. Such downscaling methods are currently tested – as outlined by *Giorgi* et al. [3] and *v. Storch* et al. [8], for example.

To illustrate what I have said, I will discuss Figure 1.

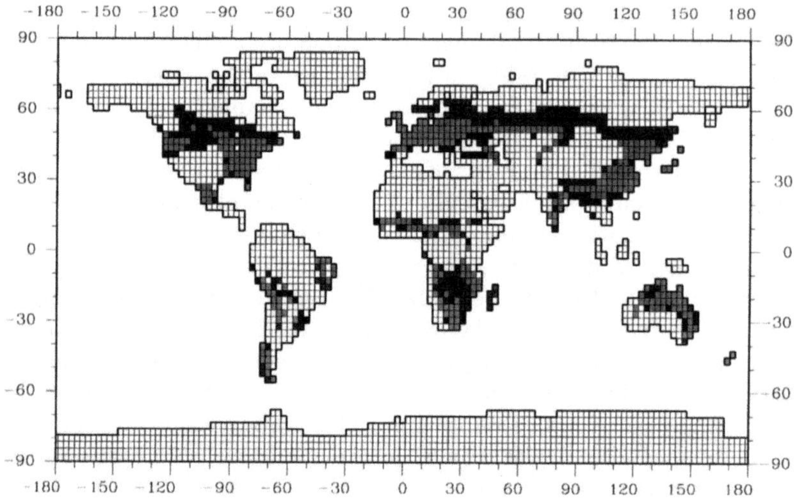

Figure 1. Change of potential agricultural land owing to a greenhouse gas induced climate change. For details see text.

In this figure, the change of potential agricultural land due to a greenhouse gas induced climate change is depicted. Light grey indicates stable agricultural land, grey, an increase, and dark, a loss of agricultural land. The model to compute the agricultural envelope is taken from *Cramer* and *Solomon* [2].

In their paper, Cramer and Solomon use the results of four GCMs which operate at a rather coarse resolution of some 5×5 degrees latitude/longitude to obtain climatic constraints needed in the land-use model. However, they present their results in the cited paper as fine-mesh, 0.5×0.5 degrees latitude/longitude maps. More precisely, they used a high-resolution map of observed present-day climate (*Leemans* and *Cramer*, 1991 [4]) and added the climate change simulated by the GCMs. In this way, the horizontal variability of climate owing to changes in small-scale orography is recaptured; however, any influence of small-scale orography on the regional climate change (for example on changes in precipitation pattern) is not included, because it cannot be simulated by the coarse-scale GCMs. Therefore, much of the information provided by Cramer and Solomon on modifications of potential agriculture induced by climate change is largely due to spatial interpolation, but not explicitly simulated by the models. This information could be considered as some scenario, but one has to be aware that the small-scale structures inferred from this scenario could be inconsistent with atmospheric dynamics.

Here, the biospheric computations are done at the same resolution (at approx. 2.8 × 2.8 degrees lat./long.) as provided by the European Center Hamburg model (ECHAM), developed at the Max-Planck-Institut für Meteorologie in Hamburg (*Roeckner* et al., 1992 [6]) which is used to drive the land-use model. The information contained in Figure 1 provides global-scale information (see below) which is consistent with GCM simulations. Of course, it is of little help for regional impact assessment, for instance concerning the change of potential agriculture in various regions of England.

3 The problem of climate variability

Figure 1 is a result of comparing two numerical realizations of present-day and future climate, respectively. However, there are, in principle, infinitely many, equally likely numerical realizations of a given climate state. These realizations differ owing to climate variability. Hence the climate impact derived from these realizations also differs. To explore the influence of climate variability on the significance of the results presented in Figure 1, three realizations of present-day climate and three realizations of future climate have been used to construct nine difference maps. A local t-test has been employed to yield the following results (for a more thorough discussion of this problem see *Claussen* [1]).

Some of the changes, depicted in Figure 1, for instance the loss of arable land in SE-Europe and in the Great Plains of N-America as well as the gain in Patagonia are locally significant. However, these changes are presumable not reliable, because they are small scale, and certainly not globally significant. The gain in arable land seen in S-Africa is a rather large-scale change, but it turns out to be not significant – it is within the natural climate variability simulated by the climate model. Only the gain in arable land in N-Europe, Siberia, and Canada constitutes a coherent, reliable, and significant feature. In fact, it is the latter change that causes a significant gain in arable land in the Northern Hemisphere. On global average, however, nothing of this significance is left as the climate variability conceals any trend.

4 Summary

In conclusion of my comment, I would strongly recommend to be very careful when using GCM results for climate impact studies. There is a scale problem and one has to bridge scales either by downscaling GCM results or by somehow upscaling climate impact models. The problem of downscaling GCM results with respect to climate impact models is currently undertaken at the Potsdam Institute for Climate Impact Research (Uwe Böhm, personal communication). Secondly, climate variability has to be taken into account to critically assess the range and the statistical significance of climate impacts.

References

1. M. Claussen. Variability of global biome patterns as a function of initial and boundary conditions in a climate model. *Climate Dyn.*, 12:371–376, 1996.
2. W. P. Cramer and A. M. Solomon. Climatic classification and future global redistribution of agricultural land. *Clim. Res.*, 3:97–110, 1993.
3. F. Giorgi, M. R. Marinucci, and G. Visconti. A 2 x CO_2 climate change scenario over Europe generated using a limited area model nested in a general circulation model. 2. Climate change scenario. *J. Geophys.Res.*, 97:10011–10020, 1992.
4. R. Leemans and W. P. Cramer. The IIASA database for mean monthly values of temperature, precipitation and cloudiness on a global terrestrial grid. Technical Report RR-91-18, International Institute for Applied Systems Analysis, Laxenburg, 1991.
5. M. L. Parry, C. Rosenzweig, P. Jones, T. Rehmam, R. B. Tranter, J. S. Marsh, J. C. Taylor, and J. E. Hossell. Climate change and land use: global and regional analyses. In H.-J. Schellnhuber and V. Wenzel, editors, *Earth System Analysis: Integrating Science for Sustainability*, Berlin, 1998. Springer. Contribution in this Proceedings.
6. E. Roeckner, K. Arpe, L. Bengtsson, S. Brinkop, L. Dümenil, E. Kirk, F. Lunkeit, M. Esch, M. Ponater, B. Rockel, R. Sausen, U. Schlese, S. Schubert, and M. Windelband. Simulation of the present-day climate with the ECHAM model: Impact of model physics and resolution. Report No. 93, Max-Planck-Institut für Meteorologie, Hamburg, 1992.
7. H. von Storch. Inconsistencies at the interface of climate impact studies and global climate research. *Meteorologische Zeitschrift*, 4:72–80, 1995.
8. H. von Storch, E. Zorita, and U. Cubasch. Downscaling of climate change estimates to regional scales: an application to Iberian rainfall in winter time. *J. Climate*, 6:1161–1171, 1993.

Chapter 3: Sustainable Development in Agricultural Landscapes

Heidrun Mühle

Abstract In modern industrial countries, intensive agriculture is one of the causes of environmental pollution and a contributing factor to global change, for instance:

- soil erosion and soil degradation;
- decreasing biological diversity;
- eutrophication of water;
- disturbance of biogeochemical and energy cycles.

It is, therefore, necessary to specify agroecological concepts – as a part of concepts for a whole landscape – which can be integrated into national and international programmes to protect the environment as well as to permit environmentally compatible production, and safeguard sustainable development in agricultural landscapes.

The basis for further investigation in the economic and social spheres seems to be the natural sphere. Before connecting different types of ecosystems it is important to know the fluxes of material end energy within a special ecosystem like an agroecosystem. Profound knowledge of the dynamics of a system is a prerequisite for ecologically sound management.

This paper gives a short survey on the development of a physiologically based detailed plant model, developed in the group "Agro-ecosystem research Quedlinburg", and the soil model "CANDY" of the section "Soil Sciences" at the Centre for Environmental Research Leipzig-Halle. The paper shows possible connections between these tasks and the issues of Potsdam Institute for Climate Impact Research (PIK).

1 Introduction

One of the most important and interesting future principles of economic development being discussed in the wake of the "Brundtland-Report" is that of "Sustainable Development".

Lively discussion is in progress about this theme at many levels and in many political, social, economic, ecological and scientific spheres. However, operationalization is very difficult for many reasons.

There is a wide range of definitions for "sustainable development" both in the natural sciences and in economics. Some want to continue real economic growth, as in their opinion only such a growth can create the capacity to solve environmental problems. Others want to create a new model of development, which considers not only limiting damage to nature but also the long-term preservation of natural potentials, both quantitatively, and qualitatively. This model advocates the integration of human activities into natural ecosystems in a way that does not irreversibly damage natural elements and cycles.

The concept of sustainable management grew up in resource management of forestry and agriculture. The concept was broadened to that of sustainable development by including the economical and social sphere. The basis for further investigation in the economic and social spheres is the natural sphere. Before connecting different types of ecosystems it is important to know the fluxes of material and energy within a special ecosystem like an agroecosystem. Profound knowledge of the dynamics of a system is a prerequisite for ecologically sound management.

2 The principle of sustainability in a cultural landscape

Haber [5] takes the view that already the phenomenon "Life" embodies the principle of sustainability. He gives an example for sustainability in different ecosystems. In his opinion the difference between so called poor and rich ecosystems in the world depends on natural resources in the earth's crust. Both of them have succeeded in evolving sustainable balances. Man has irreversibly altered the natural pattern of ecosystems in the world. Initially exploitation of resources was light. Under moderate exploitation ecosystems adapted or changed into semi natural ecosystems and became dependent on human management. Overexploitation resulted in degradation and destruction.

The pattern of a typical cultural landscape has evolved under human influence. It consists of four component ecosystems:

1. natural ecosystems;
2. semi natural ecosystems (pastured forests, scattered meadows);
3. agricultural ecosystems, consisting of fields, gardens and vineyards, consciously formed by men;
4. urban ecosystems of villages and towns with their traffic connections, businesses and industries.

Semi-natural, agricultural and urban ecosystems are stable (without great changes in a short time) only under human intervention; they are not stable (or self-restoring) without external management.

The strong interaction between the different constituent ecosystems leads to concepts for a *whole* region. Landscapes with the typical pattern have the following important functions (*Niemann*, 1985 [6]; *Haase*, 1991 [3]), and their maintaining is necessary:

- Production functions
- Cultural functions
- Human-ecological functions
- Ethical and aesthetic functions.

3 The role of agricultural landscapes within cultural landscapes

Progress toward sustainability requires intelligent management of the interacting constituent ecosystems of the cultural landscape. However, first we must begin with the assessment of interactions within an *individual* ecosystem, and so in agricultural landscapes too. We must develop strategies for their management.

We have to organize material fluxes between the component ecosystems, for instance between urban areas and agricultural landscapes or semi natural ecosystems, because urban areas are dependent on:

- food input from domestic or foreign agriculture,
- water and fresh air supply,
- energy material such as coal, oil, gas and wood,
- ability to depose of wastes to the surrounding area or in tips or in sewage treatment plants.

Agricultural ecosystems are dependent on urban markets and centres of distributions localised in towns or settlement structures. There farmers buy machines, fertilizers, pesticides, insecticides etc., and interact with central nodes of energy supply and administration. Agricultural ecosystems are distinguished from natural ecosystems in needing external regulation, because a self regulation is inoperative. Therefore the goal of ecologically oriented agriculture is to regulate agroecosystems in a sustainable way.

Agroecosystems are also dependent on inputs from natural systems, such as rainfall, groundwater, quality of soil and genetic resources.

Agricultural landscapes play an important role in cultural landscapes; they take part in preservation, cultivation and development of the whole cultural landscape, they help to preserve the multiple functions of a landscape and they interact with other types of landscapes. On the other hand agriculture both causes environmental pollution and suffers from pollution.

Sustainability in agricultural landscapes requires the protection of all the factors which are directly influenced by agricultural or farm management. However, we have also taken into account factors out of agriculture which impact on the agricultural landscape or special typical factors in it.

The prerequisites for reaching sustainability (or resource protection) are as follows:

- The soil must be protected from degradation, destruction or erosion.
- The overuse of soil is to be avoided.
- Marginal soils should not be used.
- Material cycles in farms should be as closed as possible.
- Pollution of environment, caused by agriculture is to be avoided.

- Farmers should be doing the transition from profit oriented agriculture to an ecological oriented kind of management (low doses of nutrients, pesticides and insecticides; recycling or reuse of material; in agriculture use of manure to improve the soil fertility, feeding of pigs with left-overs).
- To maintain high biodiversity it is good to raise multiple varieties or races in farms. This is one of the important prerequisites for farm stability.

Soils play an important role in reaching sustainability. Therefore we must protect the soils from sealing and the landscape from fragmentation. This is necessary to allow future generations to use the soil for their own needs or ideas.

Hey et al. [2] give recommendations for changing agricultural politics from an intensive agriculture to an ecological directed one. They propose the use of renewable energy sources in farms, the linking of production and environmental protection, regional autonomy in the goals specified by the framework of the European Union, and payment of farmers for secondary benefits of sustainable agriculture such as protection of water quality and provision of aesthetic and recreational opportunities for urban people.

How do we define boundaries of an agricultural landscape?

On the one hand there are natural differences in soil and weather conditions within a country, for instance Germany (Figure 1).

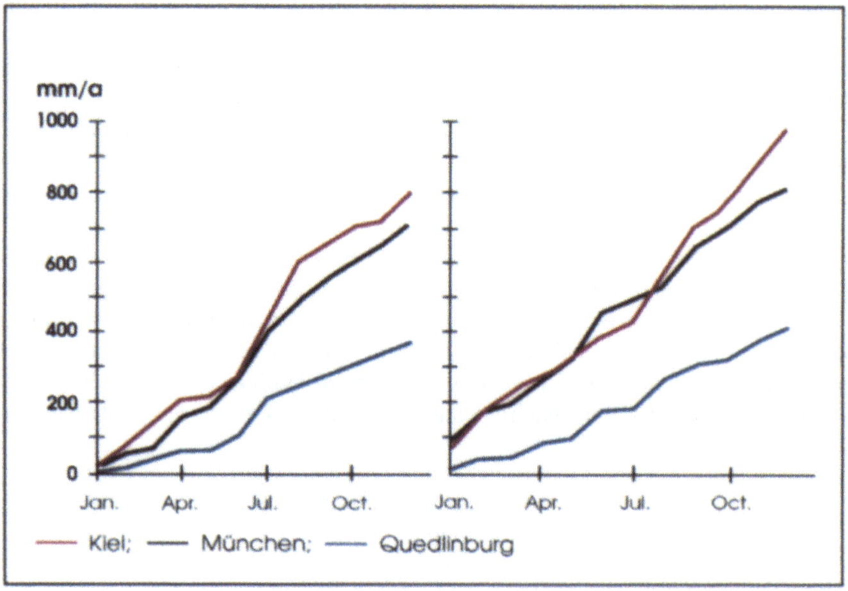

Figure 1. Cumulative precipitation in 1989 and 1990, in mm

On the other hand historical aspects of farm development play a role. Recommendations for agroecological management must take account of farming conditions, such as the sizes of farms. In the so called old states of Germany farms are rather small, in the new states they are large. Although, at present in Saxony, ecological farming practices tend to be restricted to smaller farms (area on average 20 ha), it will be also possible to manage larger farms on the same way. Large farms can also be managed ecologically. Large farms offer the following advantages:

- using the machines for seeding, ploughing, fertilizing and harvesting on larger areas is more efficient;
- dosage of fertilizer, pesticides, insecticides and herbicides (if they are used) can be made more exactly and more efficient;
- it is easier to plant more varieties or to keep more races of animals – thus lowering the risk of diseases or pest damages by broadening the genetic basis of species. This fact will promote the stability or the resilience of a farm, an ecosystem or a landscape.

4 The application of mathematical models in research of agroecosystems

The *basis* for further investigations in the fields of economy and social sphere is the natural sphere.

Before connecting different types of ecosystems it is important to know the fluxes of material (mass) and energy within component ecosystems – including agroecosystems. Profound knowledge of the dynamics of a system is a prerequisite for ecologically sound management. Research on agroecosystems must, therefore, be aimed at gaining insight into their structure, their function, regulation and stabilizing mechanisms.

Findings thus obtained are to be formalized dynamic mathematical models and made available for case studies. Using these studies, predictions can be made about the impacts of the various processes, such as changing agricultural land use, climatic changes, and contamination by pollutants.

That means research at different hierarchical levels of an ecosystem. It must be carried out for agriculturally utilized areas, and models with different degrees of details must be developed.

The methods and models to be developed have to meet the following requirements:

- They must represent the current state of the target landscape and provide quantitative information on events or conditions that may destabilize the system.
- Models must quantify the input of external flows into the system, the transformation of material and energy into system-specific material and

the output of residues from one system to another system. This must be done not only for flows between component ecosystems, but also between different compartments of a given ecosystem.
- Case studies must make it possible to derive measures for attaining stability or at least limiting the extent of damage within the system under consideration.

Figure 2 gives a survey on different hierarchical levels in a biological (or agricultural) ecosystem.

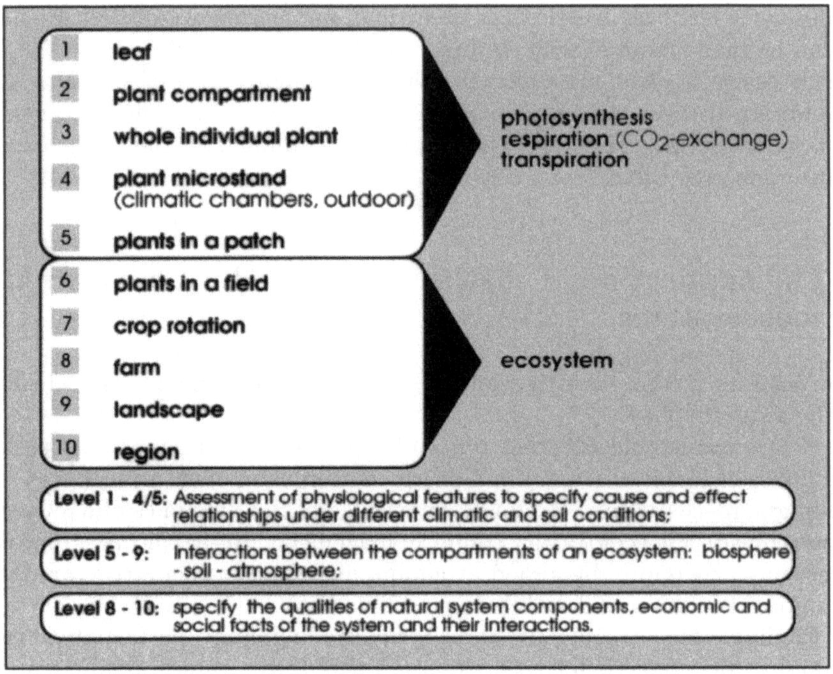

Figure 2. Hierarchical levels in a biological system

First one has to decide the way of analysis and modelling:

1. Current situation and problem: Regeneration, remediation and sustainable utilization of a landscape;
2. Definition and limitation of the problem: Examination of stress and resilience of a landscape, stress of other landscapes;
3. Definition of aims: Subdivision of natural sphere into layers and compartments, determination of material and energy fluxes and of balances within the system;

4. Development of solutions: Current state of the system, changes of the system, time series for system variables and environmental variables, interaction within the system;
5. Development of models and validation: Development of detailed models for homogeneous areas (patches), filling of a geographical information system (GIS) with information, coupling the GIS with detailed models, development of regional complex models.

A comprehensive plant and soil model is being developed which will reflect precisely the response of the plant stands to climatic and soil conditions (*Claus* et al., 1994 [1]). In Figure 3 a survey on the interaction between plant stands and the surroundings is given.

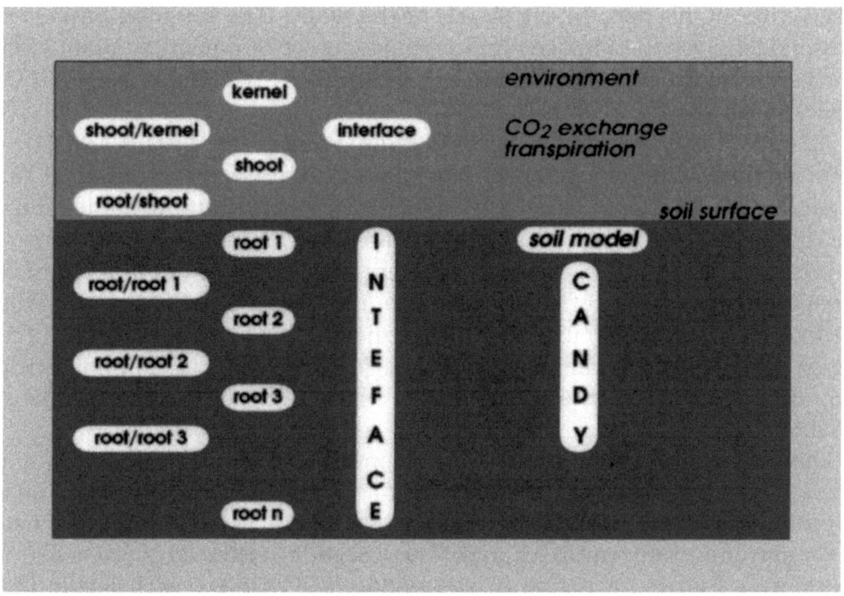

Figure 3. Plant-soil-environment interaction scheme

In this complex model several submodels describing soil, plant, CO_2 exchange in the agro-ecosystem, insect pests and predators are linked by proper defined interfaces. These models consist of compartments which are represented by instances of object types from a given object hierarchy. Further, a special object controls the interactions between the generated objects in the dynamic list of objects. The objects are able to interchange messages. All data like results, parameters and files containing values of environmental factors are managed by data bank objects. The plant model includes compartments called "shoot" and "kernel" aboveground and "root" belowground. Kernel, shoot, and root are connected by additional object types, such as

- "root/shoot"
- "root/root"
- "shoot/kernel"

corresponding to the fluxes from shoot to kernel, root to shoot, and root to root. These fluxes incorporate both phloem and xylem transport within a plant.

Environment directly affects shoot and (later) kernels through temperature – as affecting photosynthetic rates and transpiration – and through atmospheric CO_2-concentration, air humidity, radiation, and wind speed. A submodel consisting of coupled CO_2-exchange and transpiration modules is linked with an interface to the shoots of the plant. The common basis of the plant model gives the nitrogen and carbon content of a plant or plant stand. Another interface links the soil model "CANDY" (CArbon and Nitrogen DYnamics) with the root compartments.

"CANDY" is a soil model developed by *Franko* and *Oelschlägel* [4]. It exists in three different versions. Version 1 is suitable for practical farm management. Version 2 emphasizes changes of important variables during the course of crop rotations and is suitable for long-term simulations. It can also be coupled with the weather generator, a programme for deriving site-specific projections of climate change from the Large Scale Geostrophic European Center HAMburg general circulation model ECHAM/LSG of the Klimarechenzentrum Hamburg. Version 3 is suitable for coupling with a geographical information system (Figure 4). It generates time series of soil variables.

One can get synthetic data of development and growth of plant stands, on evapotranspiration and CO_2 exchange as well as data on variables of soil (N-content, water conditions and other) for many patches, and one has a survey on important variables and their variability of a large area like the "Querfurter Platte", a region in the middle of Germany with fertile black soils and dry climatic conditions (*Schenk* and *Franko*, 1993 [7]). GIS supplies the soil model "CANDY" with input data. Figure 5 shows a result of coupling GIS with the soil model "CANDY".

The plant and the soil model are the result of interdisciplinary work involving mathematicians, physicists, chemists, gardeners, agronomists and electronic engineers.

These results are the first step in the frame of the task: Creating a concept for a sustainable development of agricultural landscape as a part of a whole region.

Of course, we have to solve a lot of problems in future, for instance:

In the natural sphere:

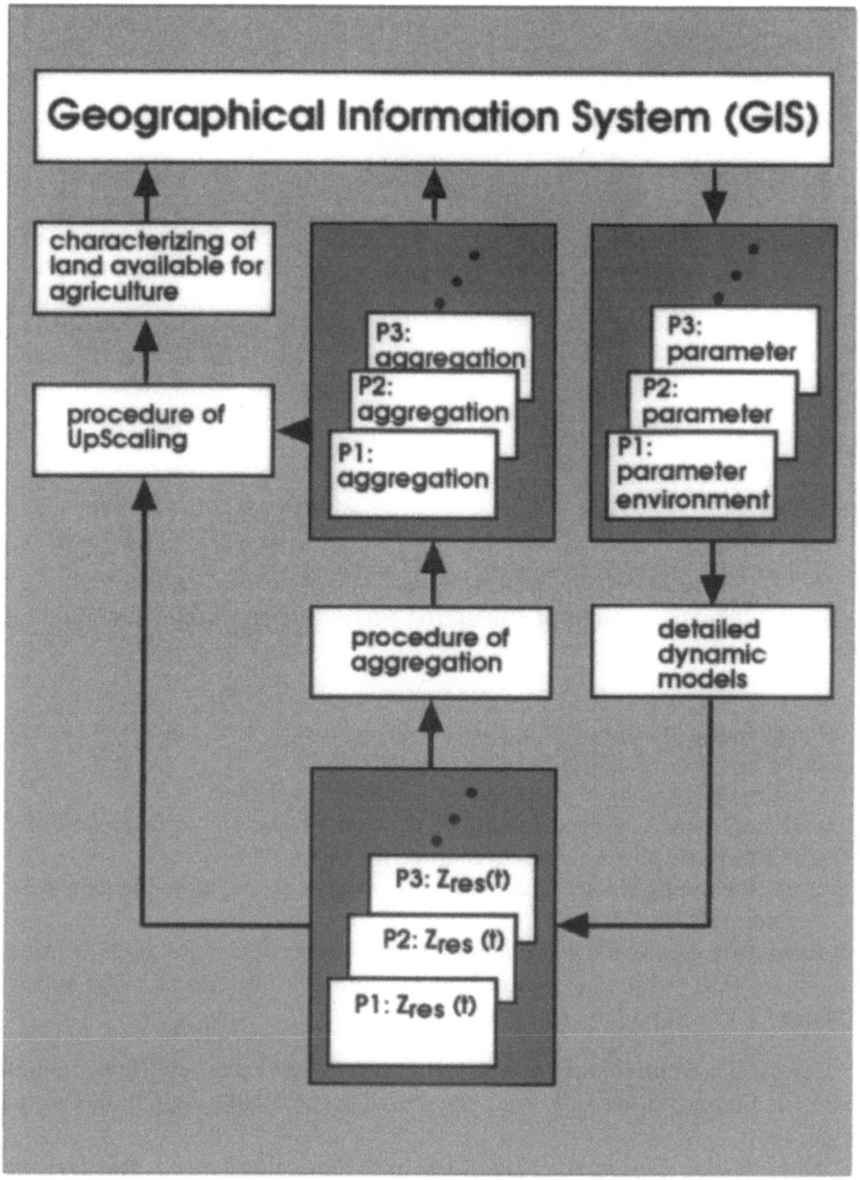

Figure 4. Inclusion of detailed models in a geographical information system (GIS)

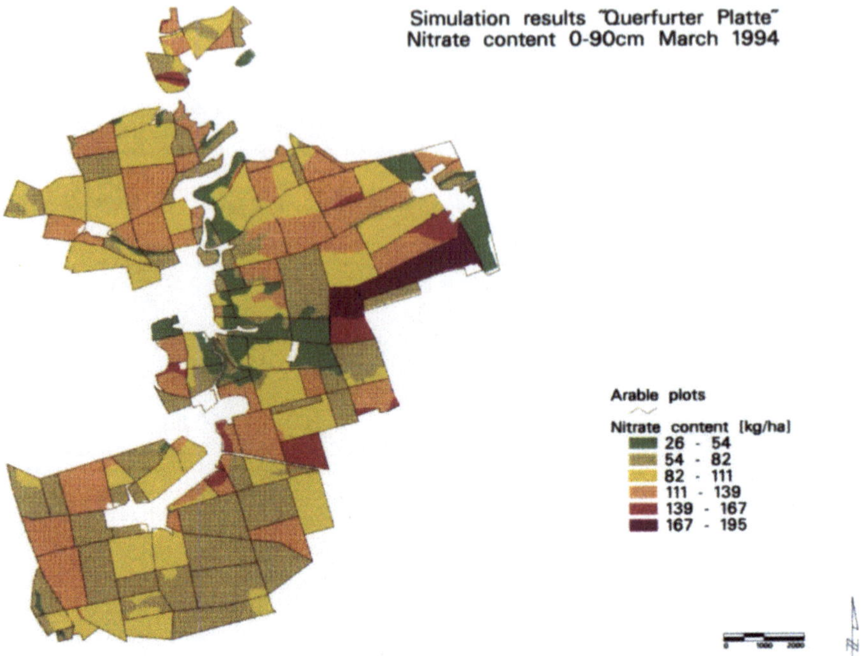

Figure 5. Simulation results of "Querfurter Platte" nitrate content 0-90 cm (March 1994)

- Development of arable crop models or adaptation of models from other groups into the modelling of crop rotations;
- Inclusion of other locations (Saxony) into the research;
- Development of scaling up methods for connecting the information of different levels within the system hierarchy;
- Search for methods to link the information from different component ecosystems of landscapes;
- Information on the cycles of material and energy fluxes in larger regions or in states (CO_2 exchange, water budget, complex mosaics of crops and wild plants).

Here will be one of the interfaces between the Centre for Environmental Research Leipzig-Halle Ltd. and the Potsdam Institute for Climate Impact Research.

We have also gaps in the field of the anthropogenic sphere. It is necessary to improve the investigations in the following way:

- Strengthening of economic research and expansion to the regional scale.

- Inclusion of social scientists in the projects and development of better understanding of the interaction between man and the regional agricultural landscape.

The problem of sustainable management and development cannot be solved in agriculture alone; solutions must encompass the entire region with urban landscapes, industry and trade. The protection of landscape functions demands good landscape planning for the whole region and not only for a few communities or municipalities.

References

1. S. Claus and H. Mühle. Complex Plant-Soil Models for Homogenous Areas - Supposition for the Quantification of Element and Energy Fluxes in an Agricultural Landscape. Conference presentation, GSF - Forschungszentrum München-Neuherberg Germany, 1994. Meeting of Terrestrial Ecosystem Research Network of Germany: "The Whole is more than the Sum of the Parts".
2. C. Hey et al. Konzepte für eine Umweltgemeinschaft Weißbuch der europäischen Umweltverbände zum EG-Binnenmarkt. *Politische Ökologie Sonderheft*, 3(22):6–19, 1991.
3. G. Haase et al. Naturraumerkundung und Landnutzung. Geochorologische Verfahren zur Analyse, Kartierung und Bewertung von Naturräumen. *Beiträge zur Geographie*, 34, 1991.
4. U. Franko and B. Oelschlägel. *CANDY*, volume 25 of *Agrarinformatik: Simulationsmodelle zur Stickstoffdynamik*, pages 99–110. Ulmer, Stuttgart, 1993.
5. W. Haber. Ökologisch nachhaltige Entwicklung von Regionen: Ansätze einer Umorientierung. *Politische Ökologie Sonderheft*, 4:13–20, 1992. Tutzinger Tagung 1992: "Sustainable Development - aber wie?".
6. E. Niemann. Ziele und Methodik einer polyfunktionalen Landschaftsbewertung. *Petermanns Geographische Mitteilungen*, 1:1–7, 1985.
7. S. Schenk and U. Franko. Kopplung des Simulationssystems CANDY an ein geographisches Informationssystem zur Bilanzierung der Stickstoffflüsse. *Berichte der Gesellschaft für Informatik in der Land-, Forst - und Nahrungsgüterwirtschaft*, 5:87–91, 1993.

Commentary on "Sustainable Development in Agricultural Landscapes"

Siegfried Bauer

1 Introductory remarks on sustainable development

The terms "sustainability" and "sustainable development" as a major policy objective have become very fashionable during the last year in the political and even in the scientific discussion. However, there is no clear and unique definition available throughout the various disciplines using these expressions. Also Dr. *Mühle* has not worked out a concept or at least a clear definition about the meaning of "Sustainable Development in Agricultural Landscape" in the previous contribution.

As early as 1991, a group of 25 agricultural scientists of different disciplines developed a policy concept, which is oriented on the future requirements of European agriculture (Wageningen Memorandum, 1991 [4]). Based on the present problems in agriculture and environment and on the future requirements, they have also developed basic policy principles for an integrated environmental oriented agricultural policy. In this connection, the term sustainable agriculture has been used for a development, which is compatible with

- environmental protection,
- reasonable income for efficient and environment-friendly farmers,
- budget control,
- stable markets,
- and international trade agreements.

This comprehensive view and the broad definition of sustainable agriculture consider the basic requirements of agriculture and the agricultural ecosystem within the global system. It is particularly important that the economic aspects have to be included in the concept of sustainability. There are many examples of concepts developed from the pure ecological point of view, which are not sustainable, because they are not economically profitable. From the environmental point of view, there is no gain, since these concepts will not be realised. For example, organic farming is only sustainable, when these farms can realise a similar per capita income due to higher producer prices. Environmental policy, as practised today, operates mainly with legal restrictions, which often do not meet the requirements of sustainability. For example, if firms go bankrupt, or if they locate their firms in other countries with lower restrictions and less environmental laws, the global environmental situation will not improve.

In most cases, sustainable development of firms in agriculture, like in other sectors, is only possible if the economic and political frame conditions are adjusted to the environmental objectives, as outlined by v. Weizsäcker in Part 3 Chapt. 3. The new economic conditions have to guarantee that an environmental behaviour and a sustainable development path of a firm (investment with low use of natural resources and low pollution) are – at least in the medium and long run – more profitable than the conventional production practice. These new conditions should basically lead to a substitution of energy by more labour intensive technologies. This would also help to solve the present unemployment problem and at least partially our global environmental problems.

2 Profit oriented versus ecological oriented behaviour

Dr. Mühle argued that there "should be a transition from profit oriented agriculture to an ecological oriented kind of management". Such arguments are very common and often expressed by ecologists. For example, the leader of the environmental expert team of the German Government has just recently similarly argued: "... it is particularly important to change the attitudes and behaviour of the farmers".

These postulates do not help as long as it is not made explicit how a change of human behaviour can be achieved, and whether this way would be successful. There are good reasons for having doubts about the seriousness of such statements. The Christian religions have tried to change the attitudes and behaviour of the people for over 2000 years but with limited success. Considering this time lag and the low efficiency, it seems that the environmental problems are too serious and that we do not have enough time to wait only for the results of moral appeals. On the other hand, it is not true that, for example, organic farmers are less profit oriented. They produce a commodity with another quality standard and consequently with other production techniques. In general they are very good managers, with good economic knowledge, and they carry out their own marketing activities. They must be profit oriented, otherwise they can not guarantee sufficient income and social security for their families. The profit oriented behaviour of these farms leads to a long run survival and thus to sustainability.

From this we can conclude that economic and profit oriented individual behaviour should be taken as a basic rule for policy concepts, which means adjusting the policy instruments with regard to the general requirements of the society.

3 Main environmental problems in agriculture and their causes

The basic principles outlined above can be exemplified very well if we look at the environmental problems of agricultural development. These problems can clearly be put down to technological, economical and political conditions under which agriculture has to operate. In the following, we will look at some typical and environmentally relevant characteristics of agricultural development, their causes and their ecological impacts (for more details see *Bauer*, 1990 [1]).

1. The increasing intensification of agricultural production, in the form of an expanded use of factors of production (fertilisers, pesticicides, concentrates, animal medicines, is mainly caused by the favourable relationship between output and input prices, and the change of the factor-price ratios. These conditions have led to a high level of input use and to an increasing use of modern technologies (high yielding varieties in plant and new breeding types in animal production). These economic conditions have also induced the development of new technologies, like the promotion of mid- and long-term breeding improvement programs. From the ecological point of view, this development has involved many problems, like residues in the food commodities, in the soil, in the ground, in the surface water, and in the air. Furthermore, another consequence of the more and more intensive agricultural practices is the decrease of the varieties and species (biodiversity), which has to be viewed very critically from the environmental point of view.
2. The agricultural development is also characterized by an increasing mechanisation, which can be explained by the increasing wage-interest ratio. In the field of crop production, mechanisation is associated with soil deformation, increasing farm sizes and larger plots as well as with faster operating times in land preparation and harvesting, partially accompanied with massive problems for flora and fauna (less protection and withdrawal possibilities for wild animals). The technical progress in animal production has resulted in larger livestock herds which in turn have negative consequences on the ecology (overstocking, manure problems). The problems which arise due to the mechanisation of agriculture can also be observed in "organic farming" (similar methods of land cultivation). The ecological problems arising from mechanisation cannot be controlled very specifically through measures of environmental and agricultural policies. One possibility would result in subsidising environmentally friendly technologies (e.g. minimum tillage machines etc.).
3. Many measures of land clearing have taken place in the 1960s and 1970s with the consequence of losing ecologically important land. Also, these land improving measures, such as removing hedges and plotting boundaries, can be put down to the high agricultural price level. The agricul-

tural price level consequently involved higher land prices in the 1960s and 1970s which induced profitable investment in land improvement measures and land reclamation. The increasing mechanisation, mentioned above, has strengthened these developments.

4. The increasing specialisation and regional concentration in agricultural production are caused by various, mainly economic effects as there are economies of scale effects, the necessity of special knowledge, modern management practices and favourable price relations between agricultural commodities and factors of production. In particular, a specialised agricultural production pattern implies a high level of input utilisation in order to compensate for decreasing yields of a short cycle crop rotation. This high level of input utilisation, in connection with a high degree of specialisation and regional concentration, has created a number of significant negative environmental effects such as high levels of animal wastes, soil erosion, less diversified crop rotation and landscape. These ecological problems are even more serious if many farms are specialised in the same direction in a given region (regional concentration). This is usually the case, since many regions have the same location advantages, e.g. soil conditions, transportation advantages, economic conditions.

The reform of the EC Agricultural Policy, carried out in 1992, has not tackled these problems. The chance has been missed to apply an integrated concept, which could have been able to solve the traditional agricultural problems together with the increasing environmental problems of agricultural development, as it has been outlined for example in the Wageningen Memorandum. In contrast, the reform was purely oriented to traditional agricultural policy problems like financial burdens, farmers' income and international trade and GATT requirements (for more details see *Bauer*, 1994 [2]). Not only does the reform not fulfil the requirements of sustainability, but also it will create some additional environmental problems (e.g. by the set aside obligation of the reform).

4 Main elements for a market oriented environmental policy in agriculture

In relation to other sectors of the economy, agriculture has negative as well as positive environmental effects. Additionally, the agricultural sector is heavily influenced by environmental pollution of other sectors. Therefore, a comprehensive environmental policy for agriculture is more complicated than for other sectors. We have to consider that this sector consists of a huge number of farmers compared to relatively few enterprises in the other sectors. Hence environmental policy for the agricultural sector should use market oriented instruments as far as possible to provide economic incentives for the single farmer for using more non-polluting production systems. As argued elsewhere

(*Bauer*, 1994 [2]) and also outlined by the Wageningen Memorandum, such a policy system should apply taxes for certain inputs and specific subsidies for the ecological output.

- Input taxes should be applied in order to internalise the negative external environmental effects of agriculture. As argued above, many environmental problems in agriculture can be put down to a high and increasing use of modern inputs such as fertiliser, pesticide, animal medicine, energy and imported feed stuff. A tax on these inputs would lead to a lower use (intensity) and also to a lower level of specialisation and regional concentration. It would become more profitable to operate a farm with less external material and energy input (more closed farm base material circle) and therefore with less pollution. In the long run, these changed economic conditions would also induce a reorientation of the technological progress (e.g. plant and animal breeding) towards input saving technologies. Environmentally oriented input taxes conforming to market principles can very easily be implemented and do not involve administrative or control problems. The massive income effect has to be compensated by higher agricultural prices and/or by payments related to the ecological output.
- The historical agricultural land use has contributed to a large extent to the biological diversity, which we want to protect today. While in traditional agriculture, biodiversity was a (unintentional) complementary by-product, in modern agriculture, there is a competition or substitution between the amount of agricultural commodities and the production of biodiversity. Since the individual price of biodiversity is zero for the farmer, he has no economic incentive to produce such an "ecological output" e.g. to apply production techniques in agriculture like crop rotation, hedges and low intensity, which are more in line with the ecological requirements from the social point of view.
- A market oriented agricultural and environmental policy should, therefore, compensate the farmer for these important amenities. The payment given to the farmer should be based on ecological criteria on the farm level and on the social value of these amenities for the society. There are many proposals and concepts available which follow this approach. The common difficulties refer to the existence of much larger problems concerning implementation, administration and control than in the case of the input tax. Additional research and experimentation is necessary in order to develop practical policy proposals.

In summary we can conclude: The pure market mechanism will not lead to a sustainable development of agriculture, because positive and negative external environmental effects are very important. Market oriented instruments, like taxes and specific subsidies should be used for internalising these effects and in order to close the gap between the individual and the social optimum as well as to keep the individual freedom as large as possible.

In the following section some more general remarks on the topic of this symposium, particularly on the relation between research and policy and on the linkage of natural science and economic models will be made.

5 Some more general remarks on research and policy

Applied environmental research and global modelling aim to gain better information for policy making. This involves some critical questions about the relation between research and policy. There are basically two extreme positions:

Position 1 states that the basic causes of environmental and climate problems are known and their knowledge is sufficient for taking the policy actions in the right direction. This position assumes that the problems are so serious that we can and should not wait until we know more details about all the causes, complex interrelationships and exact quantitative impacts through more detailed research. In contrast, position 2 assumes that first the real system's behaviour has to be clarified and that more research is needed before the right policy actions can be taken. This position has – at least implicitly - been expressed by some participants of this symposium. Policy makers may take this position as an alibi for refusing to make unpopular decisions and to apply consequent instruments. For the researcher this argument is also useful if they ask for research funds. There is the danger that this "unholy coalition" leads to a time lag in solving the real problems as well as to a cumulation of problems and even to an unsustainable development path.

How should a compromise between these two positions look like? First, we think that serious researchers should not give an alibi for delaying policy decisions, as far as some fundamentals are clear. Second, in many cases we will never know all real facts and interdependencies. Therefore, policy decisions have always to be taken in a risky and uncertain world. Third, in many fields, a policy change itself results in new information like additional variations in our data set, which may lead us to new and additional insights. For example, there are many different approaches available to estimate the impact of an energy tax on energy use and they will all lead to different numerical results. Since the principal direction and impact is clear, there is no big danger when implementing such a tax and reducing other taxes. There are many examples in historical development that trial and error processes will lead us to new insights into development. Research and global modelling can tell us more about the complexity and some interactions of the system we are dealing with, but it can not lead to political decisions without taking any risks.

6 Linking natural science with economic models

In order to analyse the impact of the climate change on the economy, linking natural science models with economic models is necessary. For an appropriate linkage, we should understand the process as follows: At first a climatic change leads to a modification of the technological conditions. For example, low rainfall may alter the yield functions between agricultural inputs and outputs. A direct impact on agricultural yields can only be analysed if we assume that all the other inputs (growth determining factors like fertiliser, irrigation, crop rotation and finally agricultural prices) remain constant. However, this is not the case. A variation of the rainfall as one growth determining factor leads to a change in the optimal level of fertiliser use and pesticide application. The farmer may also shift to other crops and finally if agricultural prices change and food stuff shortages occur, the farmer will react accordingly and apply new production techniques (e.g. *Bauer*, *Hummelsheim*, 1993 [3]). These adjustments of the economy are at the centre of economics models.

Following these arguments, we can draw some conclusions for linking natural science and economic models. First, we need a conceptual framework, which identifies exactly the linkages between natural science and economic models. Second, these linkages should be based on functional relationships like a production function rather than the level of production itself. This means that the natural science model should analyse the impact of climatic changes on the parameters of the economic models. In principle, this again requires full specification of the economic model, before the impact of climatic changes on the parameters of the economic system can be quantified. Finally, global modelling cannot just mean linking existing models or starting with an individual model, which can later be linked with other components. Instead, integrated approaches for Earth System Analysis need a conceptual outline, which identifies exactly the structure of the various model components and the linkages between variables and parameters.

References

1. S. Bauer. Landwirtschaft und Umweltpolitik: Überlegungen aus ökonomischer Sicht. *Zeitschrift für angewandte Umweltforschung*, 5:133–147, 1990.
2. S. Bauer. EC-agricultural policy and its impact on land use and environment. In R. Krönert, editor, *Analysis of Landscape Dynamics*, pages 19–29, Leipzig-Halle, 1994. 4th Seminar EUROMAB Network.
3. S. Bauer and S. Hummelsheim. Zum Forschungsbedarf und zur Forschungsausrichtung im Bereich "Klimafolgen im Agrarsektor". Unpublished, 1993.
4. Wageningen Memorandum. A view on agriculture in the European Community. Wageningen, 1991.

Commentary: Towards a Socially Accepted, Sustainable Management of Agricultural Landscapes

Felix Kienast

1 Introductory remarks

The paper of H. Mühle clearly demonstrates how agriculture is confronted with enormous uncertainties regarding the CO_2 and climate induced impacts on crop production. The autor stresses the need to combine spatially and temporally explicit vegetation or crop models that act at different levels of detail.

2 Are models of the physical system sufficient to ensure sustainable land use?

Physiologically based simulation tools shall be nested with soil models and stand simulators as input into regional risk assessment procedures. It is the output of these models that should permit policy-makers and regulatory agencies to pursue a 'minimal damage' mitigation strategy. These strategies however, whether based on elaborate impact models or simple expert judgements, often lack public acceptance since they rarely incorporate knowledge about lay-judgements of landscapes structures and processes. Thus it is questionable if models of the physical system alone are capable of convincing the broad public to support steps towards sustainable management. Hence it is decisive to couple lay-judgement and biological expert knowledge to increase public acceptance.

3 What documents are necessary to assess risk perception patterns of practitioners and lays?

It is undisputed that impact studies at the spatial level of regions or even countries are important tools for decision making at the level of governmental agencies. However, how can practitioners or lay people voice their points of view and how can they actively participate in deciding on sustainable management strategies, even if complex national or even international issues are at stake? What type of documents have to be generated that lay attitudes, beliefs and behavioural intention can be assessed properly? Do we need top-down procedures with consultation of lays or participatory, democratic bottom-up procedures?

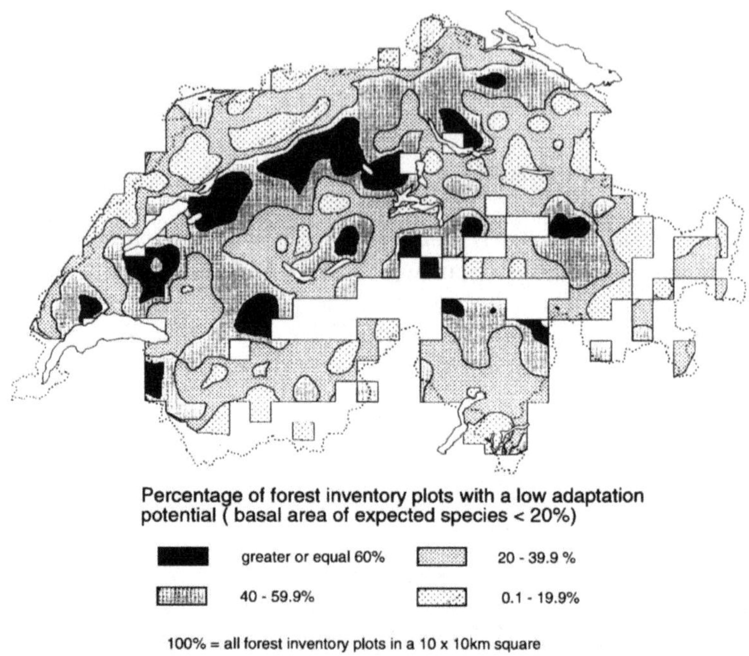

Figure 1. Risk map featuring the spatial occurrence (generalized pattern) of forests in Switzerland with a low adaptation potential to long-term species changes as expected with a $1 - 1.5$ °C warming. The isolines were generated using a 10 km $\times$10 km grid, each representing 100 km^2 of land. Squares with less than 10% forested area were excluded (white cells). The values indicate on how many forest inventory plots (relative to the total number within each 10 km $\times$10 km cell) the actual basal area of tree species that are expected under climate change is below 20% of the total basal area of the stand. Only trees with a diameter at breast height $>$ 12 cm are considered. For isoline calculations the values were concentrated geographically in the centroid of the cell.

An example of such an analysis can be illustrated for the small area of Switzerland. It aims at evaluating the long-term impacts of climate change on the forested cover at three levels of detail, namely at the national, the regional and the local scale. Numerical vegetation models and economic forestry enterprise models have been applied to perform the risk assessment (*Brzeziecki* et al., 1995 [1]; *Kienast*, 1997 [3]). Risk maps (Fig. 1) featuring the sensitivity of various regions and forest types to the potential environmental changes were used to develop mitigation strategies at the national level (*Kienast* et al., 1997 [4]). Information was predominantly generated in

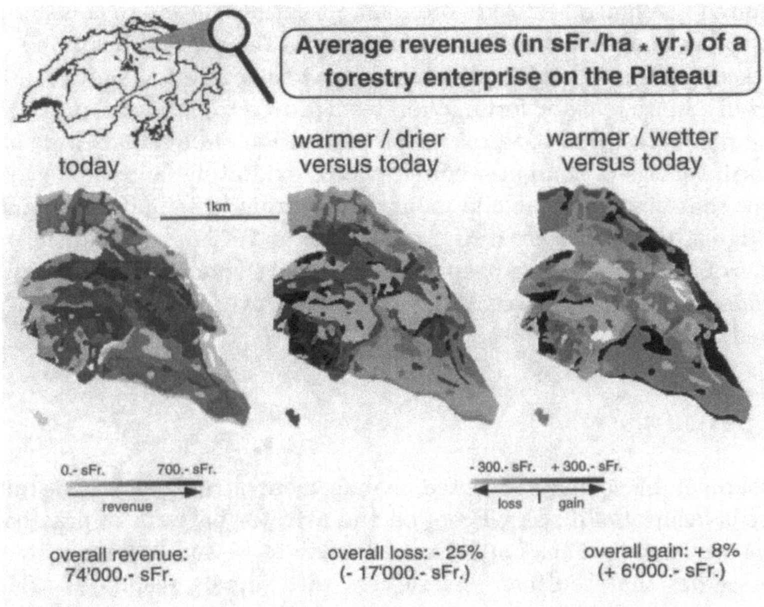

Figure 2. Potential impacts of climate change on a local forestry enterprise on the Swiss Plateau. The data are taken from a forestry enterprise model that couples potential vegetation shifts, timber production and potential yield of real forestry enterprises.

the form of maps and tables. Target groups included governmental agencies at the federal level and NGOs. Information on the intermediate scale (no Figure given) had to be more specific and regional details and cultural aspects had to be incorporated in order to assess the risk perception of stakeholders. This local aspect is of great importance at the lowest spatial scale considered where impacts of climate change had to be transformed to local economic systems such as forestry enterprises or individual farms. At this spatial scale it was important to get answers on the following questions:

- what is the threshold of vegetation change after which any further changes are perceived as a risk by the various stakeholders?
- when does the perceived risk significantly affect the decision making process?
- which innovative actions are or would be taken within the different regions and social groups to cope with the perceived risks?

Figure 2 is an example of a document that was used to assess risk perception and mitigation strategies of forestry enterprise managers at the local level (*Kienast*, 1997 [3]). It exhibits potential financial gains or losses of a specific enterprise as a result of different climate scenarios. Data were generated with the aid of a forestry enterprise model (*Lemm*, 1991 [5]). Maps such as exhibited in Figure 1 or 2 are often inappropriate means to transfer information to lay people due to the abstractness of the content displayed. Thus we checked whether complex phenomena such as vegetation shifts can be visualized in a more realistic form. Phototests proved to be a valid procedure to evaluate risk perception patterns in the population. Individuals were invited to perform ratings of computer edited photos exhibiting landscape evolution scenarios that are conceivable in mountain environments under changing climate (Figure 3), (*Hunziker*, 1995 [2], Hunziker and Kienast, submitted). It is evident from the study that people don't perceive spontaneous reforestation on abandoned fields negatively. Only completely overgrown areas seem to be perceived as a risk.

4 Conclusions

Establishment of socially accepted management strategies for sustainable land use is facilitated if risk perception and attitude patterns of practitioners and lays are known. Thus impact models have to be supplemented by social science studies that confront interviewees with impact results at a level of abstractness that fits the every-day reality of the people involved. For governmental organizations generalized maps or tables are appropriate means, whereas for practitioners or lay people a representation must be selected that is more realistic. We found that photo-edited pictures exhibiting potential impacts on landscape scenes that are familiar to the respondents, or impact results transformed to local economic systems, are most appropriate means to gather information on people's attitudes towards complex risks.

References

1. B. Brzeziecki, F. Kienast, and O. Wildi. Modelling potential impacts of climate change on the spatial distribution of zonal forest communities in Switzerland. *Journal of Vegetation Science*, 6:257–268., 1995.
2. M. Hunziker. The spontaneous reafforestation in abandoned agricultural lands: perception and aesthetic assessment by locals and tourists. *Landscape and Urban Planning*, 31:399–410, 1995.
3. F. Kienast. *Klimaänderung und mögliche langfristige Auswirkungen auf die Vegetation der Schweiz.* vdf Verlag, Zürich, 1997. Schlußbericht des NFP 31-Projektes Nr. 4031-34234: Simulating and mapping the potential impacts of increasing CO_2 and changing climate on the vegetation cover: A risk assessment study.
4. F. Kienast, O. Wildi, and B. Brzeziecki. Potential impacts of climate change on species richness in mountain forests - a risk assessment. *Biological Conservation*, 1997. in press.
5. R. Lemm. *Ein dynamisches Forstbetriebs-Simulationsmodell: Prognosen von betriebswirtschaftlichen Waldentwicklungen, Waldschäden und deren monetäre Bewertung unter variablen Einflußgrößen.* PhD thesis, ETH Zürich, Zürich, 1991.

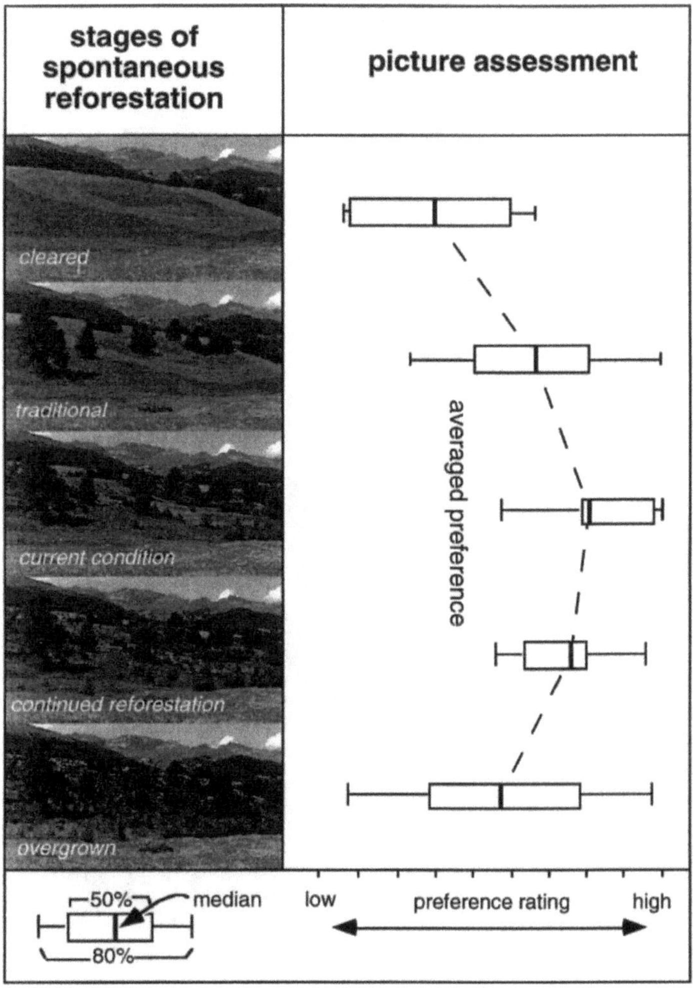

Figure 3. How do people perceive the risk of projected landscape changes? Qualitative interviews and photoseries exhibiting scenarios of landscape change may help to evaluate risk perception patterns in the population.

Part 3

Human Dimensions

Chapter 1: Modelling Global Environmental Change: Improving Human Dimensions Components

Harold K. Jacobson

Abstract[1] Modelling, which is an essential tool for improving understanding, has been a basic component of natural science global change research efforts. Even though these efforts acknowledge the importance of human interactions in global change, they have not done much to incorporate human dimensions in their models. Human dimensions are either put in a black box or incorporated by making relatively simple assumptions about human behavior. Social science modelling efforts have been limited and have focused primarily on energy production and consumption.

Efforts to improve modelling of human dimensions of global environmental change and the human dimensions components in broader models of global change must be based on a modular approach, where the modules are defined in terms of the boundaries of states. Human behavior is not uniform across the globe, but instead varies tremendously depending upon institutional context. Institutions include the structures, rules, norms, expectations, and traditions that shape human behavior. Following a modular approach in modelling human dimensions of global environmental change means that there must be modules for the largest and most important countries and political units. Other countries can be aggregated with these countries or in regionally defined units.

There are strong traditions in economics and political science that utilize rational choice and expected utility approaches on which efforts to improve modelling of human dimensions of global environmental change can be based.

1 Introduction

There is consensus that modelling is an essential tool for improving the understanding of global environmental change. There is also broad agreement that this modelling should encompass, in addition to the biological, chemical, and physical processes involved in global change, the anthropogenic factors that contribute to global environmental change and should yield forecasts of the consequences of global change for humankind. Finally, there is agreement that the models should be sufficiently comprehensive and reliable so that they can be used to clarify the likely consequences of various policy interventions. Unfortunately, such integrated, comprehensive, and reliable models of global environmental change do not exist in 1994, nor are they likely to exist in the short-term future. This chapter argues that strong efforts should be devoted

[1]Parts of this chapter are based on a paper that was presented at a Symposium on Global Environmental Change at the Autonomous University of Barcelona, 13-17 September 1993

to efforts to improving modelling of the human dimensions of global environmental change. It points out some of the serious difficulties that must be addressed, argues that a strategy based on a modular approach where the modules are defined in terms of the legal geographical boundaries of states will be necessary, and suggests that existing scholarship provides an adequate basis so that progress can be made. Though the chapter focuses particularly on issues concerning climate change, the arguments are generic to all aspects of global environmental change.

2 Modelling in current global environmental change research

Modelling is important because it is a basic tool for understanding system dynamics and thus for forecasting future developments. Models force those doing research to specify the variables that they consider to be important and to postulate the relationships that they believe exist among these variables in other words to specify their theories in terms that permit confirmation or falsification. Modelling provides a framework for research. Models guide data collection because they identify important variables, and they define and give priority to process studies by showing the relative importance of various interactions among variables and illuminating the state of knowledge about these interactions. Models forecast the future consequences of current activities, and they clarify the likely consequences of changing these activities.

Modelling has been an essential component of natural science global change research efforts and of the two international natural science programs, the World Climate Research Programme (WCRP) and the International Geosphere-Biosphere Programme (IGBP). General Circulation Models (GCMs) are central to the WCRP. In fact, describing WCRP as a global effort to improve GCMs would not be too far off the mark. The wiring diagram of bio-geo-chemical cycles developed by Francis Bretherton and his colleagues provided an overall conceptual framework for IGBP (*IGBP*, 1990 [16]; *Kuhn* et al., 1992 [9]), and considerable progress has been made in developing models for the components of this diagram and interactions among them (*Ojima*, 1992 [30]). Models are so important to IGBP that beyond including modelling in all of its core projects IGBP has a core project GAIM (Global Analysis, Interpretation, and Modelling) devoted to modelling (*IGBP*, 1994c [17]; *IGBP*, 1994d [15]).

Even though both of the international natural science research programs acknowledge the importance of human interactions in global change, neither goes very far in incorporating human dimensions in their models. Human dimensions are either put in a black box or incorporated by making relatively simple assumptions about human behavior. Modelling is a part of the younger and much less well developed Human Dimensions of Global Environmental Change Programme (HDP)[2]. One of the seven foci of HDP, the Social Dimensions of Resource Use, is defined primarily in terms of modelling demographic, economic, and technological change and resource use (*Jacobson* and *Price*, 1990 [20]), but as of 1994 this effort had barely begun. The result of this relative lack of attention to the human interactions in modelling global environmental change is apparent in the assessments of the Intergovernmental Panel on Climate Change (*IPCC*, 1990 [18]; *IPCC*, 1992 [19]).

The IPCC has used models that incorporate human dimensions to generate scenarios of climate change and its consequences for such factors as temperature, precipitation, soil moisture, snow and sea ice, and sea level. Greenhouse gas emissions are key human-induced drivers in these models, and the emissions are in turn driven by population growth, economic growth, energy intensity, and carbon intensity. In their current state, the models clearly indicate the likelihood that global warming will occur.

The world community has responded to this scientific prognosis by adopting the United Nations Framework Convention on Climate Change at the United Nations Conference on Environment and Development in Rio de Janeiro in June 1992. The convention calls for the negotiation of protocols to limit greenhouse gas emissions, and more broadly Agenda 21, which was also adopted at the Rio conference, calls for countries to undertake a wide range of activities to mitigate and adapt to global change (*UN*, 1993 [39]; *Robinson* (1993) [34]).

Unfortunately, current models cannot adequately support the next steps that the instruments adopted at the Rio conference mandate. They cannot provide a strong base for prescriptions about actions that should be taken to deal with climate change and global environmental change more generally. The models do not have adequate regional resolution, and the variance among the models concerning regional effects is much greater than that concerning global effects. These are primarily natural science issues, and considerable natural science research is in process to address them. On the social science side, the human dimensions are so primitive that they provide only a very flimsy basis for forecasting and an even weaker basis for policy intervention.

In IPCC's first scientific assessment, which was published in 1990, all of the scenarios assumed that the world's population would increase to 10.5 billion in the second half of the twenty-first century and that economic growth in the decade of the 1990s would be 2% to 3% annually in the member countries of the Organization for Economic Cooperation and Development (OECD)[3] and 3% to 5% in the Eastern European and developing countries. The economic growth rates were assumed to decrease thereafter (*IPCC*, 1990 [18]). It was assumed that the energy supply would be coal intensive and that only modest

[2] HDP has been renamed IHDP: International ...

[3] At that time the members of OECD were: Australia, Austria, Belgium, Canada, Denmark, Finland, France, Germany, Greece, Italy, Iceland, Ireland, Japan, Luxembourg, the Netherlands, New Zealand, Norway, Portugal, Spain, Sweden, Switzerland, Turkey, the United Kingdom, the United States of America.

increases in energy efficiency would be achieved. The "business as usual" scenario was generated by these assumptions. Alternative scenarios were generated by placing arbitrarily chosen limits on greenhouse gas emissions.

The business as usual scenario is not unreasonable. Population growth will continue well into the twenty-first century because there is enormous inertia in population dynamics (*USDC*, 1986 [41]). Although birth rates have fallen and continue to fall, so have death rates; consequently people live longer, expanding their total numbers. As a result of past population growth, the number of women of child-bearing age continues to increase. How rapidly global population will grow, when it will stabilize, and at what level cannot be known with certainty, but more sophisticated projections than those used in IPCC's first assessment are available. The World Bank's base case projection assumes a slow decline in population growth from 1.7% a year in 1990 to about 1% a year by 2030 (*WB*, 1992 [42]). With this assumption as a basis for projection, global population would stabilize at about 12.5 billion in the middle of the twenty-second century. Assuming a more rapid but historically plausible decline in population growth, global population would stabilize at more than 10 billion, and if the rate of population growth were to drop more slowly, global population could reach 23 billion toward the end of the twenty-second century.

Economic growth will inevitably occur, first, simply to keep up with population growth. The world's growing population will have to be fed, housed, and cared for. However, economic growth will occur for other equally fundamental reasons. More than one billion people live in abject poverty. There is global agreement, affirmed in countless declarations and resolutions of the United Nations and other national and international bodies, that this is unacceptable, that progress must be made toward the elimination of poverty[4]. There is also a connection between poverty and environmental degradation: poverty often leads to environmentally destructive behavior; economic growth can offer environmentally benign alternatives. Economic growth is seen in Agenda 21 as a proper prescription for improving the environment. The global secular trend toward democratization will also promote economic growth. There is a growing consensus, that was launched with the near-unanimous adoption in 1945 of the Universal Declaration of Human Rights, on the desirability of global democratization. As a consequence of this consensus, there is constant pressure for greater democratization. The world-wide trend toward increased educational levels is a factor in this pressure. There is a strong correlation between education and democracy. The more education people have, the more they insist on the right to participate in the processes that lead to decisions that vitally affect their lives. There is a strong, though complex relationship between economic development and growth and

[4]One of the first was *General Assembly Resolution 1710 (XVI)*, adopted 19 December 1961. This resolution proclaimed the 1960s the decade of development and set targets for growth rates.

democracy (*Huntington*, 1991 [14]). Economic performance is one of the most important criteria by which citizens evaluate the performance of their governments, and thus leaders encourage economic growth to further their tenure in office (*Nie* et al., 1976 [28]). For all these reasons then, economic growth will continue. Again, however, more sophisticated projections about economic growth and its effects are possible than those used in IPCC's first scientific assessment.

IPCC's interim assessment, which was completed in 1992, had more sophisticated assumptions about human dimensions (*IPCC*, 1992 [19]). Various UN and World Bank (WB) projections of population growth were used to generate different scenarios, and, for assumptions about rates of economic growth, the world was divided into four regions, each having different annual rates of economic growth. However, these scenarios too fall short of what would be desirable for analyzing the effects of policy intervention.

The models are most sophisticated with regard to carbon dioxide emissions. The understanding of the link between population and economic growth and CO_2 emissions, especially those resulting from energy use, is relatively well developed. Fortunately for the plausibility of the scenarios generated by the models, CO_2 emissions account for more than 50% of the human-induced greenhouse gas emissions (*IPCC*, 1990 [18]). Relatively little is known about the links between population and economic growth and other greenhouse gases, and the understanding of the link with CO_2 emissions is much better with respect to energy use than to land use. Because land use, including deforestation contributes almost a quarter of CO_2 emissions and also contributes substantially to the emissions of other greenhouse gases (*Stern* et al., 1992 [36]), the Human Dimensions of Global Environmental Change Programme has put highest priority on improving the understanding of land-use and land-cover changes (*Turner* et al., 1993 [38]).

The problems with properly integrating human dimensions inputs into the process of generating the IPCC and other scenarios, however, are deeper and more fundamental than simply the necessity of adding an understanding of emissions stemming from land use. The assumptions that are used in current modelling efforts involve relatively linear projections of past and current human behavior into the future and tend to assume that the characteristics of human behavior are relatively universal. Viewing the outputs of the models, it is easy to assume that human behavior can only be modified through regulation. Yet we know that this is not true. We know that human behavior responds to price signals and that price varies in relationship to scarcity: the scarcer the good, the higher the price, the lower the demand, and the greater the incentive to search for and discover substitutes. Human behavior also alters as a consequence of learning and the adoption of new values. Moreover, as will be discussed in detail in the following section, human behavior is not universal, but varies tremendously depending on the institutional context.

The simplistic assumptions about human behavior on which current modelling efforts rest weaken the accuracy of their projections. More complicated assumptions about human behavior are also needed for analyzing the consequences of global environmental change. As change occurs, so will human behavior, so the projected new levels of temperature and precipitation, for instance, will not apply to farming as it currently exists, but rather to farming as it has gradually adapted over time to changes in the physical environment.

To improve the models that are currently used for understanding global environmental change, not only will more aspects of human dimensions have to be included, but the assumptions about human behavior that are included will have to be made more sophisticated. The tasks will not be easy.

3 Modelling in the social sciences

Some might even argue that efforts to model the human dimensions of global environmental change will inevitably be chimerical. They can make a powerful case. Efforts that were fashionable and received considerable attention in the 1970s to predict or project world futures proved incredibly inaccurate. The pessimistic "limits to growth" argument, which was derived from some of these early modelling efforts, in retrospect seems extremely naive and biased (*Meadows* et al., 1972 [27]; *Hughes*, 1985 [13]).

Observation and analysis of human behavior risks disturbing that behavior. Just asking questions may lead humans to alter their attitudes and actions, and the publication of findings of research has an even higher probability of producing this outcome. Predictions of human behavior frequently result in modifications of behavior that nullify the predictions. Since human beings have the capacity for choice, they can change their behavior.

In this regard, the situation with respect to social science is fundamentally different from that with respect to natural science. Observation does generally not disturb the phenomena being observed, and inanimate objects do not have the capacity voluntarily to change their behavior. IGBP aims at improving understanding of important global change processes in order to improve "model-based predictions of key features of the Earth system" (*IGBP*, 1994 [17]). Comprehensive models that integrate the human dimensions of global environmental change cannot be predictive in the sense that natural scientists use the term, they can only project the likely consequences of the continuation of existing trends, or, if they are more sophisticated, forecast outcomes taking into account likely changes in human behavior.

A second problem is that social scientists are much less advanced than natural scientists in their ability to measure key variables. There are measures of national identity, but their validity and reliability is uncertain. Power, the central concept in political science, can at best be measured ordinally. Leadership, which is important in several social sciences, is basically unmeasured.

Integrating poorly measured variables into complicated models is a daunting intellectual challenge.

Finally, many crucial economic, political, and social phenomena are not well understood. Even primitive models of the interactions that produce these phenomena have not yet been developed. Key variables have not been identified, and when they have, interactions among them are often not understood.

These difficulties surely exist. Acknowledging them, however, does not imply acceptance of the argument that human dimensions cannot be integrated into models of global environmental change. Global environmental change cannot be understood without integrating human dimensions; human-induced changes are the source of concern. Because of this human dimensions have already been included in models that are currently being used. The challenge is to improve these efforts. Acknowledging the difficulties is important primarily to create realistic expectations about what can be achieved.

4 The necessity of following a modular approach

Efforts to improve modelling of human dimensions of global environmental change and the human dimensions components in broader models of global change must be based on a modular approach, where the modules are defined in terms of the legal geographical boundaries of states. This is because human behavior is not uniform across the globe, but instead varies tremendously depending upon institutional context. Rates of population and economic growth and levels and changes in energy and carbon intensity and in land use and land cover all vary and are strongly affected by institutions.

Institutions include the structures, rules, norms, expectations, and traditions that shape human behavior (*March* and *Olsen*, 1989 [26]; *Coase*, 1980 [4]; *North*, 1990 [29]; *Bogdanor*, 1991 [1]). Institutions include both formal structures and written rules and also informal and unwritten but powerful and enduring understandings. Institutions can be local, national, international, or global. Constitutions create institutional structures, such as executive bodies, legislatures, courts, and bureaucracies. They also define the mandates of these structures, determine their composition, and specify the rules they must follow for making decisions. Traditions, however, can be as influential in shaping human behavior as constitutions, so they must be included as well. Much of organized political behavior, for instance in political parties, is strongly affected by tradition. There are economic, social, and cultural institutions in addition to governmental and political institutions. Firms and markets are included within the concept of institutions as used here, as are social structures such as those relating to class. All of these institutions affect human behavior. Unfortunately, even those models that focus primarily on the social science aspects of global environmental change as opposed to the natural science aspects have devoted relatively little attention to institutional effects (*Gilbert* and *Braat*, 1991 [11]).

Because they have a monopoly of legal coercive power and competence to enact laws and regulations that determine institutional structures and patterns in other spheres, the governmental and political institutions of national states have primacy in the shorter term over other institutions. In the longer term, economic, social, and cultural institutions on a variety of levels contribute to the stability or transformation of national governmental and political institutions. How much one needs to be concerned with the impact of economic, social, and cultural institutions on governmental and political institutions depends upon the time frame of one's concerns. The longer the time frame, the greater the importance of this impact. Given the emphasis in global change studies on projections and forecasts ranging over decades and centuries, this aspect of the interaction among institutions cannot be ignored. In addition, economic, social, and cultural institutions are often the vehicles for laws and regulations, and, even when they are not, they set parameters on what laws and regulations can accomplish.

The necessity of following a modular approach to building models of the human dimensions of global environmental change or incorporating human dimensions in broader models can be stated quite simply. Although a strong correlation exists at the global level between population and economic growth and land use and land cover changes and increases in greenhouse gas emissions, such relationships vary enormously when the global totals are disaggregated by countries or regions.

Countries differ greatly in basic economic characteristics. Although this has consequences for environmental impacts such as those resulting from greenhouse gas emissions, they are not uniform.

Table 1 illustrates this point. It displays the gross national products, gross national products per capita, the real growth in GNP per capita during the decade of the 1980s, and the energy use per capita for the sixteen countries that had gross national products of more than two hundred billion United States dollars per year in 1991. Energy use is a good indicator of a country's greenhouse gas emissions and thus of some of the anthropogenic factors that contribute to global change. It is not a perfect measure, because different forms of energy production and consumption yield different types and levels of emissions, but it is a human factor for which reasonably reliable figures are available. Energy use accounts for about 50% of greenhouse gas emissions. While energy use is a convenient example, the same argument could be made with other human activities that contribute to global change, such as land use.

As can be seen, the characteristics of these countries vary substantially. Because their population ranges from China's 1 150 091 000 to Switzerland's 6 740 000 their per capita gross national products vary even more than their gross national products. The rates at which their per capita gross national products grew during the 1980s differed enormously, ranging from Mexico's low of −0.5 % per year to China's 7.4 % per year. A country's per capita GNP growth rate is a product of rates of population and economic growth. Both rates differ greatly among the sixteen countries. Brazil's rate of population growth was 2.2 % per year while Germany's was only 0.1 %; and China's rate of growth for its gross domestic product was 9.4 %, while Mexico's was only 1.5 %.

Table 1. Differences between countries in GNP and energy use (NA - not available)

Country	Gross National Product (millions of US $) 1991	GNP per capita (US $) 1991	Real growth rate (%) in per capita GNP 1980 - 1991	Energy use per capita (oil equivalent kg) 1980 - 1990
Australia	287 765	16 590	1.2	5 041
Brazil	447 324	2 920	0.4	915
Canada	568 765	21 260	2.1	10 009
China	424 012	370	7.4	598
France	1 167 749	19 590	1.8	3 845
Germany	1 516 785	23 650	2.2	3 491
India	284 668	330	3.3	231
Italy	1 072 198	18 580	2.1	2 754
Japan	3 337 191	26 920	3.7	3 563
Mexico	252 381	2 870	−0.5	1 300
Russia	479 546	3 220	1.3	NA
Spain	486 614	12 460	2.9	2 201
Sweden	218 934	25 490	1.7	6 347
Switzerland	225 890	33 510	1.6	3 902
United Kingdom	963 696	16 750	2.6	3 646
United States	5 686 038	22 560	2.1	7 822
Average		14 504	2.2	3 479
Total	17 419 556			

(Source: *World Bank Atlas*, 1993 [43], 25th Anniversary Year)

The energy use of the sixteen countries also varies greatly, ranging from Canada's high figure of 10 009 kg of oil equivalent per capita to India's low figure of 231. Because of the size of India's population, however, its total energy use is almost three-quarters of that of Canada and it is almost as much as the United Kingdom's. China's total energy use is almost three and a third times greater than the United Kingdom's and more than a third of that of the United States.

Energy use varies in a very rough fashion with income, the higher per capita income the greater the energy use, but the correlation is loose and the relationship is not linear. Canada's per capita income is less than two-thirds of that of Switzerland, but its per capita energy use is more than two and a half times as great. Obviously the physical characteristics of countries such as size, topography, and weather determine energy use, but they alone cannot completely explain all of the differences. In general, developed countries use energy more efficiently than developing countries. India used 1.71 kg of energy for every dollar of per capita gross national product, while Switzerland used only 0.12 kg. Asserted more broadly, energy efficiency is highly correlated with development.

Population and economic growth rates are the consequence of human behavior, which is conditioned by rules, norms, expectations, and traditions. Energy use too is the consequence of human behavior, and again, rules, norms, expectations and traditions affect the human behavior that produces the outcomes. Institutions are an important factor and must be brought into the analysis. The impact of institutions on population and economic growth rates and on such anthropogenic factors forcing global environmental change as energy use must be studied and modeled. The sharp differences that are so apparent in Table 1 demonstrate why these studies and models must be based on modules that are defined in terms of the legal geographical boundaries of states. These modules can be aggregated to the global level. This is the way that global models should be built. This is the only way that institutional effects can be properly incorporated.

The necessity of following a modular approach is particularly acute if models are to be used to clarify the likely consequences of strategies for mitigating or adapting to global environmental change. Virtually everyone agrees that national and international public policies will be required to deal with global change, either to mitigate it or to adapt to its consequences. There is a vast literature proposing and prescribing public policies to deal with global change (*Tuchman-Mathews*, 1991 [37]). This literature varies in analytical sophistication and quality, but almost all of it could benefit from more differentiated analyses, and the prescriptions that it offers would be more likely to achieve their intended effects if they were based on a more sophisticated understanding of differences among different institutions.

One case may be cited to illustrate this point. In 1981 the Brazilian government adopted the Northwest Brazil Integrated Development Program (*Mahar*, 1989 [25]). A main objective was to reduce forest clearing. "The plan was to conduct land use surveys to identify areas with high potential and then to concentrate new access roads, social infrastructure, agricultural research and extension, input supplies, crop storage, marketing, and farm credit in those areas" (Mahar, 1989 [25], p. 34). The program failed. The pace of deforestation did not slow, but instead accelerated, and traditional patterns of land use persisted. In the five years between 1970 and 1975, forest cover dropped from 92.1 % of the total to 86.6 %; in the five years between 1980 and 1985, it dropped from 79.4 % to 65.5 %. Pastures increased from 2.5 % of the total in 1970 to 9.8 % in 1980 and 25.6 % in 1985 (Mahar, 1989 [25], p.

35). The reason for the failure of the policies was that economic incentives encouraged people to behave in contrary directions.

Many other cases could be cited where policies designed to protect the environment failed to achieve their objective. To understand the effect of proposed policies, it is necessary to analyze how the policies interact with the other variables that comprise the system that the policies were designed to affect. The Brazilian example clearly shows that legal regulatory instruments need to be viewed in a broader context.

Often policies adopted by local and national governments to achieve non environmental goals, such as maintaining employment and stimulating economic growth, have harmful environmental side effects. Subsidies for agricultural and energy production are examples (*WB*, 1992 [42]). It is often argued that alternative policies are available that are less environmentally destructive. Again, comprehensive models are essential to clarify the effects of current policies and to evaluate alternatives.

Because of the broad differences among them, some of which are illustrated by the data in Table 1, countries are often reluctant to adopt identical policies, and even if they did, the consequences of the policies would not always be the same because of the different circumstances in which they were implemented. Most policies, particularly for adapting to global environmental change, will be formulated and implemented at the local level (*Glanz*, 1988 [12]). Agenda 21 acknowledges that most policies designed to deal with global change will by necessity be regional, national, or local.

In modelling the human dimensions of global environmental change, it is essential to pursue a modular approach in which the modules are defined in terms of the legal geographical boundaries of states because this is the only way that institutions that reflect differences among countries and regions of the world and contribute to them can be incorporated. Understanding the consequences of institutional differences is essential both to understanding the anthropogenic causes of global environmental change and to formulating strategies for mitigating and adapting to global change.

Arguing for following a modular approach in modelling human dimensions of global environmental change is not to argue that there must be a module for every country in the world. Disaggregation need not be carried that far. The sixteen countries included in Table 1 account for more than three-quarters of the world product and twelve of them, China, France, Germany, India, Italy, Japan, Mexico, Russia, Spain, the United Kingdom, and the United States account for almost three-fifths of the world's population. There must be modules for the largest and most important countries and political units: Brazil, China, the European Union (which would include France, Germany, Italy, Spain, the United Kingdom and the other members of the Union), India, Japan, Mexico, Russia, and the United States, and probably such populous countries are Indonesia, Nigeria, and Pakistan. Other countries should be aggregated with these countries or in regionally defined units. The most successful effort to model the world economy and test "limits to growth" assumptions and conclusions was based on fifteen modules (*Leontieff* et al., 1977 [22]). Surely valid and reliable efforts to model the human dimensions of global environmental change need not include more than twenty modules, but clearly they must have more than four.

5 Can institutional effects be included?

If institutions are to be a key component of a modular approach to modelling the human dimensions of global environmental change, it is essential to consider their role in more detail, assess the knowledge that exists concerning institutions, and realistically assess the possibility of modelling institutional effects. The first point is to reiterate the importance of governmental and political institutions. In his massive study of economic growth in 41 countries from 1850 through 1980 Lloyd Reynolds concluded, "For good or ill, government seems central to economic growth" (*Reynolds*, 1985 [33], p. 418). Reynolds goes on to say:

> ". . . when we ask why a particular country has grown more or less rapidly, we come back inevitably to internal factors: the resource base, the structure of economic institutions, the stability and competence of government, the wisdom of policy measures" (Reynolds, 1985 [33], p. 439).

Reynolds found that the countries that had the best growth records had strong senses of national identity, continuity in leadership, leadership committed to economic growth, competent and relatively uncorrupted administrative staffs, good institutions of higher education, favored private enterprise, and pursued astute economic policies. The character and quality of governmental institutions and policies is crucial to the economic growth rate that a country is likely to achieve.

Institutions and policies are also central to energy efficiency. Table 1 and the discussion of the data in it showed that developing countries use energy much less efficiently than developed countries. According to a World Bank study, explanations for this largely relate to institutions and policies (*WB*, 1993 [43]). The World Bank study estimated that even with existing technology in developing countries, savings of 20% to 25% of energy consumed would be possible, and with investment in more energy efficient capital stock savings of 30% to 60% could be achieved *WB*, 1993 [43], p. 13). The study maintained that the low efficiency with which energy is used is the consequence of pricing energy below the cost of its production and distribution, the tendency to treat energy production as an extension of the state, the fact that many sectors of production are dominated by a few large monopoly or state enterprises that are highly protected from foreign competition, and the failure of markets to function effectively (*WB*, 1993 [43], p. 34).

The broad point extends beyond developing countries and energy. The factor that had the most dramatic impact on energy efficiency in the developed countries was the increase in the price of petroleum in the 1970s. Energy efficiency improved dramatically in response to the price increase. In those countries that successfully delayed or prevented the price increase, energy efficiency remained constant or even declined (*Jänicke* et al., 1989 [21]).

More generally, effectively functioning markets promote efficiency, and prices affect behavior. Higher energy prices in Europe also contribute to more efficient energy use in Europe than in North America, a fact clearly illustrated in Table 1. If prices properly reflect environmental scarcities, they will promote environmentally protective behavior. However, markets do not always function effectively and prices do not always properly reflect values. This is particularly true for common or public goods such as air and water that are used or damaged in the production of goods (*Pearce* and *Turner*, 1990 [31]; *UN*, 1992 [40]). Governmental intervention is required, and governments have adopted many policies with respect to common or public goods. These policies must be incorporated into models of human dimensions of global environmental change.

This dictum needs to be supplemented by three broad cautionary points. The first is that the capacity of governments, especially in developing countries, is limited. Sir Arthur Lewis, the Caribbean-born, Nobel-laureate economist called attention to this four decades ago. His caution bears repeating because it has so frequently been ignored.

> ". . . the governments of the less developed countries are . . . less capable of taking on a wide range of functions than are the governments of the more developed. Their administrations tend to be more corrupt and less efficient, and a smaller part of the national income can be spared for government activity. . . . In fact, one cannot usefully consider in an abstract way what functions a government ought to exercise without taking into account the capabilities of the government in question. It is very easy to overload the governments of less developed economies, and it is quite clear that it is better for them to confine themselves to what they can manage than for them to take on an excessive range" (*Lewis*, 1955 [23], p. 382).

The capacities of all governments are limited. They have finite resources for monitoring and directing activity. Policies frequently fail to achieve their objectives, especially in developing countries, because governments simply do not have the capacity to implement them. Modelling cannot automatically assume that policies will be implemented fully or achieve their intended effects.

The second point is that it is difficult or impossible to achieve objectives through government regulation if economic incentives promote contrary behavior. The Brazilian efforts to protect the tropical rain forest in the Amazonian region that failed (mentioned above) illustrate this (Mahar, 1989 [25]).

The laws and regulations were well intended, but their effects were negated by other government policies that offered subsidies and tax relief and market forces that in effect encouraged destruction of the rain forest. Modelling must encompass all aspects of a country's political and economic systems.

The final point is that the effectiveness of governmental institutions is strongly conditioned by history and the social context in which they function. Alexis de Tocqueville forcefully advanced this argument more than a century ago in *Democracy in America.* More recently, Robert D. Putnam and his colleagues conducted a sophisticated and elaborate empirical analysis of the effectiveness of regional governments in Italy and confirmed and elaborated de Tocqueville's argument (*Putnam* et al., 1993 [32]). Italy established new regional governments throughout the country in 1970. Putnam and his colleagues assessed the effectiveness of these governments over the next two decades in providing services in agriculture, housing, health, and other areas. By extension their findings are applicable to environmental issues. They found that the better the "civic context" – the more local associations, engagement in community affairs, egalitarian political patterns, trust and law abidingness – the more effective the government. They also found that the civic context changes slowly.

The broad conclusions of these generalizations are:

1. though governments and policies are crucial, expectations about what they can accomplish should be realistically modest and models must not incorporate policies at simple face value;
2. the effectiveness of governments and policies depends significantly on the broader context in which they exist and this broader context must be encompassed; and,
3. by implication, since the context changes slowly, effectiveness improves slowly.

Efforts to model human dimensions of global environmental change must and can take account of these conclusions.

Although the cautions that have been stated pose formidable challenges, existing scholarship can provide a reasonable foundation for efforts to improve modelling of human dimensions of global environmental change. There are a growing number of case studies that show the effects of particular institutions and institutional arrangements on environmental policies. In the terminology of global change research, these would be regarded as process studies, studies that explore the interaction among variables so that the relationships can be understood well enough to be included in models.

One innovative study of the energy policies of Canada, France, Germany, Japan, and the United States in the aftermath of the "oil shock" carefully delineated the effects of the different institutional structures of the five countries (*Feigenbaum* et al., 1993 [10]). Although all five countries had democratic regimes and similar levels of economic development, the policy innovations that they adopted in response to the oil shock differed in direction and pace.

The study concluded first that "Policy and resource inheritances are more important in determining energy policy changes than are governmental capabilities or specific institutional arrangements" (*Feigenbaum* et al., 1993 [10], p. 99). Thus present policy is the starting point for forecasting future policy. However, having said this, the study then concluded that:

> "Political institutions play an important role in influencing government capabilities to innovate in policymaking, coordinate conflicting objectives, and implement policies. In particular, the number of effectiveness of veto points in the political system has a strong effect on each of these capabilities" (*Feigenbaum* et al., 1993 [10], p. 107).

Between 1973 and 1990, energy consumption per capita in Canada, France, Germany, and Japan increased, while in the United States it decreased. The production of nuclear energy increased in all five countries, but much more dramatically in France, Germany, and Japan, than in Canada and the United States. On the other hand, the production of coal increased substantially in Canada and also in the United States, though less substantially, while it declined in France, Germany and Japan. The case study explains the contribution of political institutions to these differences. Case studies such as this one provide essential ingredients for modelling human dimensions of global environmental change.

There are models in social science that incorporate institutional effects, and these provide another ingredient for the effort to improve modelling of human dimensions of global environmental change. Within political science, there has been a long tradition of modelling international armed conflict. Interestingly, Lewis Richardson, the same individual who developed the first rudimentary General Circulation Model, launched the tradition by creating a model of arms races. Richardson's model had an institutional component, a coefficient modifying one country's response to another's increased armament. As in the case of GCMs, Richardson's work initiated an effort to model arms races and conflict has become increasingly sophisticated (*Choucri* and *North*, 1974 [3]; *Mesquita*, 1981 [5]; *Luterbacher* and *Ward*, 1985 [24]; *Mesquita* and *Lalman*, 1992 [7]). Much progress has been made in clarifying the dynamics of international armed conflict. For instance, the circumstances under which a state would be likely to initiate an armed conflict have been relatively well defined.

The techniques that were developed to model conflicts have begun to be applied to other issue areas including the environment. Rational choice, expected utility theory models in which the actors are assumed to try to maximize the probabilities of obtaining their preferences and minimize their losses have even developed a reasonably accurate predictive capability (*Mesquita*, 1985 [6]). Various political scientists have begun to model decision making in European countries, the European Union, and the United States about environmental issues using a variety of techniques (*Brown*, 1994 [2]; *Mesquita*, 1994 [8]; *Sprinz* and *Vaahtoranta*, 1994 [35]). They have been able *ex post facto* to predict policy outcomes, and on the basis of this to project future trends. In these models, the effects of institutions broadly defined are incorporated through assumed preferences and rules that constrain decisions. Efforts have been made to compare the predictions and forecasts of the various models, and the expected utility model performs reasonably well, for instance in predicting the phasing in period for automobile emissions standards that the European Union would adopt (*Mesquita*, 1994 [8]).

6 Conclusion

To date natural scientists, demographers, and economists have assumed the predominant roles in efforts to model human dimensions of global environmental change. This chapter has argued that because of the importance of governmental and political institutions, political scientists also have a contribution to make. They are gradually becoming ready to make this contribution. Their contribution must be made in collaboration with the efforts of other social and natural scientists. What is required is a broadly collaborative effort.

This effort must be made in the certainty that the population and economic growth that drives greenhouse gas emissions and global environmental change will continue. However, the goal accepted at the United Nations Conference on Environment and Development in Rio and reaffirmed in other international fora is not to prevent global environmental change, but rather to make possible "sustainable development." The goal is to pursue development in ways that "meet the needs of the present without compromising the ability of future generations to meet their own needs" (*WCED*, 1987 [44]). Understanding institutions and the role that they play and improving modelling of human dimensions in global environmental change are instrumental to achieving this goal. It is a challenging task, but one that must and can be met.

References

1. V. Bogdanor, editor. *The Blackwell Encyclopedia of Political Science.* Blackwell, Oxford, 1991.
2. C. Brown. Politics and the environment: Nonlinear instabilities dominate. *American Political Science Review*, 88(2):292–303, 1994.
3. N. Choucri and R. C. North. *Nations in Conflict: National Growth and International Violence.* W. H. Freeman and Company, San Francisco, 1974.
4. R. H. Coase. *The Finn, the Market, and the Law.* University of Chicago Press, Chicago, 1990.
5. B. B. de Mesquita. *The War Trap.* Yale University Press, New Haven, 1981.

6. B. B. de Mesquita. *Forecasting Political Events: The Future of Hong Kong.* Yale University Press, New Haven, 1985.
7. B. B. de Mesquita and D. Lalman. *War and Reason: Domestic and International Imperatives.* Yale University Press, New Haven, 1992.
8. B. B. de Mesquita and F. N. Stokman, editors. *European Community Decision Making: Models, Applications and Comparisons.* Yale University Press, New Haven, 1994.
9. W. Kuhn et al. *Pathways of Understanding: The Interactions of Humanity and Global Change.* Consortium for International Earth Science Information Network, Saginaw, 1992.
10. H. Feigenbaum, R. Samuels, and R. Kent Weaver. Innovation, Coordination, and Implementation in Energy Policy. In R. K. Weaver and B. A. Rockman, editors, *Do Institutions Matter? Government Capabilities in the United States and Abroad.* The Brookings Institution, Washington, 1993.
11. A. J. Gilbert and L. C. Braat. *Modeling for Population and Sustainable Development.* Routledge, London, 1991.
12. M. H. Glanz, editor. *Societal Responses to Regional Climatic Change: Forecasting by Analogy.* Westview, Boulder, 1988.
13. B. B. Hughes. *World Futures: A Critical Analysis of Alternatives.* The Johns Hopkins University, Baltimore, 1985.
14. S. P. Huntington. *The Third Wave: Democratization in the Late Twentieth Century.* University of Oklahoma, Norman, 1991.
15. International Geosphere-Biosphere Programm. *Global Change, Report 30, IGBP Modelling and Data Activities, 1994-1998*, Stockholm, 1994.
16. The International Geosphere-Biosphere Programme. *Global Change, Report No. 12, A Study of Global Change, The Initial Core Projects*, Stockholm, 1990.
17. International Geosphere-Biosphere Programme. *Global Change, Report 28, IGBP in Action: Work Plan 1994-1998*, Stockholm, 1994.
18. IPCC. *Climate Change: The IPCC Scientific Assessment.* Cambridge University Press, Cambridge, 1990. J. T. Houghton and G. J. Jenkins and J. J. Ephraums, editors.
19. IPCC. *Climate Change 1992: The Supplemental Report to the IPCC Scientific Assessment.* Cambridge University Press, Cambridge, 1992. J. Leggett, W. J. Pepper and R. J. Stuart, editors.
20. H. K. Jacobson and M. Price. *A Framework for Research on the Human Dimensions of Global Environmental Change.* International Social Science Council, Human Dimensions of Global Environmental Change, Paris, 1990.
21. M. Jänicke, H. Mönch, T. Ranneberg, and U. E. Simonis. Structural Change and Environmental Impact. *Intereconomics*, (Jan.-Feb.):24–35, 1989.
22. W. Leontief, A. P. Carter, and P. A. Petri. *The Future of the World Economy: A United Nations Study.* Oxford University Press, New York, 1977.
23. W. A. Lewis. *The Theory of Economic Growth.* Richard D. Irwin, Homewood, 1955.
24. U. Luterbacher and M. D. Ward, editors. *Dynamic Models of International Conflict.* Lynn Rienner Publ., Boulder, 1985.
25. D. J. Mahar. *Government Policies and Deforestation in Brazil's Amazon Region.* World Bank, Washington, 1989.
26. J. G. March and J. Olsen. *Rediscovering Institutions.* Free Press, New York, 1989.
27. D. H. Meadows, D. L. Meadows, J. Randers, and W. W. Behrens III. *The Limits To Growth: A Report for the Club of Rome's Project on the Predicament of Mankind.* Universe Books, New York, 1972.

28. N. H. Nie, S. Verba, and J. R. Petrocik. *The Changing American Voter.* Harvard University, Cambridge, 1976.
29. D. C. North. *Institutions, Institutional Change and Economic Performance.* Cambridge University Press, Cambridge, 1990.
30. D. Ojima, editor. *Modeling the Earth System: Papers Arising from the 1990 OIES Global Change Institute*, Boulder, 1992. University Corporation for Atmospheric Research, Office of Interdisciplinary Earth Studies.
31. D. W. Pearce and R. K. Turner. *Economics of Natural Resources and the Environment.* The Johns Hopkins University Press, Baltimore, 1990.
32. R. D. Putnam, R. Leonardi, and R. Y. Nanetti. *Making Democracy Work: Civic Traditions in Modern Italy.* Princeton University Press, Princeton, 1993.
33. L. Reynolds. *Economic Growth in the Third World, 1850-1980.* Yale University Press, New Haven, 1985.
34. N. A. Robinson, editor. *Agenda 21: Earth's Action Plan, Annotated*, New York, 1993. International Union for the Conservation of Nature, Environmental Policy and Law, Oceana. Report No. 23.
35. D. Sprinz and T. Vaahtoranta. The Interest-Based Explanation of International Environmental Policy. *International Organization*, 46(1):77–105, 1994.
36. P. C. Stern, O. R. Young, and D. Druckman, editors. *Global Environmental Change: Understanding the Human Dimensions.* National Academy Press, Washington, 1992.
37. J. Tuchman-Mathews, editor. *Preserving the Global Environment: The Challenge of Shared Leadership.* W. W. Norton, New York, 1991.
38. B. L. Turner II, R. H. Moss, and D. L. Skole, editors. *HDP Report No. 5, Relating Land Use and Global Land Change.* HDP, Barcelona, 1993.
39. UN. *Agenda 21: The United Nations Programme of Action from Rio.* United Nations, New York, 1993.
40. United Nations. *Environmental Accounting: Current Issues, Abstracts and Bibliography*, New York, 1992.
41. U.S. Department of Commerce. *World Population Profile, 1985*, Washington, 1986. Government Printing Office. Bureau of the Census, Center for International Research.
42. World Bank. *World Development Report, 1992: Development and the Environment*, Washington, 1992.
43. World Bank. *Energy Efficiency and Conservation in the Developing World*, Washington, 1993.
44. World Commission on Environment and Development. *Our Common Future*, Oxford, 1987. Oxford University Press. The Brundtland Commission.

Chapter 2: Internationally Tradeable Emission Certificates – Efficiency and Equity in Linking Environmental Protection with Economic Development

Udo E. Simonis

"The neat resolution of a free market that so beautifully reconciles buyers and sellers does so far not reconcile growthists and earthists. Something new is needed."

Nathan Keyfitz

Abstract Three topics dominate the formulation of an international greenhouse-gas regime as part of an effective *global environmental policy: Efficiency, equity,* and *uncertainty*. Also three major policy instruments are discussed as regards the implementation of the 1992 Framework Convention on Climate Change: A *carbon tax/CO_2-charge, joint implementation*, and *tradeable emission certificates.*

This paper tries to answer a question that has not been rigidly asked before: How could tradeable emission certificates be tailored in such a way as to be of benefit to the *developing countries*, to facilitate global environmental protection and economic development at the same time, and to meet *both* the efficiency and the equity criterion in international relations.

Next to market organization and rules of procedure, allocation of the entitlements is crucial. The author suggests a *dynamic formula*, by which the initial allocation of certificates starts on the basis of *current greenhouse-gas emissions* but over time turns towards equity in the form of *equal per capita emissions*. In this way, making emission entitlements tradeable among countries implies not only that a globally effective limit to total emissions is attained with certainty, but also that the current unfair allocation of emission entitlements is consecutively shifted in favour of the poor countries.

1 Introduction: the idea and its political context

With respect to the formulation and implementation of a global climate policy, the "Berlin Mandate", the most important concluding document of the first Conference of the Parties to the Framework Convention on Climate Change, adopted on 7 April 1995, says the following:

> "The Parties should protect the climate system for the benefit of present and future generations of humankind, on the basis of *equity* and in accordance with their common but differentiated *responsibilities* and respective *capabilities*".

It continues:

> "The global nature of climate change calls for the widest possible *co-operation* by all countries [...]; developed countries [should] set quantified target limitation and reduction objectives within specified time frames, such as 2005, 2010 and 2020".

Finally, it states that

> "the process should begin without delay" (Berlin Mandate, 1995 [20], italics added).

As regards "joint implementation", an instrument which affects both industrialized and developing countries, the Conference of the Parties decided

> "to establish a *pilot phase* for activities implemented *jointly* among *Annex I Parties* and, on a voluntary basis, with *non-Annex I Parties* that so request".

During this pilot phase, a framework should be established

> "for *reporting* in a transparent, well-defined and credible fashion on the possible global *benefits* and the national economic, social and environmental *impacts* as well as any practical *experience* or technical *difficulties* encountered" (Berlin Mandate, 1995 [20], italics added).

The idea for the present paper arose out of this complex context of ethics and environment. The specific question to be answered is the following one: What form should a future policy instrument for the reduction of greenhouse gas emissions take if it is to enable both *global environmental protection* and *global development* while satisfying both the criteria of *economic efficiency* and *equity*? The answer, which will be explained in the following, is: by creating a market where so far no market exists, i.e. by introducing carefully designed internationally tradeable emission certificates.

2 Theoretical context

Three main issues dominate the formulation of an international greenhouse-gas regime, in the form of a "Climate Protocol" within the Framework Convention on Climate Change that came into force in 1994: *Efficiency, equity,* and decision-making under *uncertainty*. Also three policy instruments dominate the question of practical implementation of such a protocol: Introduction of an international *carbon tax* and/or CO_2 *charge, joint implementation*, and *tradeable emission certificates (emissions trading*[1]*)*. The following discussion will cover all these instruments but will focus on the interactions between the equity issue and tradeable emission certificates.

---

[1] Incidentally a concept for which twelve different names were found by the author in the English literature

2.1 International emission charges

Pearce has summed up the arguments in favour of introducing a carbon dioxide charge or carbon tax as an instrument of a global climate policy (*Pearce*, 1991 [17])[2]. As his central argument he cites Baumol and Oates, who pointed out that a tax allows total emissions to be reduced at minimum cost (*Baumol* and *Oates*, 1975 [1]). A given tax will induce emitters with low marginal avoidance costs to reduce emissions, while those with high marginal costs will find it more appropriate to pay the tax. In general terms, taxes use the market mechanism to adapt in an optimum way to the greenhouse problem, while direct government regulation can, in the individual case, be extremely expensive. In a comparative study on the USA, Tietenberg established that the average ratio of "command and control costs" to "least-cost measures" was 4 : 1 (*Tietenberg*, 1994 [19]).

Pearce adds four further advantages of a carbon tax. First, the revenue gained allows other taxes to be replaced (*neutrality of effect*). Second, the potential revenue opens up possibilities of substantial resource transfers from North to South. Third, it involves a constant inducement for industry to undergo structural change towards environment-friendly production. Fourth, if new (scientific) information about the climate problem and its impacts becomes available, the tax can be modified relatively easily.

However, there are also several disadvantages of an international carbon tax, which should not be ignored. As we have only a rough idea of the price elasticity of the supply of and demand for fuel, particularly as concerns the great dimensions we would be dealing with in practice, there is considerable uncertainty as to how large a reduction in emissions would be (two examples are shown in Figure 1). Furthermore, it is widely held that the final incidence of a carbon tax is regressive. In addition, the real distribution effect of a tax solution is usually concealed, while that of a quantitative solution – as will be shown below – is transparent, at least in the initial stage.

Probably the weightiest argument against the introduction of an international carbon tax is not concerned with economics but with organizational factors: the tax volume needed to initiate an appreciable global reduction in emissions would be so immense (the literature speaks in terms of several hundred billion dollars) that centralized administration would be unacceptable, yet an acceptable decentralized redistribution would probably be very difficult to organize (but see *Hoel*, 1991 [10] and his comments on *reimbursement parameters*).

[2]The following concrete examples of greenhouse gas emissions refer to carbon dioxide (CO_2) emissions or the equivalent amount of carbon (C) – calculated at a ratio of 3.67 : 1. If other greenhouse gases such as methane (CH_4) and nitrous oxide (N_2O) are included in the discussion about climate policy, it is recommended that they be expressed as equivalents of CO_2, in order to introduce a common "currency unit" on the emission certificates market

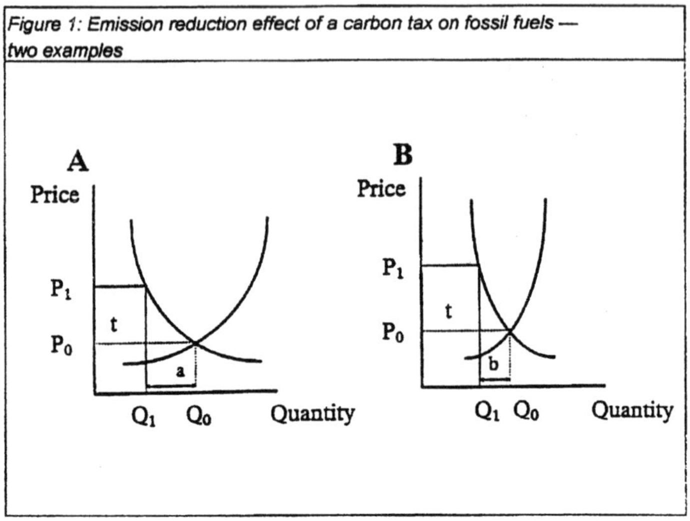

Figure 1. Emission reduction effect of a carbon tax on fossil fuels – two examples

There is also, however, an important ecological argument against an international carbon tax: in situations with rapidly increasing marginal environmental damage, taxes symbolize a possible economic compromise which the ecological system itself does not (can not) accept. In this case, quantitative (and immediate) restrictions are the only meaningful solution. Nor should one forget the problem of non-harmonized tax systems from country to country and region to region (particularly in the case of a mineral oil tax), which would be no less acute in the case of a carbon tax and, of course, the apparent reluctance of the OPEC countries (particularly Saudi Arabia) to compromise on the subject of global climate policy, especially a tax solution.

It needs also to be mentioned that a tax on *carbon* (or carbon equivalents) would be necessary as a response strategy to the climate problem, not a tax on *energy* in general. This has to do with the fact that the main task lies in extensively replacing fossil fuels by renewable sources of energy; this substitution effect would not occur if solar energy, for example, were also to be taxed on a global scale.

While efficiency and equity are central criteria of the Framework Convention and the "Berlin Mandate", it is surprising to see that in these documents no specific proposal is made of an international tax solution. This is different in the case of the two other policy instruments, which are referred to explicitly or implicitly, and will be dealt with in the following sections.

2.2 Joint implementation

As concerns the choice and structure of the policy instruments of global climate policy, Article 3, section 3 of the Framework Convention is especially relevant. This provision calls on the Parties to implement the measures agreed on in a cost-effective way: the desired reductions in emissions are to be achieved at minimum cost. In view of the ecological and economic context of a global climate policy, this efficiency clause is particularly significant: ecologically speaking – i. e. as regards the effect of greenhouse gases on the climate – it is completely irrelevant where in the world action is taken to reduce emissions, but if the costs of those reductions are to be kept at a minimum, then account will have to be taken of the fact that the marginal costs of reducing emissions (*marginal avoidance costs*) vary largely across the globe. In other words, strong economic arguments enter the climate policy arena. It was with this in mind that the instrument of "joint implementation" found its way into the Framework Convention (especially Article 4, sections 2a and 2b). At the first Conference of the Parties in Berlin in 1995, it was decided to introduce a pilot phase in order to gain respective experience.

Basically, joint implementation is an offset version of a quantitative policy with tradeable certificates: a country (a branch of industry, a company) can fulfill its reduction obligations through a combination of national (internal) reductions and international (external) offsets (offsets here means emission reduction credits which, once they have formally been certified, could be traded internationally). Until recently, this has usually been interpreted as meaning that an *Annex I* Party to the Framework Convention on Climate Change (i.e., OECD countries and countries with economies in transition) can fulfill its emissions target not only by domestic reductions, but also by investing in avoidance activities in other Annex I countries. The 1995 Berlin Conference resolved that non-Annex I states can also be included, on a voluntary basis. Joint implementation has thus become a policy instrument in the North-South context, and this provision can be seen as a first step towards a global climate policy of quantitative control and a system of internationally tradeable emission certificates.

A series of questions will have to be answered before it can be said how significant this policy instrument is or can become in the future (*Jepma*, 1995 [14]). Those questions which I feel to be most important, particularly from an ethical point of view, will now be dealt with in some detail.

The Framework Convention on Climate Change does not contain any definite target for the reduction of global emissions. According to Article 3, section 1, the industrialized countries, because of their 'historical debts' and their high emission levels, should

> "take the lead in combating climate change and the adverse effects thereof".

Along with their general commitments, i.e. developing inventories, promoting research etc., the industrialized countries assume certain additional obligations, particularly as regards the reduction of greenhouse gas emissions and the financial and technological transfer to developing countries. Since March 1994, all industrialized countries, including Eastern European States and the Russian Federation, are obliged

> "[to] adopt national policies and take corresponding measures on the mitigation of climate change, by *limiting* its anthropogenic emissions of greenhouse gases and *protecting* and *enhancing* its greenhouse gas sinks and reservoirs." (Article 4, section 2, italics added)

The exact legal scope of the term 'limiting' is difficult to assess, since the Convention does not provide any exact definition, but only a range of different aspects to be taken into account, *inter alia*, the corresponding need to maintain strong and sustainable economic growth in the developing countries.

Despite the lack of any precise timetable, Article 4, section 2a, clearly stipulates that the national policies to be adopted by the Parties have to be directed toward the 'limitation' of emissions, which should, according to the ordinary meaning of the term in the light of its object and purpose, amount to a significant modification of present emission trends, but does not have to amount to the return to 1990 levels by the year 2000. The duty under that paragraph is therefore not merely a duty of conduct – as in paragraph 1, which applies to all Parties – but a duty of *result*. However, since no definite timeframe has been agreed upon, a certain increase in carbon dioxide emissions after 1994 may be considered as still falling under the scope of 'limiting'. On the other hand, if the United States, for example, would entirely ignore their duty to *limit* their emissions for a long time, they would certainly act in breach of the Framework Convention. The agreed final aim as stipulated in the 1992 Convention is still

> "returning individually or jointly to their 1990 levels these anthropogenic emissions of carbon dioxide and other greenhouse gases not controlled by the Montreal Protocol." (Article 4, section 2b, italics added).

So far, then, there does not exist a *binding* global reduction target on absolute reductions below the (ecologically much too high) 1990 emission level nor any *country-specific* reduction target. Certain countries and groups of countries have, however, unilaterally committed themselves to definite reductions in emissions, including Germany, which, at the Berlin Conference, confirmed its assertion that by the year 2005, it would

> "reduce its emissions of CO_2 to a level 25% lower than that of 1990" (speech by the Federal Chancellor, 5 April 1995).

In this respect, joint implementation is for the time being only a policy instrument to make *unilateral* targets more flexible. However, in view of the varying marginal avoidance costs for greenhouse gases, from country to country and especially between North and South, a clear reduction in the cost of reducing emissions can be achieved. Or, to put it differently, an additional reduction in emissions can be achieved at no extra cost.

Joint implementation can also unlock positive economic effects via the transfer of low-emission technologies to developing countries. The otherwise tremendous increase in emissions that is to be expected, for example, when China and India become ever more motorized and industrialized, could be neutralized or even reduced. This positive effect of joint implementation is especially important, as no targets for the reduction of emissions have hitherto been set in these and (almost all) other developing countries.

One further important argument in favour of this policy instrument is that it can be applied without further delay, even if there is no global agreement on reduction obligations, or if no such agreement can be reached in the next few years.

Joint implementation, then, is a potentially powerful policy instrument, both for the ecologically necessary reduction of emissions and for the economically desirable transfer of efficient technology. However, its implementation is faced with potent obstacles, which can be summed up under the categories of *search costs, transaction costs* and *control costs*. These obstacles have been analyzed in quite some detail (see e.g. the 1994 Annual Report of the German Advisory Council on Global Change, 1995 [7, pp. 21 ff.]) and need not be repeated here.

The success of the joint implementation instrument will depend crucially on the institutional arrangements that are agreed upon. Several models are conceivable:

- simple bilateral systems of negotiation and information – the participating states report the reductions in emissions they have achieved to the other Parties to the Convention;
- inclusion of an international institution (such as the Secretariat of the Framework Convention); this would act as a clearing house promoting the emergence of a joint implementation market;
- in addition to that, an international institution (the Secretariat) monitors and verifies the reductions in emissions achieved as a result of joint implementatation.

One important component of these arrangements would consist in ascertaining the reduction in emissions effected by joint implementation in the form of "emission credits" for the investing country (branch, company). These credits are essential for two reasons. First, they provide the necessary incentive for investing capital abroad and, second, they must not run counter to the reduction of emissions at home (allegation of "modern sale of indulgences"). The Berlin Mandate, however, stipulates that "no credits shall accrue to any Party as a result of greenhouse gas emissions reduced or sequestered during the pilot phase from activities implemented jointly".

In order to avoid possible failure of the joint implementation policy instrument inherent in this restrictive condition, the following compromise can be suggested: Emission credits should not be credited in full to the national emissions account, but only in part – 50%, for example (as suggested by France), or 75–80% (as suggested by WBGU, 1995 [7]). In this case, if the reductions in emissions that have been achieved were used to strengthen national reduction targets (for the EU, OECD or Annex I states), then joint implementation would indirectly lead to more stringent climate protection effects.

It must be recalled that participation in joint implementation projects should not lead to any reduction in the overall financial obligations resulting from the Framework Convention or, indeed, in the actual and otherwise pledged development aid payments (Norway has submitted a proposal to this effect).

Taken together, then, the debate about joint implementation ought to be seen as an opportunity to sound the trumpet that political possibilities exist for stabilizing the climate and for coupling them with proactive development policy. Nevertheless, the quantitative significance of joint implementation in the global context should not to be overestimated. This policy instrument will only allow the industrialized countries to fulfill the minor part of their obligations to reduce emissions. Even so, in the developing countries a process can get underway which would otherwise only get off late (or even too late) – and in the end, this process could lead to a more comprehensive system of internationally tradeable emission certificates.

2.3 Internationally tradeable emission certificates

Tradeable emission certificates differ in various ways from joint implementation (or "external offsets"). Binding global obligations to reduce emissions will (must) result from the pending negotiations. The German *Bundestag's* Enquete Commission on Climate Policy, and also the Inter-governmental Panel on Climate Change (IPCC) assumes that a reduction of global CO_2 emissions by 50% compared to the 1987 levels (requiring an 80% reduction in the industrialized countries) must be reached by the year 2050 if the target of stabilizing the climate system is to be attained (see Table 1).

For the present, let us assume that an agreement of this kind (or similar) is reached in the "Climate Protocol", to be negotiated on and resolved at the Third Conference of the Parties in Kyoto in 1997, or at a later stage. What could this mean for the choice and structure of a system of internationally tradeable emission certificates?

As part of the system of agreed quantified permissible global emissions, tradeable certificates would be handed out – for the whole duration of the

Table 1. CO_2 emission plan, Enquete Commission on Climate Policy (benchmark 1987, in %)

Year	Industrialized countries	Developing countries	Whole world
1990	+5	+11	+6
1995	+6	+24	+10
2000	-4	+37	+4
2005	-20	+50	-5
2020	-40	+60	-20
2050	-80	+70	-50

Source: Enquete Commission, 1990 [5].

agreement or, better still, for certain time periods – to the participating Parties on the basis of an allocation procedure that would also have to be agreed on (which both would probably be negotiated in parallel). When the agreement comes into force, the participants would receive certificates corresponding to the emission quantities they have been allocated. If this initial allocation were insufficient for a participant (a country, a branch, a company), the participant would have to acquire additional certificates via trade. For those participating in the system, therefore, an incentive to reduce emissions would be established, be it to minimize payments for the purchase of additional certificates or to maximize earnings from the sale or lease of surplus certificates. If the initial allocation to developing countries leads to substantial quantities of surplus certificates there, a potent mechanism for the transfer of resources would be created: developing countries could sell or lease their surplus certificates to industrialized countries for money, technologies or patents.

So, there are certain similarities between an international emission charge (*price solution*) and tradeable emission certificates (*quantity solution*): the issue of the initial allocation of certificates is similar to the issue of allocating the revenue from a charge. However, there are also several differences.

The most important of these is that emission certificates exactly meet the emission reduction target in terms of quantity; the concurrent financial expenditure would be the result of the costs connected with achieving this target. This contrasts with an emission charge, which regulates financial expenditure but does not directly regulate the volume of the emissions. A further difference is that a charge generally means monetary transfers, while certificates could easily be traded for gratuities other than money (such as technologies or patents). A system of tradeable emission certificates, therefore, increases the scope of the negotiations between North and South – and might therefore be met with broad (possibly sufficient) political approval for precisely that reason.

Unlike in joint implementation projects, monitoring a system of tradeable emission certificates would be concerned with the (relatively simple) measurement of total emissions from a contracting state rather than the (more difficult) measurement of emission reductions of specific projects. The question of responsibility for adhering to the rules of procedure (i.e., compliance) is also easier, as one is not dealing with direct investments but with the sale or lease of a tradeable good (i.e., certificates).

There are other theoretical and practical issues connected with a system of emission certificates (*Victor*, 1991 [22]) – the issues connected with implementation will be dealt with in the following section. Let me end the present section with a quotation from the 1995 IPCC Report [13], which in its chapter on response strategies (Chapter 10) states the following:

> "The consequences of climate change policy will be determined by the choice of policy instruments [...] For a global treaty, a tradeable quota system is the only potentially efficient arrangement where an agreed level of emissions is attained with certainty (subject to enforcement) [...] A choice of tradeable quotas at the international level would (at the same time) provide maximum flexibility for instrument choice at the domestic level". (italics added)

3 From theory to practice

As was said at the beginning, a practicable agreement on policy instruments to limit or reduce greenhouse gas emissions has to satisfy several criteria, in particular those of *efficiency, equity*, and decision-making under *uncertainty*. The weight given to the individual criterion will determine which of the possible policy instruments or combinations of instruments is being recommended. If, unlike at the national level, the criteria of equity and uncertainty (particularly because of irreversible ecological processes) play a special role at the international level, then there is much to be said in favour of tradeable emission certificates. Yet, their practical organization entails many potential snares which can be decisive for their acceptance. The following section will look in some detail at the questions of market organization, rules of procedure, and the initial allocation of emission certificates; other issues, which, I feel, are less problematic, will not be addressed here (see *Epstein* and *Gupta*, 1990 [6]).

3.1 Market organization

Creating a market for internationally tradeable emission certificates is no easy undertaking, to put it mildly. Monitoring, certification, market access and market extension require careful management – what is more, in a highly complex area of policy. A debate about these implementation issues has at least begun, and thus one may expect the system to become established in the not too distant future. An UNCTAD study [21] had looked already into the institutional issues of market organization.

One such issue is the number of actors on this market. A "mixed" trading system in which both governments and companies participate could be optimal. Governments would remain subject to their international obligations, particularly to ensure that the certificates tally with actual emissions. Trade at the company level would enlarge the technical options of emission reduction. Yet, this could also mean that the volume of trade becomes too large. However, the major worry in the literature is that cartels could be formed. Whether or not this threat can be conquered will in the first place depend on the number of market participants, which speaks in favour of a more substantial number. If one were, for example, to start with the producers and importers of raw materials containing carbon, then, according to Maier-Rigaud, there would be about 500 actors on an EU certificates market.

There are several ways of avoiding cartelization on the emerging certificates market. Regular rounds to re-allocate or replace certificates could ensure a liquid and flexible market, and rules against hoarding and price rigging could be agreed on. In addition, the final, drastic sanction against improper conduct could be "exit", although this would have an adverse effect on all the market participants.

3.2 Rules of procedure

One question is whether emission certificates should be valid endlessly. Although any such "perpetual certificate" would not necessarily prevent the revision (and especially the tightening) of the global emission reduction target – a part of them could be withdrawn from the market or devalued regularly – much speaks in favour of certificates with only limited validity. For one thing, not all Parties to the Framework Convention will participate in the system from the start and, for another, this would counter existing fears of certificates being *bought up* by industrialized countries or multinational corporations. The other extreme version, whereby certificates would be *leased but never sold*, would only lead to more flexibility if a (more) frequent new issue was agreed on. One must say that the question of an *optimum* term for emission certificates is still an open one. In the literature, the ideas range from two to 20 years.

In this context, *Bertram* [2] has proposed an *overlapping procedure*, under which ten-year certificates would be issued and 10% withdrawn from the market every year. This proposal could be summed up by the following formula:

> "If certificates are valid for L years and a certain proportion P/L is withdrawn from the market every year, then a new tranche of certificates valid for L years can be issued."

At any given time, therefore, the market would consist of a mix of certificates, some long-term, some short-term; countries (branches, companies) could accordingly maintain a mixed portfolio and a futures market (comparable with existing other 'futures markets') could emerge.

The benefit of such a flexibilization of the system is obvious: it would be possible, on the basis of the most recent *natural* science evidence about pollution and the adaptability of the ecosystem, and of social science evidence as regards the limits to adaptability of the economic system, to hold subsequent negotiations on permissible emission limits, to include new sources and sinks of greenhouse gas emissions, and to allow additional countries to participate.

Whatever happens, this benefit should not turn into a disadvantage as a result of excessive complexity or bureaucracy. *Grubb* and *Sebenius* [8] have shown that revision periods in an overlapping system of between two and four years might be the best possible solution. Given the probable workings of the global climate regime that is to be established (such as annual conferences of the Parties, two-thirds majorities, gradual tightening of the climate protocol), this may well be a realistic assumption.

3.3 Allocation of certificates

The crucial factor for the political acceptance of tradeable emission certificates will probably be that their design be ethically based, and their initial allocation be perceived as fair. Indeed, for many authors (including myself), the system hinges on the "equity factor".

While the initial allocation of certificates does not predetermine the final distribution effect (i.e., real incidence), which ultimately is the result of market decisions (certificate price and quantity), it does predetermine the direction and possible volume of transnational resource transfer the system will involve. A certain distribution effect will, of course, result from all conceivable instruments of climate policy, especially where the dimensions are globally significant, but also in national solutions, be they price or quantity solutions.

There are different ways of effecting the initial allocation of emission certificates. One distinction is that between *burden-based* and *responsibility-based criteria*, the former laying emphasis on the burden of adjustment involved in the desired reduction in emissions, while the latter stresses the polluter-pays principle, either in terms of the current emissions or of historical and accumulated emissions.

To a certain extent, the search for an ethically acceptable allocation formula in global climate policy reflects the old debate between "realists" and "idealists" in development policy (*Sterner*, 1994 [18]).

- "Realists" would argue that certificates (and the rights to pollute that they entail) must be allocated on the basis of either the *current emissions* or of the Gross Domestic Product (GDP), since any other formula would be unacceptable (see, for example, *Pearce*, 1991 [17]). Ethically, this position is extremely weak and completely fails to satisfy any criterion of equity. In no way neither current emissions nor historical and accumulated emissions can be defined as "fair". Furthermore, this position ignores one, if not the, crucial advantage of an international system of certificates, which is that an *additional* transfer of resources can be set in motion. The "realistic" position is also unrealistic in that it almost completely ignores the developing countries, which are beginning to take an interest in global climate policy precisely because ecological necessity may turn out to be to their economic blessing.

– On the opposite side we have the "idealists", who insist, explicitly or implicitly, on the inclusion of historical and accumulated emissions (as a sign of "ecological guilt" or "historical debt"). Indeed, some of them even argue that developing countries, for the time being, should not be integrated into the global climate regime (see, for example *Hayes* and *Smith*, 1993 [9]).

Various allocation proposals have been put forward, with *globally uniform per capita allocation* presenting the strongest ethical claim. In this case, the initial allocation of emission certificates would be directly proportional to national population (in the current year or, as a softer version, in a base year, like 1992, for instance). It can be said that the practical consequences of this proposal would be considerable, if not revolutionary. Any industrialized country with above-average per capita emissions would have to purchase certificates from developing countries, not only in respect of fossil fuel consumption but also of all other sources and sinks of greenhouse gases, possibly with the exception of deforestation (slash-and-burn, clear-felling). Even under moderate projections as to the certificate's prices that would ensue, there would then be a complete reversal of the presently existing net South-North transfers. *Grubb* and *Sebenius* [8] assume that it would be at least as much as current official development assistance (ODA), which is in the order of US$ 60 billion per annum.

Of course, the alternative to any extreme position is to find a *mixed formula* – a formula which will at least guarantee that there is a net transfer from the industrialized to the developing countries (and not *vice versa*) in implementing a global climate policy. Personally, I find the formula developed by William R. *Cline* [4] extremely appealing:

$$Q_i = Q_g \left[w_h \Phi^h_{O,i} + w_y \Phi^y_{O,i} + w_p \Phi^p_{O,i} \right] \ ,$$

where Q_g is the global emission target; Q_i is the emission target of country i; w the weight of the criterion in question ($w_h + w_y + w_p = 1$); h is historical emissions, y is GDP at purchasing power parity, and p is population; Φ is the share of country i in the global total; O is the benchmark year.

This allocation formula includes the most important alternative criteria under discussion and weights them, a practice that has also been applied by the IMF (definition of country quotas), the United Nations Committee for Development Planning (definition of Least Less Developed Countries; LLDCs), and other institutions (*Levi*, 1991 [15]). The formula could be described as the necessary mixture of efficiency, equity and realism. Cline provides an illustrative example of this formula in action: the USA currently (1992) accounts for 25.7% of global GDP, 17.5% of global greenhouse gas emissions (including deforestation), and 4.8% of the world's total population. The simple average of these three figures comes to 16%. Were a global emissions target of 4 billion tons of carbon to be agreed on by the Conference of the Parties to the Framework Convention, the USA's initial share would amount to 640 million tons. Current emissions, however, total 1.2 billion tons. Accordingly, the USA would either have to reduce its emissions by 50%, or purchase an additional 100% of its initial share of allocated certificates.

By contrast, India's share would come to 8% of global emissions or 320 million tons of carbon, which would correspond to a surplus of some 50 million tons (or 17% above current emissions), which India could then sell or lease on the international emission certificates market.

As befits the logic of a mixed formula, the weighting of the *three* components could be *modified* in the course of time. For example, the weighting of the first criterion (historical, accumulated emissions) could be reduced from one-third down to zero over a period of, say, 20 years ("phasing out") and that of the second criterion (GDP) from one-third down to zero over a period of 40 or 50 years. The final result then would be an ethically strong position, namely globally uniform per capita emission rights (i.e., the population criterion).

A less formal, more 'political' solution to the allocation problem might lie in a compromise that could be described as follows:

> "The allocation of emission certificates changes over time, from a position based (more or less) on *current emissions* to a position of (more or less) *equal per capita emissions*."

A strategic compromise of this nature might be acceptable both for the industrialized countries and the developing countries, since it offers a strong incentive for a fair (a fairer) future emissions situation for the whole world in general, and for the linking of environmental protection with economic development in particular – even though this perspective was not aimed for at either the 1992 Rio de Janeiro Conference on Environment and Development or at the 1995 Berlin Climate Conference.

4 Conclusions

As far as *national environmental policy* is concerned, and in view of the high degree of institutionalization and the specific historical and cultural background of policy formulation and implementation in the industrialized countries, the author is very much in favour of a *balanced instrumental mix*, which includes market-oriented and regulatory, price-based and quantity-based policy instruments. As regards the state of the international system and the emerging contours of the structure and function of the *global climate regime*, however, I do favour market-oriented quantity solutions, especially *joint implementation* in the regime's initial phase and *tradeable emission certificates* in its final phase. If their form is prepared with the necessary care, then Peter Bohm's succinct judgement seems justified:

> "Making emission quotas tradeable among countries implies not only that a globally efficient limit to total emissions is attained with *certainty*, ... but also that the initial emission quota distribution of the treaty is shifted in favour of *the poorer countries*" (*Bohm*, 1992 [3], p. 112, italics added).

This first best solution does not exclude regulation or taxation as second best solution. In particular, if the institutionalization of the Global Environment Facility (GEF) is to be pushed forward, then it will have to be financed by taxes, as conventional multilateral development assistance would otherwise be affected negatively. In this case, other forms of taxes and bases of assessment again appear on the political agenda, such as a tax on arms exports (see *Brandt* Report, 1983 [11]), on long-distance tourism (see *Mishan*, 1970 [16]), or on international financial transactions (see *Brundtland* Report, 1987 [23], and *Qureshi* /von *Weizsäcker*, 1995 [12]).

4.1 Acknowledgements

I would like to thank Robert U. Ayres, Frank Biermann, and Carsten Helm for helpful comments on a previous draft of the paper.

References

1. W. Baumol and W. E. Oates. *The Theory of Environmental Policy.* Englewood Cliffs, 1975.
2. G. Bertram. Tradeable Emission Permits and the Control of Greenhouse Gases. *Journal of Development Studies*, 28(3):423–446, 1992.
3. P. Bohm. Distributional implications of allowing international trade in CO_2 emission quotas. In W. R. Cline, editor, *World Economy. The Economics of Global Warming*, volume 15(1), pages 107–114, Washington, 1992.
4. W. R. Cline, editor. *The Economics of Global Warming.* Institute for International Economics, Washington DC, 1992.
5. Enquete Commission "Protecting the Earth's Atmosphere". *Report of the Study Commission of 11th German Bundestag: Preventive Measures to Protect the Earth's Atmosphere*, Bonn, 1989.
6. J. M. Epstein and R. Gupta. *Controlling the Greenhouse Effect: Five Global Regimes Compared.* Washington, 1990.

7. German Advisory Council on Global Change (WBGU). *World in Transition: The Threat to Soils. Annual Report 1994*, Economica, Bonn, 1995.
8. M. Grubb and J. Sebenius. Participation, allocation and adaptability in international tradeable emission permit systems for greenhouse gas control. In OECD, editor, *Tradeable Permits for Abating Greenhouse Gases. Practical Options*, Paris, 1990.
9. P. Hayes and K. Smith, editors. *The Global Greenhouse Regime. Who Pays?*, Tokyo, 1993.
10. M. Hoel. Efficient international agreements for CO_2-control. *The Energy Journal*, 12(2):93–107, 1991.
11. Independent Commission on International Development Issues: Common Crisis, North-South. *Cooperation for World Recovery*, London, 1983. (Brandt Report).
12. Independent Working Group on the Future of the United Nations. *The United Nations in its Second Half-Century*, New York, 1995. (Qureshi/von Weizsäcker Report).
13. Intergovernmental Panel on Climate Change. *Climate Change 1995. The Science of Climate Change*, Cambridge, 1996. Cambridge University Press.
14. C. P. Jepma, editor. *The Feasibility of Joint Implementation*, Kluwer, Dordrecht, 1995.
15. M. D. Levi. Bretton woods: Blueprint for a greenhouse gas agreement. *Ecological Economics*, 2(4):253–267, 1991.
16. E. J. Mishan. *Technology and Growth. The Price We Pay*. New York, 1970.
17. D. W. Pearce. The role of carbon taxes in adjusting to global warming. *The Economic Journal*, 101:938–948, 1991.
18. T. Sterner, editor. *Economic Policies for Sustainable Development*, Kluwer, Dordrecht, 1994.
19. T. Tietenberg. Implementation issues for globally tradeable carbon entitlements. In E. C. van Ierland, editor, *International Environmental Economics*, pages 119–149, Amsterdam, 1994.
20. *United Nations Climate Change Bulletin*, Geneva, 1995. Interims Secretariat for the UN Climate Change Convention. Issue 7.
21. UNCTAD. *Combating Global Warming: Study on a Global System of Tradeable Carbon Emission Entitlements*. New York, 1992.
22. D. G. Victor. Limits to market-based strategies for slowing global warming: The case of tradeable permits. *Policy Sciences*, 24(2):199–222, 1991.
23. World Commission on Environment and Development. *Our Common Future*, Oxford, 1987. Oxford University Press. The Brundtland Commission.

Commentary: Criteria for an Equitable Distribution of Internationally Tradeable Emission Certificates

Carsten Helm

1 Introductory remarks

In his contribution (Chapt. 2), Simonis searches for a climate protection policy that equally satisfies the criteria of efficiency, equity and – one might add – political feasibility. He concludes that a system of internationally tradeable emission certificates, where the initial allocation starts on the basis of current greenhouse-gas emissions but over time turns towards equity in the form of equal per capita emission rights, would come closest to this target. As a crucial advantage of this solution he regards the additional North-South transfer it would lead to.

In my comment, I will propose a way to make the notion of equity more concrete and thereby facilitate a debate about equity issues in climate change that is less prone to two common objections: "that equity is merely a word that hypocritical people use to cloak self-interest" (*Young*, 1994 [6]) and that "it is so hopelessly subjective that it cannot be analyzed scientifically" (ibid.).

2 Structuring the equity issues of climate change

Focusing on the entitlement to emission rights, as Simonis does in his contribution, already significantly narrows the perspective on equity, because it excludes the issues of equity within countries, intergenerational equity and an equitable sharing of the damages from climate change. To be specific about what is to be shared equitably is a crucial prerequisite in applying any criteria of equity. For example, a statement like "parity is a formula for equal distribution of burdens or benefits" (IPCC, 1996 [3], p. 86) is of little help, because sharing the *burdens* of emission reductions or the *benefits* of the atmosphere's sink capacity would lead to completely different results, even if the same equity principle of parity were applied.

Figure 1 illustrates this point: that not only the choice of the appropriate equity principles is crucial, but also the choice of the appropriate perspective from which to apply them. The budget line gives the imaginary number of available permits which are to be allocated between two groups called North and South on the basis of the equity principle of parity. If the reference point, from which one equally shares the absorptive capacity of the atmosphere, is the imaginary ex ante position of a still unpolluted atmosphere, the corresponding allocation would be B. However, if the reference point is the current status quo, where the North has much higher per capita emissions than the South, the corresponding allocation would be A. These are, of course, the two extreme positions of "realists" and "idealists" as mentioned by Simonis.

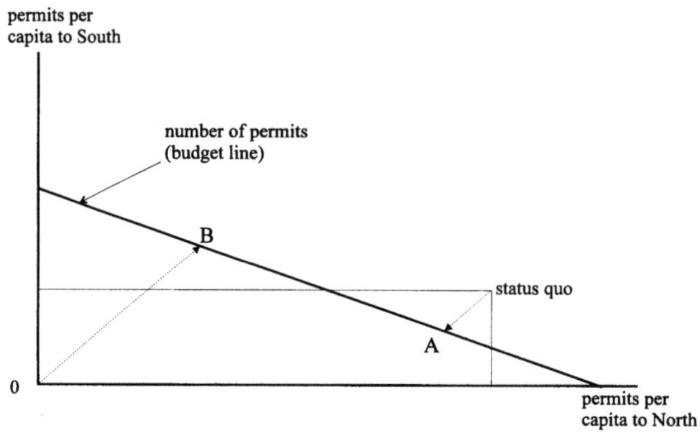

Figure 1. Fair allocation of tradeable permits

Related to the search for the right perspective is the question, to what extent the extremely unequal worldwide distribution of wealth, which is widely perceived as inequitable, should influence our perception of an equitable climate change regime. Simonis explicitly argues to link the distributional questions of tradeable emission permits with the broader level of development policy; and there are certainly strong arguments for such an holistic approach to equity, which, for example, has been followed by *Rawls* [5] and others who have tried to answer the question of the *just social order*, i.e. the just distribution of resources at large. Yet, there are also good reasons to follow the alternative approach of analyzing 'justice in the small'. In particular, agreement on what constitutes an equitable solution might be easier for an isolated issue like climate change, than if it is linked to even more controversial issues like the global distribution of wealth. Nevertheless, any equitable climate change policies should take care that at least they do not worsen global inequality.

3 An axiomatic approach to equity

Following the narrow approach of focusing solely on the equitable distribution of emissions and abatement costs in the climate change regime, some general criteria of equity can be applied, which appear as reasonable requirements for any equitable distribution of a commonly owned resource[1]. Yet, as a prerequisite to this approach the entitlements to the common resource – here the sink capacity of the atmosphere – need to be determined. Simonis regards an equal per capita entitlement as most equitable, and so do most other writers (IPCC, 1996 [3], p. 106). In fact, if one perceives the atmosphere as a global common – just as manna falling from heaven – then it is hard to argue why any one individual should a priori be more entitled to it than others. However, conceding each country equal per capita entitlements is quite different from providing them with equal per capita permits and letting them trade on international markets, as will be clear soon. As a further prerequisite – even though this could already be regarded as an equity test in itself – I demand that any allocation of emissions be efficient (Pareto-optimal), which simply says that the cake to be shared should be maximized.

Probably the most prominent equity criterion within the economic profession is the no-envy test, according to which every agent should view his share at least as valuable as anybody else's share. Unfortunately, the set of envy-free allocations of rights to pollute the atmosphere and corresponding compensatory payments is quite large, because there exists a number of different allocations where the North compensates the South for a greater share of per capita permits, while neither of the actors would prefer per capita permits and compensatory payments of the other one. More specifically, any formula that satisfies Simonis' criteria of at least guaranteeing a net transfer from the industrialized to the developing countries is envy-free as long as these transfers are not higher than the North's cost savings from the permits it receives on top of its per capita share.

However, there are two criteria of equity which offer greater guidance. The first one is the *fair share guaranteed* test (also called *acceptability*), which guarantees each actor at least the utility of his fair share of the common property resource, where, following the considerations above, fair share is defined as equal per capita entitlements. This can be stated as

$$U_i(P_i) + M_i \geq U_i(FP_i) \ , \qquad (1)$$

where $U_i(P_i)$ is actor i's utility from its permits, M_i are monetary transfers and $U_i(FP_i)$ is i's utility from its fair share of permits.

The second criterion is the *stand alone* test, which says that the benefits from common ownership of a resource should not be higher than the benefits from exclusive consumption of the resource; or formally

[1] Cf. *Helm* [2] and, for a general introduction to the equity criteria used below, *Moulin* [4]

$$U_i(P_i) + M_i \leq U_i(PG) \ , \tag{2}$$

where PG is the global number of permits. This second inequality can also be derived from the criterion of resource ***monotonicity*** that no actor should receive a higher benefit if the common resource shrinks.

Interpreting a country's benefits from permits as avoided abatement cost ($U_i(P_i) = C_i(0) - C_i(P_i)$), and taking into account that emissions in the South's business as usual path will at least initially be lower than its fair share of permits and would therefore require no abatement measures ($C_s(FP_s) = 0$), the two inequalities can be combined to yield for the South

$$C_s(P_s) \leq M_s \leq C_s(P_s) \quad \text{or} \quad M_s = C_s(P_s) \ . \tag{3}$$

In summary, as long as emissions in the South's business as usual path are lower than its fair share of permits, the two, in itself rather mild criteria of fair share guaranteed and stand alone, prescribe a single equitable solution: The South should be fully compensated for the abatement measures imposed by the criterion of efficiency, but not receive more either, as would be the case with an equal per capita allocation of internationally tradeable permits.

4 Outlook

Unfortunately, implementation of the above solution is not very straightforward. Yet, some possible modes of implementation are very close to Simonis' propositions. In particular, an initial phase of joint implementation could be succeeded by a tradeable permit system, once the South's per capita emissions in its business as usual path approach the Northern level. Alternatively, if the arguments in the *Cline* [1] formula are weighted such that initially the status quo is emphasized followed by a successive shift towards the population criterion, as proposed by Simonis, the resultant allocation of abatement measures and compensatory payments could also come close to the solution derived above.

I should close my comment with some words of caution. The solution proposed above is only equitable according to the criteria used in its derivation and given the perspective from which they have been applied. Both can be criticized with good arguments, for example by the choice of a more holistic perspective on international equity. Furthermore, a frequent characteristic of the axiomatic approach is that different criteria of equity are incompatible, even though each criterion in itself seems very reasonable – an issue not further pursued here (but see *Helm*, 1997 [2]). However, hopefully it has become clear that equity is much more than "simply a matter of the length of the judge's ears", as Elbert Hubbard once put it provocatively.

References

1. W. R. Cline, editor. *The Economics of Global Warming.* Institute for International Economics, Washington DC, 1992.
2. C. Helm. Fair play im Treibhaus. Zur gerechten Verteilung von Emissionsrechten. In F. Biermann, S. Büttner, and C. Helm, editors, *Zukunftsfähige Entwicklung. Herausforderungen an Wissenschaft und Politik,* Berlin, 1997. Edition Sigma.
3. Intergovernmental Panel on Climate Change. *Climate Change 1995. Economic and Social Dimensions of Climate Change,* Cambridge, 1996. Cambridge University Press. Contribution of Working Group III to the Second Assessment Report of the IPCC.
4. H. Moulin. Cooperative microeconomics. A game-theoretic introduction. Princeton, 1995.
5. J. Rawls. *A Theory of Justice.* Harvard University, Cambridge MA, 1971.
6. H. P. Young. Equity in Theory and Practice. Princeton, 1994.

Chapter 3: New Models of Wealth

Ernst Ulrich von Weizsäcker

Abstract Northern consumption patterns should not be copied by six billion people. Assuming that developing countries have the same right to prosperity as the North, a drastic reduction appears necessary in the ratio wealth/resource use. Fortunately, this is attainable chiefly by technological means. However, the incentive structures ought to be adjusted to invite and trigger the new generation of technologies. Lifestyles may also change to support the new trend.

1 Introduction

Dealing with global warming requires new thinking. It is a different kind of environmental challenge that we are facing. Classical environmental policy was dealing with pollution. Pollution could be dealt with by pollution control mostly at the end of the pipe.

The trouble with end of the pipe solutions, however, is that there are invariably additional costs for the producer. No product is added, but production becomes costlier. Hence industries can argue that they have to prosper in order to afford pollution control. Right they are. However, this statement is in a sense tragic. Why?

Because when we in the rich countries admonish people in the poor countries to take better care of their environment (as we did at the Earth Summit), we get their reply as a kind of *echo*: "We shall happily take care of our environment as soon as we can afford to, i.e. as soon as we prosper the way you do in the North." (See Fig. 1.)

The only way out of this situation is by considering and evaluating the link between prosperity and ecology concurrently in both directions.

2 Interdependence between prosperity and ecology

New thinking starts with and is evoked by assuming that the kind of Northern prosperity if emulated by 5 billion people, even more by 10 billion, would lead to an ecological disaster. In other words: world-wide pollution control will be seen as affordable only after our having overexploited and destroyed our ecosystems in the first place.

This may be called the pollution control dilemma. It is not in the interest of industry to make the dilemma public. It is far more comfortable to go on calling pollution control the core of the environmental question, because

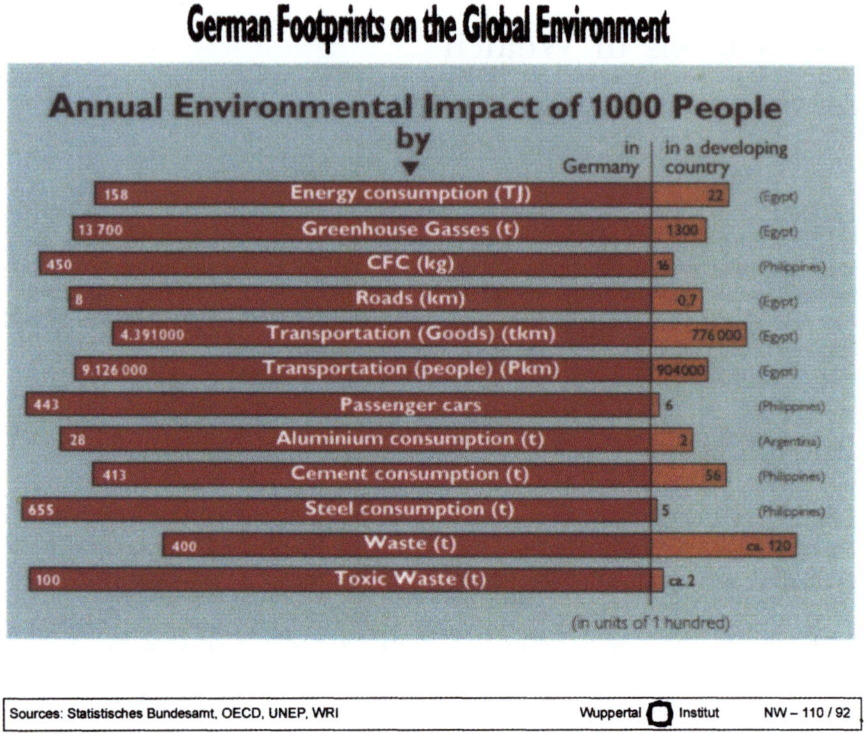

Figure 1. A thousand Germans consume roughly ten times more than the same number of people in developing countries. After *Bleischwitz*, 1992 [1]

then we can go on pretending that we have to prosper in the traditional sense before we can afford the protection of the environment.

Global warming now is the watershed. It is bound to mark the end of this irresponsible kind of thinking, because prosperity in the traditional sense implies high GHG emissions.

It is, of course, very tempting to try to integrate global warming into old thinking. This can be done by looking at existing economic activities and then calculating the costs incurred if certain percentages of the energy supplies for these existing activities were made CO_2 free. Unavoidably, these shifts to CO_2 free energy supplies turn out to be unbearably expensive. William *Nordhaus* [3] and others have demonstrated this. They are talking about trillions of dollars.

The *natural* political (and the economists') reaction to such research results is this: We may be able to calculate the costs of *adaptation* to global warming and find out if that is cheaper than the avoidance strategies. Small wonder, the answer is yes.

I must say, I am worried by the political effects this kind of consideration may have. It is just so comfortable for the combined coal, oil and gas lobbies, for car makers and the steel industry, for trade unions and others concerned with the unemployment; but why should I worry? Assuming that the calculations are correct, do I really advocate spending trillions on the avoidance?

My answer is yes and no.

Yes, because many more trillions are going to be spent anyway for this or that purpose. My hope is that a new sense of direction is going to develop for technological and civilisational progress, a sense of direction powerful enough to redirect investments and consumer spending so as to achieve all that is necessary to avoid climate disasters.

However, my answer is also no to the above question, because I do not see any reason why this redirection of investment and of consumer spending should be perceived as added *costs.* As a matter of fact I believe that it is going to be *investments* with very reasonable returns on investment rate.

3 New model of wealth

To prove this I finally have to turn to the topic I was asked to address, namely a new model of wealth.

Our present model of wealth is characterised by overuse of natural resources. We in the North, who enjoy wealth, are consuming roughly 10 times more natural resources per capita than people in developing countries (see again Fig. 1).

The Dutch concept of *ecological space* or the Canadian concept of *ecological footprints* are going a little further than this picture. They estimate the cradle to grave world-wide ecological impacts of the goods and services enjoyed by people in rich countries. And they come to the not so surprising result that extending our lifestyles to 5 or 10 billion people would require some three earths of ecological space or could leave footprints three times as large as our Earth.

This result, by the way, also proves the adaptation strategy wrong. Adaptation, too, is causing unbearably big ecological footprints.

4 Costs and quantities

Now, the obvious question is: Is there anything deserving the term wealth and leaving much smaller footprints, "sustainable footprints" if you wish? I shall outline an optimistic answer below. However, before doing so, I should like to estimate the quantitative size of the challenge. Climate protection will mean that we should reduce global greenhouse gas emissions by more than

50%. Growth aspirations, on the other hand, suggest a doubling at least of demand for energy services during the next forty years. This leaves us with a gap as large as a factor of four between what is needed and what will come if current trends continue (*Weizsäcker* et al., 1997 [7]).

The answer will be a twofold paradigm shift, one with regard to technological progress, one relating to lifestyles.

Technological progress in the past was characterised by a steady increase of labour productivity. Extended over a period of about 150 years, an increase by a factor of 20 is perhaps a fair estimate. This resulted in an increase by a factor of 20 in GDP (to be multiplied by the number of people living in the respective economies). Energy consumption went hand in hand with GDP growth.

It is worth mentioning that during this period, labour became more expensive every consecutive decade during these 150 years, while energy and raw materials became cheaper every consecutive decade (except for the short period from 1973 - 1982). A hundred years ago, the costs of one kilowatt-hour were roughly as high as the costs of four hours of human labour. Today, a kilowatt-hour costs perhaps as much as twenty seconds of human labour.

Technology went with profitability. Increasing labour productivity became ever more profitable, while increasing energy productivity remained a low priority affair. The technological reaction to the oil price shock after 1973 proved that there is much scope for improvement of energy productivity.

However, this was a very unsatisfactory experience still, from the point of view of redirecting progress and investments.

- It came as a shock and therefore caused disruption and big economic losses;
- Nearly all the additional money spent on oil and gas was additional costs, not benefits, to the OECD countries;
- As a result, the oil crises resulted in widespread anxiety of high energy prices and massive attempts world wide to pull them down again, successfully, as the reader knows.

Nevertheless, there is an element of lasting encouragement in the oil crises experience: It proved that decoupling is technically possible.

A *second look* at the economic effects of high energy prices seems to show that high energy prices per se do not harm the economy. In fact, countries maintaining high energy prices against world market trends seem to have had a positive, not a negative effect on the respective economy's performance (Fig. 2).

5 "Factor Four"

Let me now turn to the challenge of closing the gap between the goal of climate stabilisation and the trends of growing energy demands. As indicated

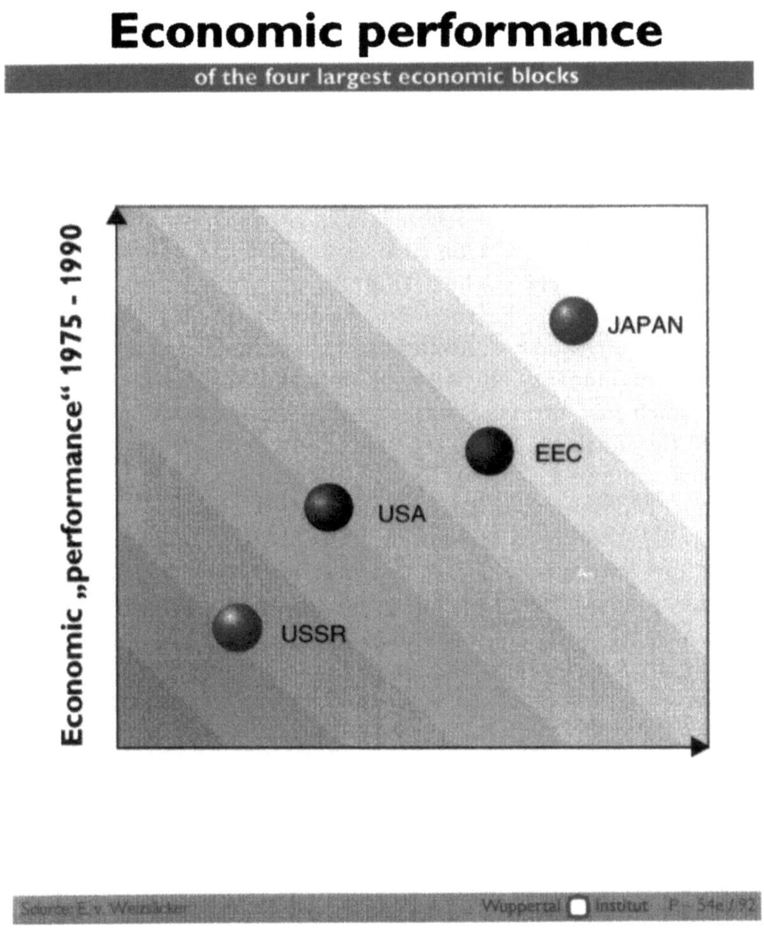

Figure 2. A positive, rather than a negative correlation seems to exist between a composite index of economic performance and energy prices. After *Rechsteiner*, 1993 [4]

above, the gap is as large as a factor of four. Some contributions towards closing the gap can be expected from an increase of carbon-free energy sources. However, what we at the Wuppertal Institute consider is most promising is a strategy of systematically increasing energy productivity – together with other resource productivity in general.

We are looking at various sectors of energy and resource consumptions and find that increases by a factor of four are generally possible under the condition that systemic changes are allowed. An increase in the recycling rate of aluminium from 15% to 85% results in a factor of four in energy productivity.

A factor of four or more is attainable in CO_2 efficiency in private homes and office buildings. Amory *Lovins* (see *Weizsäcker* et al., 1997 [7]) claims that for most household appliances a factor of four in energy efficiency can be reached. He is even proposing a passenger car achieving 150 miles per gallon (1.5 litres per 100 km) which again is easily a factor of four better than today's fleet. This book offers fifty examples of quadrupling resource productivity. Some are pure technical innovations, while others relate to logistics and even lifestyles. One such example is the transport intensity of manufacturing strawberry yoghurt (Fig. 3).

Other examples from "Factor Four" relate to the durability of consumer goods. This requires new attitudes on the part of consumers; low transport intensity, high energy efficiency homes, and durable goods – that is not something which just springs from industrial laboratories and automatically spreads into the markets. It is a true paradigm shift for a civilization that is determined to achieve a *sustainable* development.

How can this determination be achieved? Is it a matter of brain-washing? Rather not. What we need is a major shift in the rules and incentive structures. Let me only mention one instrument serving to induce this shift: ecological tax reform (1992 [5]). By gradually reducing the fiscal burden on human labour and capital and simultaneously raising taxes on resource use, the incentive structure would be changed. Slowly but steadily the incentive would increase to rationalise energy; and the pressure to rationalise labour will be slowed down.

New models of wealth will also require new perceptions and new measurements of wealth. As Wouter *van Dieren* [6] has outlined in a Report to the Club of Rome, the conventional measures of growth are fallacious and misleading. They do not truly distinguish the creation of added value from the using up of non-renewable resources. The very notion of the "Gross National Product" appears to be unsustainable.

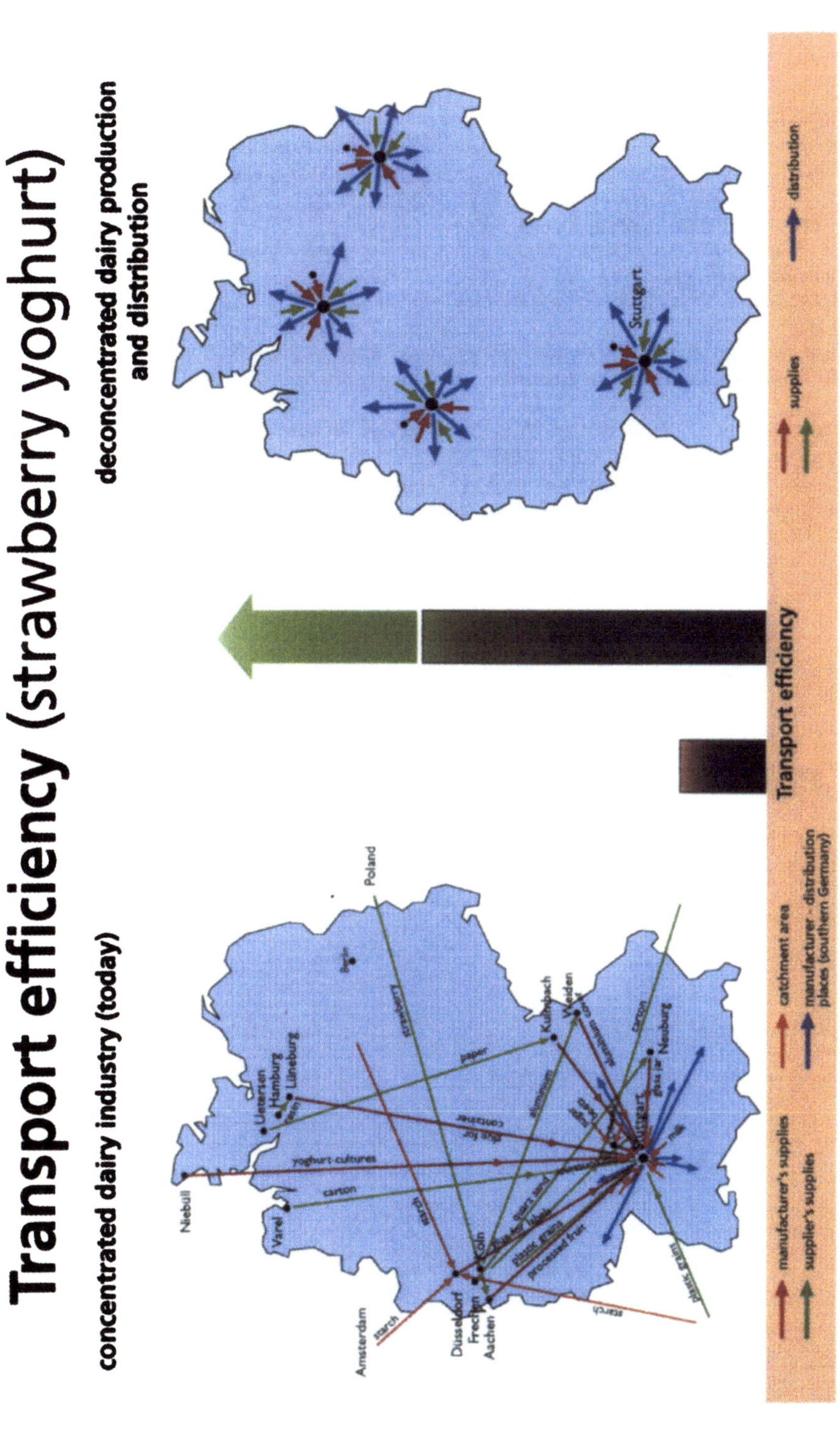

Figure 3. As Stephanie *Böge* [2] has found out, the typical manufacturing of strawberry yoghurt in Germany requires lorries to drive a total of 8000 km of road length, including for the shipment of supplies to suppliers of manufacturers.

References

1. R. Bleischwitz and H. Schütz. *Unser trügerischer Wohlstand. Ein Beitrag zur deutschen Ökobilanz.* Wuppertal Institute, Wuppertal, 1992.
2. S. Böge. Erfassung und Bewertung von Transportvorgängen: Die produktbezogene Transportkettenanalyse. In D. Läpple, editor, *Güterverkehr, Logistik und Umwelt.* Edition Sigma, Berlin, 1993.
3. W. D. Nordhaus. An optimal transition path for controlling greenhouse gases. *Science*, 258:1315 pp., 1992.
4. R. Rechsteiner. Sind hohe Energiepreise volkswirtschaftlich ungesund? *GAIA*, 2:310–327, 1993.
5. R. Repetto and R. Dower et al. *Green Fees. How a Tax Shift Can Work for the Environment and for the Economy.* World Resources Institute, Washington, 1992.
6. W. van Dieren. *Taking Nature Into Account. A Report to the Club of Rome.* Copernicus-Springer, New York, 1995.
7. E. U. von Weizsäcker, A. Lovins, and H. Lovins. *Factor Four. Doubling Wealth While Halving Resource Use. A Report to the Club of Rome.* Earthscan, London, 1997.

Commentary on "New Models of Wealth"

Hans-Jürgen Bolle

1 Introductory remarks

In his programmatic lecture Professor Weizsäcker envisaged the possibility of generating in people's minds a new attitude towards the environment through a revaluing of products and services, enforced by taxation and price policies. A major problem in this context obviously is the intelligent regulation of matter and material fluxes without damaging prospering economies. Both the enhancement of welfare as well as the conservation of natural resources must figure among the aims of future environmental policies.

2 Two questions

It is undoubtedly true that our economic thinking requires a new system of values, which takes into account the limitations of our natural resources. Two questions arise: Firstly, to what extent are political and economic measures able to encourage such a reorientation of thinking and, secondly, is there congruence between the time scales of the growth of public awareness on the one hand and the urgent need for the stabilization of the environmental conditions on the other?

3 Categories of threats

The global threats as well as regulation mechanisms to counter these threats can be divided into three categories:

- Direct effects of anthropogenic activity, which occur regionally but synchronously at different points around the globe. This concerns primarily the pollution of the atmosphere, water, and soils.
- Indirect global impacts on the environment due to climatic change caused by the change of atmospheric optical depth in the thermal infrared part of the spectrum due to increasing concentrations of absorbing gases.
- Changes induced by worldwide economic factors like energy consumption, population density, changes of land-use including such consequences as erosion, fires and run-off of fertile soils through the rivers into the oceans, and, last but not least, changes in human understanding and behaviour.

4 Global vs. regional "logics"

With regard to the last category, it must be borne in mind, that economical regulation mechanisms may have far-reaching effects; measures that are required in one part of the world may have substantial impact on the ecology of another part of the world.

Ecological changes that occur around the globe are normally experienced as regional effects and are scarcely recognized in their full global dimension. Studies are frequently made of isolated subsystems such as cities, specific landscapes (mountainous, coastal, semi-arid regions), tropical forests, or business enterprises. Their global interdependencies often cannot be analysed because no universal model exists as yet in which such investigations can be imbedded. Even if such a model existed, it would be difficult to derive results from it since man's actions and reactions to global threats are to a large extent unpredictable.

5 Synergetic approach for research and management

The problems involved in repairing our distorted relationship with nature were presented and discussed last year at the Free University of Berlin in a series of lectures held by representatives of all pertinent disciplines. From these lectures it became evident that changes in our economic axioms can only be successfully brought about if based on globally accepted "new ways of thinking". These cannot be imposed by new taxation and price policies that may take effect in one part of the world only. Attitude must change in a broad stream of "philosophical" understanding of the interdependency between man and nature. This requires a synergetic approach of all the disciplines concerned. Research findings of the natural sciences must also be scrutinised by other disciplines before passed on to the decision makers at a practical-normative level. Philosophy, for example, can contribute by clarifying the responsibility of the individual not only for the society, but also for the environment in which he lives. This may also have a bearing on our moral code, for which theology is, in part, responsible. Psychology must play a role in examining the motives behind human actions and reactions. Global environmental laws must be drafted. The education system must pass a new knowledge to the next generation. Moreover, interdisciplinary discourse about issues of mutual importance must improve and a change of paradigms must take place in our universities to foster the understanding of complex systems and to further interdisciplinary research. Only a broad consensus can lead to better planning concepts and less uncertainty in the drawing of hypotheses.

Such a broad approach – the harmonization of global and regional "logics" – may ultimately lead to changes in our management of both ecology and economy, and thus to a change of what we presently understand under

"Global Change". The time factor involved in achieving this goal is most critical. The degradation of our natural environment is continuing apace. Do we have the time and the means to begin from the very roots of our civilisation to convey the idea of a "New Deal" between man and nature? On the other hand, acting without sufficient knowledge of the whole system might cause more damage then good.

6 The possible role of a global model

Do we know at all or can we derive the criteria for a change towards a "better world"? Furthermore, if we know or can derive them, are these criteria the same for both regional and global environments, or do different logics apply in different subsystems? The definition of such criteria would necessitate a universal view and the inclusion of interactions between subsystems. This calls for a global model that improves our understanding of the interaction between man and nature. In this respect such a model would be of great help, but it would never be complete. Errors and bifurcations in temporal developments are inevitable, so complex and unpredictable is the evolution in time of human action and reaction. A model must therefore never be understood as a tool with which these interactions can be managed.

7 Concluding remarks

To try to draw a conclusion under these very complex circumstances is difficult and maybe not appropriate. It seems to be clear that the research results of the natural sciences as well as political actions based upon this new knowledge are not sufficient to reestablish balance between man's action and nature. A broader approach including all disciplines is needed to adjust the axioms of civilisation to the new situation on our planet.

Commentary on "New Models of Wealth"

Gyula Bora

1 Introductory remarks

It is almost a common view nowadays that the existence of human civilization depends upon a single earth and on its biosphere. The countries and communities have cared about each other very little so far in their struggle for survival or welfare, and even less about nature, of which all of us are only parts and not masters. In many parts of the world economy is governed by narrow-minded production practices, aimed at yielding profit in the short run, which have, therefore, shaky grounds as we look to their future.

In order to solve contradictions between nature and human society, to do away with uncertainties of our future the philosophy of "sustainable development" came to life, the importance of which has already been accepted by many; moreover it has gained grounds in science and, to some extent, also in policy. A number of publications on the subject describe at length and analyse the problems from different aspects, while exploring relationships. However, a few ideas, hardly attainable, have been suggested for the ways and means of putting into effect the goals. In my opinion, the most important aim, that is searching for a new way for the future and progress of mankind, acknowledges that this is an extremely difficult case to solve. Nor should we forget that one of the central issues is, among other things, how to balance the dissenting interests of countries of different development and consumption levels.

Professor von Weizsäcker has outlined a new model of searching possible solutions, especially to avoid, or to reduce, the risk of global warming. The emphasis has been placed on energy and taxation factors; for this reason I am to discuss mainly the proposed economic means. In my view it is the area in which the key to the solution should be looked for.

2 Ecological taxes

In the ultimate analysis what we are talking about is the necessity of evolving a new form of behaviour, and within this of consumption. The facts of history indicate that many aims and efforts can only be put into effect if adequate economic means are available. It has been proved also by environmental policies of different countries in the initial stage, where legal means alone might have turned out to be powerless.

According to the proposal a well-considered ecological tax reform in the field of resources will result in a gradually increasing tax level, while the simultaneous decrease of other forms of taxes is aimed at achieving a new equilibrium. Raising resource prices through taxation means to put constraints on consumption on the one hand and, in this price system, to lay foundation for the better use of energy and materials on the other.

3 Energy efficiency in Hungary

Professor von Weizsäcker mentioned the problems of Eastern Europe, where bad energy efficiency is a general problem. The long-lasting low prices created crucial squandering in the use of energy, e.g. in the case of Hungary, this can be proved by the so called specific energy consumption. Before the collapse of the planned economy, for the production of one dollar per capita GNP three or four folds higher energy input was necessary; therefore, neither the economy, nor the population were interested in the energy saving. In reality, our energy system has so great a reserve that with the help of energy rationalization long-run economic growth could continue with the current capacity of the energy system. The Hungarian government recognized the importance of higher energy prices, but this process was not followed by decrease of other taxation; quite to the contrary more new taxes were introduced in many fields. The general bad economic conditions made necessary these steps. It does mean that firm economic stabilization could open ways for the implementation of Professor von Weizsäcker's idea.

The study discusses actions to be taken at national level; however, their effect on the world economy should also be pointed out. As it is well-known developed countries exploit to a high extent the resources of developing countries with complete disregard for damaging local environment. Thus, a decrease in consumption serves, besides resource conservation, environmental protection too.

4 Accelerated technology transfer

The proposals also concern increasing the durability of products, diminishing wastes etc. that can be fostered by higher resource prices. That is to say they may contribute to change over from quantitative to qualitative development. It is hoped that policy makers, who keep in mind the principles of sustainable development, will inevitably strive to connect economic development more closely to the environment.

The same applies to technical progress. Many notable scholars believe that technical innovations did not meet necessarily actual social requirements in the last decades. The new policies should enhance, through economic means, the development of technologies which might serve the faster spread of modern technologies of less specific use of energy and materials.

A series of examples shows that the results of technological progress can reduce the specific use of energy and materials. One example of them is the development of computers. The comparison is very easy between a first generation and an up-to-date personal computer, in size and in energy consumption. I suppose the differences are undisputable.

5 Range of commodity flow

The presentation of Professor von Weizsäcker also mentioned a very important goal for the sake of reduction of energy use, i.e. to shorten the long distances of commodity flow. According to the theory and practices of conventional division of labour, the energy, material and labour inputs necessary per unit of products can be reduced by modern technological processes by the application of great series of production, by the so called economies of scale. As a result, the costs of production can be decreased which allow the increasing transportation distances of the products. The share of transportation costs had a decreasing tendency in the final market prices which contain also some other elements (like taxes); therefore the border of the market of a given product has been broadened in physical and in economic term. However, at a global level the flow of big masses of commodities needs an extremely high dimension of energy use in the world economy. In the future, a better balanced practice will be necessary in production and transportation to reduce the energy consumption; probably this goal also can be achieved by higher energy prices.

6 Limitations

While admitting the justification of the proposals reference should be made to the likely limits, thus:

- in the present world economy of complex interrelations, it opens the serious question whether a country alone would be able to adopt the solutions mentioned above, while not being put at disadvantages in international competition.
- can a government succeed in, simultaneously with raising taxes, effectuating tax cuts in a way that there would be no time intervals when the conditions of certain economic sectors become worsened?
- reactions of micro-economy can be considered uncertain (the presentation mentioned these problems concerning the attitude of different industrial lobbies).

- for lessening unemployment, which possibly presents itself in the course of transition, will the more flexible treatment of worktime, reduced work hours etc. be sufficient?
- finally, there is no way of knowing how people having been accustomed to other circumstances respond, and whether their resistance will jeopardise progress.

7 Concluding remarks

There is no single model for sustainable development since economic and social systems as well as ecological conditions are diverse. Each country must elaborate the concrete policy on its own. Apart from the differences, sustainable development should be viewed as a global aim.

The way to sustainable development can only be found through reshaping international relations. This requires far-reaching changes in trade, flow of capital and technologies, which will bring better consistency to environmental needs. However, highly differentiated solutions are required, for which Professor von Weizsäcker's initiatives may provide a basis to start from.

Chapter 4: Urban Metabolism and Disaster Vulnerability in an Era

James K. Mitchell

Abstract Research on the human dimensions of global change typically examines the vulnerability of sociotechnical systems to anticipated environmental stresses, most often in non-urban settings. This paper takes a somewhat different track; first it argues that urban areas should receive high priority in global change research and second, it highlights an expanded view of "urban metabolism" as a neglected set of variables that mediate between environmental hazards and human responses. Other researchers have defined "urban metabolism" as the production and consumption of natural resources and physical environments by cities (*Stren*, *White*, and *Whitney*, 1992 [16, p. 9]). In the brief overview that is presented here the term has broader connotations; it includes various means – social as well as biophysical – by which cities acquire or lose the capacity for sustainability in the face of diverse and competing problems. Chief among these is the process by which opportunities for reducing natural disasters are increased or decreased as a result of interactions among different urban issues.

1 Introduction: Global Change hazards and disasters

If global climate changes in the ways projected by most General Circulation Models (GCMs). the earliest and most discernible effects are likely to include increased storms, floods, droughts, and other short-term weather extremes (*Mitchell* and *Ericksen*, 1992 [15, p. 142]). These hazards will add further burdens to the rising global potential for disaster that is mainly fueled by increasing exposure and vulnerability of human populations as well as lack of investment in improved disaster management.

Risks of potential climate change are not spread evenly across the Earth's surface. Low-lying coasts are believed to be particularly susceptible to flooding associated with sea level rise; polar regions are thought to be disproportionately at risk to small shifts in mean temperatures; tropical islands might feel the effects of precipitation changes more than adjacent mainland areas; high mountains could witness significant altitudinal shifts in vegetation zones and the limits of agriculture and pastoralism would probably fluctuate most along the margins of hot deserts and other semi-arid areas. As research on the ecological implications of climate change continues, it will doubtless be possible to refine these generalizations and to identify local ecosystems and specific places that are most at risk.

However, information on the susceptibility of natural environments to atmospheric perturbations provides only limited guidance about the likely human impacts of global change risks. To understand enough to make appropriate management decisions we must examine the effects of those risks on the places that people live, work or use and on the activities that they pursue. To a large and expanding degree, this means that *cities* – especially very large cities – should become be a primary focus of global change research.

2 Cities and natural disasters in a changing world

Researchers who study the hazards of environmental change have good reasons to focus on cities. Humans are transforming the biosphere in unprecedented ways and cities are the most intensely human-created of all environments. Moreover, cities will soon be the dominant form of human settlement throughout the world. In addition, compared with other types of land use, urbanization tends to increase disaster potential. The process of urbanization is also changing in ways that have far-reaching implications for hazards management. Finally, urban areas are affected by unusually complex mixes of overlapping and synergistic human problems that act to broaden or narrow opportunities for resolving hazards issues. Different constituencies and different agendas for different emerging problems collide and interact in ways that are only poorly understood, especially for non-western cities (*McGranahan* and *Songsore*, 1994 [13]). Each of these factors deserves additional comment and the last of them requires considerable emphasis.

3 Cities and the expanding human role in environmental transformation

During the past century scientific conceptions of Mankind's role as an environmental modifier have broadened considerably. In the late 1800s G.P. *Marsh* [11] suggested that people *modified* environments. By the early twentieth century Russia's V.I. Vernadsky judged that Mankind had begun to act on a scale that was comparable to natural processes. In the 1950s Carl Sauer believed that humans had already substantially *changed* the face of the Earth (*Thomas*, 1956 [17]). Now, at the close of the twentieth century, other scholars are arguing that the biosphere has been fundamentally *transformed* by humanity (*Turner* et al., 1992 [18]). There is no mistaking the implications of these shifts in word usage; they reflect a growing awareness that human capabilities to alter the world have expanded enormously. Nowhere is this more obvious than in cities.

The fabric of every city is thoroughly marked by intentional human designs and unintended outcomes. Although natural features may be visible and natural processes are still essential to the city's survival, both are usually heavily modified by human actions and becoming more so. This situation is clearly evident in a contemporary megacity like Hong Kong. There, most of the colony's estimated six million people are crammed into an area of about 50 square miles that is mainly composed of small rocky islands and peninsulas. Essentially unoccupied a century ago, the city grew up on a dramatic site, exposed to tropical cyclones and lacking significant natural resources.

Today, Hong Kong Island and the Kowloon Peninsula are fringed with filled and bulkheaded land, created by planing off adjacent hilltops and dumping rock into the ocean. The resulting building sites are occupied by closely packed high rise apartment blacks and office towers, cheek by jowl with industrial facilities and an international airport whose main runway projects far out into a busy shipping channel. Many residents live permanently on boats moored in typhoon shelters or on adjacent waters. Ravaged stumps of hills and water laden with garbage from ships, onshore developments and upstream China form a backdrop to multinational trading corporations. Bays are dammed to create fresh water reservoirs for a city that lacks major surface streams and significant aquifers. Steep slopes are veneered with concrete, festooned with fences and planted with trees to retard erosion and mass movement. Watersheds are protected against fire and clearance both by public laws and controls on access. The few remaining streams are channeled, diverted and culverted. Road and rail tunnels carry heavy traffic under the harbor; elevated moving sidewalks transport pedestrians between the glass and concrete citadels of commerce, past the second storeys of adjacent homes and businesses; a rack railway climbs the steepest slope and scenic highways are cantilevered out from nearby hillsides. All of this overlooks a roadstead filled with hundreds of deep sea freighters and near shore junks that serve one of the world's most buoyant free trade centers. Despite reflections of the non-human environment, Hong Kong is an intensely humanmade place! At one time such cities contained only a minority of humanity; this is no longer the case.

4 An approaching urban millennium

The 1990s will witness one of the world's great demographic transitions. If current trends continue the global population will become predominantly urban during this decade. At present, 47.1 % of world population can be classified as urban and 69 countries are predominantly urban (*Fellmann* and *Getis*, 1992 [6, p. 371]). Most of the latter are located in Europe (26), followed by Asia (18), North America (9), South America (9), Africa (5) and Oceania (2) (Note 1.). In the year 2020 it is estimated that 62.5% of the world's people will live in cities. By itself, the onset of an urban millennium implies that public policies for hazards management will be strongly influenced by the experience of urban populations (*Anderson*, 1992 [1, p. 77]).

5 Disaster-forcing attributes of urbanization

Urbanization, especially rapid large-scale urbanization, increases disaster susceptibility in various ways and the disaster vulnerability of megacities has been identified as a high priority topic for research during the United Nations-sponsored International Decade for Natural Disaster Reduction (National Research Council; Science Council of Japan; U.S. Office of Science and Technology Policy). First, urbanization concentrates people and investments, thereby raising the potential for losses. For example, urban areas take up about 3% of the land surface of the United States but host 75% of the national population (*Berry*, 1984 [3, p. 180]). Given the expansive nature of North American cities, the proportion of urban land is probably smaller in most other countries (Note 2.).

Second, growing cities typically contain high numbers of recent arrivals many of whom are poor or otherwise ill-prepared for city living and therefore particularly vulnerable to urban disasters. For example, the proportion of Third World urban dwellers who occupy the sub-standard housing of "informal" settlements in megacities typically varies from 30-85% (World Resources Institute, 1990 [21, p. 76]): similar proportions of the poorest urban people do not have access to piped water, indoor sanitation and basic health care facilities (World Resources Institute, 1992 [22, p. 87]). Not only are the new immigrants separated from known rural environments and traditional disaster response systems, they are also exposed to a variety of new types of disasters- including complex blends of technological and natural risks (*Anderson*, 1992 [1]). These are the people who suffer most when Calcutta's streets are flooded by the monsoon, or Khartoum's mud houses crumble as the rising Nile waterlogs their foundations.

Third, as cities expand, competition for land encourages the use of previously bypassed hazardous areas such as exposed coasts and steep slopes (*Gilbert* and *Gugler*, 1992 [8, p. 193]). Mexico City's long history of accommodating growth by filling shallow lakes that once provided defense, food and water and by expanding up the slopes of nearby volcanoes is an excellent example. Increased risks associated with using hazardous land can be offset by judicious adjustments. The early development of Moscow is instructive. Initially, settlement spread downslope from the Kremlin and extended along the banks of the Moscow River. Repeated flooding persuaded Catherine the Great to authorize construction of a diversion channel (i.e. the Vodootvodnyi Canal) that cut through the loop of a large meander and funneled excess water south of the growing city center. Fortunately for Muscovites this structural engineering project appears to have mitigated subsequent flood losses.

Fourth, the process of land conversion frequently worsens existing risks: hillsides are undercut, floodplains are filled in, fire-resistant native vegetation is replaced with inflammable exotics, unstable sites are occupied by housing tracts. The development of greater Los Angeles exhibits all of these activities and others besides but most megacities face similar problems (*Cooke*,

1984 [5]). Bangkok klongs (canals) that used to accommodate overflow from the Chao Praya river have been filled in to create streets that are now chronically floodprone. Heavily developed slopes above Rio de Janeiro, Caracas and Tijuana give way in destructive landslides triggered by rainstorms.

Fifth, most *cities age faster than they can be replaced.* Hence, the urban fabric deteriorates and its ability to resist damage declines; wood rots, concrete cracks, water systems spring leaks, dams silt up, disaster-resistant construction practices are undermined by unregulated building extensions and conversions. The plight of people who live in the decaying tenement-style housing of Lima's inner urban neighborhoods is a representative example (*Maskrey*, 1989 [12]); (Note 3.).

Of course, the organization and operation of cities vary widely. For example, differences in the internal structure of an urban area and its relationship to the surrounding rural hinterland can have significant effects on hazards and hazard management. Consider, for example, the concentration or dispersal of buildings and other facilities. The degree to which essential life-support functions and services are concentrated in tight central nodes or dispersed widely has a direct bearing on the vulnerability of lifelines and infrastructure systems and the costs of hazard mitigation. Dispersed low density cities like Los Angeles are less likely to be immobilized during a disaster than are more compact European or Indian cities because multi-nodal transportation and utilities networks contain greater built-in redundancy in the face of disruptive events. On the other hand, the costs of providing effective warning and evacuation systems or sophisticated structural protection works may be prohibitive in low density urban areas. This helps to explain why outer suburbs of Oakland, California and Sydney, Australia have found themselves facing catastrophic losses from wildfires in recent years.

Transportation and land use patterns are also important. Many western cities have orderly street patterns, regulated densities and strict segregation of noncompatible land uses whereas cities in developing countries or Japan often contain neighborhoods that exhibit convoluted street patterns, self-built "informal" housing, and radically different uses in close juxtaposition. As a result the same type of extreme event may have contrasting consequences in different cities. For example, limited amounts of public open space in Lima and Tokyo restrict the availability of emergency refuges during earthquakes and lack of effective zoning in Bhopal (India) contributed to the heavy loss of life in adjacent shantytowns when a toxic gas release occurred at Union Carbide's methylisocyanate plant in 1984.

6 Changes in the process of urbanization

The process of urbanization appears to be undergoing substantial changes. Some of these pertain to the numbers and scale of urban forms; others to the internal arrangement of urban functions and land uses; yet others to the convergence or divergence of different styles of urbanism.

Over the next two or three decades, the number of very large cities will grow substantially. There are now somewhere between 300 and 400 megacities (i.e. cities with populations of at least 1 million) and they contain approximately one quarter of the world's population (*Fellmann* and *Getis*, 1992 [6, p. 371]). Megacities are now more or less ubiquitous (Table 1). At least 80 countries possess them – including 28 where rural populations are in the majority (Note 4.). The latter include places like Vietnam, Bangladesh, Mozambique and Ethiopia (Note 5.).

Not only are there more megacities, megacities are growing bigger. Trends that affect the world's 30 largest cities are representative (Table 2). Taken together, these cities contained approximately 286 million people in 1985; by 2000 the total will be around 388 million – a 36% increase. In 1985 the world's largest city (Mexico City) housed around 18 million people; by 2000 it is likely to have about 26 million. Regional concentrations of large megacities will, however, show significant change.

Most (20) are already located in developing countries and this pattern will intensify throughout the rest of the decade, but South Asia will gain the most new large megacities while there will be relatively fewer large megacities in the Middle-East, Europe and North America. However, except for a handful (e.g. Paris) all of today's big cities will grow substantially over the next several decades. With increasing size comes increasing physical and social complexity and increasingly difficult problems of urban environmental management (Note 6.).

How far will the trend towards global megacities proceed? Are we headed towards increasingly similar urban forms and functions? Current trends suggest that population will continue to concentrate at the global and national scales of analysis but that there may also be regional and localized dispersals. There is no doubt that massive agglomerations are springing up in many parts of the world (e.g. Megalopolis; the Tokyo-Osaka corridor; the Rio de Janeiro-Sao Paulo focus) and the growth of most megacities shows no signs of ceasing.

Other trends point towards a convergence and homogenization of urban forms. For example, the architecture and building materials of cities are beginning to converge on certain "international" styles. While the downtown office buildings under construction in large Asian cities may be sheathed in bamboo scaffolding, the final products – highrise glass, steel and concrete edifices – are not substantially different in appearance from the metal-scaffolded office buildings that are going up in North American or European cities. Bangkok's new high level river bridge and Istanbul's Bosphorus bridge are both built to designs that can be found throughout most of the western world. The underground passenger railroads of Hong Kong, Mexico City and Calcutta are designed, constructed and operated on similar principles.

Table 1. Megacity states

State	Urban pop.(%)	Pop. in megacities(%)	Urban pop. in megacities(%)
AFRICA			
Algeria	51.7	12.2	23.6
Angola	26.3	17.1	65.0*
Côte d'Ivoire	40.4	18.1	44.8
Egypt	46.7	24.3	52.0*
Ethiopia	12.9	3.8	29.5
Ghana	33.0	7.9	23.9
Guinea	25.6	22.5	87.9*
Kenya	23.6	6.3	26.7
Libya	70.2	45.4	64.7*
Morocco	48.0	17.1	35.6
Mozambique	26.8	10.1	37.7
Nigeria	35.2	8.3	23.6
Senegal	38.4	20.4	53.1*
South Africa	59.5	17.6	29.6
Sudan	22.0	7.7	35.0
Tanzania	32.8	6.1	18.6
Tunisia	54.3	20.0	36.8
Zaire	39.5	9.9	25.1
NORTH-CENTRAL-AMERICA			
Canada	77.1	29.8	38.9
Costa Rica	47.1	33.7	71.5*
Cuba	74.9	19.8	26.4
Dominican Rep.	60.4	30.7	50.8*
Haiti	28.3	15.8	55.8*
Mexico	72.6	32.5	44.7
Nicaragua	59.8	26.1	43.6
United States	75.0	36.3	48.4
SOUTH AMERICA			
Argentina	86.3	42.5	49.2
Bolivia	51.2	16.9	33.0
Brazil	74.9	35.2	44.0
Chile	85.9	35.9	41.2
Colombia	70.0	27.3	39.0
Ecuador	56.0	27.5	70.1*
Peru	70.2	29.0	41.3
Uruguay	85.5	38.7	45.3
Venezuela	90.5	26.6	29.4

Table 1. *(Continued)*

State	Urban pop.(%)	Pop. in megacities(%)	Urban pop. in megacities(%)
ASIA			
Afghanistan	18.2	9.5	52.1*
Bangladesh	16.4	7.7	47.0
China	33.4	9.1	27.2
India	27.0	8.6	31.9
Indonesia	30.5	10.0	32.8
Iran	56.7	23.3	41.1
Iraq	71.3	21.4	30.0
Israel	91.6	40.9	44.7
Japan	77.0	27.5	35.7
Jordan	68.0	25.6	37.6
North Korea	59.8	10.2	17.1
South Korea	72.0	49.9	69.3*
Kuwait	95.6	52.9	55.3*
Malaysia	43.0	9.6	22.3
Myanmar	24.8	7.9	31.9
Pakistan	32.0	13.4	41.9
Philippines	42.6	13.6	31.9
Saudi Arabia	77.3	22.6	29.2
Singapore	100.0	100.0	100.0*
Syria	50.4	30.1	59.7*
Thailand	22.6	12.8	56.6*
Turkey	61.3	21.6	35.2
Vietnam	21.9	6.5	29.7
EUROPE			
Austria	58.4	27.6	47.3
Bulgaria	67.7	13.2	19.5
Czechoslovakia	77.5	8.3	10.7
Denmark	87.0	26.9	30.9
Finland	59.7	20.3	34.0
France	74.3	19.2	25.8
Germany, W.	87.4	13.0	14.9
Germany, E.	77.2	7.8	10.1
Greece	62.5	34.2	54.7*
Hungary	61.3	20.0	32.6
Italy	68.9	25.4	36.9
Netherlands	88.5	14.0	15.8
Poland	61.8	17.5	28.3
Portugal	33.6	15.6	46.4
Romania	52.7	9.4	17.8
Spain	78.4	22.1	28.1
Sweden	84.0	19.7	23.5
U.K.	89.1	23.4	26.3
Yugoslavia	56.1	6.6	11.6
Former USSR	65.8	15.3	23.3
AUSTRALIA	85.5	50.6	59.2*

Data source: WRI, 1992 [22, pp. 264-5].

*States with more than half the urban population in megacities

Table 2. The world's 30 largest metropolitan areas ranked by population (in millions, 1985 and 2000)

Rank	Location	1985 size	Location	2000 size (projected)
1	Mexico City	18.1	Mexico City	26.4
2	Tokyo/Yokohama	17.2	Sao Paulo	23.6
3	Sao Paulo	15.9	Tokyo/Yokohama	21.3
4	New York/NJ	15.3	New York/NJ	16.1
5	Shanghai	11.8	Calcutta	15.9
6	Calcutta	11.0	Bombay	15.4
7	Buenos Aires	10.9	Shanghai	14.7
8	Rio de Janeiro	10.4	Teheran	13.7
9	Seoul	10.2	Jakarta	13.2
10	Bombay	10.1	Buenos Aires	13.1
11	Los Angeles	10.0	Rio de Janeiro	13.0
12	London	9.8	Seoul	13.0
13	Beijing	9.2	Delhi	12.8
14	Ruhr	9.2	Lagos	12.5
15	Paris	8.9	Cairo	11.8
16	Moscow	8.7	Karachi	11.6
17	Cairo	8.5	Manila	11.5
18	Osaka/Kobe	8.0	Beijing	11.5
19	Jakarta	7.9	Dhaka	11.3
20	Tianjin	7.8	Osaka/Kobe	11.2
21	Delhi	7.4	Los Angeles	10.9
22	Baghdad	7.2	London	10.8
23	Teheran	7.2	Bangkok	10.3
24	Manila	7.0	Moscow	10.1
25	Milan	7.0	Tianjin	10.0
26	Chicago	6.8	Lima	8.8
27	Istanbul	6.8	Paris	8.8
28	Karachi	6.8	Milan	8.7
29	Lima	5.7	Madras	7.9
30	Bangkok	5.5	Bangalore	7.7

Source: U.N. 1987; *Jones* and *Kandel* [9]

Yet, convergence and homogenization are not the only characteristics of contemporary urbanization. In the next century some old contrasts among cities may remain but some new ones also appear to be emerging. For example, today the environmental concerns of residents in different cities often vary widely (*Kates*, 1994 [10]). and the internal organization of cities reflects differences in culture and societies. The range of expressions of both indices of urbanization is likely to remain wide. Latin American cities usually have a single downtown business district surrounded by well-constructed wealthy neighborhoods with poor shantytowns on the periphery whereas North American cities typically contain a financially troubled central business district (CBD), multiple vigorous suburban business districts, poor inner city slums and outer fringes occupied by the middle class and the wealthy. Large African and Asian cities often lack major public open spaces but include substantial tracts of garden land or agricultural holdings; many exhibit twin urban foci – one an indigenous informally laid out core and the other a planned colonial settlement. The former socialist cities of Eastern Europe and Russia have high percentages of land devoted to (decaying) industries and few socioeconomic contrasts among neighborhoods. When a disaster strikes, buildings and human activities may be permanently altered but the underlying patterns of land use and urban organization quickly reassert themselves, thereby preserving the contrasts of form and function.

Among the new contrasts that may be emerging are trends towards different settlement densities and the appearance of wholly novel urban forms. For example, the emptying out of western cities into low-density exurbs contrasts with the continuing high population densities of most Third World cities. Among the new urban forms are "edge cities" and "cyburbias" that are functionally disconnected from the places they occupy but plugged into vast frictionless electronic networks. Such places offer one vision of the future that has no historic precedent (*Fishman*, 1987 [7]; *Chudacoff* and *Smith*, 1994 [4]). Another emerging trend is the contrast between cities of the "polycenter" and cities of the periphery. This distinction echoes the well-known center-periphery models of global political and economic analysis with the difference that there is no single core city of the global urban system. Instead, a group of 19 cities make up a polycenter that directs and controls most international economic activities (*Berry*, 1990 [2]). Tokyo, New York and London are the dominant cities of the polycenter. It is these cities that now account for the vast majority of the world's electronic data flows and their dominance is likely to grow even stronger in upcoming decades. Polycenter cities are wealthy but periphery cities occupy a wide range of economic niches including many that are poor. Undoubtedly the polycenter may expand to include more megacities but prospects for closing the gap between these fortunate few and the burgeoning megacities of the periphery are not encouraging.

7 The emergence of complex urban hazards

Thus far, it has been argued that cities are sensitive indicators of global environmental change, particularly insofar as this involves the exacerbation of natural hazards and disasters. It has also been suggested that cities are changing in ways that raise questions about the applicability of previous human experience with hazards to cities of the future. A further development must now be taken into consideration – the genesis of urban hazards is also changing.

In the past urban floods, storms, earthquakes, droughts and other hazards killed people and destroyed property chiefly by direct physical means (e.g.

inundation, dessication, collapse of structures etc.). Now, the materials that make up the city, the resources that it consumes, the energy that powers it, and the wastes that are dumped from it are all likely to be hazards in their own right. The so-called "natural" hazards are increasingly likely to involve technological and other risks such as explosive substances or flammable materials and radiological, biological or chemical contaminants. Some examples follow.

Floods carry chemical leachates out of toxic sites and into adjacent neighborhoods. Storms cut electricity supplies to storage plants that keep liquid ammonia safe under refrigeration. Pumping of aquifers (or oil wells) triggers subsidence of urban districts and incapacitates existing dykes and bulkheads. Rising waters float compressed gas tanks off their foundations and send them careening into bridges and among riverbank houses. Droughts cause lakes to recede thereby uncovering old waste disposal sites. A flash flood saturates old mine tailings and precipitates a devastating mudflow. Decayed urban infrastructure falls, admits water from an adjacent canal into the basements of highrise commercial buildings and brings metropolitan businesses to a standstill. A storm surge picks up chemically treated timber in a waterfront storage yard and scatters it among nearby houses and freshwater sources. A reservoir cracks during an earthquake and threatens to release floodwaters into downstream communities. A natural gas pipeline explodes and burns after being cut by earth moving machinery that is building a temporary floodwall. Most of these events have already happened in large North American and Asian cities and others have only been narrowly averted. Moreover the possible combinations of natural extremes and hazardous technologies in cities are very large in number; further surprises lie ahead if one also takes account of the so-called instrumental hazards, such as arson, sabotage and warfare. Thus permutations of unprecedented urban environmental hazards grow even larger.

The combination of worsening familiar hazards, new forms of hazard, changing urban contexts, and the possibility of more surprises poses formidable problems for urban managers and populations at risk. In highly uncertain circumstances like these there is a temptation to be pessimistic about the prospects for avoiding or preventing emerging urban hazards before they arise and to concentrate instead on dealing with their effects by improving generic programs of emergency management and postdisaster recovery. Such a view ignores the potential for new approaches to hazard reduction that comes into view if one broadens the analytic focus to take account of the full range of interacting issues and problems that occur in modern megacities. Because these problems are so tightly intermeshed it is no longer possible to treat them separately. They are an essential part of urban metabolism. This is an arena that has not been well explored by researchers and, one suspects, that its full potential has not been appreciated by hazards management professionals either.

8 Synergistic problems of disaster in modern megacities

Earlier I called for a broader interpretation of urban metabolism that takes account of neglected factors which enhance or retard urban sustainability. It is now time to return to these ideas of sustainability and metabolism.

The sustainability of cities is an offshoot of the "sustainable development" concept that was popularized by the Brundtland Commission Report on Environment and Development (World Commission on Environment and Development, 1987 [20]). Sustainable Development is based on the notion that management of the global economy and management of the global environment are interdependent activities that should be pursued together in the interests of maximizing long term returns. Unfortunately, neither the sustainability of cities nor the role of natural hazards in achieving or retarding sustainability is well developed in the professional literature.

Mostly natural disasters are viewed as proof of unsustainable development; the implicit assumption is that disasters will disappear if sustainable development is achieved. Clearly, this is not so; sustainable development does not equal safe development. Natural disasters will still continue to occur, if for no other reason than the advent of unprecedented – and therefore unplanned for – forms of hazard such as those associated with global climate change (e.g. sea level rise). and new environmental technologies. Therefore, it is necessary to rethink the role of natural hazards in urban sustainability. It is also necessary to rethink the nature of urban sustainability itself.

As presently formulated, the concept of urban sustainability reflects a western, developed world perspective; it is largely concerned with retaining the already secured "goods" of urban life and keeping threatening "bads" at bay (*McGranahan* and *Songsore*, 1994) [13]). We need to remember that for many of the world's urban residents – perhaps a majority – the day to day situations they face reverse the equation. The "bads" (e.g. natural disasters, poor environmental quality, unhealthiness) are the norm; the "goods" must be sought. However, it is naive to believe that these two properties lie at opposite ends of a spectrum and that the principal task of Mankind is to move the human condition away from one and towards the other. The reality is more complex; both goods and bads exist together in different combinations in different places and at different scales of analysis. In other words, the concept of urban sustainability should incorporate hazards as well as resources.

This leads to reconsidering the subject of urban metabolism. As originally formulated, the definition of urban metabolism refers only to flows of materials and energy – the "getting and spending" of resources that support the city. I suggest that issues, conflicts and other expressions of interaction among urban environmental problems are also part of urban metabolism. There phenomena are the behavioral counterparts of natural resources. A city becomes more sustainable if it increases the range of available human choices as well as meeting the material needs of its population. In the context of this paper, a move toward increased urban sustainability can be equated with the broadening of choice for urban populations at risk through the discovery or development of alternatives for reducing hazard. An outline of the reasoning for this assertion follows.

Scholars of natural disaster have long recognized that the optimal time for enacting new hazard management polices and for taking related initiatives is soon after a major disaster occurs. The conventional explanation for this is that many factors come together to favor action in the wake of disaster. Among others, the immediate stimulus of loss, post-disaster altruism and camaraderie, the availability of external resources designated to assist relief and recovery, and the glare of publicity combine to increase the likelihood that inertia or opposition can be overcome in pursuit of improved collective security.

However, it would be a mistake to believe that only factors of hazard are of importance in determining which ameliorative actions, if any, are undertaken in the aftermath of disaster. Quite the contrary; it is the potential for focusing jointly on nonhazard issues as well as hazard ones that makes the post-disaster period so rich in opportunities for changing policies and management systems. Moreover, it is the megacity environment that offers the largest suite of overlapping problems and, hence, the greatest scope for new policy departures that seek to address cross-cutting issues. Of course, the very diversity and scale of megacity problems may also raise insurmountable barriers to hazard management innovations. This paradox, the fact that the same situation can present both great opportunities for change and major barriers to change is a subject that is worthy of close study. Let us examine some examples drawn from recent North American megacity disasters.

Evidence from the Loma Prieta earthquake and other recent megacity disasters in the United States suggests that management problems are becoming considerably more complex than in the past and that both the scope and context for hazard policy making are now much wider than heretofore. Thus, big city disasters possess features that have not been common in smaller cities and that raise new problems of disaster management (*Mitchell*, 1994 [14]). For example, disaster impacts are likely to be extensively, continuously and obsessively reported by the mass media whereas impacts on smaller places are frequently overlooked or underestimated. The consequences for skewing disaster relief are considerable. The complex social mixes of megacities affect the delivery of emergency response services and disaster relief; in increasingly polyglot big cities linguistic differences between victims and helpers and among different groups of victims are becoming marked thereby adding yet another layer of social variables to the public policy mix. Third, the infrastructure of megacities has become so vast and so intricate that it is particularly liable to disruption. Finally and in part because of the foregoing changes, recovery from megacity disasters appears to be significantly slower than in smaller places. These and other pieces of evidence are indicators that the practical problems of disaster management may be changing in megacities. However, so too is the canvas of possibilities for action. In the big cities, more and more issues are becoming intermixed in the aftermath of disaster.

9 Complex urban issues in disaster's wake

After hurricane Andrew devastated south Dade County, Florida in 1992, disaster recovery became enmeshed with concerns about urban poverty, immigration, migrant labor, crime, national defense, expanded roles for a peacetime military; public transportation, building code enforcement, intergovernmental relations, tourism and a variety of other issues. It is not impossible that some of these issues might also have arisen in a non-megacity community but it is inconceivable that there would have been so many overlapping, contentious and difficult-to-solve problems in such a setting.

The proposed closure of Homestead Air Force is a representative example. Here was a facility that was scheduled for termination in the wake of post-Cold War military reorganization plans but whose continuance became central to the recovery program for much of the hurricane impacted area. Had the base been located in a lightly populated region of the United States, the victims of disaster would have been fewer and conventional relief programs might well have been sufficient for local needs. Instead, at a time when tens to hundreds of thousands of people had suffered heavy economic losses in the wake of Andrew, the prospective closure of the air base was portrayed as a body-blow to the regional economy. Those who wanted to keep Homestead open as a contribution to disaster recovery found themselves making common cause with many constituencies whose agendas were quite different. Among others these included anti-drug groups who pressed for policies designed to bolster the base's role as a smuggling surveillance center and south Florida's large Cuban exile community who viewed the base as a bulwark against the ostensibly hostile regime of Fidel Castro. That both of these overlapping issues – drugs and the Castro regime – have broad national significance takes nothing away from the fact that they are also deeply embedded in the Greater Miami megacity. In other words, the Homestead example shows how a megacity disaster can crystallize a variety of broad location-specific issues; in this case – recovery, crime, and national security. It also demonstrates how the coming together of the constituencies that are affected by those issues opens up possibilities for addressing megacity disasters in ways that transcend the usual forums and procedures of hazard management decison-making.

The recent Los Angeles (Northridge) earthquake provides more vivid examples of synergisms among megacity issues in the aftermath of disaster. Many of the victims who remained longest in the tent cities established after the quake were recent immigrants from Central America. These extended family groups were typically unfamiliar with the Los Angeles area, often spoke little English, sometimes included undocumented (illegal) aliens, and manifested a deep mistrust of assurances about the safety of undamaged buildings that were given by local emergency managers (Note 7.). The sociocultural visibility of these victims made them a focal point of disputes among various disaster interest groups and they became pawns in the California politics of multiculturalism, immigration and economic recession. However, in the end, the very fact that all these victims were embroiled in such contentious issues helped to ensure that their plight was not ignored. All were able to leave the tent cities and take up residence in better-serviced more permanent structures.

Other overlapping issues associated with the Northridge earthquake included: tensions between the continuing needs of already poor people and those who were impoverished by the earthquake; grievances about unequal treatment of several sets of victims of different disasters that had occurred in the greater Los Angeles region (i.e. earthquake victims, wildfire victims, riot victims); debates about the feasibility of rebuilding the Los Angeles transportation network to make it more hospitable to mass transit carriers and less burdensome to air quality; and counterpressures of (a) accountability for the management of relief during a period of national fiscal restraint and (b) expediency to meet the needs of a vocal and politically strategic constituency in southern California.

Now, seven months after the Northridge earthquake, it is not yet known how these issues and controversies have worked themselves out. Detailed studies of the recovery process have yet to be completed. However, the possibilities that are generated by overlapping and interacting problems are striking. For example, the role of victims becomes quite ambiguous when there are so many different types (e.g. "newly impoverished" and "old impoverished"; earthquake victims, fire victims and riot victims). Is one kind of victim more deserving of assistance than another? How shall donors and helpers adjudicate among multiple miseries? Are there overlaps among the recovery phases of different kinds of disasters that affect the same area? How do the victims of later disasters fare compared with those of earlier ones? Is there more sympathy and more aid or less? Is it likely that megacities will face so many disasters that specialized agencies will be developed to focus attention on them or will most of the assistance still come from outside (national) sources?

10 Conclusions

The fact that terms like coalitions, constituencies and agendas are an implied part of the foregoing discussion should not mislead the reader. This is not simply a call for increased attention to the fluidity of urban politics in the wake of megacity disasters. It is more a recognition that the context of disaster decision making is changing in big urban areas. Where once a rare disaster might prompt disparate interest groups to come together on an ad hoc basis in pursuit of unusual policy changes, now the megacity setting provides enough disasters and such a large number of interacting issues that different arrangements for addressing hazards management in the context of other problems are both possible and likely.

This transition is a further stage in the development of environmental management practices and policies. Initially, such practices and policies tended to focus on achieving single purposes by single means. (e.g. controlling floods by building dykes). Later, they moved through intermediate stages. Multiple purposes (e.g. water supply, water quality, electric power generation, flood control, recreation) were addressed by single means (multipurpose dam projects) or single purposes (e.g. flood control) were addressed by multiple means (e.g. dykes, warning and evacuation, insurance, land use zoning). Finally, at least in more developed countries, we entered an era of policy-making and management for multiple purposes by multiple means (e.g. integrated river basin management); (*White*, 1969 [19]).

Now, we are in the course of a further transition that involves even broader contexts. Hazards, resources and economic development are becoming bound together under the heading of "sustainable development". Moreover, the types of issues that arose in the aftermath of Loma Prieta, hurricane Andrew, the Northridge earthquake and other megacity disasters illustrate that social equity and ethics as well as other matters are routinely a major part of the urban disaster panorama. Perhaps we should now speak of "multiple contexts-multiple purposes-multiple means" policies and programs as the next philosophical and managerial horizon?

These developments suggest that global change-driven hazards and disasters are caught up in a complex little understood "brew" of other issues and problems that is bubbling up the settlement hierarchy from the level of evolving megacities towards the world scale. The ensuing situation is rife with possibilities for innovative policies and programs. There is considerable untapped potential for bringing new thinking to bear on a wide range of environmental concerns if we can clarify the processes by which such issues emerge and interact at the megacity scale, and if we can develop the appropriate management procedures and tools. In other words, global environmental change researchers must remember that broad societal changes are also afoot and that the two types of change constitute a potent mixture.

11 Notes

1. Most of the highly urban (i.e. 80% + urbanized) large countries are located in Europe, Latin America and Australasia while small heavily urban countries can be found in places like Singapore, Djibouti, Israel, Lebanon and the Persian Gulf states (e.g. Bahrain, Kuwait, Qatar, United Arab Emirates). Urban states (i.e. 50% + urbanized) include: *Algeria*, Djibouti, **Libya**, South Africa, Tunisia, Canada, Cuba, **Dominican Republic**, Jamaica, Mexico, Nicaragua, Panama, Trinidad

and Tobago, United States, Argentina, Bolivia, Brazil, Chile, Colombia, **Ecuador**, Peru, Uruguay, Venezuela, Bahrain, Cyprus, Iran, Iraq, Israel, Japan, Jordan, North Korea, **South Korea, Kuwait**, Lebanon, Mongolia, Qatar, Saudi Arabia, **Singapore, Syria**, Turkey, United Arab Emirates, Austria, Belgium, Bulgaria, Czech Republik, Slovakia, Denmark, Finland, France, Germany, **Greece**, Hungary, Iceland, Ireland, Italy, Luxembourg, Malta, Netherlands, Norway, Poland, Romania, Spain, Sweden, Switzerland, U.K., Yugoslavia, Commonwealth of Independent States, **Australia**, New Zealand. (States that contain some megacities in italics; so-called "megacity states" in boldface; see Note 3.) All data for 1990. Source: World Resources Institute, 1992 [22].

2. Americans typically live in very low density cities. The spectrum of population densities (per square mile) of megacities includes: Calcutta - 62 200; Sao Paulo - 46 500; Mexico City - 46 000; Tokyo - 25 700; New York - 11 500; Los Angeles - 9 400.
3. Old buildings are not always the most vulnerable. Hundreds of Cairo residents died in hastily constructed mudbrick buildings that were not designed to withstand even the modest seismic shaking that caused them to collapse during a recent earthquake.
4. Rural states with megacities include: **Angola**, Ivory Coast, **Egypt**, Ethiopia, Ghana, **Guinea**, Kenya, Morocco, Mozambique, Nigeria, **Senegal**, Sudan, Tanzania, Zaire, **Costa Rica, Haiti, Afghanistan**, Bangladesh, China, India, Indonesia, Malaysia, Myanmar, Pakistan, Philippines, **Thailand**, Vietnam, Portugal. (States that contain some megacities in italics; so-called "Megacity states" in boldface; see Note 6.) All data for 1990. Source: World Resources Institute, 1992 [22].
5. Megacity states (i.e. states with >50% of urban population in megacities) include: Angola, Egypt, Guinea, Libya, Senegal, Costa Rica, Dominican Republic, Haiti, Ecuador, Afghanistan, South Korea, Kuwait, Singapore, Syria, Thailand, Greece, Australia.
6. The connection between city size and environmental problems is complex. McGranahan and Songsore's recent survey of several Third World cities notes that "City size is especially important to city- and regional-level problems" but they go on to point out that "The environment of a smaller city may be just as unhealthy as that of a megacity if inadequate household water supplies, bad sanitation, insect infestation, indoor air pollution, and local accumulations of solid waste are common." They also suggest that neighborhood environmental problems (e.g. sewers) may be more characteristic of cities in poor countries, whereas city-wide or region-wide problems (e.g. smog) may be more prevalent in the cities of wealthy countries. In general, with increasing wealth, urban problems become spatially more extensive and priorities shift from specific issues of environmental health to concerns about broad-based sustainability (*McGranahan* and *Songsore*, 1994 [13]).

7. Some observers have interpreted the reluctance of Spanish-speaking Los Angeles residents to return to earthquake-rocked buildings as a carry-over of behavior that might well have been appropriate in the towns and villages of Central America where non-earthquake resistant adobe brick construction is widespread.

References

1. M. B. Anderson. Metropolitan areas and disaster vulnerability: A consideration for developing countries. In A. Kreimer and M. Munasinghe, editors, *Environmental Management and Urban Vulnerability*, World Bank Discussion Paper No 168, pages 77–92, Washington, D.C., 1992. The World Bank.
2. B. J. L. Berry. "Urbanization": The earth as transformed by human action. In B. L. Turner II, W. C. Clark, R. W. Kates, J. F. Richards, J. T. Mathews, and W. B. Meyer, editors, *Global and Regional Changes in the Biosphere over the Past 300 Years*. Cambridge University Press, Cambridge, 1990.
3. D. Berry. Threats to American cropland: Urbanization and soil erosion. In R. H. Platt and G. Macinko, editors, *Beyond the Urban Fringe: Land Use Issues of Nonmetropolitan America*, pages 179–192. University of Minnesota Press, Minneapolis, 1984.
4. H. P. Chudacoff and J. E. Smith. *The Evolution of American Urban Society*. Prentice Hall, Englewood Cliffs, 1994.
5. R. U. Cooke. *Geomorphological Hazards in Los Angeles*. Allen and Unwin, London, 1984.
6. J. Fellmann and A. Getis. *Human Geography: Landscapes of Human Activities*. Brown Publishers, Dubuque, 1992.
7. R. Fishman. *Bourgeois Utopias: The Rise and Fall of Suburbia*. Basic Books, New York, 1987.
8. A. Gilbert and J. Gugler. *Cities, Poverty and Development: Urbanization in the Third World*. Oxford University Press, Oxford, 1992.
9. B. Jones and W. A. Kandel. Population growth, urbanization, disaster risk, and vulnerability in metropolitan areas: A conceptual framework. In A. Kreimer and M. Munasinghe, editors, *Environmental Management and Urban Vulnerability*, World Bank Discussion Paper 168, pages 51–76, Washington, 1992. The World Bank.
10. R. W. Kates. A tale of three cities. *Environment*, 36(6), 1994.
11. G. P. Marsh. *Man and Nature: The Earth as Modified by Human Action*. Belknap Press of Harvard University Press, Cambridge, 1864. 1965 edition.
12. J. Maskrey. *Disaster Mitigation: A Community Based Approach. Development Guidelines No. 3*. Oxfam, Oxford, 1989.
13. G. McGranahan and J. Songsore. Wealth, health, and urban household: Weighing environmental burdens in Accra, Jakarta, and Sao Paulo. *Environment*, 36(6):4–11 and 40–45, 1994.
14. J. K. Mitchell. Negotiating the contexts of disaster prevention. In J. K. Mitchell, editor, *Society and Disaster Prevention*, Mexico City, 1994. Mexican Council for Social Science. in press.
15. J. K. Mitchell and N. Ericksen. Effects of climate changes on weather-related disasters. In Irving Mintzer, editor, *Confronting Climate Change: Risks, Implications and Responses*, pages 141–152. Cambridge University Press, Cambridge, 1992.

16. R. Stren, R. White, and J. Whitney, editors. *Sustainable Cities: Urbanization and the Environment in International Perspective.* Westview, Boulder, 1992.
17. W. L. Thomas Jr. *Man's Role in Changing the Face of the Earth.* University of Chicago Press, Chicago, 1956.
18. B. L. Turner II, W. C. Clark, R. W. Kates, J. F. Richards, J. Tuchman-Mathews, and W. B. Meyer, editors. *The Earth as Transformed by Human Action: Global and Regional Changes in the Biosphere over the Past 300 Years.* Cambridge University Press, Cambridge, 1990.
19. G. F. White. *Strategies of American Water Management.* University of Michigan Press, Ann Arbor, 1969.
20. World Commission on Environment and Development. *Our Common Future,* Oxford, 1987. Oxford University Press. The Brundtland Commission.
21. World Resources Institute. *World Resources 1990-91*, Oxford, 1990. Oxford University Press.
22. World Resources Institute (WRI). *World Resources 1992-93*, Oxford, 1992. Oxford University Press.

Part IV

Integration for Sustainability

Chapter 1: Multi-actor Optimization of Greenhouse Gas Emission Paths Using Coupled Integral Climate Response and Economic Models

K. Hasselmann and S. Hasselmann

Abstract The simple Structural Integrated Assessment Model (SIAM) applied previously to optimization studies for the abatement of CO_2 emissions for an integrated global economy with a single decision maker is extended to the multi-actor case. The general non-cooperative multi-actor optimization problem is formulated both for the case of non-trading actors, in which the coupling between actors is limited to the jointly modified global climate, and for trading actors, where the control variables of individual actors also directly affect the welfare of other actors. Numerical examples are presented for both classes of interaction. In the non-trading case, the optimal CO_2 emission paths are less affected by the change from a single-actor cooperative strategy to a non-cooperative multi-actor strategy than may have been anticipated intuitively. For a modest number of identical actors (5 - 20), the abatement of the individually optimized emission paths is generally weakened but is of the same order as in the single-actor case. For the case of a single actor undertaking mitigation measures in the presence of $(n-1)$ actors pursuing a 'business-as-usual' policy, the lone mitigator even enhances his (or her) abatement measures for $n < 10$. To illustrate the impact of trade, a two-actor fossil fuel supplier-consumer model is considered. The conflicting goals of the two actors can lead to an effective neutralization of the consumer's attempts to mitigate climate change through reduced fossil fuel consumption by the supplier, who has a motivation to stimulate consumption by reducing the fossil fuel price.

1 Introduction

A major challenge facing mankind today is the development of an effective climate protection strategy to avert or mitigate a major global climate warming. There is a broad scientific consensus that an unimpeded increase in greenhouse-gas emissions at current growth rates will produce a global mean temperature increase within the next century of the order of 3 °C, leading to a climate regime beyond the historical experience of mankind (cf. reports of the Intergovernmental Panel on Climate Change, IPCC, 1990, 1992, 1994, 1995 [10, 11, 12, 13]). If emission levels continue to rise, climate model projections predict a still more drastic global warming in the following centuries of at least twice this magnitude (cf. *Cline*, 1992 [3], *Hasselmann* et al., 1997 [8] – referred to in the following as HHGOS).

Since it will be impossible to completely avoid greenhouse warming – indeed, an estimated global warming of the order of $0.5 - 1$ °C has already taken place today – an optimal climate-protection strategy must seek to minimize the net impact of climate change by minimizing the total costs incurred through the abatement of greenhouse-gas emissions and the adaptation to the residual climate warming. 'Costs' are interpreted here in the general sense of 'welfare loss', including both direct economic costs and quality-of-life factors which, although difficult to quantify economically, must nevertheless be considered in political value judgements and trade-off decisions.

The rational determination of an optimal greenhouse-gas emission strategy requires the construction of integrated 'Global Environment and Society' (GES) models describing the interactions between the climate and the socio-economic system (cf. *Hasselmann*, 1991 [7], HHGOS). Both the climate and the socio-economic system are governed by a complex set of processes extending over a broad range of dynamical time scales, including the time scales relevant for anthropogenic climate change. It follows that the derivation of optimal greenhouse-gas emission paths must be formulated as a dynamical optimization problem.

A number of cost benefit studies using dynamical models have been carried out (e.g. *Reilly* et al., 1987 [25], *Nordhaus*, 1991 [20], 1993, *Nordhaus* and *Yohe*, 1983 [22], *Manne* and *Richels*, 1991 [17], *Peck* and *Teisberg*, 1992 [23], *Michaelis*, 1994 [19], *Tahvonen* et al., 1994 [29], *Tahvonen* et al., 1995 [30], Beltratti, 1995, *Richels* and *Edmonds*, 1995 [26], HHGOS, and also the more complete list of references and discussion in *Cline*, 1992 [3], or *Fankhauser*, 1995 [5]). Most were based on strongly idealized climate and socio-economic models and served mainly to demonstrate the method and investigate the sensitivity to various parameters and assumptions, rather than to provide quantitative data as a basis for political decisions (although, for lack of other data, they have also been evoked for this purpose). A majority of the studies were scenario investigations. Relatively few (e.g. *Nordhaus*, 1991 [20], *Peck* and *Teisberg*, 1992 [23]) carried out explicit optimization analyses.

A basic simplification of the first optimization studies has been that the world could be viewed as a single economic region. In effect, it was assumed that there existed an international agreement on the global abatement-plus-adaptation cost function which was to be minimized. The optimization problem was thus reduced to the task of determining the path of the globally integrated greenhouse-gas emissions which minimized the total, time-integrated (suitably discounted) abatement plus adaptation costs.

Although providing a useful reference baseline, this idealized picture of an effectively single decision maker world economy does not represent a particularly good approximation to the real situation. In practice, the efforts at arriving at a gobal climate protection strategy are better described as a multi-player game, in which each player, representing a particular political-economic region or sector, is engaged in trying to optimize his (or her) individual welfare function on the basis of policy decisions made under the constraints of local-interest pressures, international trade relations and various other political side conditions.

Recently, a few authors have addressed the multi-actor problem using generalizations of single-actor models used in earlier studies (e.g. *Nordhaus* and *Yang*, 1995 [21], *Manne* and *Richels*, 1995 [18]). In this paper we similarly consider the multi-actor problem using an extension of the idealized single-actor Structural Integrated Assessment Model (SIAM) used in HHGOS. We adopt the same philosophy as in HHGOS: in view of the large divergences of the conclusions inferred from existing single-actor models, we regard it as premature to embark on a multi-actor analysis using a relatively complex economic model. The goal of our analysis is to understand the basic system-analytical implications of the multi-actor problem, and to generalize the conclusions of our previous sensitivity study of the single actor system to the multi-actor case, using a model designed for maximal structural transparency.

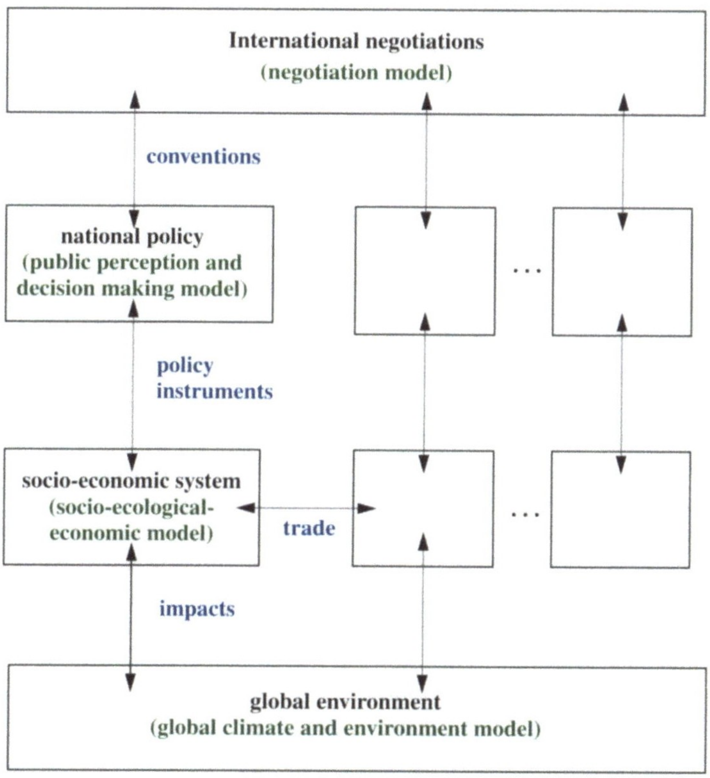

Figure 1. Interactions and sub-systems of an integrated Global Environment and Society (GES) model, as in Fig. 1 of HHGOS, but broken down into separate interacting actors.

The basic structure of such a multi-actor GES model is indicated in Fig. 1. It describes the same set of interactions between the climate system, the socio-economic system and the policy making sub-systems depicted for the single-actor case in Fig. 1 of HHGOS (see also *Hasselmann*, 1991 [7]), but broken down now into individual columns representing different political-economic regions or sectors. The disaggregation can refer to either different geographical regions or different sectors within the economy (e.g. consumers and suppliers of fossil fuels, or private and government sectors).

Different actors interact through trade and negotiations, and are coupled through global climate change, to which all contribute, but which affects each actor differently. Each actor seeks to optimize his individual welfare function. Whether the outcome is a non-cooperative Nash equilibrium (assuming such an equilibrium exists) or an optimal cooperative solution is an open question and depends on the type of interactions between the various sub-systems, the negotiation rules (possibilities of retribution, reward, etc) and the negotiation strategies of the individual actors. We will consider various alternatives.

As basic building blocks for our multi-actor GES model we use the climate impulse-response and structurally simplified economic-cost modules of the single-actor SIAM model developed in HHGOS. Since realistic climate models based on coupled ocean-atmosphere general circulation models (CGCMs) (cf. IPCC, 1992 [11], *Cubasch* et al., 1992 [4]) and three-dimensional ocean-atmosphere carbon cycle models (*Maier-Reimer* and *Hasselmann*, 1987 [16], *Maier-Reimer*, 1993 [15]) require too much computing time to be used for greenhouse-gas emission studies involving iterative optimization computations, simpler dynamical climate models must be developed. The impulse response climate model used in SIAM represents a linearized approximation of the response computed with a realistic, fully nonlinear climate model. It captures the principal features of the response of the complete model to external greenhouse forcing, but is valid only for relatively small perturbations of the climate system. For the examples considered in this study the linear response approximation should be adequate. Economic models suitable for optimization studies of greenhouse gas emissions studies similarly need to be more strongly aggregated than most global macroeconomic models. The resulting reduced GES models nevertheless still remain too complex for an analytical solution of the optimization problem, so that the optimal emission paths need to be determined numerically, for example by conjugate gradient minimization methods.

The paper is structured as follows: The reference HHGOS single-actor model SIAM used as the basis for our study is presented briefly in Section 2. As in most other studies, we consider as greenhouse gas only CO_2 (the influence of the non-CO_2 gases is discussed briefly in HHGOS). In Section 3 we consider two examples of the generalization of the single-actor case to n non-trading actors. The coupling between different actors occurs here only through the impact of the emissions of each actor on the global climate. The general fully-interacting multi-actor problem, including coupling through trade, is presented in Section 4. As example, we consider then in

Section 5 the optimization of greenhouse-gas emissions for a two-actor system consisting of a cartel of fossil-fuel suppliers and a single consortium of fossil-fuel consumers. As discussed in *Blank* and *Ströbele* [2] (see also *Richels* et al., 1996 [27]), the pricing strategy of the fuel suppliers is shown to have a strong impact on the CO_2 emissions path resulting from the conflicting efforts of each actor to maximize his individual welfare expression. The results are summarized in the concluding Section 6.

2 The SIAM single-actor model

The Structural Integrated Assessment Model (SIAM) used in HHGOS consists of two modules: a climate module, which determines the evolution of climate for a given CO_2 emissions path, and a costs module, which evaluates the climate damage and CO_2 abatement costs for given climate and CO_2 emissions paths.

The climate module was constructed by projecting the simulation results of a state-of-the-art coupled ocean-atmosphere general circulation model and a three dimensional global carbon cycle model onto a linear impulse response model. In principle, an impulse-response climate model can contain the same detailed climate change information as the sophisticated models against which it is calibrated, but in HHGOS only the global mean (surface) temperature was retained as output climate variable.

The costs module contains two components: a simple but widely used expression for the climate damage costs, composed of a term which depends quadratically on the change in temperature and a similar second term which depends quadratically on the rate of change of temperature; and a structurally highly simplified abatement costs expression. The costs incurred at any given time (specific costs) are integrated over all time, from the present to infinity, to determine the total costs. Future specific costs are discounted at rates which may be chosen differently for climate damage and abatement costs.

In the following we summarize the principal features of the SIAM model, together with the main conclusions of the single-actor HHGOS study relevant for the present n-actor investigation. A more detailed description is given in HHGOS.

2.1 The general impulse-response climate model

The climate module of the SIAM model is based on the principle that, although the climate system and sophisticated climate models are inherently strongly nonlinear, the response of the climate system, as of any differentiable nonlinear system, to small external forcing is to first order linear. As external greenhouse forcing the model considers only the annual emissions $e(t)$

of CO_2. This represents about 60% of the total anthropogenic greenhouse forcing today, and – in the absence of abatement measures – is projected to contribute a larger fraction in the future.

In the linearized approximation, the change $\mathbf{x}(t)$ of the climate state relative to an initial state at some time t_0 (which we take as some time in the pre-industrial period) to an arbitrary, sufficiently small emission function $e(t)$ can be represented in the general integral form

$$\mathbf{x}(t) = \int_{t_0}^{t} \mathbf{R}(t - t')e(t')dt', \tag{1}$$

where the climate impulse-response function $\mathbf{R}(t - t')$ represents the climate response at time t to a unit δ-function emission at time t' and the climate state consists, in a discretized model representation, of the vector of all climate variables at all model gridpoints.

Since a doubling of the CO_2 concentration corresponds to an increase in radiative forcing of about 4 W/m^2, or little more than 1% of the global mean incident solar radiation of 340 W/m^2, the linear form is adequate for most applications concerned with optimal emission scenarios which lead to an acceptable stabilized climate. However, for extreme climate change scenarios, corresponding to unregulated or only weakly regulated CO_2 emissions, the linear approximation is no longer applicable and can be used only as a rough guide (after application of first-order nonlinear correction factors) to the order of magnitude of the predicted climate change. Since we shall be concerned mainly with optimized emission scenarios which avoid unsustainable climate change regimes, the limitations of the linearization approximation will not be serious.

We note that the dimension of $\mathbf{R}(t)$ in eq.(1) is the same as that of $\mathbf{x}(t)$: the linearized form of eq.(1) implies no loss of information in the representation of the climate state relative to the complete nonlinear system, either in terms of geographical resolution or with regard to the set of climate variables (temperature, humidity, precipitation, ocean currents, sea level, etc). However, we shall restrict the analysis in the following, as in HHGOS, to a reduced version of the general impulse response climate model in which only one component of the full climate state vector $\mathbf{x}$ is considered, namely the global mean temperature T.

The response function can be determined empirically by fitting simple analytical forms, consisting, for example, of sums of exponentials, to numerical climate response experiments with realistic three-dimensional carbon cycle and coupled ocean-atmosphere CGCM models. In HHGOS, the calibration was performed against the carbon cycle models of *Maier-Reimer* and *Hasselmann* [16] and *Maier-Reimer* [15] and the CGCM model simulations of *Cubasch* et al. [4], as analyzed by *Hasselmann* et al. [9].

The impulse-response formulation of simplified climate models has several advantages over the traditional approach of constructing simple climate models in the form of empirical box models with only a small number of degrees of freedom: there is no loss of information; the model can be readily calibrated against realistic CGCM and carbon cycle models; and the model dynamics is not limited to simple low-order differential equations, but applies generally for differential equations of arbitrary order – indeed, for any dynamical system, regardless of how the dynamics is formulated, provided only that the system is differentiable.

2.2 Reduction to global mean temperature

In the following we replace the climate state vector $\mathbf{x}$ in the general climate response relation of eq.(1) by the single climate variable T representing the change in global mean temperature. The net response T to the CO_2 emissions is constructed from two sub-systems:

1. A *carbon cycle model*

This describes the evolution of the change w of the atmospheric CO_2 concentration relative to the pre-industrial value in response to the CO_2 emission $e(t)$,

$$w(t) = \int_{t_0}^{t} R_w(t-t')e(t')dt', \tag{2}$$

where $R_w(t-t')$ denotes the impulse response of the concentration at time t for a unit δ-function emission pulse at time t'. We shall choose t_0 later as the pre-industrial date 1800 (the exact date is immaterial, since $e(t) \equiv 0$ in the pre-industrial era).

Time throughout is in units of years. To retain the same carbon units GtC (giga tonnes of carbon) for w and the emissions e (in GtC/yr), w is expressed in all equations in terms of the total carbon in the atmosphere. However, we shall present results for w in the figures later in the usual units of ppm. The conversion factor is w [GtC] $= 2.123w$ [ppm]. The present atmospheric CO_2 concentration of 358 ppm corresponds to an atmospheric carbon content of 760 GtC, while the pre-industrial concentration is taken as $w_0 = 280$ ppm $= 594$ GtC.

Initially, all of the emissions enter the atmosphere, so that

$$R_w(t_0) = 1. \tag{3}$$

$R_w(\infty)$ defines the fraction of the emissions which is retained in the atmosphere in the asymptotic equilibrium state. If the ocean sink alone is considered, the retention factor is approximately 15%; if the dissolution of CO_2 in the upper layers of the ocean sediments is also included, the long-term atmospheric retention factor may fall (*Maier-Reimer*, 1993 [15]) to about 7%. The increased storage of CO_2 in the terrestrial biosphere through CO_2 fertilization and the significantly slower loss of CO_2 through sedimentation in the ocean are not included in these estimates.

Invoking eq.(3), the time derivative of eq.(2) (which will be needed to couple the CO_2 model to the temperature response model defined below) is given by

$$\frac{dw}{dt} \equiv \dot{w}(t) = \int_{t_0}^{t} \dot{R}_w(t-t')e(t')dt' + e(t). \tag{4}$$

In an analysis of the response of a nonlinear three-dimensional global ocean carbon cycle model to various CO_2-emission levels, *Maier-Reimer* and *Hasselmann* [16] found that the model response could be fitted to a linear relation of the form of eq.(1) quite well for an increase in the CO_2 level up to a factor of two. For stronger emission levels, producing a four-fold increase in the CO_2 concentration, the linear response underestimated the atmospheric concentration predicted by the full model by about 30%. This was due primarily to the nonlinear decrease of the solubility of CO_2 in sea water with increasing CO_2 concentration. A relatively simple nonlinear extension of the linear response form to allow for the nonlinearities associated with the solution of CO_2 in sea-water has recently been proposed by *Joos* et al. [14].

2. A *global temperature response model*

This describes the change $T(t)$ of the global mean temperature induced by the rate of change $\dot{w}$ in the CO_2 concentration,

$$T(t) = \int_{t_0}^{t} R_T(t-t')\dot{w}(t')dt', \tag{5}$$

where the temperature impulse response function $R_T(t-t')$ represents the change in the global mean temperature produced at time t by a unit step-function increase in the atmospheric CO_2 concentration at time t'.

Since the climate system has inertia, the instantaneous climate response to a sudden change in CO_2 concentration is zero,

$$R_T(0) = 0. \tag{6}$$

At the other end of the time scale, $R_T(\infty)$ represents the asymptotic equilibrium response of the ocean-atmosphere system to a unit increase of the atmospheric CO_2 concentration.

The generalization of this simple one-parameter climate model to more complex climate-state models, including, for example, regional temperature distributions, modified precipitation patterns and sea level change is basically straightforward. Such models could be readily constructed, in accordance with the general form of eq.(1), from existing data generated by CGCM climate-response simulations. However, for illustrative purposes we restrict the model here to a single climate variable, the global mean temperature.

In applications to multi-actor optimization problems, one needs generally to consider different climate impact functions for different actors. This can be achieved within the framework of the present one-parameter climate model by simply applying different regional impact factors to the global mean temperature T. Assuming that the climate impact for a given region can be characterized by the average temperature change over the region, this can be justified by the results of numerical global warming simulations with coupled CGCMs (*Cubasch* et al., 1992 [4]). The response of the global temperature distribution is dominated in these simulations by a single pattern, the first EOF (empirical orthogonal function), implying that the average temperature response for any region can indeed be related to the global mean temperature by a time-independent scale factor.

The linear response relation between the temperature change and the change of the CO_2 concentration can be modified to accommodate the more realistic logarithmic dependence between the radiative greenhouse forcing and the CO_2 concentration by replacing $\dot{w}$ by $d(\ln w)/dt$ in eq.(5). This introduces no significant complications in the numerical examples considered in the following section. However, the difference between the linear and logarithmic formulation is small for small forcing (which we assume), and for the present illustrative purposes, the linear relation of eq.(5) has the advantage of yielding a net linear climate response to the emissions in accordance with eq.(1) [cf. eqs.(9, 10) below].

Approximations of global warming CGCM simulations in terms of linear response relations (*Hasselmann* et al., 1993 [9]) suggest that, as in the case of the linearized carbon cycle model, the linearized temperature response relation is applicable for climate changes associated with CO_2 concentration increases up to about double the pre-industrial level, i.e. for a temperature rise up to about 3 °C. Apart from the inherent individual nonlinearities of the coupled ocean-atmosphere circulation system and the carbon cycle system, the linear response relations also become inaccurate beyond this range because the temperature feedback on the CO_2 model (increasing temperature decreases the CO_2 solubility of sea-water) has not been included in the CO_2 response relation of eq.(2).

Combining the carbon cycle and global temperature response models, the net response of the 'climate' T to the emission $e(t)$ can now be written

$$T(t) = \int_{t_0}^{t} dt' R_T(t-t') \left\{ e(t') + \int_{t_0}^{t'} dt'' \dot{R}_w(t'-t'') e(t'') \right\}. \tag{7}$$

Noting that

$$\int_{t_0}^{t} dt' \int_{t_0}^{t'} dt'' = \int_{t_0}^{t}, dt'' \int_{t''}^{t} dt', \tag{8}$$

this may be expressed as

$$T(t) = \int_{t_0}^{t} R(t-t')e(t')dt', \tag{9}$$

in accordance with the form (1), where

$$R(t) = R_T(t) + \int_{0}^{t} R_T(t-t')\dot{R}_w(t')dt'. \tag{10}$$

The net temperature impulse response or *global warming response* function $R(t)$ represents the temperature increase at time t due to a unit δ-function CO_2 input into the atmosphere at time $t = 0$. It reflects the net effect of both the thermal inertia of the ocean-atmosphere climate system and the decrease of the atmospheric CO_2 concentration through the transfer of CO_2 from the atmosphere to other components of the carbon cycle. [$R(t)$ should not be confused with the "global warming potential" or "commitment" defined in IPCC [10], as a measure of the time integrated radiative forcing.]

2.3 Numerical values

The response functions R_w and R_T have been determined empirically from numerical response experiments using realistic three-dimensional models of the global carbon cycle (*Maier-Reimer* and *Hasselmann*, 1987 [16], *Maier-Reimer*, 1993 [15]) and the coupled ocean-atmosphere climate system (*Hasselmann* et al., 1993 [9]). A number of different models were intercompared in HHGOS. It was found that the differences had little impact on the computed optimal emission scenarios. We shall accordingly use only their baseline model R00 (cf. Fig. 2):

$$\begin{aligned} R_w &= 0.07 + 0.648\exp(-t/258.5) + 0.101\exp(-t/71.9) \\ &\quad + 0.097\exp(-t/17.6) + 0.084\exp(-t/1.6), \end{aligned} \tag{11}$$

$$\begin{aligned} R_T &= w_0^{-1}\{1.21(1-\exp(-t/2.1)) + 0.759(1-\exp(-t/12)) \\ &\quad + 0.531(1-\exp(-t/138.6))\} \\ &= w_0^{-1}R'_T. \end{aligned} \tag{12}$$

The normalized function $R'_T = w_0 R_T$ represents the temperature response to a CO_2 doubling relative to the pre-industrial CO_2 concentration w_0. The response curves in Fig. 2 demonstrate [in accordance with the analytical expresssions of eqs.(11, 12)] that the net climate response to a δ-function CO_2 emissions pulse cannot be characterized by a single time constant. Following a rapid temperature rise in the first few years as the upper mixed layer of the ocean warms, the net response function for the global mean temperature

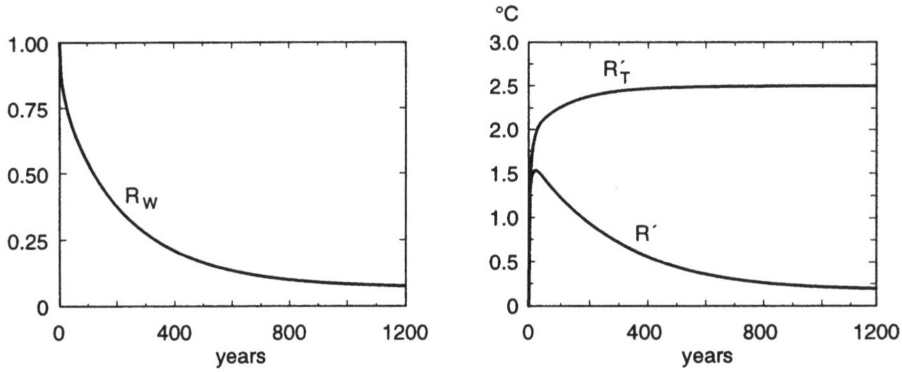

Figure 2. *Left panel*: Response function R_w representing the atmospheric retention factor for a unit δ-function emission of CO_2 at time $t = 0$, as given by the CO_2-response model of eq.(11). *Right panel*: Temperature response functions $R'_T = w_0 R_T$ and $R' = w_0 R$ for a step-function doubling of the CO_2 concentration at time $t = 0$. The temperature response R'_T describes the global warming which would occur if the initial CO_2 input were completely retained in the atmosphere [eq.(12)], while the net global warming response R' [eq.(10)], allows for both the thermodynamic temperature response and the uptake of the initial CO_2 input by the ocean.

increases more slowly as the warming penetrates into the deeper ocean. After a few decades, the CO_2 transfer from the atmosphere into the ocean causes the temperature to gradually relax back, over a period of several hundred years, to an asymptotic equilibrium value of $2.5 \times 0.07 = 0.175$ °C. For the optimization of greenhouse-gas emission paths, both the near-time and far-time climate response characteristics must be considered.

2.4 The cost function

In the SIAM model the impulse-response climate model is coupled with a simple economic costs module. This consists of the sum of the costs C_a incurred through the abatement of CO_2 emissions and the climate damage costs C_d. The optimal emissions path is the path which minimizes the total climate induced costs

$$C = C_a + C_d \tag{13}$$

or maximizes the total welfare

$$W = W_A - C, \tag{14}$$

where W_A represents the global welfare for a reference non-regulated 'business as usual' (BAU) emissions path $e_A(t)$ (e.g. Scenario A of IPCC, 1990 [10]) in which climate damages are ignored. The determination of the reference path $e_A(t)$ is irrelevant for the following; we regard $e_A(t)$ as given. Welfare is equated here simply with negative costs, but it is assumed that costs include monetarized values of quality-of-life factors such as the protection of species or the quality of the environment. Nonlinearities in the relation between welfare and costs are irrelevant for the single-actor optimization problem as long as the relation is monotonic, but could play a role in the multi-actor problem. However, for the present general systems-analysis approach, these effects will be ignored.

It is assumed that the net abatement and climate damage costs can be expressed as integrals over the *specific* costs $c_a(t)$, $c_d(t)$ in the form

$$C_a = \int_{t_0}^{\infty} c_a(e(t), \dot{e}(t), \ddot{e}(t), t) dt \tag{15}$$

$$C_d = \int_{t_0}^{\infty} c_d(T(t), \dot{T}(t), t) dt. \tag{16}$$

The integrals converge for the assumed infinite time horizon provided suitable discount factors are introduced.

Cross-coupling of the climate and emission variables in the cost expressions is not considered. Although a change in the socio-economic system induced by a change in emissions will presumably modify the sensitivity of the system to climate change, and a change in climate will conversely have some impact on the effectivity of abatement measures, these effects are ignored.

The first and second time derivatives $\dot{e}$ and $\ddot{e}$ of the emissions are included in the specific abatement-cost function in order to parametrize the effects of economic inertia.

As simplest mathematical expression which captures the principal properties of the abatement costs anticipated from a more detailed economic model the SIAM model sets

$$c_a = \left\{ (\frac{1}{r} - r)^2 + \tau_1^2 \dot{r}^2 + \tau_2^4 \ddot{r}^2 \right\} D_a(t), \tag{17}$$

where τ_1 and τ_2 are time constants, $r = e/e_A$, and

$$D_a(t) = \exp(-t/\tau_a) \tag{18}$$

is an abatement-cost discount factor, characterized by an abatement-cost discount time constant τ_a (inverse annual discount rate).

The first term in the form of eq.(17) has the property that any positive or negative departure from the reference BAU emissions path e_A incurs costs which are quadratic in the deviations $\delta r = r - 1$ for small δr, $(\frac{1}{r} - r)^2 \approx 4(\delta r)^2$, and approach infinity for $r \to 0$ and $r \to \infty$. The second and third terms depending on the time derivatives of $e(t)$ penalize rapid changes in the emissions relative to the BAU emissions curve $e_A(t)$. They prevent the occurence of discontinuities in $e(t)$ and $\dot{e}(t)$. For $e_A(t)$, a simple linear growth form from a 1995 emission level of 6.3 GtC/yr up to a level of 38 GtC/yr in the year 2200 was assumed, with frozen emissions at this level thereafter (cf. Fig. 3).

For the climate damage costs the SIAM model assumes

$$c_d = \left\{ \left(\frac{T}{T_c} \right)^2 + \left(\frac{\dot{T}}{\dot{T}_c} \right)^2 \right\} D_d(t), \tag{19}$$

where

$$D_d(t) = \exp(-t/\tau_d) \tag{20}$$

is the climate damage costs discount factor, with discount time constant τ_d, and T_c, $\dot{T}_c$ are constants. Climate damages are assumed to arise not only through a change in the temperature itself but also through the rate of change of temperature. The quadratic dependencies reflect the general view that climate damage costs increase nonlinearly with climate change and that costs are incurred through climate changes of either sign.

For the optimization problem, only the ratio of the cost functions is relevant. The freedom to choose an arbitrary normalization constant has been used to set the coefficient of the first term of the abatement cost function of eq.(17) equal to unity. This scales the constants T_c, $\dot{T}_c$ in the damage cost function in relation to the abatement costs: the climate damage and abatement costs become comparable when $T = O(T_c)$ or $\dot{T} = O(\dot{T}_c)$ and $r = O(0.5)$. Thus the parameters T_c and $\dot{T}_c$ define a critical (soft shouldered) elliptical window or corridor in the climate phase space $T_c, \dot{T}_c$ within which the climate damage costs remain smaller (in order of magnitude) than the mitigation costs at a 50% abatement level. Outside the critical corridor, the climate damage costs begin to exceed the mitigation costs at this reference abatement level.

The minimal-cost solution can be found numerically by a method of steepest descent (e.g. a conjugate gradient technique, cf. *Press* et al. [24]). This requires computing the (functional) gradient

$$g(t) = \delta C/\delta e, \tag{21}$$

defined by the integral relation

$$\delta C = \int g(t)\delta e(t)dt, \tag{22}$$

of the cost with respect to the control variable $e(t)$. For a linearized climate model expressed in integral response form, the gradient can be computed explicitly (cf. HHGOS). However, for the numerical results presented in HHGOS and below, the gradient was computed automatically using a general numerical functional derivative compiler developed by *Giering* and *Kaminski* [6]. The computations of HHGOS were referred to a baseline case $S0$ using the climate model defined in eqs.(7-12) and the following cost function parameters:

$T_c = \sqrt{2}$ °C ≈ 1.4 °C, $\dot{T}_c = 2\sqrt{2}$ °C/yr ≈ 0.028 °C/yr; $\tau_1 = \tau_2 = 100$ yr; $\tau_a = 50$ yr, $\tau_d = \infty$ yr. We shall use the same climate model and cost function parameters for our multi-actor computations later.

2.5 Results of the one-actor SIAM model

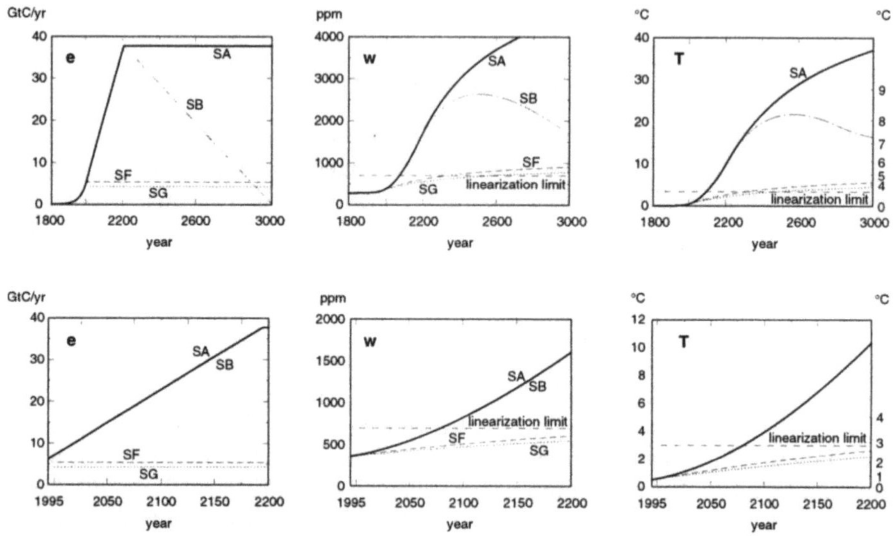

Figure 3. CO_2 emission paths, computed CO_2 concentrations and global warming (*from left to right*) for the time periods 1800-3000 (*top*) and 1995-2200 (*bottom*) for the BAU scenario (SA, *full curves*, cf. HHGOS), modified BAU scenario (SB, *dashed-dotted curves*), frozen emissions at 1990 levels after the year 2000 (SF, *dashed curves*) and 20% reduced emissions relative to the 1990 level after 2000 (SG, *dotted curves*). The linear model is not applicable above the indicated levels (*dashed lines*). The logarithmic T scales on the right ordinate axes of the panels on the right indicate the order-of-magnitude temperature response allowing for the logarithmic dependency of the radiative forcing on the CO_2 concentration. (Reproduced from HHGOS.)

The principal conclusions of the HHGOS sensitivity study may be summarized as follows:

1. The long response time of the climate system implies that optimal emission strategies must be conceived over time scales of several centuries, far exceeding normal political and economic planning horizons. Short period analyses can seriously underestimate the long term climatic impact of CO_2 emissions (compare the lower panels of Fig. 3 for the period 1995-2200 with the upper panels covering the full period 1800-3000).
2. To constrain future global warming within acceptable bounds, CO_2 emissions must be drastically reduced in the long term. However, the required

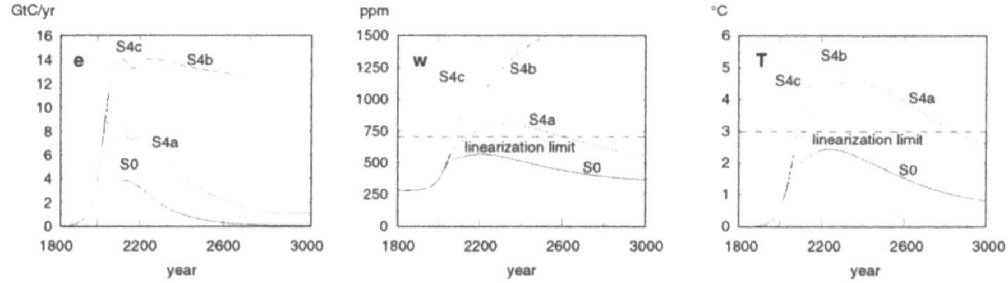

Figure 4. Optimal CO_2 emission paths, CO_2 concentrations and global warming (from *left* to *right*) for baseline optimal emissions scenario $S0$ climate damage-cost discounting (*full curves*) and scenarios assuming finite discount time constants $\tau_d =$ 100 yr (S4a, *dotted curves*), $\tau_d = 50$ yr (S4b, *dashed curves*) and $\tau_d = 25$ yr (S4c, *dashed-dotted curves*). In the last two cases the temperature rises monotonically over the entire 1200-year period. In all runs the abatement costs discount time constant was set at $\tau_a = 50$yr. (Reproduced from HHGOS, where further details are given.)

transition to carbon-free energy technologies can be achieved gradually over many decades.

3. Economic inertia and the dependence of climate damages on the rate of change of temperature have little impact on the computed optimal emission paths for the long time scales characterizing these solutions.
4. Optimal emission paths yielding acceptable global warming limits are obtained only if the discount rate for climate damages is set at a significantly lower level than the discount rate for abatement costs (cf. Fig. 4). Discounting climate damages at normal economic discount rates implies that negligible present value (in the sense of willingness to pay today) is attached to sustainable development over time scales of a few centuries, and that there is therefore no incentive to avoid long term global warming.

In the following sections we investigate the impact of multi-actor interactions on these results.

3 Non-trading multi-actor models

We consider first the simplest case in which the interdependence between actors occurs only through global climate change, with each actor seeking to optimize his (or her) individual welfare function, independently of the actions of the other actors. The more general case of actor interdependence through trade and optimization strategies based on foresight of the anticipated actions of other actors is considered in the subsequent section.

3.1 Identical actors

In general, the optimal solutions for non-cooperative (or cartel-type) strategies are inferior to the optimal cooperative solution of the single-actor model. We illustrate this first for the simplest example of n identical actors representing different economic regions, each of which is characterized by the same per capita abatement and damage costs as in the single-actor case. We use the same cost expressions as in the single-actor SIAM model. Since the normalization of the cost functions in the SIAM model was arbitrary, we may assume that they represent per capita costs.

The non-cooperative Nash equilibrium for this problem is obtained by minimizing the cost function C_i for each actor i with respect to his own emissions e_i, keeping the emissions of the other actors fixed. From the symmetry of the problem it follows that all emission functions e_i will be equal for the Nash equilibrium solution. Thus

$$e = \sum_{j=1}^{n} e_j = ne_i. \tag{23}$$

The minimization problem for a given individual actor i is then essentially identical to the previous single-actor case, with the control path $e(t)$ replaced by $e_i(t)$.

To compute the variation of the abatement costs C_{ai} of actor i with respect to his emissions e_i, we need to write the SIAM expression (eqs. (15, 17)) for the per capita abatement costs $C_{ai} = C_a$ as a function of e_i instead of e: $C_{ai}(e_i, \dot{e}_i, \ddot{e}_i, t) = C_a(e/n, \dot{e}/n, \ddot{e}/n, t)$. Thus

$$\frac{\delta C_{ai}}{\delta e_i} = \frac{\delta C_a}{\delta e}\frac{de}{de_i} = n\frac{\delta C_a}{\delta e}. \tag{24}$$

To determine the variation of the per capita climate damage costs $C_{di} = C_d(T, \dot{T}, t)$ induced by a variation δe_i, we need to determine the variation δT of the global mean temperature (and similarly $\delta \dot{T}$) induced by δe_i. Here the partial derivative $\partial e/\partial e_i = 1$ of the total emissions with respect to the emissions e_i is relevant, as given by the left hand part of eq.(23):

$$\frac{\delta T}{\delta e_i} = \frac{\delta T}{\delta e}\frac{\partial e}{\partial e_i} = \frac{\delta T}{\delta e}. \tag{25}$$

The net result of both transformations is that in the expression for the gradient with respect to e_i of the total per capita costs, the ratio of the climate damage to abatement costs is reduced by a factor n^{-1} relative to the single-actor case. Thus the Nash equilibrium for the case of n identical non-cooperating actors is identical to the single-actor solution with the climate damage costs reduced by a factor n^{-1}. In effect, each actor considers only his own contribution to the climate damage costs and thus acts as though the climate damage costs are a factor n^{-1} smaller than they in fact are when

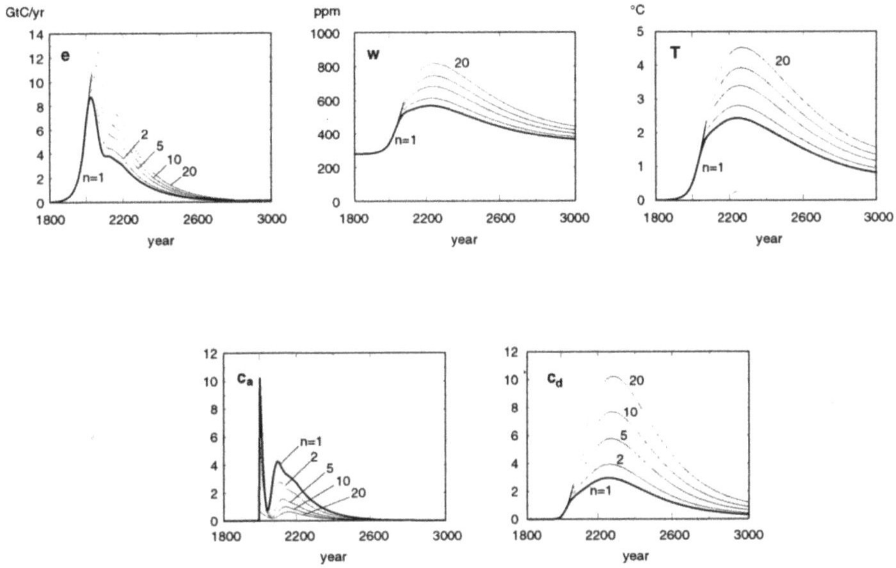

Figure 5. Optimal CO_2 emission paths, CO_2 concentrations and global warming (*top, left to right*) and specific abatement and (total) climate damage costs c_a, c_d, respectively, (*bottom, left to right*) for the n-identical actor Nash equilibrium solution. The effective damage-cost function of an individual actor is reduced by a factor n^{-1} relative to the single actor case (computed for the parameters of the baseline reduced-emissions scenario $S0$ of HHGOS, *full curves*, also shown in Fig. 4).

summed over all actors. In the limit of a large number of actors, one obtains the free-rider solution: no one carries out abatement measures.

However, for a moderate number of actors, of the order of 10 or 20, the non-cooperative Nash equilibrium emission paths do not differ as strongly from the cooperative solution as one may have expected intuitively. Fig. 5 shows the impact of a non-cooperative strategy, compared with the cooperative single-actor solution, for the baseline reduced-emissions scenario reproduced in Fig. 4, for different numbers of actors.

Similar results showing the effect of a change in the ratio of climate damage to abatement costs (but with an enhancement rather than a reduction of the ratio) were presented in HHGOS in a discussion of the role of non-CO_2 gases. The relatively minor impact can be explained by two effects: Firstly, a decrease of the climate damage costs by a factor n^{-1} relative to the abatement costs implies an increase of the critical climate temperature T_c (and rate of change of temperature $\dot{T}_c$) by a factor of only $\sqrt{n}$ [cf. eq.(19)]. Thus the climate damage costs computed by the individual actor in the non-cooperative case are raised already to the level of the true climate damage costs, as computed in the cooperative case, when the emissions are increased by a factor of only $\sqrt{n}$. However, while the computed climate damage costs for the n-actor problem correspond then to the original climate damage costs for the single-actor case, the computed abatement costs, because of the higher emission levels, are lower. In the optimal n-actor solution, a balance is attained between the abatement and damage costs. Thus the abatement costs will be higher and the emission levels lower than these values. Hence the increase in emission levels for the non-cooperative Nash equilibrium solution will be even smaller than the factor $\sqrt{n}$.

Because of the relatively small differences between the cooperative and non-cooperative optimal emission solutions for a moderate number of actors, it could be argued that there is little incentive for trying to persuade all actors to adopt a cooperative strategy. However, an essential assumption in this scenario is that all actors share the same assessment of the climate damage and abatement costs. Once this consensus has been achieved, it is only a small additional step to achieve a cooperative climate-protection strategy. The more difficult (and relevant) multi-actor situation arises when different actors have different climate damage and abatement-cost perceptions, as illustrated in the following example.

3.2 The single mitigator problem

A more realistic description of the n-actor climate-protection problem is that some actors regard the risks of adverse effects due to climate change as high, while others are more concerned with the negative effects of possible mitigation measures on the economy. To study the impact of different cost assessments on the individually optimized non-cooperative solutions, we consider a particularly simple example. We assume that one actor assigns the same values to climate damage and abatement costs as in the SIAM model, while the remaining $(n-1)$ actors ignore all climate damage costs, as in the business-as-usual (BAU) scenario. For simplicity, all actors are again assumed to represent identical economic regions, the only difference being in their assessments of the climate related costs. Since the emissions $e_A(t)/n$ of the BAU actors are prescribed, the problem reduces again formally to a single-actor problem for the first actor 1.

For the $(n-1)$ BAU actors with prescribed emissions $e_A(t)/n$, the distinction between different actors is irrelevant. We may thus replace the set of $(n-1)$ BAU actors simply be a single 'rest of the world' BAU actor whose emissions are given by $[(n-1)/n]e_A(t)$. However, to retain an analagous notation to the previous example, we shall continue to refer to '$(n-1)$ BAU actors'.

Regarding costs again as per capita costs, the expression for the abatement costs for actor 1 remains the same as in the global single-actor case of eq.(17), with the global emission abatement factor $r = e/e_A$ replaced by the individual abatement factor $r_1 = e_1/e_{1A} = ne_1/e_A$ of actor 1. However, the climate damage costs are now dominated by the prescribed emissions of the other $(n-1)$ BAU actors.

The total emissions are given by

$$e = \frac{n-1}{n} e_A + e_1. \tag{26}$$

Since the climate change in our model depends linearly on emissions, the global mean temperature T and rate of change of temperature $\dot{T}$, which determine the climate damage costs, are given by

$$\begin{aligned} T &= \frac{n-1}{n} T_A + T_1, \\ \dot{T} &= \frac{n-1}{n} \dot{T}_A + \dot{T}_1, \end{aligned} \tag{27}$$

where T_A and T_1 represent the climate response to the emissions e_A and e_1, respectively, as defined by eqs.(9, 10). Substitution of eq.(27) into eq.(19) defines the damage costs as a function of $e_1(t)$, for given $e_A(t)$. Adding to these the abatement costs of actor 1, which are also defined as a function of e_1, the total costs of actor 1, given the BAU emissions of the remaining actors, are thus expressed in terms of the emissions of actor 1 and can be minimized with respect to $e_1(t)$.

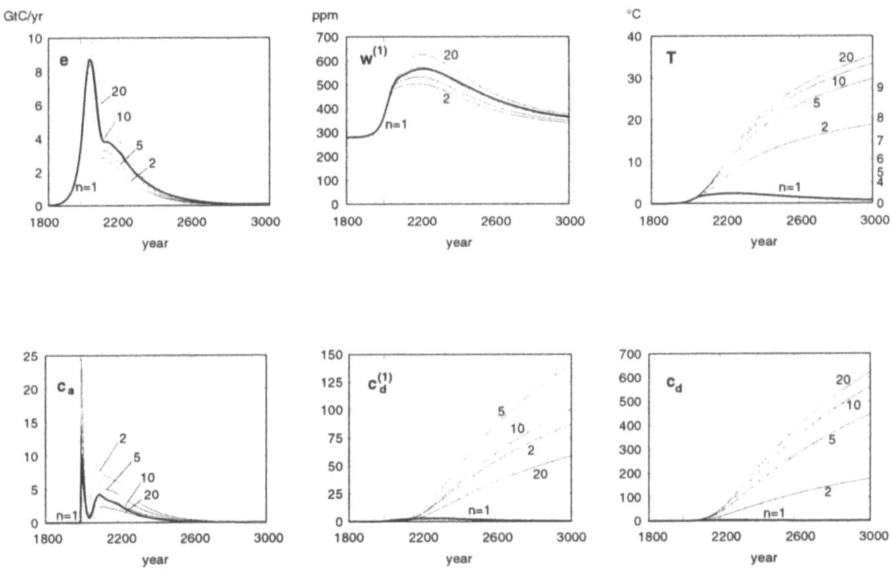

Figure 6. Optimal emissions solution for a single mitigator responding to climate damage costs in the presence of '$(n-1)$ BAU actors', ignoring climate damage costs. Panels on the *left* and in the *middle* refer to the single mitigator, panels on the *right* to the total contributions from all actors. For ease of comparison, the emissions and concentrations of actor 1 have been multiplied by n and thus represent the equivalent contributions if all actors had adopted the same emissions as actor 1. The computations were made for the parameters of the baseline single-actor scenario, *full curves*, shown also in Fig. 4 (layout analogous to Fig. 5).

The resulting optimal paths are depicted in Fig. 6 for various values of n. If the number of non-cooperative BAU actors is less than 10, the single mitigator compensates for the lacking abatement policy of the other actors by enhancing his own abatement measures. However, as n increases beyond 10, it becomes more difficult for the single mitigator to influence the climate damage costs, and ultimately, for very large n, actor 1 resigns and also undertakes no abatement measures. However, the asymptotic solution is approached more slowly than may have been anticipated intuitively.

It is surprising also, in view of the relatively small contribution of the single mitigator to the total climate damages, that for $n < 10$ it is economical for him to enhance his abatement efforts. This is a consequence of the assumed nonlinear dependence of the climate damage costs on the climate change. The background temperature increase caused by the BAU actors amplifies the climate change impact of the single mitigator. This can be seen by separating the total climate damage costs into the contributions from the $(n-1)$ BAU actors and the residual contribution from the mitigator (including the nonlinear interaction term with the BAU actors).

According to eq.(27), the total climate damage costs (ignoring, for simplicity, the $\dot{T}$-terms, which can be treated similarly) are given by

$$c_d = T_c^{-2}\left(\frac{n-1}{n}T_A + T_1\right)^2 D_d(t) = c_d^{(A)}\frac{(n-1)^2}{n^2} + c_d^{(1)} D_d(t), \tag{28}$$

where $c_d^{(A)} = c_d(T_A)$ is independent of e_1 and

$$\begin{aligned} c_d^{(1)} &= \left(\frac{2(n-1)}{n}\frac{T_A T_1}{T_c^2} + \frac{T_1^2}{T_c^2}\right) D_d(t) \\ &\approx \left(\frac{2(n-1)}{n}\frac{T_A T_1}{T_c^2}\right) D_d(t) \quad \text{for} \quad T_1 \ll T_A \end{aligned} \tag{29}$$

represents the climate damage costs $c_d^{(1)}$ incurred by actor 1. These may be compared with the climate damage costs

$$c_d^{(n)} = \left(\frac{nT_1}{T_c}\right)^2 D_d(t) \tag{30}$$

incurred if all actors adopted the same emissions path as actor 1 in a cooperative scenario. The ratio

$$\frac{c_d^{(1)}}{c_d^{(n)}} \approx \frac{2(n-1)}{n^2}\frac{T_A}{nT_1} \tag{31}$$

contains a first factor $2(n-1)/n^2 \approx 2/n$, which becomes small for large n, as anticipated, and a second amplification factor $T_A/nT_1 = T_A/T_0$, where $T_0 = nT_1$ is the temperature change for the equivalent cooperative emissions path $e = ne_1$. For the baseline cooperative optimal emissions path, a typical value of T_A/T_0, averaged over the first part of the emissions path, is of order 5. Thus the cross-over point at which the additional damage costs incurred by a single mitigator become comparable to the total damage costs of the cooperative solution is near $n = 10$, as indeed found.

For policy makers this simple result has important implications. Although actors who regard climate change as a potential hazard which should be mitigated through appropriate abatement measures will naturally strive to achieve binding international agreements on joint actions, the frequently heard argument that there is no incentive for reducing emissions on an individual basis does not appear to be valid. Our model suggests that self-interest should motivate at least the larger industrial countries to undertake unilateral mitigation actions which are comparable to the measures they wish to realize ultimately in an international agreement. However, this conclusion is strongly dependent on the assumed nonlinear nature of the climate damages and could be modified if trading interactions are taken into account (e.g. problems of competition and leakage), as discussed in the following section.

4 Multi-actor models with trade

Although the non-trading multi-actor model discussed in the previous section is useful for investigating certain aspects of the multi-actor problem, it clearly represents an unrealistic simplification of the true situation. The coupling between different actors will in general be governed not only by changes in global climate, but also by trade. Moreover, in attempting to maximize his individual welfare expression, each actor will normally not ignore the actions of other actors. We consider again only the non-cooperative case, so that direct negotiations are excluded. Each actor will nevertheless endeavour to anticipate the response of other players to his own actions, and will develop a foresighted strategy accordingly.

To model the general case, we assume that there exists a general multi-actor economic model in which the actions of each player i are described by a set of control variables $v_{i\alpha}(t), \alpha = 1, 2, \ldots$ including, in addition to the CO_2 emissions, various other economic control factors such as the prices of commodities, tariffs or, in place of CO_2 emissions, a CO_2 tax or some other regulatory instruments. In the following, it will be convenient to discretize the time variable t and introduce the control vector $\mathbf{v}_i = (v_{i,\alpha,t})$ representing the set of paths of all control variables of actor i.

The individual welfare functions W_i which each actor i strives to maximize will depend generally not only on the individual control parameters $\mathbf{v}_i$ of the relevant actor, but on the complete set of control variables $\mathbf{V}$ of all actors. This interdependence arises both directly through trade and indirectly through the dependence of W_i on the change in global climate $\mathbf{x} = \mathbf{x}(\mathbf{V})$ produced jointly by all actors: $W_i = W_i(\mathbf{V}, \mathbf{x}(\mathbf{V}))$. For the following, the distinction between coupling through trade and climate change will be irrelevant, however, and we shall therefore write simply $W_i = W_i(\mathbf{V}) = W_i(\mathbf{v}_i, \bar{\mathbf{v}}_i)$, where $\bar{\mathbf{v}}_i = (\mathbf{v}_1, \mathbf{v}_2, \cdots, \mathbf{v}_{i-1}, \mathbf{v}_{i+1}, \cdots, \mathbf{v}_n)$ is the complementary set of control variables to $\mathbf{v}_i$, i.e. the set of all control variables $\mathbf{v}_j$ with $j \neq i$.

Assume now that for given values of the control variables $\bar{\mathbf{v}}_i$ of the other actors, actor i will adjust his control variables $\mathbf{v}_i$ (by some strategy, which we discuss below) such that his welfare W_i is maximized. The resulting optimal control vector $\mathbf{v}_i^{opt}$ will then be a function

$$\mathbf{v}_i^{opt} = \mathbf{f}_i(\bar{\mathbf{v}}_i) \tag{32}$$

of the complementary set of control variables. It follows that a change $\delta\mathbf{v}_j$ in the control variable of actor j, $j \neq i$ implies a change

$$\delta\mathbf{v}_i^{opt} = \frac{\partial \mathbf{f}_i(\bar{\mathbf{v}}_i)}{\partial \mathbf{v}_j}\delta\mathbf{v}_j \tag{33}$$

in the optimized control variable of actor i, or

$$\delta\mathbf{v}_i^{opt} = \mathbf{M}_{ij}\delta\mathbf{v}_j, \tag{34}$$

where the *response matrix*

$$\mathbf{M}_{ij} = \mathbf{M}_{ij}(\bar{\mathbf{v}}_i) = \frac{\partial \mathbf{f}_i(\bar{\mathbf{v}}_i)}{\partial \mathbf{v}_j} \tag{35}$$

characterizes the marginal response of the optimal control variables of actor i to a marginal change of the control variables of actor j.

We note that the matrix $\mathbf{M}_{ij} = M_{i,\alpha,s;j,\beta,t}$ depends not only on the actor indices i, j but also on the control-variable indices $(\alpha, s); (\beta, t)$, which are suppressed in the matrix notation. Thus matrix products in eq.(34) and below, representing contractions over the compressed control variable indices α, s, correspond in continuous notation to contractions with respect to α in combination with time integrals over the control variable path.

Depending on how one interprets the response strategy of the actors, one can distinguish between three forms of Nash equilibrium as outcome of the multi-actor optimization problem.

4.1 The Cournot-Nash equilibrium

In text-book treatments of coupled economic optimization problems (e.g. *Binmore*, 1992 [1]), the interactions between players are often regarded as single-shot games. Each player i specifies his (or her) control variables $\mathbf{v}_i$ once and for all. His choice is based on some assumption about the control variables of the other players, but the possible 'response' of the other players to his own choice is irrelevant, since the other players have no opportunity to respond, once they have chosen, independently, their control variables $\bar{\mathbf{v}}_i$.

Thus the Nash equilibrium (also termed the Cournot-Nash equilibrium in this context) is given by the solution of the simultaneous set of equations

$$\frac{\partial W_i}{\partial \mathbf{v}_i} = 0 \tag{36}$$

for all actors i.

This approach corresponds to the examples discussed in the previous section and other recent analyses of the multi-actor problem (*Tahvonen*, 1993 [28], *Nordhaus* and *Yang*, 1995 [21]). However, as pointed out, this model is in general not very realistic. For example, in the two-player fossil-fuel supplier-consumer model considered below, it will be found that there exists no Cournot-Nash equilibrium. In a time dependent interactive dynamic control problem, the actors clearly do have an opportunity to react to the choice of control variables of the other actors, and will adjust their control variables accordingly. In chosing their control variables, they will furthermore anticipate the response of the other actors, and will optimize their strategy taking this response into account.

4.2 The self-consistent interactive Nash equilibrium

If the response of the other players is included in the optimization analysis, the simultaneous solutions of eqs.(36) no longer represent simultaneous local extrema (or turning points) of W_i with respect to the control variables $\mathbf{v}_i$. Although a variation $\delta \mathbf{v}_i$ induces no direct variation of W_i, since $(\partial W_i / \partial \mathbf{v}_i) \delta \mathbf{v}_i = 0$, it generates a variation $\delta \bar{\mathbf{v}}_i$ in the other control variables through the coupling relation of eqs.(34). This produces an indirect variation in W_i through terms $(\partial W_i / \partial \mathbf{v}_j) \delta \mathbf{v}_j$. The necessary condition for a maximum of W_i, considering the response of the other actors, is therefore the stationarity condition

$$\frac{dW_i}{d\mathbf{v}_i} = \frac{\partial W_i}{\partial \mathbf{v}_i} + \sum_{j \neq i} \frac{\partial W_i}{\partial \mathbf{v}_j} \mathbf{M}_{ji} = 0. \tag{37}$$

Thus the Nash equilibrium in the interactive case is given by the solution of the set of eqs.(37), which must be solved simultaneously with the defining eqs.(35) for the response matrix. The two sets of equations are coupled and can normally be solved only iteratively: to determine the solutions of eqs.(37), one needs the response matrix, but the response matrix can be evaluated from eqs.(35) only when the solutions of eqs.(37) have been found.

4.3 The conjectured response Nash equilibrium

If the actors have no reliable information on the welfare functions of the other actors – or are not able to carry out the fully coupled optimization analysis for all actors simultaneously – they will be unable to determine the true response of the other actors to their own actions. To optimize their welfare, they will therefore be forced to make some assumption regarding the reactions of the other actors. If each actor i has some conjecture about the form of the components $\mathbf{M}_{ji}$, $j \neq i$ of the response matrix relevant for his own optimization strategy, the resulting Nash equilibrium for all actors can be determined by solving the set of eqs.(37), without regard to the consistency relation of eqs.(35).

If all actors succeed in modifying their control variables simultanously such that the gradients $dW_i/d\mathbf{v}_i$ of their welfare functions vanish simultaneously (and the zero gradient points do indeed correspond to local maxima), a mutual (pseudo-) maximum welfare solution has been found. However, although each actor believes that he has maximized his welfare given the strategy of the other actors, the resultant set of solutions $\mathbf{f}_i(\bar{\mathbf{v}}_i)$ does not in fact represent a true joint optimum: the conjectured response matrix will in general not be consistent with the true response matrix resulting from the optimization strategy, as defined by eqs.(35). Nevertheless, the solutions represent a real Nash equilibrium, given the imperfect information available to the actors.

We note here a basic conceptual difference between interactive Nash equilibrium solutions, of either type, and the non-interactive Cournot-Nash equilibrium. An interactive equilibrium – if it exists – is automatically realized by the actors, who systematically adjust their control variables until all actors have jointly achieved an optimal solution (or possibly one of a set of locally optimal solutions). In contrast, in the one-shot Cournot-Nash equilibrium, the solution represents only a possible outcome. Each actor is unaware of the control variables of the other actor, and the Nash equilibrium is realized only if all actors are versed in game theory and assume that the Cournot-Nash equilibrium is indeed the strategy which everyone will adopt.

4.4 Numerical determination of interactive Nash equilibrium solutions

Both the self-consistent and conjectured response interactive Nash equilibrium solutions may be constructed numerically by iteration methods, the self-consistent solution requiring an additional external iteration loop to satisfy the self-consistency condition. The conjectured response Nash equilibrium may be derived by a multi-component method of steepest ascent, applied simultaneously to all welfare functions. Alternatively, the solution can be determined by maximizing the welfare for each actor in turn, keeping the control variables of the other actors fixed (but following the gradient direction defined by the total rather than the partial derivative), and then iterating the procedure.

The self-consistent interactive Nash equilibrium may then be constructed from the conjectured response solution by requiring further that the derivatives $\partial\mathbf{f}_i/\partial\mathbf{v}_j$ inferred from the resulting solutions $\mathbf{v}_i = \mathbf{f}_i(\bar{\mathbf{v}}_i)$ are consistent with the conjectured response matrices, in accordance with the definition of eqs.(35). This can be achieved by successively replacing the conjectured response matrices by the response matrices computed from the derivatives of the conjectured response solutions. If the iterative adjustment procedure converges, one arrives at the internally consistent interactive Nash equilibrium.

Intuitively, one may expect such straightforward iterative constructions to converge for both the conjectured response and the internally consistent Nash equilibrium solutions, provided the solutions exist. However, general conditions for the existence of Nash equilibrium solutions in these cases, and the convergence properties of appropriate numerical algorithms for their construction have not, to our knowledge, been studied.

Numerically, the derivation of internally consistent response matrices for the dynamical interactive optimization problem is not a simple exercise. The computation of the response matrix $\mathbf{M}_{ij}$ involves the determination of the second-derivative matrix $\partial^2 W_i/\partial\mathbf{v}_j\partial\mathbf{v}_i$ obtained by differentiating the partial gradient $\partial W_i/\partial\mathbf{v}_i$ of the welfare function W_i with respect to the set of control vectors $\mathbf{v}_j$, $j \neq i$, to be carried out in the high-dimensional space resulting from the discretization of the time axis. The computation must then be repeated many times to obtain the internally consistent response matrices through the iterative adjustment of the conjectured response matrices to the response matrices computed from the resultant solutions.

In practice, however, the determination of the internally consistent interactive Nash solution is probably an academic exercise. The construction of the solution – if it indeed exists – will be as elusive for the actors in a real situation as for the theoretical analyst. To define an iterative joint optimization algorithm (which may be regarded as a mathematical proxy of the real joint adjustment procedure, which also may or may not converge), it is sufficient that each actor has a model of the response of the other actors to his own strategy, regardless of whether or not the model corresponds to the actual response computed a posteriori for the resultant jointly optimized solution. The real situation is probably best studied by numerical experiments, for example, by considering the sensitivity of the mutual optimization exercise with respect to the conjectured marginal response matrices.

5 Two-actor interactions between fossil-fuel suppliers and consumers

A realistic multi-actor game theoretical model of global climate-protection strategies must consider not only different climate damage and abatement cost functions for different fossil fuel consumers, but also the different roles of fossil fuel suppliers and consumers. Strategies of fuel consumers for reducing climate damage costs by reducing emissions can well be counteracted by fuel suppliers, who may respond by reducing fuel prices to enhance fuel consumption (cf. *Ströbele*, 1994 [2], *Richels* et al., 1996 [27]). As a simple but illustrative example of an interactive multi-player model we consider a two-player system consisting of a single world fossil fuel producer and a single world fossil fuel consumer. Thus we assume that all economic actors generating greenhouse gas emissions agree on a joint optimal emission-path strategy and act as a single player, while all fossil-fuel producing actors form a cartel (e.g. OPEC), also acting as a single player.

To describe the interactions between fossil fuel suppliers and consumers we need to introduce as control variable, in addition to the consumption of fossil fuel (expressed as before in terms of the emissions), the fuel price. This did not appear in the previous single-actor considerations of HHGOS, since the activities of fossil fuel production and consumption were aggregated, and in a cooperative scenario the fossil fuel price affects only the distribution of welfare, not the total welfare itself.

The fuel supplier, actor 1, is assumed to have no other function than providing fuel and to be impervious to climate change. His (or her) goal is to maximize his welfare, which he equates with his total earnings

$$W_1 = \sum_t \alpha(t)(p(t) - p_e(t))e(t)D_a(t))\Delta t, \tag{38}$$

where $p(t)$ is the price of a unit of fossil fuel corresponding to a unit of emissions (we make no distinction between different types of fossil fuel), $p_e(t)$ denotes the price of extraction, and the notation is otherwise as in Section 2, except that we have used, as elsewhere in this section, a discretized time variable t, with discretization interval Δt. The profit $p(t) - p_e(t)$ from the sale of a unit of fossil fuel is discounted with – for simplicity – the same discount factor $D_a(t)$ of eq.(18) as used in the abatement cost function in the single-actor case. A normalization factor $\alpha(t)$ of dimension $[(\text{emissions} \times \text{price}]^{-1}$ has been introduced to relate W_1 to the dimensionless climate damage and abatement costs in Section 2. Since W_1 represents normal economic costs, we have committed ourselves at this point to equating 'welfare' with negative costs. However, as pointed out, this represents no basic restriction within the qualitative framework of the present optimization analysis, provided that the 'true welfare' is a monotonic function of cost.

The welfare function of the fossil fuel consumer, actor 2, is given by the difference

$$W_2 = W - W_1, \tag{39}$$

where W is the total welfare function defined for the single-actor case. In the cooperative case, the dependence on price of the total welfare $W = W_1 + W_2$ vanishes, as pointed out: the supplier's gain is the consumer's loss.

In the non-cooperative case, each actor tries independently to maximize his individual welfare expression: the fuel supplier by selecting a price path

$\mathbf{v}_1 = (p(t))$ which maximizes his earnings, the fuel consumer by chosing an emission path $\mathbf{v}_2 = (e(t))$ which minimizes the sum of his abatement, climate damage and fuel costs. It can immediately be seen that there exists no non-interactive Cournot-Nash equilibrium solution for this problem: one can find no pair of control variable paths $p(t), e(t)$ such that the total costs of the fuel consumer are minimized with respect to variations of $e(t)$ for fixed $p(t)$ and, conversely, the earnings of the fuel supplier are maximized with respect to $p(t)$ for fixed $e(t)$: for given $e(t)$, the earnings of the fuel supplier are unbounded, increasing linearly with $p(t)$.

In practice, of course, the fuel supplier will hardly be induced to raise his prices indefinitely, since he knows that the response of the fuel consumer to rising prices is to reduce consumption, resulting ultimately in a point of diminishing returns. His pricing policy will depend on his assessment of the marginal response of the fuel consumer to marginal price changes. Conversely, the fuel consumer, in determining the optimal emission path which minimizes his total costs, will attempt to make some assessment of the impact of marginal changes in his emission path on marginal changes in the price path of the fuel supplier. Thus the appropriate definition of the Nash equilibrium in this case is the interactive Nash equilibrium, based on the conjectured or internally consistent interaction response matrices. We shall consider only the simpler case of the conjectured response Nash equilibrium.

Assume, for example, that the fossil fuel supplier 1 anticipates that the response $\delta e(t)$ of the fuel consumer 2 to a change $\delta p(t)$ in fuel price can be represented by an instantaneous marginal response coefficient γ,

$$\delta p(t) \to \delta e(t) = -\gamma\,\delta p(t). \tag{40}$$

Equation (40) corresponds to a diagonal response matrix $M_{2,s;1,t} = -\gamma\delta_{ts}$ with respect to the suppressed time indices t, s in eq.(35), where δ_{ts} denotes the Kroneker symbol. The marginal response coefficient γ has the dimension [emissions/price] and can be a function of time, $\gamma = \gamma(t)$, but is assumed otherwise to be independent of p and e. (It is more convenient for the present example to assume a state-independent response coefficient γ rather than a constant elasticity coefficient $p\delta e/e\delta p$.)

The maximization of the fuel supplier's earnings, eq.(38), yields then the condition

$$\begin{aligned}\delta W_1 &= \sum_t \alpha\,\{\delta p(t)e(t) + \delta e(t)(p(t) - p_e(t))\}\,D_a(t)\Delta t \\ &= \sum_t \alpha\delta p(t)\,\{e(t) - \gamma(p(t) - p_e(t))\}\,D_a(t)\Delta t = 0,\end{aligned} \tag{41}$$

or

$$e(t) - \gamma(p(t) - p_e(t)) = 0, \tag{42}$$

so that

$$p(t) = p_e(t) + e(t)/\gamma. \tag{43}$$

Thus the fossil-fuel supplier can immediately determine his maximal-earnings price from eq. (43), given the level of emissions (fuel consumption) $e(t)$ of the fuel consumer. His earnings for this optimal price is then given by [cf. eqs.(38, 43)]

$$W_1 = \sum_t \alpha\gamma^{-1} e^2 D_a \Delta t, \tag{44}$$

or, expressed in terms of non-dimensional variables,

$$W_1 = \sum_t \beta r^2 D_a \Delta t, \tag{45}$$

where $r = e/e_A$ and

$$\beta = \alpha\gamma^{-1} e_A^2. \tag{46}$$

The coefficients α and γ occur in eq.(44) and in all following dimensionless relations only in the non-dimensional combination $\alpha\gamma^{-1}e_A^2 = \beta$. In the numerical examples presented below we shall therefore assume that the dimensional parameters γ and α scale with the BAU emissions (the only relevant externally prescribed dimensional variable) such that the non-dimensional parameter β remains constant.

Assume now that the fuel consumer is aware of the fuel supplier's strategy. He therefore knows that the marginal response of the fuel supplier to a marginal change in emissions will be given, according to eq.(43), by

$$\delta e(t) \rightarrow \delta p(t) = \delta e(t)/\gamma. \tag{47}$$

Thus the coefficient characterizing the fuel supplier's marginal price response to a marginal change in consumption (emission) is just the negative inverse of the fuel supplier's conjectured response coefficient of eq.(40) describing the marginal emission response of the consumer to a marginal change in price.

Substituting eq.(47) into eqs.(38, 39), we obtain for the gradient of the cost C_2 for the fuel consumer (negative welfare gradient)

$$g_2(t) = -\frac{\delta W_2}{\delta e} = g(t) - g_1(t) \tag{48}$$

where $g(t) = -\delta W/\delta e$ is the gradient of the total costs (negative total welfare) for the single-actor case of eqs.(21, 22) and

$$g_1(t) = -\frac{\delta W_1}{\delta e} = -\alpha\left(p(t) - p_e(t) + \gamma^{-1}e(t)\right) D_a(t)\Delta t \tag{49}$$

is the additional cost gradient term arising from the price interaction with the fuel supplier. The term proportional to $p(t) - p_e(t)$ in eq.(49) represents the incremental costs transferred to the fuel supplier for an incremental increase in consumption (emission), while the term proportional to $\gamma^{-1}e(t)$ represents the feedback term resulting from an increase (decrease) in fuel price by the fuel supplier in response to an increase (decrease) in consumption.

Applying eq.(43) to eliminate $p(t)$, eq.(49) may be written more simply:

$$g_1(t) = -2\alpha\gamma^{-1}e(t)D_a(t)\Delta t, \tag{50}$$

or, in terms of non-dimensional parameters,

$$g_1(t) = -2\beta e_A^{-1} r D_a \Delta t. \tag{51}$$

Since both gradient contributions g and g_1 in eq.(48) for the fuel consumer cost gradient g_2 are given as functions of the emissions $e(t)$, the minimal cost $C_2 = min$ can be determined by varying $e(t)$ using a method of steepest descent. Through the analytical determination of the optimal response in eq.(43) of the fuel supplier to the emission strategy of the fuel consumer, the two-actor case has thus been reduced computationally to a single-actor problem.

However, before carrying out the computations we must first modify our definition of the abatement cost function. In the HHGOS global single-actor case, the abatement costs were defined as the additional costs incurred through a deviation from the BAU emissions path $e_A(t)$, where $e_A(t)$ maximized the total welfare W in the absence of climate damage costs at the value W_A. A necessary condition that W_A represents a maximum is that the functional derivative $\delta C_a/\delta e$ with respect to the emissions path $e(t)$ vanishes for the BAU path $e = e_A$. However, in the fossil fuel supplier-consumer problem, the emissions are controlled by only one of two actors, the fossil fuel consumer, who will chose the BAU emission path e_A to maximize his own welfare $W_2 = W - W_1$, rather than the total welfare W. If we retain the same expression of eq.(17) for the abatement costs as in the single-actor case, W_2 will not be maximized for $e = e_A$, since for this path $\delta W_2/\delta e = -\delta W_1/\delta e = -2\beta e_A^{-1} D_a \Delta t \neq 0$ [cf. eq.(45)].

Through the presence of the additional fuel cost term W_1 in the welfare expression of the fuel consumer, the optimal emissions path is depressed relative to the single-actor BAU path $e_A(t)$ in the unregulated emissions case. If we wish to recover the original single-actor BAU path also for the two-actor case, this depression factor must be appropriately neutralized. This can be achieved, for example, by rescaling the ratio $r = e/e_A$, $r \to \tilde{r} = \lambda r$, in the abatement costs expression of eq.(17). The rescaling factor $\lambda\,(< 1)$ can be chosen such that one recovers $r = 1$ as the optimal solution for the BAU case.

Introducing the scaling factor λ into the climate damage costs, the net costs for actor 2 in the unregulated case, i.e. without climate damage costs, are given by [cf. eqs.(17, 45)]

$$C_2 = C_a(\lambda r) + W_1 = \sum_t \left\{ \left(\frac{1}{\lambda r} - \lambda r \right)^2 + \tau_1^2 (\lambda \dot{r})^2 + \tau_2^4 (\lambda \ddot{r})^2 + \beta r^2 \right\} D_a \Delta t. \tag{52}$$

The necessary condition for the minimization of C_2 is accordingly (noting that the first and second time derivative terms of r do not contribute to the gradient if λ is adjusted such that $r = 1$, i.e. $\dot{r} = \ddot{r} = 0$, for the optimal solution)

$$\frac{\delta C_2}{\delta r} = 2\lambda \left(\lambda r - \frac{1}{\lambda r} \right) \left(1 + \left(\frac{1}{\lambda r} \right)^2 \right) + 2\beta r = 0, \tag{53}$$

or

$$(\lambda r)^4 - 1 + \lambda^2 r^4 \beta = 0, \tag{54}$$

which yields

$$r = \left(\lambda^4 + \beta \lambda^2 \right)^{-1/4}. \tag{55}$$

Thus to recover the BAU solution $r = 1$ we must set

$$\lambda^4 + \beta \lambda^2 = 1, \tag{56}$$

or (taking the relevant positive real root)

$$\lambda = \left\{ -\frac{\beta}{2} + \left(\frac{\beta^2}{4} + 1 \right)^{1/2} \right\}^{1/2}. \tag{57}$$

With this modification, the welfare of the fossil fuel consumer is given by

$$W_2 = W_A - \tilde{C}_a - C_d - W_1, \tag{58}$$

where W_A is the total welfare in the BAU case, $\tilde{C}_a(r) = C_a(\tilde{r})$ [eqs.(17, 57)], C_d has the same form of eq.(19) as in the single-actor case, and the fuel costs W_1 are given by eq.(45). Equation (58) may be written in the form

$$W_2 = W_{2A} - C_{2a} - \Delta W_1 - C_d, \tag{59}$$

where

$$W_{2A} = W_A - \tilde{C}_a(1) - W_1(1) \tag{60}$$

denotes the welfare value of actor 2 in the BAU case, $r = 1$, without consideration of the damage costs,

$$C_{2a} = \tilde{C}_a(r) - \tilde{C}_a(1) \tag{61}$$

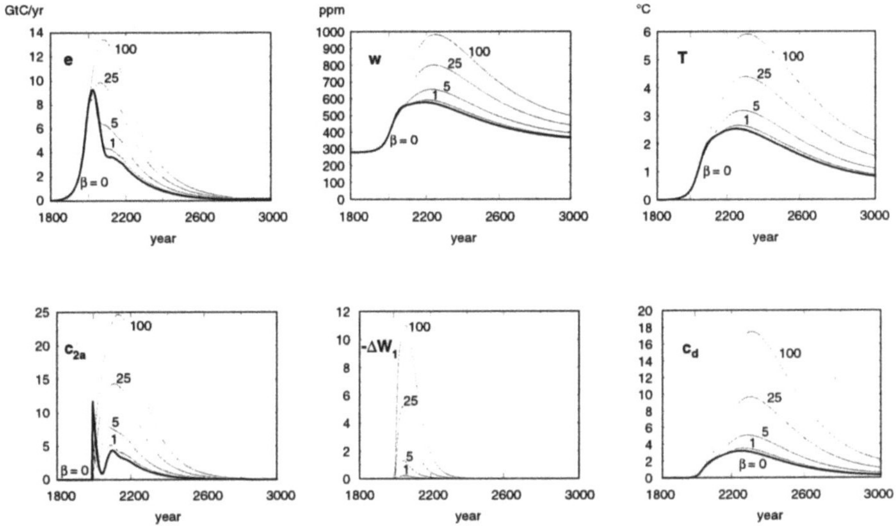

Figure 7. Impact of the price-consumption interaction parameter β on the simultaneous optimal-emissions, optimal-price solution for the non-cooperative two-actor consumer-supplier problem (computed for the parameters of the single-actor baseline scenario $S0$ shown in Fig. 4 and reproduced here as *thick curves* for comparison). Layout analagous to Figs. 5, 6.

represents the direct abatement costs for actor 2 arising from a deviation from the BAU path, and

$$\Delta W_1 = W_1(r) - W_1(1) \tag{62}$$

denotes the change in fuel costs arising from the change in emissions relative to the BAU path. All cost terms are defined as functions of the emissions path $e(t)$, so that the welfare W_2 can be maximized with respect to $e(t)$.

Fig. 7 shows the computed optimal solutions for the two-actor case for different constant values of the non-dimensional price-consumption coupling parameter β. The smaller the marginal response coefficient γ, i.e. the larger the parameter β, the larger the impact on the optimal emissions path. Small values of γ imply that fossil fuel suppliers respond to reductions in fossil fuel use with large decreases in the fossil fuel price. This stimulates consumption, thereby counteracting the abatement measures.

The impact of the fuel costs on the computed optimal emissions paths is seen to increase with β, but at a less than linear rate. This is an artifact of our side condition that in the absence of climate damage costs, the BAU optimal emissions curve $e_A(t)$ should always be the same, independent of the interaction parameter β. Our adjustment of the abatement cost expression to satisfy this condition [eqs.(52-62)] has the effect that, as β is increased, the abatement costs grow more rapidly with decreasing emission reduction factor $r = e/e_a$. Thus the economic system becomes 'stiffer', and the costs of counteracting the negative effects of climate change by reducing emissions are increased. This cross-coupling of two opposing effects would presumably be avoided in a more realistic economic model, including a specific description not only of the costs incurred through a deviation from the BAU path, as in the present analysis, but also of the basic economics determining the reference economic BAU path itself.

Specific implications for policy can clearly be derived from interactive optimization analyses of the type presented here only if they are based on more realistic economic models than we have used. However, we anticipate that the general qualitative conclusions of the present study, as well as the general methodological approach, will carry over to more detailed quantitative models.

6 Conclusions

The implementation of a global climate protection strategy with optimized emissions of CO_2 and other greenhouse gases in the real world of many interacting, interdependent decision makers with diverse interests and different assessments of climate change impact is a complex multi-actor problem. The optimization problem has been investigated with quantitative models so far only in two limiting cases, both of which are rather far from reality: full cooperation, assuming an agreement has been reached on joint mitigation goals, for which the problem reduces to the single-actor case; and the fully non-cooperative n-actor problem, which ignores all negotiatory aspects, including the various options of forming partial alliances. Despite these limitations, the two limiting cases are useful in identifying basic features of the problem and defining a space of possibilities which may span some of the key conclusions of more realistic multi-actor models.

The principal results of single-actor optimization computations, using the same basic model as in this study, have been summarized in HHGOS and need not be repeated here. The main conclusions of the present n-actor investigation is that, while non-cooperative strategies generally yield reduced abatement measures and a stronger global warming, the impact of non-cooperative optimization on the individual emissions paths is smaller than may have been anticipated intuitively. In particular, there is no reason, based on our model simulations, to postpone mitigation measures until an international climate protection agreement has been achieved; individual emissions abatement, although in a modified form, is cost effective also in the non-cooperative case.

Thus in the case on n identical non-trading actors sharing the same assessment of climate change damages and mitigation costs, non-cooperation has little effect for small n (less than about 10), the limiting solution of zero abatement being approached only for large n (greater than about 100).

A similar picture emerges in the case of a single mitigator faced with (n-1) actors who ignore climate impacts and pursue the business-as-usual emissions path. Although the single mitigator is unable to control the large climate damage costs incurred by the BAU actors, he can nevertheless still reduce his own climate damage contribution. If the climate damages depend quadratically on the temperature change, these are amplified by the climate damages incurred by the BAU actors. For $n < 10$, it is cost effective for the single mitigator to enhance his abatement efforts beyond the normal cooperative level. However, as n increases, his impact on the climate damage costs becomes smaller, and he reduces his mitigation measures accordingly. However, the fall-off is relatively mild, the limiting zero-abatement solution for $n \to \infty$ being approached at a similarly slow rate as in the case of n identical actors pursuing the same non-cooperative strategy.

In the general multi-actor case of different players interacting through trade, the non-cooperative Nash equilibrium solution depends on the marginal interaction matrix characterizing the trade coupling between different actors. The Nash equilibrium can be formulated either as a conjectured response problem, if the actions of other players in response to the actions of an individual player are simply assumed by the individual player, or as an internally consistent solution, if the additional side condition is introduced that the assumed response is consistent with the computed true response inferred from the equilibrium solution.

In analyzing the impact of trade interactions, it is important to consider not only different fossil fuel users with different climate damage and mitigation cost assessments, but also fossil fuel suppliers. A simple two-actor supplier-user model illustrates that for suitable consumption-price marginal response coefficients, reductions of the fossil fuel price by the suppliers can effectively counteract efforts of fossil fuel users to mitigate climate change by curtailing consumption.

The purpose of the present exploratory paper was primarily to clarify some of the concepts and problems posed by multi-actor optimization problems, rather than to offer concrete solutions for particular practical situations. In addition to improvements of the basic climate-socio-economic model outlined in HHGOS, the present analysis clearly needs to be extended to include negotiations between actors in order to bridge the gap between purely cooperative and purely non-cooperative optimization strategies.

6.1 Acknowledgements

We are grateful to Ralf Giering for providing his adjoint model and functional derivative compiler, and his assistance in its application, which greatly simplified all computations. The constructive discussions with *v. Storch*v. Storch were also highly appreciated. The optimization analysis was stimulated in part by a contribution in collaboration with Hans-Joachim Schellnhuber for a report of the Wissenschaftlicher Beirat Globale Umweltveränderungen (German Advisory Council on Global Change).

References

1. K. Binmore. *Fun and Games. A Text on Game Theory.* D. C. Heath and Co., Lexington, 1992.
2. J. E. Blank and W. J. Ströbele. The economics of the CO_2 problem. What about the supply side? Rep., Univ. Oldenburg, 1994.
3. W. R. Cline, editor. *The Economics of Global Warming.* Institute for International Economics, Washington DC, 1992.
4. U. Cubasch, K. Hasselmann, H. Höck, E. Maier-Reimer, U. Mikolajewicz, B. D. Santer, and R. Sausen. Time-dependent greenhouse warming computations with a coupled ocean-atmosphere model. *Climate Dynamics*, 8:55–69, 1992.
5. S. Fankhauser. *Valuing Climate Change.* Earthscan, London, 1995.
6. R. Giering and T. Kaminski. Recipes for adjoint code construction. *ACM Transactions on Mathematical Software*, 1996. submitted.
7. K. Hasselmann. How well can we predict the climate crisis? In H. Siebert, editor, *Symposien- und Konferenzbände des Instituts für Weltwirtschaft an der Universität Kiel*, pages 165–183, Mohr, Tübingen, 1991. Conf. on Environmental Scarcity: The International Dimension.
8. K. Hasselmann, S. Hasselmann, R. Giering, V. Ocana, and H. von Storch. Sensitivity study of optimal CO_2 emission paths using a simplified structural integrated assessment model (SIAM). *Climate Change*, 37(2):345–386, 1997.
9. K. Hasselmann, R. Sausen, E. Maier-Reimer, and R. Voss. On the cold start problem in transient simulations with coupled atmosphere-ocean models. *Climate Dynamics*, 9:53–61, 1993.
10. IPCC. *Climate Change: The IPCC Scientific Assessment.* Cambridge University Press, Cambridge, 1990. J. T. Houghton and G. J. Jenkins and J. J. Ephraums, editors.
11. IPCC. *Climate Change 1992. The Supplementary Report to the IPCC Assessment.* Cambridge University Press, Cambridge, 1992. J. T. Houghton, B. A. Callander and S. K. Varney, editors.
12. IPCC. *Climate Change 1994. Radiative Forcing of Climate Change and an Evaluation of the IPCC IS92 Emission Scenarios.* Cambridge University Press, Cambridge, 1994. J. T. Houghton et al., editors.
13. IPCC. *Scientific Assessment of Climate Change.* Cambridge University Press, Cambridge, 1995. B. Bolin, J. T. Houghton and L. G. Meir, editors.
14. F. Joos, M. Bruno, R. Fink, U. Siegenthaler, T. F. Stocker, C. Le Quéré, and J. L. Sarmiento. An efficient and accurate representation of complex oceanic and biospheric models of anthropogenic carbon uptake. *Tellus*, 48B(3):397, 1995.
15. E. Maier-Reimer. The biological pump in the greenhouse. *Global and Planetary Climate Change*, 8:13–15, 1993.
16. E. Maier-Reimer and K. Hasselmann. Transport and storage of CO_2 in the ocean - an inorganic ocean-circulation carbon cycle model. *Climate Dynamics*, 2:63–90, 1987.
17. A. S. Manne and R. G. Richels. Global CO_2 emission reductions: The impacts of rising energy costs. *The Energy Journal*, 12:88–107, 1991.
18. A. S. Manne and R. G. Richels. The greenhouse debate: Economic efficiency, burden sharing and hedging strategies. *The Energy Journal*, 16:4, 1995.
19. P. Michaelis. The cost of stabilizing the atmospheric concentration of greenhouse gases. In *EU workshop: Designing Economic Policy for Management of Natural Resources and the Environment.* Rethymon, 1994.
20. W. D. Nordhaus. To slow or not to slow: The economics of the greenhouse effect. *The Economic Journal*, 101:920–937, July 1991.

21. W. D. Nordhaus and Z. Yang. A regional dynamic general equilibrium model of optimal climate-change policy. Draft report, Yale University, 1996.
22. W. D. Nordhaus and G. W. Yohe. Future carbon dioxide emissions from fossil fuels. In Carbon Dioxide Assessment Committee National Research Council, editor, *Changing Climate: Report of the Carbon Dioxide Assessment Committee*, pages 87–153. National Academy Press, Washington, 1983.
23. S. C. Peck and C. J. Teisberg. CETA: a model for carbon emissions trajectory assessment. *Energy Journal*, 13:55–77, 1992.
24. W. H. Press, B. P. Flannery, S. A. Teukolsky, and W. T. Vetterling. *Numerical Recipes.* Cambridge University Press, Cambridge, 1986.
25. J. M. Reilly, J. A. Edmonds, R. H. Gardner, and A. L. Brenkert. Uncertainty analysis of the IEA/ORAU CO_2 emissions model. *The Energy Journal*, 8:1–29, 1987.
26. R. G. Richels and J. Edmonds. The economics of stabilizing CO_2 emissions. *Energy Policy*, 23:4/5, 1995.
27. R. G. Richels, J. Edmonds, H. Gruenspecht, and T. Wigley. The Berlin Mandate: The design of cost-effective mitigation strategies. Draft report, Stanford University, Energy Modeling Forum-14, 1996.
28. O. Tahvonen. Carbon dioxide abatement as a differential game. Paper in Economics and Buisiness Studies No. 4, University of Oalu, Oalu, 1993.
29. O. Tahvonen, H von Storch, and J. von Storch. Economic efficiency of CO_2 reduction programs. *Clim. Res.*, 4:127–14, 1994.
30. O. Tahvonen, H. von Storch, and J. von Storch. Atmospheric CO_2 accumulation and problems in dynamically efficient emission abatement. In G. Boero and Z. A. Silbertston, editors, *Environmental Economics.* St. Martin's Press, New York, 1995.

Commentary: Short Remarks on the Problem of Integrated Modelling

Arkadii Maltsev

1 Introductory remarks

First, I would like to thank the organizers for inviting me to this highly interesting and important meeting. Since the summer 1993, IIASA and PIK have maintained close contact trying to coordinate joint efforts, particularly in the field of global modelling.

Please let me now briefly comment on Professor Hasselmann's presentation, referring to the presentations of Professor Golitsyn and Professor Schellnhuber as well.

2 Large vs. simplified climate models

Everyone knows what fascinating work is done in Hamburg on their gigantic model. Quite often, the progress on Climate Change Impact Assessment is connected with the progress of work on that model, and this is well justified. However, my view is that it cannot be considered as a signal to stop developing simplified climate models like the one which was initially developed at Professor Golitsyn's Institute of Atmospheric Physics and then finalized and refined at IIASA. Such models seem to be useful tools for the further development of global modelling. And, it is not only the question of differences between climate models and social-economic models Professor Hasselmann was speaking about. He also pointed out the fact that for huge multidimensional models it has become a non-trivial time-consuming operation to trace the influence of a given parameter on a given trajectory.

3 The human ecological niche

My second short comment is just based on my wish to underline the importance of the human health-environment dimension for any serious work on global modelling. What does the "ecological niche" mean for the *Homo sapiens*? If we have no answer to this question, all global models, no matter how sophisticated they might be, will be useless.

4 The example of the Caspian Sea

My third and final comment is on the closing remark of Professor Schellnhuber's presentation, i.e., that to shape the future is often simpler than to predict it. I consider this remark as a warning, and Professor Golitsyn's recent comment gives a very good example. He has just mentioned that the Caspian Sea level rise between 1977 and 1993 was more than 2 m. However, between 1930 and 1977, the Caspian Sea level fell by 2.75 m. This fact was attributed to the excessive water withdrawal from the Volga, the Ural and Tereck rivers, which are the main tributaries of the Caspian Sea.

Attempting to save the lake, the Government of the former USSR approved an ambitious plan to divert water from catchment areas of some of the northern rivers to the Volga river. The logic behind this decision was based on straightforward engineering thinking: let's take the water from the place where it is abundant (the north slope of the continental platform) to the place where there is a shortage (the south slope).

The possible consequences of the accepted plan were analyzed unofficially by a group of scientists (mathematicians as well as other specialists). It was clearly proved that the application of the plan would result in a regional environmental disaster. Based on the findings of this group of experts, the project was abandoned. Shortly thereafter, and contrary to the expectations of the project's authors, the Caspian Sea level began to rise.

Commentary: Global Environment and Society (GES) Models, a Problem of Multiple Control Theory

Otto Fränzle

1 Introduction

Professor Hasselmann has demonstrated the difficulties of modelling complex systems such as climate systems in evolution and socio-economic systems. The problems involved may be considered from the viewpoint of scaling, where one is faced with the fact that – dependent on scale – different variables have different relative importance.

Thus it appears necessary to make some complementary remarks. Localized and regionalized instabilities of variable extent and duration, e.g. the patch-dynamics of forest ecosystem complexes, appear as essential elements of larger-scale system stability. This means that natural systems are (normally) affected by non-linear changes at uncertain intervals and at different scales. Basically and in retrospective analysis the importance of such events can be assessed in terms of recurrence intervals, as for excessive rainfalls, hurricanes or high floods.

2 Calibration and validation of models

Comparative analysis of climatic events has not furnished evidence for a global increase in frequency and intensity, but there is much evidence for regional displacements and especially for more devastating effects. They are likely to induce a sharp increase in material damage in the First World, and a concomitant increase in the number of victims in the Third World due to overpopulation (*Richter*, 1993 [5]). In the light of these facts the usefulness of global climate models for prognostic purposes may be more precisely assessed or questioned. GCMs have predicted a considerable temperature increase for the high latitudes and a small one for the tropics (cf. *Cubasch*, 1992 [1] and *Lohmann* et al., 1992 [4]), but hitherto the effect could not be corroborated by measurements (see e.g. *Flohn* et al., 1992 [2], *Kahl* et al., 1993 [3]). In other words, an actualistic calibration and validation of these models has not been possible on the basis of available data. Thus it may appear worthwhile paying more attention to simpler models of the scenario type for regionalized prognostic purposes.

3 Scaling problems in ecological modelling

The appropriate consideration of scaling problems is also constitutive of ecological modelling. In view of the overwhelming complexity of major ecosystems, i.e. landscapes, German ecosystem research has attributed highest priority to a geostatistically founded selection of spatially representative ecosystems for comparative analyses, in order to obtain data which faithfully represent site qualities and related functional interrelationships of larger areas. This applies in particular to ecotones, i.e. the interfaces between adjacent ecosystems which are characterized by increased gradients and consequently also enhanced fluxes of energy and matter and thus they play a particular role in the framework of modern dynamic ecology. Ecology thus operates at several scales, starting with the biodynamics of individual leaves or needles or microbial synusia of even smaller dimensions proceeding in a stepwise methodology to larger-scale entities such as stands and ecosystem complexes. The results of these interrelated bottom-up and top-down analyses and subsequent analogous modelling approaches permit the more precise definition in the realm of validity of the formalized descriptions of systems behaviour at each scale considered. This is the only way to discover emergent properties as related to the hierarchic structure of living systems.

4 Modelling of social systems

Human systems – or to phrase it more correctly – systems with man as predominant structural factor or key species are clearly the most complicated to model. Two basically different albeit convergent approaches may be distinguished:

- the first views man as a "zoon politicon" whose actions and reactions may be defined by means of empirical sociological, psychological, philosophical and theological analyses of social groups acting under more or less well defined economic and political boundary conditions.
- in the second focus is on the so-called decision makers, who are then considered as paradigms of groups acting and interacting in the sense of central elements of behavioural models.

This implies that the calibration and validation problems mentioned in the climate and ecosystem context are much higher still. Already relatively simple models like those presented by Meadows are necessarily characterized by a bewildering number of positive and negative feedback loops whose parameter sensitivity is necessarily extremely high. Chaotic behaviour is consequently (almost) inevitable, and this applies the more if models of this type are coupled with complex other models like comprehensive climate models.

5 Conclusion

The conclusions to be drawn are (i) that models should be considered as optimally precise descriptions of the results of systems analysis, but beyond the scope of comparative elementary and selective interrelationships they hardly claim to have further-reaching prognostic qualities. Yet they constitute most valuable tools for the reproducible construction of scenarios. (ii) Cooperation between PIK and other Federal Research Institutions and Ecosystem Research Centers is recommended for the aim that valid data for modelling purposes can be obtained as a result of integrated systems analysis on various levels of generality. Rationally handled, the different approaches mutually reinforce each other, thus providing reciprocally new and deeper insights.

References

1. U. Cubasch. Das Klima der nächsten 100 Jahre. Szenarienrechnungen mit dem gekoppelten globalen Ozean-Atmosphärenmodell aus Hamburg. *Physikal. Blätter*, 48:85–89, 1992.
2. H. Flohn et al. Water vapour as an amplifier of the greenhouse effect: new aspects. *Meterolog. Zeitschr.*, 1:122–138, 1992.
3. J. D. Kahl et al. Absence of evidence for greenhouse warming over the Arctic Ocean in the past 40 years. *Nature*, 361:335–337, 1993.
4. U. Lohmann et al. The Köppen climate classification as a diagnostic tool for general circulation models. Report, Max-Planck-Institut für Meteorologie, Hamburg, 1992.
5. M. Richter. "Gaia" und "Surprise" - Dimensionen zwischen globalem Klimawandel und klimabedingten Katastrophen. *Petermanns Geograph. Mitteil.*, 137:325–338, 1993.

Chapter 2: Global Change and Sustainable Development: Towards an Integrated Conceptual Model

Jan Rotmans

Abstract A global integrated assessment model TARGETS is being developed dedicated to analysis and assessment of the linkages among social and economic processes, biophysical processes, effects on ecosystems and humans.

The objective of this global scale model is to better understand the complex interrelations between increasing pressure on the environment, the impacts on biogeochemical cycles and the human society. Strategies for a sustainable development can be derived, based on sustainability indicators to be defined or selected, when using this integrated assessment model.

The presentation of information generated by the model is of crucial importance for the communication between science and policy-making and between natural and social sciences as well.

Various scientific and cultural perspectives are incorporated into the model to reflect different understandings of sustainability.

1 Introduction

In the international decision-making community there is a growing need for an operationalisation of the notion of sustainable development. Therefore, the National Institute of Public Health and Environmental Protection (RIVM) in The Netherlands launched the research programme "Global Dynamics and Sustainable Development". This contribution represents a set of ideas and concepts that formed the foundation of this research programme as initiated in 1992. During the past 5 years these concepts have been worked out through the realization of the TARGETS model, however, in a somewhat different way than originally was intended. The model is now operational and extensively described in *Rotmans* and *de Vries* (1997) [34]. The model is also available on CD-ROM as published by Baltzer Science Publishers. So the reader should be prepared that the conceptual description of the TARGETS model and its related uncertainties as presented below may significantly differ from the realized version as described in the TARGETS book.

The main objective of the above mentioned research programme is to operationalise, and to render applicable, the concepts of global change and sustainable development in such a way, that these notions can be used in the decision making process. To this end, a global, integrated assessment model is being developed, the TARGETS model, by means of which an analysis and assessment can be performed, on a global scale, of the quantitative and qualitative linkages among social and economic processes, biophysical processes and effects on ecosystems and humans. Such a global, integrated assessment model can yield insights into the complex interrelations in time and space between increasing worldwide pressure on the environment, pollution and disruption of essential biogeochemical cycles, and the effects of anthropogenic influences on the environment, human health and society as a whole, in both the short and the longer term. Furthermore, an integrated assessment model can be used to develop concrete strategies for sustainable development, a process which requires the formulation of sustainability indicators. The referential framework for such sustainability indicators is given by the elements of the causality chain which the model links together. Therefore, the model should be regarded as an aid to the formulation of possible projections for the future, and not as a means of generating predictions as such.

A great deal of attention will be paid to the presentation of the relevant information generated by the model in an insightful manner, which is of crucial importance in the communication process between science and policy-making. This necessitates the "opening" of models, an exercise which ranges from the construction of a model which can be used interactively, to the creation of strategic planning exercises, whereby models serve to provide guidelines in the background. Because it is becoming increasingly evident that there is no unique definition of sustainable development, and that the interpretation of this concept depends strongly on the perspective of a specific actor, an attempt is to be made to incorporate various scientific and cultural perspectives in the integrated assessment model.

A top-down approach is chosen, whereby the analysis starts at the highest aggregation (abstraction) level, i.e. the global level, considering the globe as a whole. In the next phase the model will be disaggregated to the level of major world regions, using regional data sets.

The research programme will be performed by a multidisciplinary core research team consisting of about fifteen full-time researchers. The main scientific disciplines involved in the research programme are: mathematics, economics, ecology, hydrology, physics, epidemiology and philosophy/sociology. Not only the model-building itself, but also the use of the model are two of the focal points of the research programme. Therefore, it is expected that the programme will be able to add a new dimension to the use of integrated assessment models: namely, the use of such a model as means of communication not only between scientists and policy makers/analysts, but also between exponents of the natural sciences and the social sciences.

2 Global Change

Human activities have significantly affected the structure and the functioning of the Earth system. Our use of land, water, minerals and other natural resources has increased more than ten-fold during the past 200 years, and future increases in population and development will only intensify this pressure.

Large-scale chemical transformations and natural transfers of energy and materials around the world are involved. During the last few decades we have seen increasing acceptance of the insight that the increase in magnitude and complexity of such large-scale problems causes serious harm, varying from a disruption of local ecosystems up to global disturbance of the biosphere as a whole. Here, the totality of changes on planet Earth, including all human interventions and alterations, is considered as constituting global change. The concept of global change is broader than the concept of global environmental change, since the latter only refers to the human-induced biophysical changes to the dynamics of the Earth system, while global change refers to changes of both the biophysical and the human system. One of the major difficulties with respect to global environmental change is to disentangle the natural changes to which the global Earth system is subjected, these being part of a continuous disturbance or variability, from the anthropogenic changes.

In this study global change is, in contrast to earlier studies, considered from an integrated perspective. This means that the classical way of splitting up environmentally related problems into themes, functions and scales (*Langeweg*, 1989 [19]) will be abandoned. This means that the starting point here is to consider the common causes, mechanisms and impacts of a number of coherent themes, functions and scales, and to translate them in terms of Pressure, State, Impact and Response (P-S-I-R approach). The starting point chosen here is more in line with the universal principle of approaching environmentally related problems, which assumes that many of those problems are generic in nature. In this study the integrated (P-S-I-R), universal approach is a *leitmotiv* for analysing global change.

3 Sustainable Development

In order to withstand the worldwide threats to the global system, the notion of sustainable development was introduced during the 1980s. Since then, the concept of sustainable development has mainly been used as a conceptual guide and target for environmental policy making. However, the operationalisation of sustainable development appears to be extremely difficult, and during recent years, several (competing) definitions of sustainable development have been put forward. The World Commission on Environment and Development (WCED, 1987 [50]) played a key role with their widely-cited anthropocentric definition of sustainable development: "a development that fulfils the needs of the present generation without endangering the future needs of future generations". An alternative definition of sustainable development is presented by the IUCN/UNEP/WWF [15] from an ecocentric perspective: "improving the quality of human life while living within the carrying capacity of supporting ecosystems". In view of its normative character, there is no unambiguous definition of sustainable development that can fit with everybody's conception of the notion. It appears that the concept of sustainable development can be understood in many different, subjective ways. The concrete elaboration of the notion seems to depend on the actor's cultural perspective. Therefore, different cultural perspectives will be elaborated, which will be translated into different preferences for the operational definition of sustainable development. Hence, instead of putting forward another definition of sustainable development, both the anthropocentric and ecocentric approach (and other perspectives as well) towards the notion of sustainable development will be elaborated. In this study, sustainable development will be considered on a global scale, linking the notions of global change and sustainable development. Moreover, as opposed to previous studies dealing with this concept, it will not be considered from a monodisciplinary viewpoint, but it will be a multi- and interdisciplinary effort which allows for an integrated vision of what sustainable development is, and how it can be achieved.

4 An integrated systems approach

One of the main experiences in environmental research thus far is that the reductionistic approach based on aspect-compartment oriented research methods has failed in analysing adequately complex, multidisciplinary, large-scale global phenomena. A more promising way seems to be the holistic, integrated approach, based on a systems-oriented analysis, which concentrates on the interactions and feedback mechanisms between the different subsystems of cause-effect chains rather than focusing on each subsystem in isolation. The justification for adopting a systems approach to operationalise the concepts of global change and sustainable development is based on the following line of argument: sustainable development is closely allied to the natural resilience and buffer capacity of the biosphere in relation to anthropogenic disturbances. This disturbance on a global scale is denoted as global change, and can be represented by a set of interrelated cause-effect chains. To operationalise global change and sustainable development the cause-effect chains are chosen as a starting point in this project, which requires an integrated approach. The inextricably interconnected cause-effect chains form an organized whole, a complex system, the properties of which are more than just the sum of its constituent parts, the subsystems. The subject of systems analysis is not only to study the particular system structure and to classify and describe the entities (components) of the system, but also to understand the processes, interactions and feedback mechanisms within the system that generate changes in its dynamics and structure. It enables a synoptic approach that addresses the interdependencies between the cause-effect chains. Given the complexity of the system under consideration, and the relative ignorance about the basic processes and interactions that determine its dynamics, the systems approach can help to foster understanding of the causal relationships that are responsible for changes in the structure and dynamics of the Earth system.

It then provides the opportunity to conceptualize the issues of global change and sustainable development in a holistic, integrated way.

An important feature of the research strategy is the top-down approach which will be chosen, which has implications for several aspects of the research. One major aspect is the degree of generic specificity aimed at, while developing an overall system for the analysis of global change and sustainable development. The subsystems to be built are developed in as generic a form as possible, i.e. irrespective of aggregational, regional or temporal differences. This means that the theories used must be applicable at different levels of spatial aggregation and for different regions in different periods.

Adopting the systems-based approach as a guiding principle, the biosphere can, in global terms, be considered as a system of reservoirs and processes connecting the reservoirs; many of these processes are cyclic, nonlinear or otherwise qualitative in nature. Both are influenced, directly and indirectly, by human interventions. Exogenous and endogenous developments perturbing the system, as well as interventions due to human activities, can lead to irreversible changes in the system. Although aggregation is possible at each conceivable level, a plausible division of the cause-effect chains into subsystems on the global scale would be the following, as represented in Fig. 1:

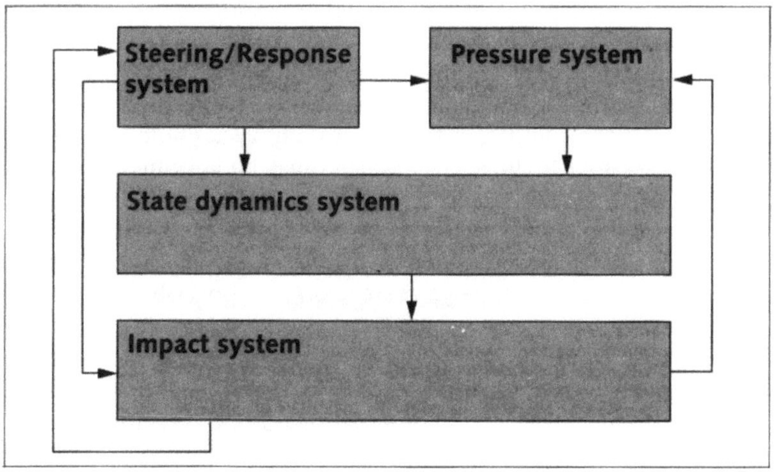

Figure 1. System diagram of the Steering-Pressure-State-Impact chain

Steering subsystems

These systems are designed to describe the user's possibilities to influence (change) either human activities and/or the environment, which also includes the response to societal and ecological impacts.

Pressure subsystems

These systems are designed to describe the driving forces underlying the continuously changing pressure on the environment, characterized by three crucial aspects: population, economy/technology and resources.

State dynamics subsystems

These systems are designed to describe the changes in the physical, chemical and biological state of the biosphere, as well as the change in human population and resources/capital stock. The biogeochemical cycling of elements and the accumulation of chemical agents are major foci of interest.

Impact subsystems

These systems are designed to describe the effects of human interventions on the global environment (global cycles); the effects of anthropogenic multiple stress on ecosystems; the environmentally related consequences for human health; and finally, the socio-economic effects on various sectors.

All these subsystems have their own characteristics within the four-dimensional space built up from spatial and temporal scales. For the economy/technology subsystem the circulation time may be determined by the useful life of investments or the time required to introduce a new technology; for the biosphere subsystem the response time may be a characteristic which is dependent on biogeochemical cycling; for ecosystems the time scale may be determined by the multiple stress on the ecosystems, varying from several weeks to hundreds of years. The characteristics of the subsystems may change as a result of anthropogenically induced or evolutionary processes.

For each subsystem, and also for the system as a whole, the boundaries and the interactions with the surroundings must be specified. Furthermore, for each subsystem and the whole system, the system structure, aggregation level, uncertainties and sensitivities must be investigated. As stated above, dynamic behaviour can be analysed by means of model experiments, to be performed in a systematic and consistent manner, upon which representative indices can be designed.

It is important to note that a full systems analysis is the only analysis which makes sense. This is because a certain development within one or more subsystems can tend towards a sustainable state, but may be unsustainable for the whole system, or even worse, may lead to a shift from a sustainable to an unsustainable state in other subsystems. Using some principles of the system dynamics, the whole system and the various subsystems identified can be analysed in terms of the basic concepts and definitions.

In order to investigate the systems concepts and characteristics defined above, a series of experiments are to be performed, using simulation models. For each subsystem, the perturbation of the natural state is to be determined, and the vulnerability, resilience, buffer capacity and recovery capability are to be defined by model experiments. By performing various model experiments in a systematic manner, the behaviour of the subsystems and of the complex system as a whole can be analysed. Similarly, the carrying capacity of the system can be determined.

5 The TARGETS model: an integrated framework for appraising global change and sustainable Development

5.1 Basic philosophy

Based on the various subsystems of the whole system portraying global change, a series of highly-aggregated modules are being built, interlinked and ultimately integrated. This results in an overall integrated assessment framework, TARGETS : *T*ool to *A*ssess *R*egional and *G*lobal *E*nvironmental and Health *T*argets for *S*ustainability. The TARGETS model serves to explore long-term, to some extent inherently unknowable, dynamics of global change which may shape the Earth system over the next 100 years. It consists of metamodels, which are simplified and condensed representations of coherent subsystems and which are continuously evaluated vis-à-vis expert models.

A two-dimensional integration approach is incorporated in TARGETS in the following way. The TARGETS integrated framework basically consists of a population and health model, a resources/economy model, a biophysics model, a land use model and a water model, which are all interlinked. All types of models comprise a linkage of causative, state-descriptive and impact modules, in this way representing the – vertical – integrated cause-effect chain. By integration of cross-linkages and interactions between various causes, various state-dynamics descriptions and various impacts, the horizontal integration comes into play. This enables the representation of TARGETS as a two-dimensional integration matrix, as shown conceptually in Fig. 2. The time horizon for the TARGETS model will span about two centuries, starting at the beginning of this century, the year 1900, symbolizing the end of the pre-industrial area, until the end of the next century, the year 2100, with time steps varying from one month to one year.

Although there are some serious attempts being made to construct an integrated model of the Earth's atmosphere, hydrosphere and terrestrial biosphere (*Fisher*, 1988 [9]; *Krapivin*, 1993 [17]), it is conceptually and technically not yet (or may never be) possible to link, let alone integrate a variety of complex, detailed and three-dimensional models. Therefore, the TARGETS model inevitably consists of simpler versions of the more complex models, denoted here as metamodels. Basic characteristics of metamodels are that they are reformulated, simpler versions of more elaborate and complex aspect-compartment models (expert models) which have previously been constructed and described in the literature (*Rotmans*, 1990 [32]). A prerequisite for these metamodels is that they should be tested, verified, calibrated and validated thoroughly against expert models and observational data.

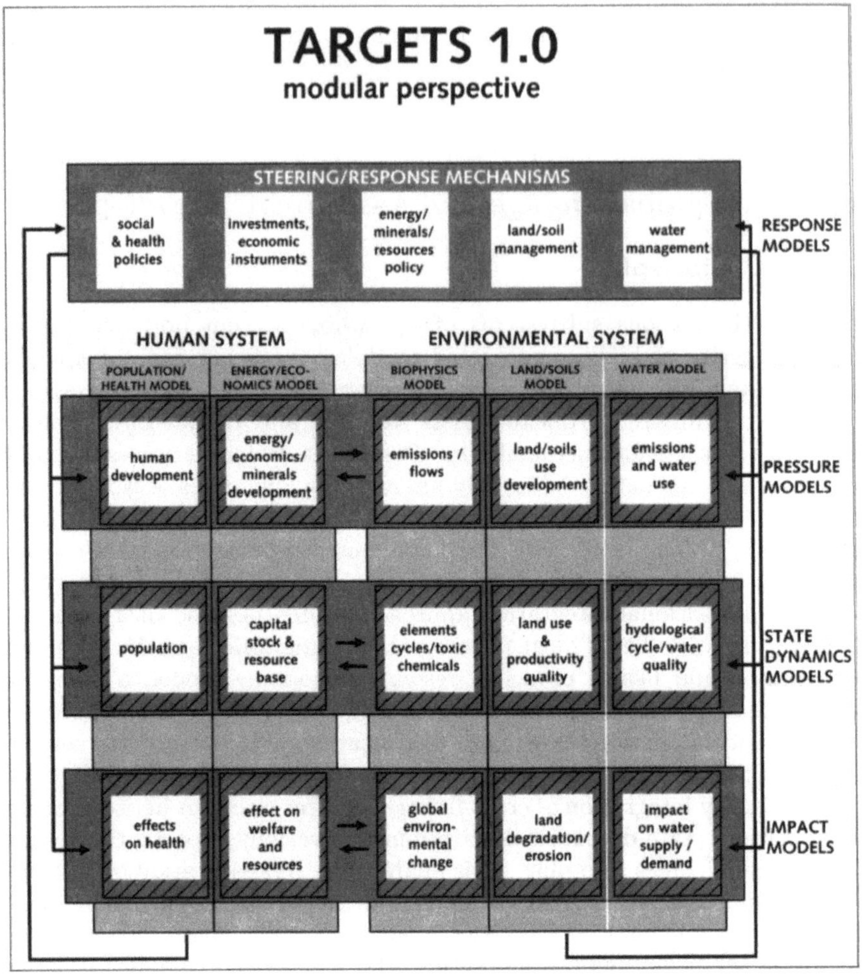

Figure 2. Modular perspective of TARGETS.

The TARGETS model does not pretend to offer a comprehensive picture of all of the relevant processes of global change. In view of the accumulation of uncertainties, which is inevitable in integrated assessment modelling, the interpretative and instructive value of the TARGETS model is far more important than its predictive capability, which is limited by the incomplete science upon which it is constructed. Therefore, rather than a prediction tool, the TARGETS model is an interpretative tool, whereby its predictive value is rather low. The challenging aspect of building the TARGETS framework is to find the right balance between simplicity and complexity; aggregation and realistic outcomes; stochastic and deterministic elements; qualitative and quantitative linkages; transparency and uncertainty. Furthermore, it is fundamentally important to face the limitations of the model to be built and to recognize the type of issues/questions that *cannot* be addressed by and what falls *beyond* the scope of the model.

5.2 Global models

The TARGETS model belongs to the class of global, integrated assessment models and elaborates on a tradition that was founded in the early 1970s by the Club of Rome (*Meadows* et al., 1972 [25]). A global, integrated assessment model is defined here as a model that is designed to analyse the phenomenon of global change from an integrated perspective. Over the past 20 years, numerous global models have been built (*Brecke*, 1993 [2]) all of which were burdened by an excessive number of processes and feedbacks, and all suffered from a number of drawbacks, namely:

1. the too complex and opaque structure;
2. the unacceptably high aggregation level for basic processes and feedbacks;
3. the production of sometimes disputable results which should be treated with great care;
4. an inadequate understanding of the system being modelled;
5. absence of stochastic behaviour;
6. an inadequate treatment of uncertainty;
7. the poor documentation and difficulty to run, although there are some exceptions like the WORLD3-model (*Meadows* et al., 1972 and 1992 [25, 24]), which is well documented;
8. the limited possibility of verification and validation of modelling structure and results.

Thus, in view of the above, we must learn from the experiences with global models (*Meadows* et al., 1982 [23]). Therefore, in developing the TARGETS model, a great deal of attention is paid to the issues of uncertainty, stochastic behaviour, aggregation levels, transparency, and how to deal with incomplete knowledge. In addition, the TARGETS model will be well documented, and, if the model is to be useful to decision makers, it should have a reasonably quick turnaround time.

The TARGETS model is not a traditional model in the sense that all global models developed so far focused on the social and economic components of the human system. Since the 1970s, the knowledge about and the political importance of environmental issues increased considerably, which allows for a more balanced treatment of the human and environmental systems. Therefore, the TARGETS model aims at keeping the balance steady between social, economic and environmental processes in order to quantify the global dynamics of the real world in a simplified fashion.

The TARGETS model is also not a traditional model in the sense that it assumes that incremental changes in parts of the global change system will cause gradual and incremental changes in the system as a whole. The real world does not function in such a simple, linear way. Therefore, TARGETS will be a composite framework of simple systems – represented by metamodels – which may show nonlinear and adaptive (self-learning), perhaps even chaotic, behaviour, resulting in a complex system. This means that incremental changes in conditions of subsystems may result in considerable changes in the results of the overall system, which may not always be predictable.

In the light of far from sufficient knowledge in natural and social sciences pertaining to global change, it is absolutely clear that it is impossible to model all entities, processes and feedbacks in a quantitative sense. Therefore, there is a need for a new class of global, integrated assessment models that, on one hand try to capture the essential dynamics of complex discipline-oriented expert models in meta-form (metamodels), but on the other hand are intended to promote an understanding of the qualitative interactive linkages, which are poorly understood in a quantitative sense. This implies that when it comes to implementation of the insufficiently understood aspects of global change, subjective expert judgement is a prerequisite (*Dowlatabadi* and *Morgan*, 1993a and 1993b [7, 8]).

5.3 Top-down approach

An important aspect of the research is the top-down approach which has been chosen, and which also has implications for the order in which the simulation model will be implemented. Firstly, a prototype of the TARGETS model will be made operational on a global level, with global data sets. In this global version (version 1.0) of TARGETS, both the distributions on a scale below the temporal resolution in the model and the heterogeneities below the spatial resolution level of the model are dealt with by introducing classes and spatial distribution functions. In Fig. 3 different types of models developed by or used at RIVM are set alongside each other pyramidally. As this picture shows the TARGETS model is placed on top of this model pyramid, because it has the highest level of aggregation and integration, which is, as mentioned above, only possible when using the simplicity principle. The TARGETS model utilizes aggregated data and processes which are provided by theme-specific integrated models such as the IMAGE (climate assessment) and the RAINS (acidification assessment) models. Theme-specific integrated models in turn make use of data and processes generated by aspect-compartment models such as air pollution models or groundwater models on the local or national scale.

On the other hand, integrated models may reveal weaknesses in aspect-compartment models, which may lead to improvements of these discipline-oriented models. After operationalisation for the globe as a whole, the model will be disaggregated to the level of major world regions, using regional data sets: Western Europe, Eastern Europe, North America, Latin America, Africa and Asia. For each of the world regions, one representative country or

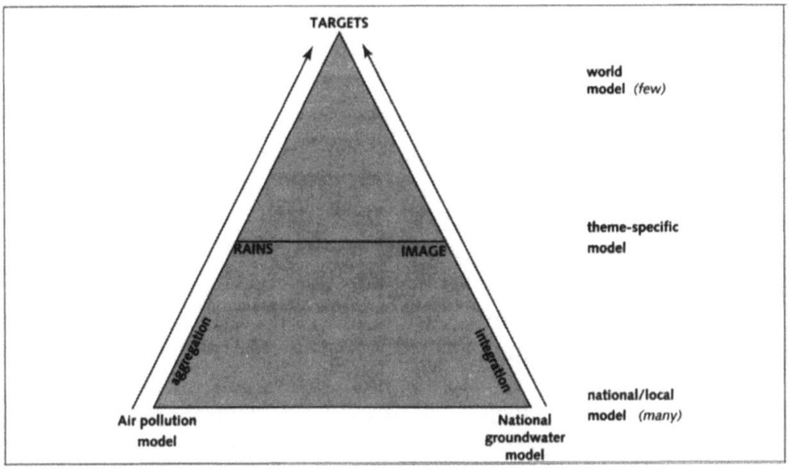

Figure 3. Pyramid of models according to different levels of aggregation and integration

area will be selected for which an integrated assessment study will be carried out. In practice, this means that such case studies pursue the question of what sustainable development on the regional level means.

5.4 Horizontal integrative description of TARGETS

5.4.1 Steering models. Horizontally, the TARGETS model can be subdivided into steering, pressure, state dynamics and impact submodels, which are described below. All steering models are fully interlinked, just as all pressure, state dynamics and impacts models. The steering model contains those variables by which the user can influence human activities or/and the environment. The steering variables are clustered into financial and legislative measures. An essential part of the steering model is the allocation model which distributes the total amount of money towards the different sectors, and controls it so that the entire stream of money forms a closed system, like the physical flows do in the pressure, state dynamics and impact systems. Three types of investments are distinguished: investments in communication and education, investments in infrastructure, and investments in science and technology. The rate of change in technology is one of the major determinants of sustainable development. Therefore, in the steering model, investments in science and technology will steer the development of new technologies and transfer of technologies from developed to developing countries for the energy sector (energy submodel), the winning of resources (minerals submodel), agriculture and food production (agricultural and food submodel), chemical technology and the transportation sector (economy submodel). The steering variables are all information variables which influence the physical or material variables within the pressure, state dynamics and impact systems.

5.4.2 Pressure models. A comprehensive, predictive model of humanity's interaction with the natural world still lies far beyond our reach. The pressure models are only intended to shed some light on the interdependencies which exist among factors such as growth in population, developments in the stock of capital goods, use of resources, growth in production and technological development. Important intermediary factors are the agricultural and food-provision system, the infrastructure, including health and education provision, rural vis-à-vis urban systems, income distribution and the stocks of natural resources. In TARGETS, all causative or pressure models describe the developments in human population, resources/economy, land use and water use, respectively, where all developments are fully interlinked. In practice, this means that the pressure system within a specific module is composed of pressure elements arising from a number of other modules. As a starting point, two separate submodels of the WORLD3-models (*Meadows* et al., 1992 [24]) are here used as a foundation: a conceptually improved model version of the natural resources and the population dynamics model, both described in *de Vries* et al. [3]. These models are coupled to an economic model, a land use development model and a water use model.

5.4.3 State dynamics models. The state dynamics models describe the biogeochemical status of the environmental system and the social and economic status of the human system, which are of course strongly interrelated. The biogeochemical status of the Earth is characterized by an interconnected complex of transport mechanisms and transformation processes, many of which are of a cyclic nature (*Wollast* et al., 1993 [49]). An integrated model is being developed that delineates the physical, chemical and biological state of the biosphere. An essential part of the model is the integration of the element cycles (C, N, S and P). Other cycles implemented are the global hydrological cycle and global cycles of some representative heavy metals, pesticides and organic micropollutants. All these cycles are related to important universal issues: climate change, soil degradation, acidification, stratospheric ozone depletion, and toxification. This global disturbance of the biospheric system interferes with the changing state of the population and resources reservoirs of the human system. The size of the demographic reservoirs subsystem is determined by the fertility rates, and disease and age-specific mortality levels. The mineral and fossil resources reservoirs are determined by human decisions and behavioural rules, such as resource exploitation through investments by private companies or by government planning.

5.4.4 Impact models. The impact models to be designed can be divided into three types of strongly interrelated generic models:

1. Models which describe the effect of anthropogenic multiple stress on the functioning of ecosystems (ecological succession, structure and resilience);

2. Models which describe the impact of determinants such as socio-economic developments and changing global environmental conditions on human health;
3. Models which specify the direct socio-economic effects (monetary and physical damage) and indirect socio-economic effects (multiplier effects in the economy resulting from direct losses in sectors) for the public health sector, water management sector and the coastal defence sector.

5.5 Vertical integrative description of TARGETS

Vertically, the TARGETS modelling structure can be subdivided into various autonomously functioning models, called modules, each module representing and covering as much as possible of the cause-effect relationship with respect to a particular aspect or theme of global change. The modules describing the human subsystem are the population/health module and the energy/economics module, whereas the environmental subsystem is described by the biophysics, toxics, land use and water modules.

5.5.1 The integrated population and health model. The global population and health model includes a demographic subsystem, two causal subsystems, i.e. the health services and determinants, and two effect subsystems, i.e. diseases and deaths, as shown in Fig. 4. The aim of the model is to simulate the changes in morbidity and mortality levels under varying socio-economic and environmental conditions. A prototype of the model is described in *Niessen* and *Rotmans* [27]. The steering/response system consists of social and health policies which may lead to interventions in the various sectors, modulating the forces in the pressures (driving forces) compartment. The pressures part consists of forces depending on the state of socio-economic development determining water and food supply, income and literacy level and preventive and curative health services level. Global environmental change influences exposure to health risks in a direct way, affecting e.g. the presence of micro-organisms causing diarrhoea and vector-borne diseases (*Martens* et al., 1994 [22]), and indirectly, via the world food supply (*Rosenzweig* and *Parry*, 1994 [31]), which may lead to premature death, infectious diseases or certain degenerative diseases. In the dynamics component, the demographic changes take place under the various environmental and socio-economical pressures that lead to changes in birth and mortality rates. Fertility rates are determined by income expectation, life expectancy/infant mortality rates, desired family size and contraceptive prevalence. Mortality rates are determined by age-specific mortality functions which in turn depend on average life expectancy. This, in turn, depends on the levels of food, industrial and service output per capita. The effects are reflected in the health impact part, that distinguishes various disease states and disease-specific mortality rates.

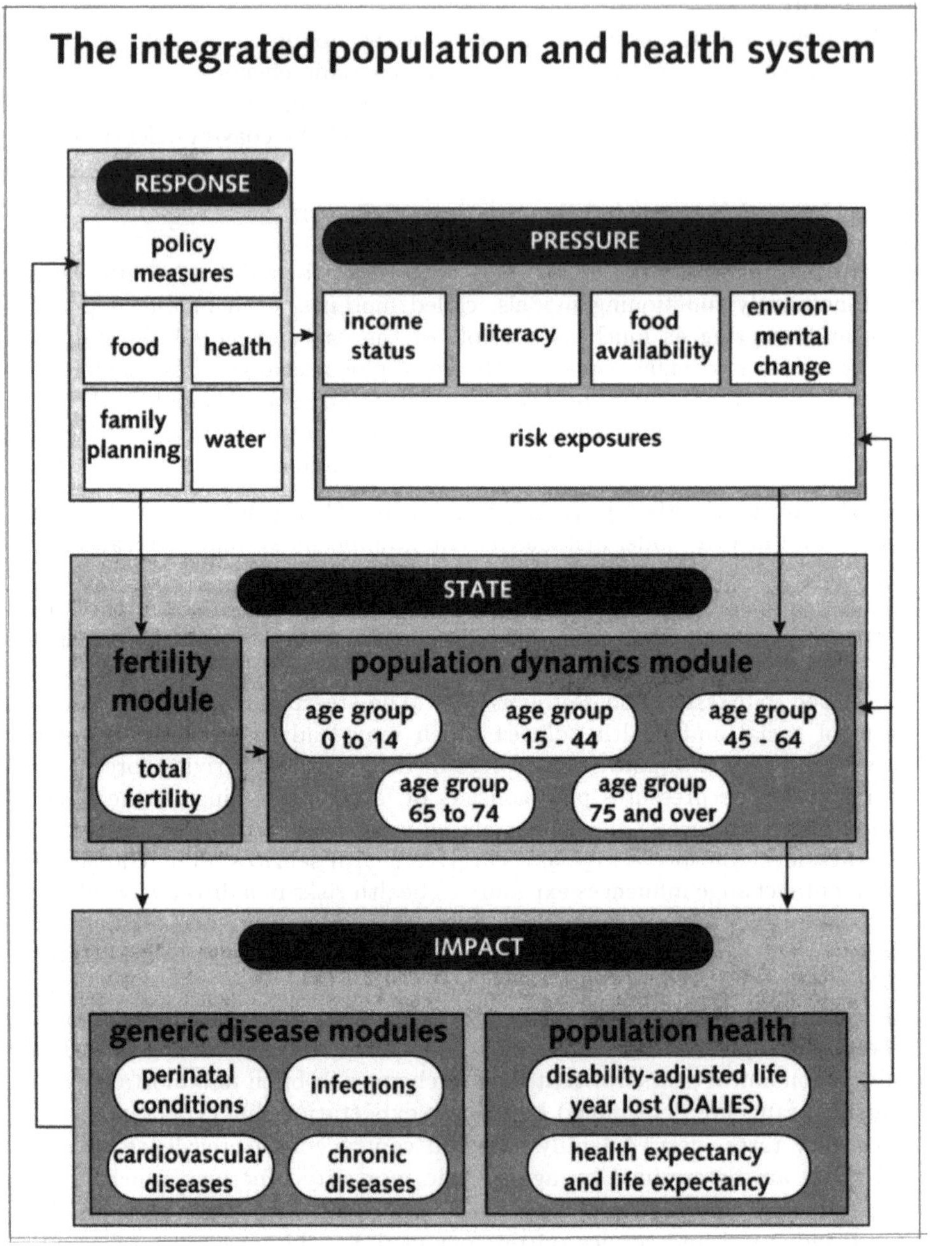

Figure 4. Integrated population and health module

5.5.2 The energy/economics/resources models. The model is divided into an energy submodel, an economic submodel and a minerals submodel. The energy submodel simulates the use of fossil fuels and the eventual introduction of renewable and nuclear energy sources, as well as ways to substitute energy end-use for capital. The energy submodel is divided into four sections: the solid fuels model, the liquid fuels model, the gaseous fuels model and the electric power model. The three fuel supply models have a similar structure. Depletion is governed by long-run cost-supply curves for the resource base. New investments are based on profitability. Prices are related to costs through a mechanism which reflects the impact of capacity shortages in the production of capital stocks. If an alternative (solid/liquid/gaseous) enters the market, its relative price will determine the degree of market penetration. The minerals submodel simulates the processes of discovering, exploiting and partly recycling a finite resource base of moderately scarce metals. It also simulates the usage patterns of the bulk of relatively abundant metals and their eventual recycling. The very simple economy submodel consists of an industry, services and consumers part. Industrial output is generated at a fixed capital-output ratio; labour is not taken into account. The service sector is expanded on the basis of an assumed relationship between industrial output per capita and desired service output per capita. Consumption is as yet determined by a constant savings rate. For a more elaborate description see *de Vries*, et al. [3].

5.5.3 The biophysics model. The biophysical part of TARGETS consists of the following parts: a global elements cycle module, a climate assessment module, an ozone assessment module, an acidification assessment module, and a toxification assessment module which are interlinked. The global elements cycle model simulates the cycling of carbon, nitrogen, phosphorus, and sulphur between the atmosphere, terrestrial biosphere, lithosphere (soils) and hydrosphere (rivers and oceans) and integrates the interactions between the various compartments. The global elements cycle model triggers, amongst others, the climate assessment model in TARGETS, which comprises an atmospheric chemistry model, a radiation model and a climate model, which are described in *den Elzen* [4]. The global-mean temperature projections are then used to generate regional patterns of mean monthly and seasonal changes in future temperature and precipitation for the world, using results of coupled-atmosphere GCM transient experiments, based on a method proposed by *Santer* et al. [37], as already applied in the ESCAPE model (*Rotmans* et al., 1994 [36]).

The ozone assessment model (or the UV-B chain model) comprises a halocarbon model and a UV-B impact model. The halocarbon model simulates halocarbon concentrations, and the resulting atmospheric chlorine concentration and the overall (direct and indirect) radiative forcing. The UV-B impact model uses as input the atmospheric chlorine concentrations derived from the halocarbon model, and calculates stratospheric ozone depletion, increased UV-B radiation, and non-melanoma and melanoma skin cancer risks due to increased UV-B radiation associated with ozone depletion. For a detailed description of the UV-B chain model see *den Elzen* [4].

The acidification assessment model is a meta-version of the RAINS acidification model for Europe, described in *Alcamo* et al. [1]. The model is fed with emissions of SO_2 and NO_x, which provide the input to the atmospheric transport model, which uses a transfer matrix (source receptor matrix) that expresses the relationship between country emissions and local deposition. The depositions form the input to the soil model, which calculates the soil acidity, taking into account acid load and the soil's buffering capacities. To integrate effects of soil acidification with soil effects of nutrients cycling and climate change, a single integrated soil model has to be developed. It is intended to develop versions for three specific regions: East Asia, Northern America and Europe.

The global toxic substances model will be based on the nested version of the SIMPLEBOX model (*Van de Meent*, 1993 [43]), which is a generic box model that describes transport and degradation processes within the atmosphere, soil and water boxes. The global distribution model for toxic substances focuses on those toxic micropollutants which, on a global and a regional scale, may adversely affect human health and ecosystem functioning, or may adversely affect agricultural production, or may result in a deterioration of drinking water resources. At most ten different "aggregated" toxic chemicals will be selected, which are supposed to represent the major classes of heavy metals, persistent organic chemicals and some non-persistent chemicals.

5.5.4 The global land model. The global land model deals with the dynamics of the land system (land use and quality) and the food system. The model portrays the complex land dynamics from an integrated perspective, simulating and coupling the various causes (social, demographic and economic processes resulting in food and wood demand, as well as global environmental change), the mechanisms (biophysical land use and fertility processes), the effects (on society in terms of food supply changes, and on ecosystems in terms of deforestation, erosion, degradation and desertification) as well as steering mechanisms (land and soil management practices). The model basically consists of a land/soil management and investment module (steering part), a land development module (which forms together with outer pressure factors such as population, economic and energy development the pressure part), a land use and land productivity module (state dynamics part, focusing on the main agricultural processes of irrigation, land clearing and intensification, food production – based on agricultural inputs like fertilizer use – and degradation) and a supply/demand module for food, wood and land use (impacts part). Although the prototype acts on a global scale, spatial heterogeneities are introduced by disaggregation into specific classes for soil, climate and land use, as described in *den Elzen* et al. [5].

5.6 The global water model AQUA

The global, integrated water model AQUA describes the global water system (both hydrological and qualitative aspects), the human pressure on this system and the effects of changes in the water system on human health, food supply and socio-economics, as well as options to draft and evaluate water management policies (*Hoekstra*, 1994 [12]). The hydrological submodule of AQUA is divided into eleven compartments, where each compartment is considered as a reservoir in which water is stored. Fig. 5 shows the compartments and the water fluxes between them, where each arrow represents a water flux that is modelled separately, preferably on a physical basis. AQUA pretends to be a generic model for an arbitrary river basin. To test this hypothesis the generic model has to be validated against some of the 20 major river basins or watersheds in the world. In *Kwadijk* and *Rotmans* [18] a methodology is described to estimate the impact of a human-induced climate change on stream flow changes in large river basins. To validate parts of the AQUA model, this method is applied to the Ganges/Brahmaputra and the Yangtze. In these case studies we focus on two water-related problems: on the one hand, local and seasonal water scarcity, which comprises both the problems of water absence and inferior water quality, which are strongly interrelated and affect population (domestic water supply), agriculture (irrigation water supply) and industry (industrial water supply); and on the other hand, local and seasonal over-abundance of water, which causes regular floodings in the delta area of the Ganges/Brahmaputra and the Yangtze.

6 Calibration and validation

There are many definitions and interpretations of the term calibration and validation. Moreover, complete calibration and validation of simulation models is impossible, because the underlying systems are never closed (*Oreskes*, 1994 [30]). Within the framework of this study calibration is defined as the procedure for comparing the model results with the results of the real system (historical output, for instance observational data). The question to answer is how close the model output approximates to the observational data. In spite of the implied vagueness in the term "as close as possible", statistical techniques can be used to quantify the difference between the model output and measured data.

Validation is defined here as the procedure for testing the adequacy of a given mathematical model. Validation can be divided into two different types. The first type is practical validation, which concerns the validity of the outcomes of the model. In fact, this implies that the outcomes of the model are compared with observational data. The major difference with calibration is that the comparison should be based on a new set of data, which lies outside the calibration pathway.

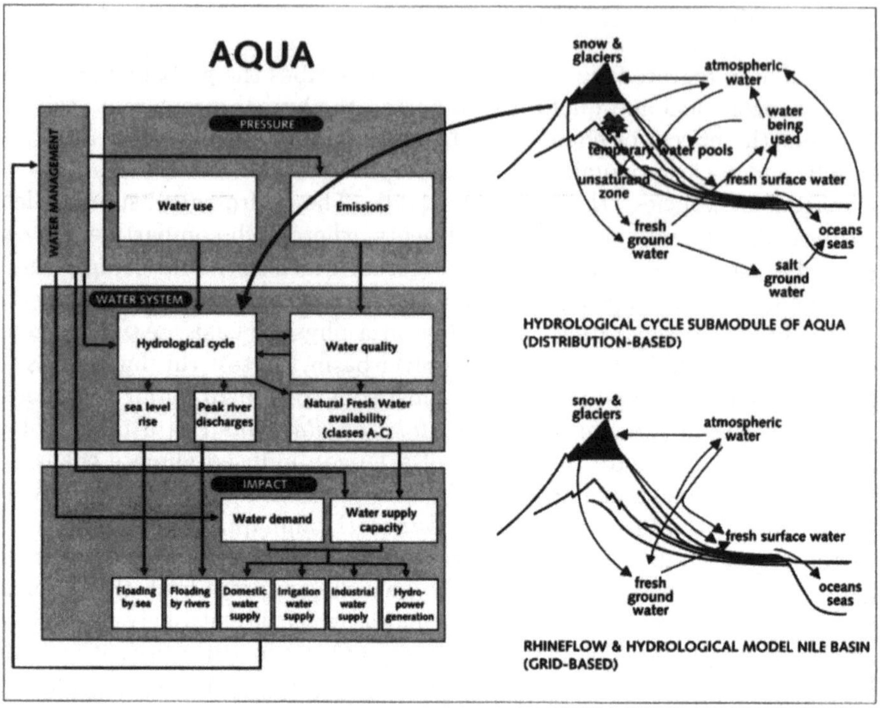

Figure 5. Various components of the AQUA model

The second type of validation is conceptual validation, which concerns the test of whether the model represents the real system. This implies that the internal structure of the model is tested, by testing whether the concepts and theoretical laws of the system under consideration are interpreted and represented in a sound way. Conceptual validation is often carried out by applying statistical techniques to test the estimates of parameter values; to test the distribution functions of parameters; and to test the coherence of relations (the latter mainly by regression and correlation analysis). As a matter of fact, conceptual validation is often problem- and domain-dependent. In general, the fundamental difficulty in validation is that, if the validation test is too stringent, or if it is not stringent enough, the model may be an inadequate description of the real system. The difficulties in validating a global, integrated model are discussed below.

Each submodel should be scientifically valid in the sense that the model structure, relations, parameters and dynamic behaviour over the period 1900-1990 reflect the prevailing theoretical insights and the key facts pertaining to that part of reality which the model is supposed to represent. In the case of submodels, it will not be easy to collect reliable data sets for the validation period 1900-1990, and failing this, a more recent and briefer period of a few decades will be taken, e.g. 1970-1990. It is of major importance that the various submodels are validated at the regional or country level. Therefore, experts in the field of small-scale, detailed models in a specific region or country will be asked to analyse the model results and to validate regionalized model results for a specific world region against regional data subsets. In view of the generic structure of the modules (submodels), which makes them (as far as possible) applicable at different spatial aggregation levels and for different regions in different periods, the experts should indicate the extent to which the generic structure of the module(s) should be adapted to regional conditions and dynamics. A second way of validating the submodels is by means of the intercomparison with expert models which they are supposed to represent at the meta-level.

An example of validating a simple carbon cycle model (based on *Goudriaan* and *Ketner*, 1984 [11]) against observational data and more complex two- and three-dimensional carbon cycle models is presented in *Rotmans* and *den Elzen* [35] and *Wigley* [48]. Having validated the submodels, the overall validation of the modelling framework should be performed, whereby special attention should be devoted to the validation of the coherence and internal consistency of the model.

7 A hierarchical framework of indicators

One of the possibilities to operationalise the concept of sustainable development is to design sustainability indicators. Here an indicator is defined as a characteristic of the status and the dynamic behaviour of the system under concern. From the systems-based definition of an indicator, it follows that an indicator is a one-dimensional systems description, which may consist of a single variable (absolute indicator) or of a set of variables (relative indicator). Indicators can monitor the pressure on, the status of, and the impact on the global environment, which enables us to anticipate an impending global change, to establish priorities, to formulate adequate strategies and to test the effectiveness of those strategies.

While integrated assessment models are important for analysing the phenomenon of global change, indicators serve as the vehicles for the communication of the model results, on the basis of which sustainable routes can be mapped out. In particular there is a strong need for highly aggregated and composite indicators or indices, in which condensed information is assembled. The distinction between indicators and indices is based on a difference in aggregation level. An index is here defined as a multi-dimensional composite made up from a set of indicators. As a result of their composite, and therefore abstract, character the requirements mentioned above do not necessarily hold for an index. A prerequisite for indices of sustainable development i.e. sustainability indices is that they represent environmental pressure, and the state of, and impact on environmental conditions. In other words, they should capture as much as possible of the cause-effect chains they represent and relate pressure and effects to criteria for sustainable development. This causal chain approach is comparable with the pressure-state-response indicator mechanism proposed by the OECD [29]. So far, compared with an economic index such as GNP and a social index such as HDI (Human Development Index, produced by the United Nations Development Programme) which are recognized worldwide as useful one-dimensional measures, no environmental indices have been developed. Because of the lack of comprehensive, generally accepted measures of the global environment, which can facilitate decision making within the context of sustainable development, there is thus a need for innovative ideas and indicator concepts that are both comprehensive and achievable.

The method proposed here attempts to develop a hierarchical framework of indicators/indices which is linked to the integrated modelling framework TARGETS. The indicator/index framework serves as a communication layer between modellers and decision makers. Similar to the hierarchy of models in TARGETS, the associated set of indicators should also be hierarchical, and consists of a system of pressure, state, impact and response indicators and indices. The indicator framework can be considered as a layer on top of the model layer (see Fig. 6).

Within the hierarchical framework of indicators, four levels of aggregation are distinguished. The first level represents the model-aggregated indices; the second level denotes aggregated indices for the various submodels; the third level denotes relative indicators (absolute indicators per unit); while the fourth layer represents data/statistics. In this way a tree diagram of pressure, state, impact and response indicators and indices is built up (Fig. 7), with at the top-level the overall sustainability index, capturing the pressure, state, impact and response dynamics for the model as a whole. One level below, the sustainability indices for the various submodels can be composed. The different aggregation levels within the indicator framework enable the building of the bridge between the advocates of using highly composite indicators and the traditional statisticians.

The main advantage of linking a set of indicators to an integrated modelling framework is that it yields insight into the complex dynamics of the system under concern. This enables the production of coherent information about linkages between causes and effects (vertical integration) and the addressing of cross-linkages between different issues (horizontal integration). This coherent and integrative information can only be generated by an interconnected framework of indicators, and not by separate indicators. The ultimate goal of the hierarchical framework is to create an overall index which captures the key characteristics of the global environment in a single measure, and to demonstrate the dynamic interrelations between this abstract index and the real-world indicators. In constructing the hierarchical framework of indicators and indices the following steps can be distinguished: experimen-

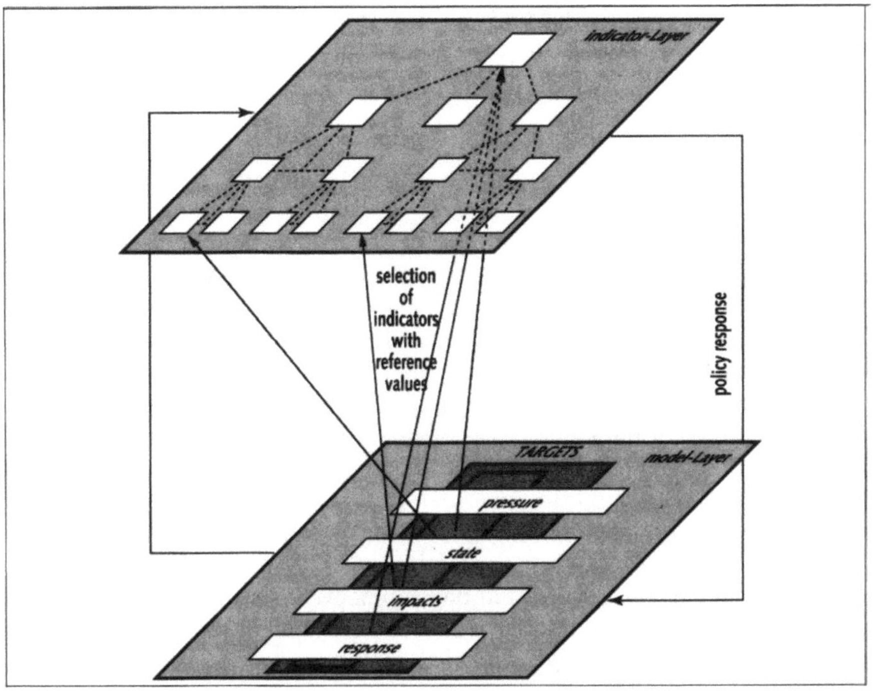

Figure 6. Indicator-layer versus model-layer

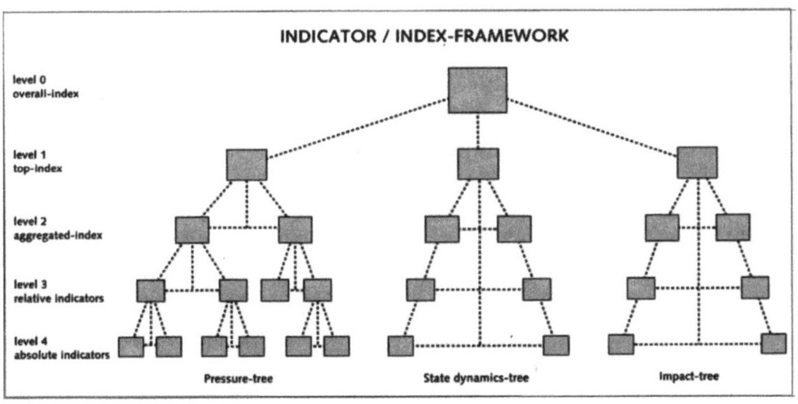

Figure 7. Hierarchical framework of indicators/indices

tation, selection, scaling, weighing and aggregation. The various steps described above can be based upon a combination of expert judgement, delphi-techniques, multi-criteria analysis, public opinion polls, value-based decisions and modelling experiments. This requires frequent and intensive interaction between decision makers and modellers.

8 Strategies for Sustainable Development

The hierarchical framework of indicators and indices will be used to develop and evaluate alternative strategies for sustainable development. A common and useful distinction is drawn here between "scenarios" and "strategies". Scenarios reflect both various exogenous developments, i.e. "uncontrollable" or autonomous developments outside the system considered, and various endogenous developments, i.e. developments within the system considered that cannot be steered or influenced. Various strategies (to be decided upon by decision makers) are analysed under various scenarios which are not controlled by decision makers.

Because extrapolation of current trends and conditions is an inadequate method for making long-term projections, it will not be used here in developing strategies for sustainable development. Instead, we first determine sustainable states and then indicate whether and how these sustainable states could be reached. Such a pathway may include surprises, and we are aware that no methodology has yet been developed which enables the systematic incorporation of surprises into scenarios and strategies.

Sustainable states can be determined by introducing target values which are based on empirical knowledge or on modelling results. Then it will be indicated whether and how these sustainable states could be reached, by calculating the components of the indices in the hierarchical framework and comparing these values with the target values. Using this method of "backcasting" to develop coherent and consistent response strategies for sustainable development, the techniques of optimization (search strategies) and fuzzy logic might serve as a helpful guide. Although there is no optimal solution possible for sustainable development, optimization techniques can be used to explore the solution space in search of specific strategies for sustainable development. However, while integrated assessment simulation models cannot be used for optimization goals in the usual way – apart from simple climate assessment models like that of *Nordhaus* [28] – it is possible to make use of heuristic methods which employ metamodels and optimization techniques in an attempt to find local optima by starting the search in several reference scenarios. A heuristic method is described in *Janssen* et al. [16].

The concept of fuzzy programming is introduced to reflect the uncertainty and vagueness associated with the optimization of sustainable development. The reason for using fuzzy logic programming is that we wish to satisfy all kinds of imprecisely known constraints as well as possible, without achieving a maximum performance. By using fuzzy approaches, decision makers are exposed to a broader range of solutions.

9 Uncertainties

The concept of uncertainty encompasses a multiplicity of terms. The variety of types and sources of uncertainty can cause considerable confusion. Several attempts have been made to classify the different types and sources of uncertainty (*Morgan* and *Henrion*, 1990 [26]; *Funtowicz* and *Ravetz*, 1989 [10]). *Morgan* and *Henrion* [26] distinguish three kinds of uncertainty in models: uncertainty about technical, scientific, economic and political quantities; uncertainty about the appropriate functional form of technical, scientific, economic and political models; and disagreements among experts about the value of quantities or the functional form of models. An alternative classification is the distinction made by *Funtowicz* and *Ravetz* [10] in technical uncertainties (concerning observations versus measurements), methodological uncertainties (concerning the right choice of analytical tools) and epistemological uncertainties (concerning the conception of a phenomenon).

The uncertainties in the TARGETS model can be aggregated into two categories, based on the subdivision of the Earth's system into the human and environmental system (*Rotmans*, 1994 [33]):

1. scientific uncertainties: those occurring in the environmental system which arise from the degree of unpredictability of global environmental change processes;
2. social and economic uncertainties: those occurring in the human subsystem which arise from the degree of unpredictability of future geopolitical, socio-economic and demographic evolution.

Note on 1: Scientific uncertainties originate from lack of measurements; inconsistency of measurements; unknown emission factors; deficient knowledge of the key physiological, chemical and biological processes; inadequate understanding of the many geophysical and biogeochemical feedback responses which can amplify or dampen the response of the biosphere system; the poor knowledge of the ecological and socio-economic effects and the effects on human health. There is thus a very high degree of inherent uncertainty in the whole complex system portraying global environmental change.

Note on 2: Social and economic uncertainties are related to the set of behavioral rules and human decisions that try to describe the dynamics of the socio-economic system. The information on which these rules and decisions are based is often surrounded with uncertainty. At the same time, because these rules and decisions are loaded with value judgements, they are under continuous change due to changes in cultural value orientations. One of the consequences is that there are no unanimously accepted models to describe the socio-economic system.

A significant number of the uncertainties within the concept of global change originate from structural misconceptions (scientific uncertainties) and from "unpredictable" geopolitical, socio-economic and demographic processes (social and economic uncertainties). In this research programme an attempt is made to make these uncertainties explicit. On the one hand scientific uncertainties will be related to different scientific paradigms, and on the other social and economic uncertainties which are not based on scientific theories will be coupled to cultural perspectives.

10 Various perspectives

10.1 Introduction

No unique and correct description of the real world exists. The problem is not that there are no consistent descriptions of the world, but rather that there are too many (*Wallace* and *Norton*, 1992 [46]). Because of the complexity of global change, here considered as the totality of changes on planet Earth, including all human interventions and alterations, there are numerous valid perspectives from which to describe and evaluate the phenomenon of global change. Choosing a perspective or paradigm for global change automatically implies a specific way of describing global change, so that each scientific description of global change is value-based. This means that the integrated systems approach to global change and sustainable development, chosen in this study as the starting point for analysing the phenomenon of global change, is just one of the many possible representations.

The various perspectives considered here are based on scientific hypotheses or theories (scientific perspectives), or originate from differences in cultural perspectives (cultural perspectives). An approach is introduced by which uncertainties in models can be made explicit by relating different types of uncertainty to different types of perspectives, as is shown in Fig. 8. This is done by creating perspective-based alternative modelling routes, some examples of which will be discussed below.

10.2 Scientific perspectives

Scientific perspectives, some examples of which will be given below, relate to the appreciation of the interacting biological, physical, chemical and social processes. Vernadsky was the founder of the scientific perspective which has it that the biosphere acts as an interacting, organic-geologic physiological whole (*Vernadsky*, 1945 [44]). Vernadsky introduced the term "noosphere", which represents the amalgamation of the biosphere and the technosphere (UNEP, 1992 [42]). If you are a "Vernadskyite" you assume that the wholeness of all physical and socio-economic processes on Earth cannot be modelled in an

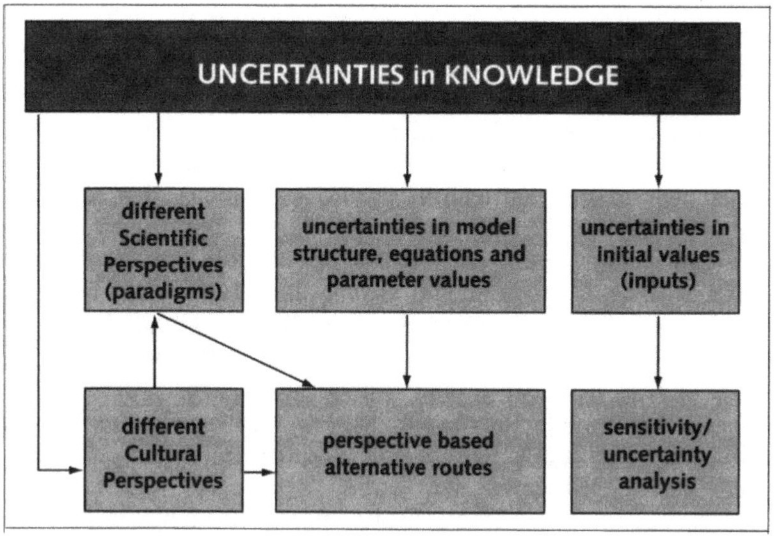

Figure 8. Perspectives related to different types of uncertainties

adequate way. In fact, the modern viewpoint of interacting socio-economic and ecological systems is based on Vernadsky's perspective.

One of the most inspiring, but also controversial theories in recent environmental science has been the "Gaia-hypothesis" developed by James *Lovelock* (1979 and 1988 [20, 21]). The Gaia theory can be considered as a new scientific paradigm, which looks upon the Earth system as a cybernetic whole, as a coherent system of life which is self-regulating and self-changing. If you are a "Gaian" you consider that the complex interactions among biogeochemical processes constitute a self-correcting global organism on different temporal and spatial scales. Lovelock has illustrated his ideas with a simple model, Daisyworld, a planet in which the environment is represented by the temperature and the biota of a single species, namely daisies.

Gaia-adherents consider the Gaia theory to be a major scientific revolution of our time, but sceptical researchers refer to the counterexamples which have been postulated, and more generally regard the Gaia theory as an untestable hypothesis. A basic issue in a vivid scientific discussion was the question of whether the Gaia theory should be interpreted literally or as a metaphor (*Wallace* and *Norton*, 1992 [46]).

A third new scientific perspective deals with the "expectation of the unexpected" (*Timmerman*, 1986 [41]). This perspective is very topical and focuses on the concepts of chaos, risk, discontinuity, disequilibrium and surprise. This scientific referential framework fits in well with the concepts of the carrying capacity of ecosystems, which is of a cyclic nature, as well as resilience of ecosystems, introduced by *Holling* [13, 14]. Not until the past couple of years has the idea gained acceptance that the concept of chaos, which is a measure of the sensitivity of a system to initial conditions, works not only in physical systems but also in human systems.

The last scientific perspective treated here is that of "evolutionary and adaptive behaviour" as, amongst others, described in *Waldrop* [45]. This perspective abandons the deterministic, pseudo-mechanical representation of the Earth dynamics. The evolutionary perspective assumes that the Earth system cannot be described by deterministic laws only, because living organisms such as human beings, animals, ecosystems and societies can respond, react, learn, change, adapt and influence each other. Historical evidence shows that both on longer and shorter time scales life has adapted to a changing environment. The question is whether, and if so, what species may adapt to the unprecedented fast rate of human-induced changes in the global Earth system. Therefore, modelling the global Earth system requires an approach which facilitates the implementation of adaptive and structural changes in subsystems and the whole Earth system.

In the TARGETS model, three scientific perspectives will be implemented, i.e. the "Gaia perspective", the "Expectation of the unexpected perspective" and the "Evolutionary and adaptive behaviour perspective". The numerous "scientific uncertainties" within the TARGETS modelling framework will be investigated from the viewpoints of the three apparently antithetical scientific perspectives. In the first step of the approach, these uncertainties need to be related to the various perspectives. To this end, subjective probability distributions will be selected or designed for the major scientific uncertainties. This can be done in such a way that, for instance, from the "Gaia" perspective, the negative feedbacks dominate the system (dampening effect), while in the "Expectation of the unexpected" perspective the positive feedbacks dominate (accumulative effect). To avoid obtaining unrealistic results from these approaches, the outcomes have to be validated. It is therefore proposed to start such an analysis with a detached part of the TARGETS model, namely the coupled carbon cycle/climate model, for which extensive data records are already available.

More realistic experiments, however, would imply the reformulation of subsystems of the TARGETS framework. This is because representative perspective-based systems can only be realized by adding or omitting certain entities, processes, interactions and feedbacks to the system. In *Schneider* and *Boston* [38] examples are presented of Gaia-model representations of the global carbon cycle, which differ in dynamic structure from conventional global carbon cycle models. Similarly, *Westbroek* [47] offers an interesting prototype of a world model in which a Gaia-system is incorporated.

The coupled carbon cycle/climate model will thus be adjusted to the various scientific perspectives by

(a) changing the dynamic structure of the underlying system; and

(b) including the natural variability in the coupled carbon cycle/model, to test the sensitivity of the system to fluctuating inputs.

After having performed a limited number of experiments using experimental designs, the outcomes can again be validated against the data records for atmospheric carbon and global mean temperature. If this restructuring and reformulation of the coupled carbon cycle/climate model yields satisfactory results, such an approach can be extended to the human system part of the TARGETS framework. In the human system intriguing matters have to be questioned such as the reaction of human behaviour to policy measures by adaptation, learning and rejection, as well as how do people respond to different kinds of measures and how do different social structures develop?

10.3 Cultural perspectives

Human attitudes towards the environment have evolved during history and are determined by, among other things, cultural perspectives, which have been classified in various ways. During the last couple of years there has been an increasing recognition of the usefulness of the various cultural perspectives within the context of sustainable development, but, alas, only in a qualitative sense. An interesting contribution has been made by *Schwartz* and *Thompson* [39] and *Thompson* et al. [40]. In their cultural theory they combine anthropological insights from e.g. *Douglas* [6] with recent ecological knowledge developed by e.g. *Holling* [13, 14]. The crux of their theory is that societies can be characterized along two axes: groups and grids. The group axis measures the degree to which individuals consider themselves part of a group of individuals with whom they share values and beliefs. The grid axis indicates the extent to which individuals are subjected to role prescriptions within a larger structural entity. The group-grid characterization yields different perspectives from which people perceive the world and behave in it. *Thompson* et al. [40] argue that the viability of a perspective depends upon a mutually supportive relationship between a particular cultural bias and a particular pattern of social relations. They claim that five and only five perspectives – hierarchy, egalitarianism, fatalism, individualism and autonomy – meet the conditions of viability. These perspectives are in competition for adherents, but, on the other hand, are interdependent. Here the autonomous perspective will not be elaborated, because the "hermit" according to *Thompson* et al. [40] is not interested in what happens in the world, but only wants to dislodge himself from the world.

Thompson et al. [40] claim that the notions of human and physical nature are socially constructed, and that the four myths of nature, derived from ecologists (*Holling*, 1986 [14]; *Timmerman*, 1986 [41]) closely coincide with the ideas of nature. The different characteristics concerning nature, myth of nature and human nature of the four perspectives are summarized in Table 1.

As a matter of fact, the four perspectives sketched above are merely indicative and theoretical constructions. In reality, perspectives consist of

Table 1. Characteristics of the various cultural perspectives

Perspective Notion	egalitarian	individualist	hierarchist	fatalist
nature	accountable	skill controlled cornucopia	isomorphic	lottery
myth of nature	ephemeral	benign	perverse	capricious
attitude towards nature	great care	laissez-faire	regulate	just cope
human nature	born good	self-seeking	evil	unpredictable
attitude towards humans	construct egalitarian society	channel rather than change	restrict	distrust

Table 2. Cultural biased preferences in terms of strategies

Perspective Strategy	egalitarian	individualist	hierarchist	fatalist
needs and resources	can manage needs but not resources	can manage needs and resources	can manage resources but not needs	can manage neither resources nor needs
economic growth	not preferred	preferred: create personal wealth	preferred: create collective wealth	preferred: chance may bring it
attitude towards risk	risk-aversive	risk-seeking	risk-accepting	risk-avoiding

combinations of the perspectives stated above, and will develop dynamically, shifting from one perspective to another, in a process of mutual competition and addition. In spite of the caricatural character of the perspectives sketched above, they can serve as a useful framework with which to make more explicit the role of such perspectives in the complex problem of global change. For example, by creating a sequence of culturally biased preferences with respect to how to make a living, different strategies can be developed for different perspectives. *Thompson* et al. [40] deduced four strategies for "making ends meet", of which some basic features are shown in Table 2. In general, each of these strategies aims at upholding the corresponding perspective.

In a manner analogous to that used for the scientific perspectives, the cultural perspectives will be related to the concept of uncertainty. Social and economic uncertainties, which are often due to subjectivity and disagreement, can be made explicit by filling them in differently for the various cultural perspectives. To this end, a consistent mode of conduct or strategy is to be developed for each perspective, based on its biases and preferences, some of which are given in Table 2; others can be found in the literature (*Thompson* et al., 1990 [40]; *Schwartz* and *Thompson*, 1990 [39]). For each module, a limited number of crucial uncertainties will be selected, each of which will be modelled according to the biases and preferences of the perspective-based strategy under concern. Levers for alternative modelling routes can be derived from the perspective-based strategies. An alternative modelling route can be considered as a model interpretation in which a sequence of crucial uncertainties is "opened" and modelled explicitly in accordance with the biases and preferences of a particular perspective. Thus, a separate modelling route or pathway will be generated for each perspective. This implies that the fatalistic perspective is of no use to us, because from this perspective everything is a lottery; thus the appropriate model would consist of random relations and parameters. Therefore, although fatalists are an essential element of the total system – *Thompson* et al. [40] consider them as the cultural equivalent of compost – this "passive perspective" will not be implemented in the TARGETS model.

After having framed alternative modelling routes in a qualitative sense, the perspective-based sequence of uncertainties needs to be mathematically translated and quantified. To this end, various mathematical techniques are available: parametric analysis, probability distribution functions, fuzzy programming, delphi-techniques, multi-criteria analysis and expert judgement, each method with its own pros and cons.

11 Conclusions and expected results

The integrated systems and modelling approach proposed here will enable the consequences of several types of human influences to be evaluated simultaneously. It is hereby envisaged that synergetic effects, which are currently beyond the horizon of predictive competence, may be brought into view. Such a global, integrated assessment model should be regarded as an aid to the formulation of possible projections for the future, and not as a means of generating predictions as such. As said above, the interpretative and illustrative value of such a modelling framework is therefore much more important than its predictive potency.

A great deal of attention will be paid to the presentation of the relevant information generated by the model in an insightful manner. This necessitates the "opening" of models, an exercise which ranges from the construction of a user-friendly model which can be used interactively, to the creation of strategic planning exercises oriented towards strategies for sustainable development, whereby models serve to provide guidelines in the background. It is therefore expected that it will become possible to add a new dimension to the use of integrated assessment models: namely, the use of a model as a medium for communication between exponents of the natural sciences and the social sciences. Although this overall goal may seem rather ambitious, the experience gained within the IMAGE-project may be useful in avoiding major pitfalls, in order to realise the objective stated above as far as possible.

The user groups that could possibly profit from the integrated assessment framework TARGETS include researchers as well as policy analysts and policy makers. Even though each different user requires a different functionality from TARGETS, the choice made here is nevertheless to reckon with all potential users. A prerequisite for the usage by dissimilar user groups is that the model should be flexible and transparent, and lend itself to tailoring to the particular needs of a user group. The development of TARGETS is thus directed by the demands of the targeted user groups, albeit anticipated, rather than revealed demands. At the institutional level, the three user groups can be connected via the following institutions, in decreasing order of priority:

1. United Nations Environment Programme (UNEP). The TARGETS model could be used to provide integrated environmental assessments as support for environmental policy. In addition, the TARGETS model can serve as a conceptual "thinking" framework to structure the discussion concerning global change and to demonstrate the coherence between the various items of the AGENDA 21. Fig. 9 shows which items of the AGENDA 21 are, in a highly aggregated and simplified way of course, covered by the TARGETS framework.
2. World Health Organization (WHO) and the World Bank. The TARGETS model could be used to give a quantitative demonstration of the numerous interrelations between the economy, environment and population.
3. International Geosphere-Biosphere Programme (IGBP) and the Human Dimensions of Global Environmental Change Programme (HDGECP). The TARGETS model could prove especially worthwhile for establishing one of the possible links between IGBP and HDGECP.

References

1. J. Alcamo, R. Shaw, and L. Hordijk, editors. *The RAINS Model of Acidification: Science and Strategies in Europe.* Kluwer Academic Publishers, Dordrecht, 1990.
2. P. Brecke. Integrated global models that run on personal computers. *SIMULATION*, 60(2), 1993.
3. H. J. M. de Vries, T. Fiddaman, and R. Janssen. Strategic planning exercise about global warming. RIVM-Report 461502002, National Institute of Public Health and Environmental Protection, Bilthoven, 1993.
4. M. G. J. den Elzen. *Global Environmental Change: An Integrated Modelling Approach.* PhD thesis, University of Maastricht, Maastricht, 1993.

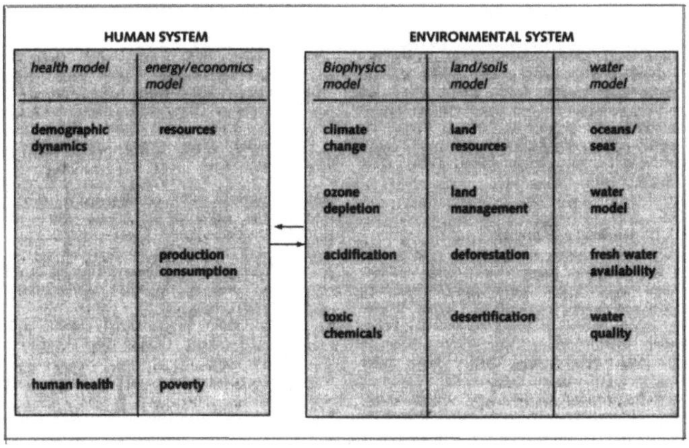

Figure 9. TARGETS framework versus AGENDA 21 issues

5. M. G. J. den Elzen, J. Köster, W. J. M. Martens, and M. Heil. The global land model of TARGETS. RIVM-Report 461502004, National Institute of Public Health and Environmental Protection, Bilthoven, 1994.
6. M. Douglas, editor. *Essays in the Sociology of Perception.* Routledge, London, 1982.
7. H. Dowlatabadi and G. M. Morgan. Integrated assessment of climate change. *Science*, 259:1813–1814, 1993.
8. H. Dowlatabadi and G. M. Morgan. A model framework for integrated studies of the climate problem. *Energy Policy*, pages 209–221, 1993.
9. A. Fisher. One model to fit all. *MOSAIC*, 19(34):53–59, 1988.
10. S. O. Funtowicz and J. R. Ravetz. Managing uncertainty in policy-related research. In *Les Experts sont Formels: Controverses Scientifiques et Décisions Politiques dans le Domaine de l'Environment*, Arc et Senans, 1989. 11.-13. September 1989.
11. J. Goudriaan and P. Ketner. A simulation study for the global carbon cycle including man's impact on the biosphere. *Climatic Change*, 6(167-192), 1984.
12. A. Y. Hoekstra. AQUA: an integrated tool for global water policy making. RIVM-Report 461502005, National Institute of Public Health and Environmental Protection, Bilthoven, 1994.
13. C. S. Holling. Resilience and stability of ecological ecosystems. *Ann. Rev. Ecology and Systematics*, 4:1–23, 1973.
14. C. S. Holling. The resilience of terrestrial ecosystems: local surprise and global change. In W. C. Clark and R. E. Munn, editors, *Sustainable Development of the Biosphere*, Cambridge, 1986. International Institute for Applied Systems Analysis, Cambridge University Press.
15. International Union for Conservation of Nature and Natural Resources, United Nations Environment Programme, World Wildlife Fund. *Caring for the Earth: a Strategy for Sustainable Living*, Gland, 1991.
16. M. A. Janssen, J. Rotmans, and O. J. Vrieze. Climate change: optimization of response strategies. Report M 94-01, University of Limburg, Maastricht, 1994. Reports in Operations Research and Systems Theory.
17. V. F. Krapivin. Mathematical model for global ecological investigations. *Ecological Modelling*, 67:103–127, 1993.

18. J. Kwadijk and J. Rotmans. The impact of climate change on the River Rhine: a scenario study. *Climatic Change*, 30(40):397–425, 1995.
19. F. Langeweg. *Concern for Tomorrow: A National Environmental Survey 1985 - 2010.* National Institute for Public Health and Environmental Protection, Bilthoven, 1989.
20. J. E. Lovelock. *Gaia: A New Look at Life on Earth.* Oxford University Press, Oxford, 1979.
21. J. E. Lovelock. *The Ages of Gaia: A Biography of our Living Earth.* Norton and Company, New York, 1988.
22. W. J. M. Martens, J. Rotmans, and L. W. Niessen. Climate change and malaria risk: an integrated modelling approach. RIVM-Report 461502003, National Institute of Public Health and Environmental Protection, Bilthoven, 1994.
23. D. Meadows, J. Richardson, and G. Bruckmann. *Groping in the Dark - the First Decade of Global Modeling.* John Wiley & Sons, New York, 1982.
24. D. H. Meadows, D. L. Meadows, and J. Randers. *Beyond the Limits.* Chelsea Green Publishing Co., Post Mills, 1992.
25. D. H. Meadows, D. L. Meadows, J. Randers, and W. W. Behrens III. *The Limits To Growth: A Report for the Club of Rome's Project on the Predicament of Mankind.* Universe Books, New York, 1972.
26. G. M. Morgan and M. Henrion. *Uncertainty: A Guide to Dealing with Uncertainty in Quantitative Risk and Policy Analysis.* Cambridge University Press, Cambridge, 1990.
27. L. W. Niessen and J. Rotmans. Sustaining Health: Towards a Global Health Model. RIVM-Report 461502001, National Institute of Public Health and Environmental Protection, Bilthoven, 1993.
28. W. D. Nordhaus. An optimal transition path for controlling greenhouse gases. *Science*, 258:1315 pp., 1992.
29. OECD. Environmental indicators: basic concepts and terminology. In *Proceedings of the Workshop: Indicators for Use in Environmental Performance Reviews*, Paris, 1993. Organization for Economic Cooperation and Development. Background Paper no. 1, by Group on the State of the Environment.
30. N. Oreskes, K. Shrader-Frechette, and K. Belitz. Verification, validation, and confirmation of numerical models in the Earth sciences. *Science*, 263:641–646, 1994.
31. C. Rosenzweig and M. L. Parry. Potential impact of climate change on world food supply. *Nature*, 367:133–138, 1994.
32. J. Rotmans. *IMAGE: an Integrated Model to Assess the Greenhouse Effect.* Kluwer, Dordrecht, 1990.
33. J. Rotmans. Global change and sustainable development: a modelling perspective for the next decade. RIVM-Report 461502000, National Institute of Public Health and Environmental Protection, Bilthoven, 1994.
34. J. Rotmans and H. J. M. de Vries. *Perspectives on Global Change: The TARGETS Approach.* Cambridge University, Cambridge, 1997.
35. J. Rotmans and M. G. J. den Elzen. Modelling feedback processes in the carbon cycle: balancing the carbon budget. *Tellus*, 45B, 1993.
36. J. Rotmans, M. Hulme, and T. Downing. Climate change implications for Europe: an application of the ESCAPE model. *Global Environmental Change*, 1994.
37. B. D. Santer, T. M. L. Wigley, M. E. Schlesinger, and J. F. B. Mitchell. Developing climate scenarios from equilibrium GCM results. Report 47, Max Planck Institute für Meteorologie, Hamburg, 1990.
38. S. H. Schneider and P. J. Boston, editors. *Scientists on Gaia.* MIT Press, Cambridge, 1992.

39. M. Schwartz and M. Thompson. *Divided We Stand: Dedefining Politics, Technology and Social Choice.* Harvester Wheatsheaf, New York, 1990.
40. M. Thompson, R. Ellis, and A. Wildavsky. *Cultural Theory. Political Cultures Series.* Westview Press, Boulder, 1990.
41. P. Timmerman. Mythology and surprise in the sustainable development of the biosphere. In W. C. Clark and R. E. Munn, editors, *Sustainable Development of the Biosphere*, IIASA-Book, Cambridge, 1986. Cambridge University Press.
42. M. K. Tolba and O. A. El-Kholy, editors. *The World Environment 1972-1992: Two Decades of Challenge*, London, 1992. United Nations Environment Programme, Chapman & Hall.
43. D. van de Meent. SIMPLEBOX: a generic multimedia fate evaluation model. RIVM-Report 672720001, National Institute of Public Health and Environmental Protection, Bilthoven, 1993.
44. V. Vernadsky. The biosphere and the noosphere. *American Scientist*, 33(1):1–12, 1945.
45. M. M. Waldrop. *Complexity: The Emerging Science at the Edge of Chaos.* Contact Publishers (Dutch version), Amsterdam, 1992.
46. R. R. Wallace and B. G. Norton. Policy implications of Gaian theory. *Ecological Economics*, 6:103–118, 1992.
47. P. Westbroek. *Life as a Geological Force: Dynamics of the Earth.* Norton, New York, 1991.
48. T. M. L. Wigley. Balancing carbon budget: implications for projections of future carbon dioxide concentration changes. *Tellus*, 45B, 1993.
49. R. Wollast, F. T. Mackenzie, and L. Chou. *Interactions of C, N, P and S: Biogeochemical Cycles and Global Change*, volume 4 of *NATO ASI Series I.* Springer Verlag, Berlin, 1993.
50. World Commission on Environment and Development. *Our Common Future*, Oxford, 1987. Oxford University Press. The Brundtland Commission.

Commentary on "Global Change and Sustainable Development"

Max Tilzer

1 Introductory remarks

The presentation of Jan Rotmans has beautifully demonstrated the complexity which has to be tackled in order to define sustainable development strategies. In particular the aspect of stochastic and unpredictable processes deserves further consideration.

In the following I will propose six theses which attempt to demonstrate that the global system is even more complex than stated, and that some basic concepts have yet to be defined before we can hope to be able to develop models with the potential of providing the necessary foundations for decision making.

2 What is global change?

The earth system undergoes dynamic changes over a wide range of time scales ranging from minutes to millions of years. These changes can either be cyclic or directional. Effects of various causes of change are superimposed upon each other, thus leading to chaotic behaviour of the system, or at least to patterns that cannot be analyzed with respect to their relative significance in influencing the system as a whole. Anthropogenic changes can be viewed as *just one* set of effects caused by *just one* influencing factor. This view implies that "global change" has existed throughout earth history, long before the advent of humankind.

3 Natural versus anthropogenic changes

Our present knowledge of short-term changes occurring over time scales from years to decades in most cases is insufficient to separate natural from anthropogenic effects. This is because our data sets do not cover sufficiently long time spans, and because our knowledge of effects generated by certain influencing factors as yet is inadequate.

4 The role of humankind in global change

The concept of *sustainable development* at least implicitly is based on the notion that we can influence our own decisions and hence can control anthropogenic effects on the earth system. The difference between natural and anthropogenic change in the global system is that humans know (or at least believe they know) of their role in the global system. This knowledge is an indispensable prerequisite for any measures with the intention to alleviate or avert adverse anthropogenic effects on the environment. In many cases we at present do not yet have this knowledge.

5 What is sustainable development?

The term *sustainable development* in my view is somewhat misleading in that it implies that it is based on objective facts. In reality, however, this concept is based on *value judgements.* I therefore believe that the term *acceptable development* would be more appropriate.

6 Which future is acceptable?

What we consider acceptable is based on *cultural and ethical values.* These values vary between individuals within one culture, and even more so between cultures and socio-economic systems. The latter is because the needs of humans vary widely depending on a region's wealth and/or socio-economic philosophy. In this context even the question is unresolved whether conservation of nature is only for our own needs, or whether we have the obligation to preserve species of plants and animals for their own sake. When we share this latter view the protection of species requires no additional justification.

7 Ecosystems and socio-economic systems

The TARGETS approach attempts to model a *"grand unified earth system"* which includes the bio-geochemical and the human components. Both subsystems are complex and highly interactive. It is essential to know whether both subsystems obey identical or at least similar principles.

Commentary on "Global Change and Sustainable Development"

W. Windhorst

1 Introductory remarks

Looking at the expected results, as they have been formulated by Jan Rotmans, the conceptual model TARGETS aims to serve researchers, policy analysts and policy makers as an integrated assessment framework for questions connected with Global Change and Sustainable Development. TARGETS therefore shall cover quantitative and qualitative linkages among social, economic and biophysical processes to elucidate effects on ecosystems and humans on the global scale. Although the disaggregation to the level of "major world regions" is scheduled for the next phase of model development, the first version will have to take scale dependencies into account, just as, for example, the first version of the integrated soil model which is part of the biophysical model. However, having already entered this topic, it has to be questioned whether there is additional need to classify and to describe the interactions between human activities and global phenomena in order to identify those problems and actions which are predestined to be treated in the presented TARGETS approach. From the perspective of possible users of TARGETS the value of the model will depend on the extent, the model structure and model results will reveal the users' range of action and responsibility. If for example UNEP will use the model, the developed strategies will have to include a valuation of the actions being within the range of possibilities on the national level. In order to underpin the success of TARGETS it will be helpful to illuminate the presented approach with the following questions and remarks.

2 Are global approaches necessary?

As *Ravetz* [5] pointed out the prospects for our civilisation will be dependent on the answer to whether the technological development can be ecologically sustainable or not. Moreover, taking into account that meanwhile mankind is facing processes of globalisation in environmental as well as in economic affairs (*Henderson*, 1990 [2], *Laszlo*, 1992 [4]), it is obvious that methodological approaches to tackle global questions with a global concept, as presented by Jan Rotmans, are urgently needed.

3 How to enhance the acceptability of global approaches

The presented method of analysis is an excellent offer to enhance the communication process between different scientific disciplines, in order to elaborate harmonised statements, the TARGETS approach will help to answer the questions society is raising in a better way than was possible before now. There is considerable disagreement between the members of various constituencies as to what constitutes a technological risk, as well as about how such risks should be prioritised for regulatory attention (*Rayner*, 1986 [6]). Having reached a harmonised scientific statement the broad acceptance of the regulatory necessities is undoubtedly dependent on the acceptability of the aims and values applied (*Röglin*, 1990 [7]). So it will be necessary to couple the top-down approach with a bottom-up validation strategy. This includes the development of a harmonised framework of values that allow the balance of ecocentric and anthropocentric targets or goal functions, as well as nested measuring approaches covering results from the patch scale to the global scale. Otherwise it will not be possible to prove the relevance of the projections generated with the TARGETS approach and to achieve the desired acceptability.

4 Are there additional conditions for success?

At least at this point ecosystem theory will take the floor. It will be necessary to operationalise the fact that the whole is more than the sum of the parts, which means that we have to identify and to deal with emergent properties which describe the dynamics of the present (eco)systems as well as evolutionary processes. It will be of evident importance not only to identify the required aggregation level for the whole system or for the subsystem but to name the emergent property that reveals the behaviour of the (sub)system selected. An important question in this context is, if the Sustainable Development Index (SDI) proposed by Jan Rotmans meets this in a complementary way to existing measures like ascendancy, stability, efficiency, etc., to what extent will the SDI as a single index be able to assess the (sub)system behaviour as adequate, as is possible with the six basic orientors (existence, effectiveness, freedom of action, security, adaptivity, consideration of other systems) *Bossel* [1] proposed.

5 Communication and co-operation

In order to deal successfully with holistic questions such as Global (environmental) Change it will be necessary to develop a new quality of communication between the scientific disciplines involved; it will not be enough to read and to hear it will be necessary to experience the synergistic gains from interdisciplinary exercises. Scientists, scholars and policymakers *Ravetz* [5] will need to open up and share their genuine but limited insights of science, so that a common understanding, enriched and enhanced by dialogue, can emerge. Therefore both methodological approaches, like Jan Rotmans' presented, and even more interdisciplinary institutions like the Potsdam Institute for Climate Impact Research (PIK) or Ecosystem-Research centres are necessary if scientific progress is to cope with human-induced changes of the environment. Examples of the necessary co-operation have been mentioned during the symposium "Earth-System- Analysis". Thus Hans-Joachim Schellnhuber demonstrated the comparison between the so called modular and the parallel modelling approach, and Richard C. Rockwell opened the discussion about the adequate spatial resolution of analysis in Earth-System-Analysis. In all these cases I am almost sure that there will not be final and/or general solutions, because the complexity of the system(s) under investigation will force us to develop case and question oriented strategies with limited validity. To meet this limitation as efficiently as possible it will be necessary to use and to develop technical instruments to give full transparency and to allow quick access to those solutions found. As an example of an ongoing activity in this field I want to mention the UFIS-Project (*Knorrenschild* et al., 1993 [3]) which developed by close co-operation of several ecologists a documentation scheme for ecological models as a basis for an environmental research information system. All together, the approach of Jan Rotmans, institutions like PIK, Ecosystem-Research centres and instruments like UFIS are promising and necessary steps to overcome the structural amnesia in science described by *Rayner* [6]. However, the success of all participants will be dependent on the quality of co-operation they achieve at the national and international level.

References

1. H. Bossel. Real-structure process description as the basis of understanding ecosystems and their development. *Ecological Modelling*, 63:261–276, 1992.
2. H. Henderson. Der Einfluß gewandelter Paradigmen auf eine post-industrielle Welt. In M. Schüz, editor, *Risiko und Wagnis - Die Herausforderung der industriellen Welt 1*, pages 276–294, Neske Verlag, Pfullingen, 1990.
3. M. Knorrenschild, R. Lenz, E. Forster, and C. Herderich, editors. *UFIS - A Database of Ecological Models. Talk Presented at the International Congress on Modelling and Simulation*, Perth, Australia, 1993.
4. E. Laszlo. *Evolutionäres Management - Globale Handlungskonzepte.* Paida Verlag, Fulda, 1992.
5. J. R. Ravetz. Usable knowledge, usable ignorance: incomplete science with policy implications. In W. C. Clark and R. E. Munn, editors, *Sustainable Development of the Biosphere*, pages 415–432, Cambridge University Press, Cambridge, 1986.

6. S. Rayner. Commentary to Ravetz, J. R. In W. C. Clark and R. E. Munn, editors, *Sustainable Development of the Biosphere*, pages 432–434, Cambridge University Press, Cambridge, 1986.
7. H.-C. Röglin. Der chaotische Weg zur Vernunft - Risiko und Kommunikation in der Industriegesellschaft. In M. Schüz, editor, *Risiko und Wagnis - Die Herausforderung der industriellen Welt 2*, pages 32–45, Neske Verlag, Pfullingen, 1990.

Chapter 3: From a Fictional Globe to POETic Ecosystems: Modelling Human Interactions with the Environment

Richard C. Rockwell

Abstract Models at the global scale for the interactions of humans with the natural environment are criticized as insufficiently realistic and specific for either social science research or policy analysis. The author suggests building social-science models at a meso-scale level, which might be as small as a city or county, or in some cases as large as a continent. An example of how meso-scale modelling forces consideration of factors not usually taken into account at the global level is provided by the case of demographic modelling of the growth of the population of sub-Saharan Africa. Units of data collection and sampling are discussed, with consideration given to the nation-state as the unit most commonly employed in the social sciences and its probable inappropriateness for many applications. Finally, a unifying concept first advanced in the 1950s is put forward as often being appropriate for meeting social-science data collection and modelling needs – the concept of the "community," which is defined as a complex of population, environment, technology, and social organization.

1 Introduction

This chapter considers the question of which scale of quantitative models is most appropriate and powerful for research on human interactions with the biogeochemical environment. One possibility being actively explored today, including at the Potsdam Institute for Climate Impact Research, is the construction of a unified global-scale model. Constructed primarily in the interests of elegance, parsimony, tractability, and isomorphism with global General Circulation Models, this global social science model would employ one set of quantitative functions to describe all human impacts on the environment, all human adaptations to environmental changes, and all impacts of environmental changes on humans and their societies, wherever and whenever they occur. The classic example of such a model is World3, the famous foundation for *Limits to Growth* (*Meadows* et al., 1972 [9]) and its successor, *Beyond the Limits* (*Meadows* et al., 1992 [8]). World3 is perhaps the leading contender for the presumptive global model in the social sciences, but there are other serious contenders for what the author sees as a dubious goal. There are reasons to doubt that global-scale models can be sufficiently realistic and specific to support either social science research or policy analysis.

The alternative proposed by this chapter for the social sciences is the construction of numerous smaller-scale or meso-scale models, each tailored to a particular human population, that can be linked into one unified global *system* of social-science models. This system of models may then be linked to General Circulation Models at either the global level or regional levels. Possible intermediate or "meso" scales include continents (e.g., Australia), regions of continents (the Legal Amazon), nations (India), sub-national regions (the U.S. Desert Southwest), or even smaller units (the metropolitan area of Belo Horizonte, Brazil).

Social science theory and methodology are sufficiently powerful to construct models for meso-scale areas, which can be described in detail and have often been studied for years. Data collection at such scales can be done efficiently and at a reasonable cost. Social scientists know how to draw samples from such areas, and they have years of experience in analyzing such data, especially in the sub-discipline of sociology known as "human ecology"[1]. Before exploring the suggestion of meso-scale models, this article states methodological reasons for holding that building global-scale models is unlikely to be a fruitful endeavor in the social sciences. An example of fallacious application of a global model to a continent is provided.

2 The concept of a global system

To the natural scientists who formulated the objective of the International Geosphere-Biosphere Programme, it was unproblematic to speak of "the interactive physical, chemical, and biological processes that regulate the *total Earth system...*" (*IGBP*, 1990 [6]). It was clear that such a global system does exist and that the processes regulating it need to be understood at the global level as well as at regional levels. The idea that changes in the whole earth system can be induced by human action was arguably one of the prime intellectual accomplishments of the natural sciences in the late nineteenth century. It represented a substantial shift from the older viewpoint that, on a large scale, nature is impervious to human action.

Research spurred by this hypothesis has led to documentation of the fact that today's human beings are not only capable of causing systemic changes at the level of the entire globe over periods of decades to centuries but also that they have already caused such changes. It is now widely recognized that human actions are affecting the global climate system, the processes for cleansing the atmosphere, the transport and deposition of acids and particulate contaminants, and the penetration of Earth's atmosphere by ultraviolet radiation. These environmental changes are systemically global and are modelled as such. In addition, through a diverse set of local actions humans are effecting other changes that accumulate to global-scale changes: lessening

[1] What no one seems yet to know is how to devise a worldwide sampling design for selecting these areas, a problem apparently shared by ecologists designing sample surveys to measure biodiversity.

biodiversity worldwide, reducing forest cover, changing marginal lands into deserts, diminishing wetlands and coastal marshes, reducing the quantity and quality of water supplies, and depositing nuclear and other toxic wastes.

These systemic and cumulative biogeochemical changes are occurring in the context of sweeping changes in human societies that are also having global impacts. Industrialization has been changing the face of the earth for more than three centuries, but the pace of industrialization and economic development has quickened in many developing countries in just the past decade. Global trade and the ensuing global inter-dependence are becoming realities for even the least-developed societies. The telecommunications revolution, coupled with the increasing ease of intercontinental transportation, supports not only global trade but potentially also global learning and consensus-formation. This increased exposure to world society also supports a "revolution of rising expectations" by displaying the lifestyles of the older industrialized countries as models for the industrializing countries. The human population continues to grow but at sharply different rates around the world, increasing pressure on the environment. Large cities will become the predominant form of human habitation sometime around the millennium, continuing a trend towards urbanization that began with the advent of agriculture. Privatization of economic activity has, in one form or another, displaced other economic systems in most of the world, and its calculus is affecting environmental decision-making everywhere. Democracy has spread into every continent, even though it is still fragile in many countries. These worldwide changes are of intrinsic interest to social scientists, but their interactions with global-scale changes occurring in the natural environment are especially interesting to many of us.

The need to understand the multifarious interactions of humans and nature has opened the door to unprecedented cooperation between social and natural scientists. However, if that cooperation is to yield solid and timely results, many conceptual and methodological problems must be sorted out, ranging from technical measurement issues to variant meanings of fundamental terms such as "prediction" and "model". One of the more prominent issues concerns the appropriate scale of models for the social sciences and, subsequently, how social science models at one scale can be linked to natural science models at another scale.

The temptation demonstrably exists in the social sciences to mimic the natural sciences and to construct social-science analogues of their global-scale models (*vide* the Club of Rome). However, this choice implicitly assumes the invariance of human activity around the world, just as biogeochemical processes are invariant around the world. It assumes that a single process of agricultural production or human reproduction operates without variation around the world, or at least that the processes are sufficiently similar worldwide that it is useful to model the processes as globally invariant. That is arguably not the case for the phenomena studied by the social sciences, while it is often or always the case for the natural sciences.

To be sure, the biology of human reproduction or the biology of the growth of plants do not change from society to society. Invariant biology is of far less importance to the population growth rate or to whether people are adequately fed than is the human context in which biology operates. Biology sets the limits; humans choose where their societies will perform within those limits.

The same physical, chemical, and biological processes operate in, say, Southern California as operate elsewhere. The culture and politics of Southern California, no matter how different from the rest of the world, do not influence the nature of the biogeochemical processes being modelled there by natural scientists. The carbon cycle does not have a different chemistry in Santa Barbara than in Rome. Nor do the special characteristics of the Southern California coast create a demand for new "laws" pertaining to photosynthesis and plant respiration – although the particular species involved, the settings in which the processes take place, and their contributions to composition of the global atmosphere do differ. To be sure, a global model in the natural sciences could be refined at a regional level to provide additional information about specific ecosystems or specific climates, but that would simply be an enhancement built upon a common biogeochemical foundation.

3 The lack of comparable global processes in the social domain

Is there anything equivalent to "the total Earth system" for the social scientists who are seeking to understand the human contributions to global changes and the effects of these changes on humans? Is there a solid foundation upon which to build a quantitative model that pertains to all of the Earth's peoples and societies? Are there economic, social, cultural, or political processes that operate at the level of the entire globe or invariantly around the globe, and that can be usefully modelled at that level?

This chapter answers "no" to these rhetorical questions by addressing two issues. First, it considers whether a model with a single functional form can be globally applied to human systems. Second, it proposes the alternative of meso-scale models. The fundamental point is that humanity and its activities are extremely heterogeneous, and that the modelling enterprise must reflect that heterogeneity. Because of this heterogeneity, it is argued that social science models cannot realistically treat the globe as a single functional system[2]

[2]There has been a movement, based primarily in sociology and political economy, for studies of "the world system," with Immanuel M. Wallerstein as one of its leaders. Focusing largely on trade and exploitative economic and political relationships, it has had little to say about how humans affect the environment or how humans are affected by the environment. (See *Wallerstein*, 1974 [17].)

This point of view is consistent with prevailing trends in the social sciences, in which history and cultural context have assumed new importance, and "Grand Theory" and the search for "universal laws" correspondingly less importance. The thesis of this article cannot be proven, for it is a negative: global-scale models will not work in the social sciences. It is logically impossible to prove the truth of such a statement. However, it is possible to provide reason for doubt and to seek to direct valuable time and resources to what the author believes to be more likely to be a fruitful endeavor.

This is not to say that human actions do not have global effects. Quite the contrary: human actions or inactions are the proximate cause of most of the global changes observed today. For example, the human role in altering the carbon cycle clearly has global implications. However, the ways in which humans alter the carbon cycle are not global systems in and of themselves. All human activities occur in local context, and that context is conditioned by local history, economics, culture, politics, and social organization. The human activities that alter the carbon cycle range from the burning of biomass for land clearing to the combustion of coal for energy production, and all of these activities occur at quite local levels. The cultural, historical, political, social, and economic contexts in which these activities occur differ widely around the globe. There are obvious differences in these processes between the advanced industrial societies and the developing and largely agrarian societies, and less obvious differences within these broad groups[3].

Perhaps more importantly, the underlying causes of the proximate human actions or inactions that cause global changes are themselves heterogeneous. These underlying causes include growth in the human population and its uneven dispersion around the planet, rising levels of consumption in many societies, the adoption of sustainable or unsustainable life-styles, the utilization of efficient or inefficient energy technologies, the presence or absence of "green politics," whether a population is generally wealthy or impoverished, etc. Models for these underlying causes of environmental change must reflect the heterogeneity of human society.

4 Demographic projections of global population growth

Arguably the most important underlying cause of environmental change is the growth of the human population, which is by no means an invariant process around the globe. This section considers a quantitative model that assumes

[3]In *Beyond the Limits* [8], the 1992's successor to *Limits to Growth* [10], the model consists of five sectors: persistent pollution, nonrenewable resources, population, agriculture (food production, land fertility, and land development and loss), and economy (industrial output, services output, and jobs). All of these sectors are modelled on a global basis, using global statistics and global estimates. Even the differences between developed and developing societies are elided in this formulation.

a single set of equations for population growth throughout the globe. Demographic models are always deeply informed by data, and these data were generated from collections of local measurements. In this regard, such models resemble biogeochemical models, which also depend on local measurements. However, the suggestion here is that demographic phenomena are not adequately modelled if it is assumed that the same simple processes operate in exactly the same way throughout the world. In particular, the argument is that the carrying capacity of the land where the people actually live – not the carrying capacity of the entire globe – imposes constraints on population growth that are ignored in global-scale modelling.

Population modelling is primarily a science of very few variables, which may account for much of its strength as a rigorous discipline. Population demographers are concerned with birth, maturation, reproduction, migration, and death. People are born, they mature, about half of them are capable of reproducing, many of these people do reproduce, some of them move, and then all die. When in the life span these events occur and at what rates are the crucial factors in population projections. It is obvious that a society increasing at a net growth rate of four percent a year will in a few years look very different from a population of the same initial size that is increasing at the rate of only one percent. However, these growth rates are very blunt summaries of more subtle differences among societies. The same four percent growth rate could result from several different combinations of birth and death rates. It could be found in populations with different compositions of sex and age. Holding birth rates constant, a society with an age composition tilted towards the younger ages has a far different fertility future than does a society not so tilted. This is why simple exponential growth projections that ignore age and sex composition are not useful for human populations and are commonly so recognized as inadequate.

Population demographers are, of course, concerned with more than the four or five vital events. They have also long contemplated the classic Malthusian factors of war, pestilence, famine, and "moral restraint". One reason for their attention to these factors (the first three of which increase death rates and can decrease birth rates) was that only such large-scale perturbations seemed capable of disturbing what has been called the "incredible inertia" of demographic change. This inertia is the result of the momentum of growth that is locked into place by the future reproductive potential of a population that, because of its earlier growth, is tilted in the direction of younger age groups, and specifically towards women still in their childbearing years or soon to come into their childbearing years.

This inertia resembles the inexorable phenomena noted in studies of global change, in which, for example, the concentration of CFCs in the stratosphere will apparently increase for about another fifty years, even if all emissions of CFCs stop immediately. Assume that fertility everywhere instantaneously drops today to the replacement level of about 2.1 children per woman that is now required for developed countries[4]. If fertility dropped that fast and that far, the global population would nevertheless continue to grow for about another 50 years because most of the women of childbearing age have already been born. It is, of course, unreasonable to expect such a dramatic decrease in fertility, leading to the firm expectation that – *sans* Malthusian events – the global population will continue to grow throughout our lifetimes and most of the lifetimes of our children.

The United Nations "medium-fertility" projection of global population as of 1989 was about 8.2 billion people by the year 2020. The lowest projection was a population of 6.5 billion. There are three strong assumptions behind the medium projection: (1) the continuation of favorable mortality trends that will produce a life expectancy in developing countries of about 65 years, which is about the same as that experienced in North America 70 years ago; (2) the leveling-off of mortality improvement in developed nations, at about 80 years life expectancy (as opposed to a continued increase in human life spans); and (3) a decline in fertility rates so that the number of children per woman in Africa declines to the level of about 3.5 children, which is about the same as in North America during its "Baby Boom" years[5]. Demeny reports United Nations projections as of 1989. Those projections have since changed slightly but without effect on the arguments in this article. In addition, updated rates of change and projections have been drawn from Bureau of the Census, (U. S. Department of Commerce, September 1992 [16]). Questions can be raised about whether these assumptions are reasonable, particularly the assumption of continued declines in fertility rates in developing countries. If that assumption is not made, the global population could be as much as three billion persons greater in 2020 than in the 1989 medium projection.

Since this 1989 projection, world population growth has slowed. The United Nations medium-fertility projection as of 1996 is that "only" some 9.4 billion people will inhabit the earth in 2050, substantially below the forecasts of just two years preceding and only 1.2 billion people more than previously forecast for 2020. The years 1985-1990 may have been the peak years of absolute population growth, with some 87 million persons added each year. Today, the earth is absorbing around 81 million persons each year. For those concerned with over-population, this is welcome news. However, the news is not all good, nor is it everywhere the same. The number "81 million" is still a large increment to the population, and what is most disturbing about the addition of these 81 million people is that many of them will be born into those regions of the world that are least able to support them.

---

[4]The replacement level is the fertility that is required for a woman to replace herself and her partner in the population; it is slightly higher than 2.0 in developed countries because of the need to account for failure to reproduce for a variety of reasons, including deaths at young ages. It is much higher in developing countries with high mortality.

[5]This discussion is largely based upon Paul *Demeny*, "Population," in B. L. *Turner* II et al. (eds.) [13, pp. 41–54].

There are huge regional differences in rates of population growth. Despite fertility declines in some countries, Sub-Saharan Africa is still the fastest-growing region of the world, at 3.1% each year, compared to 1.9% each year in Asia (excluding Japan). Even within regions, there are large differences. The birth rate in Cuba is 1.9%, while across the channel in Haiti the birth rate is 6.3%. These regional differences are subsumed into global projections and blurred accordingly.

It is not only in growth rates and birth rates that the regions of the world differ. Regions also differ in their economies, cultures, politics, and histories. This section focuses on Africa as a regional instance of a projection made fallible because these other differences are ignored. The proposition is that Africa will not be able to sustain the population growth that is forecast for it in a simple model consisting only of birth, death, and migration rates. The carrying capacity of sub-Saharan Africa is probably inadequate to support this population, and this deficiency of carrying capacity has little to do with arable land, soil nutrients, or water. This case is illustrative of the hazards of ignoring regional and sub-regional differences.

5 Global projections are fallible in regional context

In the period leading up to 2020, Africa has been projected to at least double its population density, growing to more than 140 persons per square mile in the UN medium projections model. Under this model, the number of persons added in Africa between 1985 and 2020 will be about 884 million. By 2020 about eighteen percent of the world's population will live in Africa, or 1.5 billion people. Can Africa support this additional population?

There is superficial reason to believe that it could do so. Malthus underestimated the capacity of agricultural technology to increase productivity in order to feed growing populations. The famines that he had feared in Europe did not occur on the scales he had foreseen. Wars have indeed killed large numbers of people, but never enough (not yet) to slow growth appreciably over the long term. Until this century, wars primarily killed the most dispensable portion of the human population, young males, of whom society needs relatively few in order to maintain high levels of reproduction. Disease continues to wrack human populations, but even the African AIDS epidemic is still forecast, on a continental scale, only to slow the rate of population growth, not to result in population decline. On the scale of villages and perhaps even of nations, AIDS can cause population declines, but diseases are not, in general, very efficient means of reversing population growth patterns. A far more pervasive disease, malaria, co-exists with very high levels of population growth virtually wherever it is found. Perversely, disease can be a spur to increased fertility and population growth, as parents ensure the survival of at least some of their children by having many of them.

For many generations the classic Malthusian factors have not played a large role in controlling the growth of the human population. Since the beginning of the Industrial Revolution, technology, buttressed and facilitated by social organization, has usually risen to the occasion when growing populations put greater demands on agriculture, water supplies, and energy resources. It was deliberate interference with agricultural technology and the social organization supporting it which caused the most devastating contemporary famine, that in China during its Great Leap Forward.

Over the next quarter of a century, will agricultural technology again rise to the occasion so that these projected additional 884 million people in Africa can be supported? There is surely space in Africa into which to expand agriculture and animal husbandry, because Africa presently has low continental population density. In fact, as is true almost everywhere, the African population tends to be concentrated on the continental margins. After the uncultivated arable land of the margins is brought into production, there remains much land in the interior that is ostensibly open for major agricultural expansion. However, much of the interior has long been the domain of the tsetse fly, which is the carrier of a sleeping sickness to which indigenous African wildlife is resistant but to which humans and their herds are not resistant. The problem of the tsetse fly is now being resolved and soon the interior can indeed be opened to production – at great loss, to the concern of many, to the species, genetic, and ecosystem biodiversity of Africa. This great expansion into the interior assumes a considerable degree of economic development. Where will Africa find the capital needed to develop first its coastal margins and then its great interior? Where will it find the funding for irrigation, fertilizers, food storage, food processing, and distribution systems? Note that Africa will be investing in this production technology at the same time that its growing population is in need of other services, such as housing, health care, and education. Moreover, history is not encouraging that this capital will be available. For two decades Africa experienced negative economic growth of about one percent each year, with upward trends only very recently. It would take a confirmed optimist to deny this history's relevance for the future.

Assuming that the capital is available for these investments in agriculture, there remain other questions. Where will Africa find the energy that is needed to grow the energy-intensive crops of the Green Revolution and the energy to process and distribute all this new food? Where will it find the clean water for humans and their herds? Where is the new form of social organization that can create the cultivation, harvesting, storage, and distribution systems that would be needed to feed a doubled population? Noting the persistence of food habits, we must also wonder whether Africans will eat the crops that can be grown in profusion; all over the world, people have starved in the presence of food that they did not know how to eat or would not eat. Cultural habits persist even in trying circumstances. This vast agricultural expansion will have to occur in an historical context in which per capita food production in Africa declined in most years since the 1980s, shifting some African countries from net food-exporters to net food-importers.

One possibility is that most of these 884 million additional people will till family farms, living directly off the land on traditional crops at a subsistence level and using traditional methods of farming. Subsistence farming requires little or no capital and social organization, so many of the concerns raised here are not germane. However, subsistence farming is just that – it produces no food for the cities. Moreover, subsistence farming will require that much larger portions of the continent be brought into cultivation (largely through deforestation), because capital-poor farming is less efficient than capitalized farming in terms of productivity per acre. Most of the people added to the population will be young and thus will be less productive as workers on these subsistence farms; the oldest will be 35.

Further, the major migrations required to colonize the interior with subsistence farmers seem unlikely. A rural pattern of population growth would be a reversal of present trends in Africa and elsewhere, for it is the cities that are growing most rapidly. Reversal of this pattern of urbanization could have undesired side effects. Urbanization has been key to reducing levels of fertility through its provision of job opportunities and education for women. However, it might happen that Africa becomes a largely agrarian continent – assuming that the soils, rainfall patterns, and endemic diseases permit the land to be farmed so extensively. The effects of the necessary widespread deforestation on regional climatic conditions and on the expansion of African deserts would have to be evaluated. It is conceivable that these new farms would last no longer than the short-lived farms and ranches of the Amazon.

If Africa does not grow all the food it newly needs, perhaps it could import the large quantities of food and energy required to support this additional population. Where would it find the money or the goods to trade for these enormous imports? What does Africa have of which the world wants enough that trade in that commodity will annually earn the costs of feeding a doubled population – and are those resources distributed throughout Africa as the population is distributed? Diamonds, chromium, uranium, cocoa, and timber exports may not suffice to feed a continent. For African participation in the global economy on this scale – and this rapidly seems highly unlikely – this change would have to occur in less than a quarter of a century. Even if it did occur, within little more than two decades Africa would have to create an exceptional system for storage and distribution of this vast quantity of imported food.

If purchases cannot occur, perhaps the wealthier countries of the world will provide large food subsidies to Africa on a very much larger scale than they do today. Or perhaps these countries will go further and provide the capital needed to permit Africa to develop agricultural self-sufficiency. However, it seems unreasonable to believe that the industrialized countries will engage in that large a transfer of resources during the next 25 years, given present retrenchments in bilateral and multilateral aid programs in many developed nations. What would be required from the wealthier countries is on the scale of transferring to Africa the equivalent of the infrastructure presently supporting the entirety of the food, water, and energy systems of India, itself hardly a model for how to feed a population. This may well be beyond the capability of the world community, some of which will be otherwise occupied in dealing with its own population growth.

Emigration from Africa might be a substantial part of the answer. If emigrants from Africa are partitioned among other countries on the basis of those countries' present population size so as to hold Africa's population at 1985 levels, of the 884 million the U. S. would accept some 41 million immigrants over the next 25 years. Germany would accept 13 million; and China, about 188 million. The latter figure can be put another way: in 25 years China would have to absorb migrants numbering the equivalent of the total population of the U. S. in 1963 – and it would have to augment its own agricultural system to feed those additions to its population. Environmental refugees are certain to be a prominent aspect of world politics under a scenario of population growth and environmental changes, but it seems unlikely that emigration on such a scale would occur. Some significant fraction of the projected additional 884 million Africans would be likely to remain in Africa.

This discussion of capital and infrastructural needs has ignored a larger question behind this volume: what might be the effects of climate change on the productivity of African crops? Of depletion of the ozone layer? Of carbon dioxide fertilization? The early part of the next century will see the convergence of environmental trends with which humans have no experience and over which they apparently have little immediate control. At the same time that the global population is growing at a high absolute rate, the world is entering into a period with many unknowns about whether the human-constructed systems on which we have depended for food and water will continue to function as well as at present. If climate change were only to shift precipitation patterns in Africa just a bit, so that the same amount of rain fell annually but at less desirable times during the growing seasons, the impact on food production could be enormous. If ozone depletion increased ultraviolet radiation damage to a few crops, this impact could be exaggerated. Carbon dioxide fertilization might not offset this damage, and there is some evidence that plants grown under conditions of carbon dioxide fertilization are less nutritious to humans.

6 Carrying capacity and human will

The above is basically an argument based upon the idea that the "carrying capacity" of the African continent is more a function of human social organization, culture, and economics than it is of biological and physical environmental factors. Good land, water, and sun cannot offset inadequate organization, culture, and economics. Social organization, culture, and economics that were sufficient for a smaller population could fail when confronted by population increases of the size projected for Africa. Limits to growth will result from human action or inaction well before they are imposed by physical and biological parameters, and failures of carrying capacity will occur at local and regional levels even when the globe is theoretically capable of supporting a larger population.

The concept of carrying capacity has been based in the biological study of "natural" ecosystems in which human culture and politics play no salient role. Conventionally defined, an ecosystem consists of abiotic elements (air, water, minerals, energy) and biotic elements (plants, animals, microorganisms) that interact and are inter-related. An ecosystem can be as small as a forest tract or (in loose usage) as large as the entire planet. Ecosystems are open on both the input and the output sides, in that organisms migrate, materials flow in and out, and energy enters and leaves the system (*Eblen* and *Eblen*, 1994 [2]). The volume and content of the trans-border flows reveal whether an ecosystem can sustain itself without drawing down the capacity of other ecosystems to sustain themselves.

Human-influenced ecosystems have all these characteristics and more. It is both the deliberate organization of biotic and abiotic elements through purposive human action and the human management of flows across borders that transform the concept of carrying capacity from a purely biological concept to a biosocial concept. Whether in natural or human-influenced ecosystems, carrying capacity is limited by the capacity of the ecosystem's weakest component. In many cases the weakest component is the food supply, and humans control that supply through both willful and inadvertent acts.

As previously admitted, the historical experience has been that the carrying capacity of the planet – viewed from the standpoint of how many humans it can support, not how many wolves or birds – can be endlessly expanded by technological and social changes. We should question whether this historical observation is forever a basis for sound modelling or sound policy at the levels of particular peoples in particular situations. The cornucopians may be right for much of the world, but the immediate question is whether the cornucopia also overflows for Africa in the next quarter of a century. Even if the world should produce enough food to feed Africa from its surplus, it is not a given fact that this food will find its way into African mouths. It might just be added to the waste already produced by developed nations – waste that is already not finding its way into starving mouths.

Note carefully that this argument is not challenging the capacity of the globe to carry a population of 8.2 billion. It challenges only the capacity of one continent at a particular time in history to carry a population of 1.5 billion. Under optimistic scenarios, the world could reach an astonishing population level while maintaining or even improving the average quality of life, but what is true for the world as a whole need not be true for Africa or for any other particular place. Much of that worldwide population growth will occur in Asia, and the quality of life in Asia might continue to improve. Although in 2020 more of humanity will live in Asia than in the rest of the world combined, the Asian population will have grown at the leisurely rate of 1.2 %. This is like the growth rate in North America in the early 1970s. Normal economic growth could accommodate this rate of population increase, and much of Asia is experiencing far better than "normal" economic growth. Within Asia, it is primarily in India that population growth may raise some of the same concerns as in Africa, but even there growth rates are falling (*Freedman*, 1995 [3]). The Asian population is growing at a far less rapid rate than in Africa, and Asian economies are growing at a far more rapid rate. The African experience may be far different from the Asian experience.

The concept of limits to carrying capacity may be largely irrelevant at foreseeable population levels when applied to the entire globe, but at the geographic levels at which humans actually produce and consume food, it may be all too meaningful. We must therefore consider and plan for the possibility that the African population will exceed the carrying capacity of the continental ecosystem in the medium-term future. If that dire possibility seems far fetched when applied to the projected African population of 1.5 billion in 2020, perhaps it will seem less far fetched for the African population of about 3.5 billion projected for 2050. There must be limits at some point. We as a species have continually extended our ecological niche through technological change, but perhaps *for some of us* the niche has little more room for expansion. We cannot plan for the future on the basis of the ridiculous assumption that all humans everywhere have equal access to the entire globe's productivity.

This concern about the carrying capacity of the African continent is far from an optimistic vision of a world saved from a burden of 8.2 billion people. It is a specter of starving people, particularly starving children, elderly people, and women; of life-shortening diseases associated with malnutrition; of strife as societies fall apart under the combined pressures of growing populations and declining living standards; and perhaps of the crash of this large part of the human population. To be sure, the 1.5 million Africans projected for 2020 would represent only about 18% of the total human population. Whatever happens in Africa would not have a major effect on global population, compared to the effect if similar events occurred in Asia. That does not mean that the fate of Africa can be ignored, if for no other reason than that this is one world: the smoke from burning African savannas and the dust from wind-eroded African fields already form plumes reaching across the Atlantic towards the Americas.

7 On building a less-fallible population growth model for Africa

Population projections that have assumed a growth of the human population unbridled by considerations of carrying capacity, because all the demographic factors point in that direction, must begin to take account of whether humans have in some places finally increased to such numbers that technology cannot practically sustain us. The mathematics of population growth must soon confront the reality of the carrying capacity of specific ecosystems. Demographers will have to learn some ecology, returning in a sense to the discipline of "territorial demography" advocated by C. H. Cooley at the end of the nineteenth century. Demographic models must become much more complex[6].

More complex demographic models must take account of variations in carrying capacity at meso-scale levels. They cannot attend only to demographic vital events and ignore food, water, energy, social organization, and capital requirements. What kind of African regional model(s) ought to be constructed on the way to a global system of models of population growth? Whether a single regional model is appropriate for Sub-Saharan Africa or whether several sub-continental or even smaller models are required should be evaluated. Among other factors, projections should take account of:

- Agricultural productivity and the extent to which it can feasibly be increased by technological improvements.
- Availability of arable land.
- Barriers to expansion into available land.
- Capabilities of social and political organizations for harvesting, storing, processing, and transporting food.
- Capital resources and competition for those resources from other sectors.
- Cultural preferences regarding the organization of farming and herding.
- Food preferences and their mutability.
- Labor force composition effects.
- Labor force preparation for employing advanced technologies.
- Other natural resources, including energy and soil nutrient levels.
- Political stability and efficacy of governments as a facilitator or barrier to improvement of agricultural production.
- Technology already in place or transferable to the continent.
- The effects of climate change, ozone depletion, and carbon dioxide fertilization on agricultural production.
- Water supply for agriculture and human consumption.

Building these models is not a job for demographers alone. It requires the participation of both ecologists and Africa specialists. The involvement of African social scientists – indeed, their leadership – is key to producing

[6] An excellent discussion of the building of regional-scale models integrating physical, biological, and social sciences is given by *Groffman* and *Likens* [4].

models that not only make sense but also that will carry weight with Africans and their governments.

8 Linking meso-scale models into a global system of models

Across the seven continents and the many island systems, few population growth models will have exactly the same content as an African model, and the functional forms of models may well differ among regions. That is not in itself a problem for the goal of linking these models into a global system of models. Meso-scale models may be linked into a global system if each regional model produces as its output projections of the growth, composition, and distribution of population, and each is capable of accepting, as input, exogenous shocks such as the supposed effects of climate change.

An exemplar for constructing a global system of population-growth models is that of Project Link, a non-governmental, cooperative consortium of more than 250 researchers that links 80 diverse national econometric models into a world econometric model. Its most important feature is that the national models are constructed by economists resident in each country to represent their particular economic conditions. Models range from systems of a few dozen equations to systems of several thousand equations. The models are linked into a worldwide system of models through "trade flows, price linkages, capital flows, interest rates, exchange rates, migration, technology transfers, and global commodity markets" (*WWW*, 1997 [18]). Similar linkages among meso-scale models can be constructed for demographic models, and a similar consortium can be created. The Potsdam Institute could take a lead in establishing such a consortium.

The argument for building a global system from differing parts is not simply that global models lack credibility and specificity when held up against the local context and the imperative for local action. More importantly, global-scale models can be misleading and untrustworthy for both scientific and policy purposes. Global models can provide useful alerts or wake-up calls, as perhaps did the Club of Rome reports, but they are unlikely to be useful for policy formulation and evaluation. It is even conceivable that these models provide too mild an alert, because the vagueness required for global modelling may reduce the scale and scope of the changes that they forecast.

Consider the relationship of the growth of the human population to deforestation of the land surface. Across many millennia, the decline in forest cover is correlated with the growth of the human population. Most studies of the relationship assert that the driving force of deforestation is either a high rate of population growth or high population density. Alberto Palloni quotes a leading researcher as asserting: "The slow and almost imperceptible increase in population during premodern times and its rapid rise from roughly

1600 onwards has led to a steady decrease of the world's forests" (1994 [12]). *Palloni* finds this causal assertion to be far too simple, noting that there are "important social institutions that create an environment where population pressure may or may not affect deforestation" (*Palloni*, 1994 [12]). This is precisely the argument made in this chapter.

Alternatively, consider the potential to mislead of the graph that opens the International Geosphere-Biosphere Programme's major report on its initial core projects (*IGBP*, 1990 [6]). This graph shows world population from 1800 through 2000, superimposed upon graphs of the concentrations of carbon dioxide and methane in the atmosphere for the same period. The curves follow strikingly similar paths. It is tempting to draw the conclusion that something causal is shown here, and the impression left behind is probably that the growth of the human population drives everything else. The policy implications are starkly clear in this simplistic understanding. Yet only a bit of digging will uncover the fact that it is principally the increased burning of fossil fuels that accounts for these two changes in atmospheric composition, and that the nations (and continents) with the highest population growth rates are not where the use of fossil fuels most accelerated over three centuries. In fact, the fastest-growing parts of the human population still contribute little to changes in atmospheric composition. Birth control programs in Africa will do little to reduce the atmospheric load of greenhouse gases.

9 Units of data collection, sampling designs, modelling, and the scope of generalizations

To recapitulate the fundamental recommendation of this chapter: social science models should be designed at the level of analytical geography that makes sense for the phenomena being modelled. However, this recommendation begs the question of what level of analytical geography does make sense for a given model. What is the appropriate political, economic, or geographical unit for modelling purposes? Do we treat nation-states as our modelling units, or do we use some other geographical/economic/political unit? While offering no definitive answer, I do propose desiderata for choosing these units, beginning with consideration of the drawing of samples.

Sampling designs are of tremendous importance for the utility of results from research. No one would believe that a study of Southern California residents could be extrapolated to the entire United States, just as no one ought to believe that a study of methane production in one paddy in Southeast Asia can be extrapolated to methane production from rice agriculture worldwide. If one wishes to describe the culture of the United States, then one must sample the entire United States.

However, this presents a serious problem for research on humans and their environments. There is no sense in which the people of the United States can reasonably be treated as if they live in a common natural environment – exposed to the same levels of air and water pollution, contributing the same amounts of carbon monoxide to the atmosphere, experiencing the same effects of ozone depletion. To each individual observation we must attach observations from the local environment in which the individual lives; this is the difficulty, because in order to say anything with confidence about how the residents of Southern California interact with their environment, we need a good sample of the residents of that region.

The scope of the generalizations that we wish to draw is intricately related to the sampling design. If we have no interest in describing relationships at the level of Southern California, we do not have to have a representative sample of Southern Californians, but if Southern California is what we want to study, then it turns out that we need a sizable sample of Southern Californians, ranging from 500 to 1 500, depending upon how much error due to sampling we are willing to tolerate[7]. In fact, we need samples of around 500-1 500 residents of any area that we wish to describe, regardless of its population size. If we wish to study the relationships between humans and their activities at the level of the United States, we need one sample of about 500-1 500 people that is representative of the entire population. If we wish to study those relationships within each state, then we need 50 such samples. That aggregate sample would be quite expensive, probably demanding levels of science funding that will not be available.

The problem is compounded by the fact that many natural science measurements make no sense at anything other than a local level. If we are to connect human activities to environmental measurements, we must find a way to accommodate the measurement systems of the natural sciences while maintaining the ability to pursue the goals of the social sciences.

An example of both the potential and the difficulty of connecting natural and social science measurements is research on air pollution. One could do a national survey of attitudes to air pollution without regard to actual levels of exposure to pollution, but the results would be of dubious value. Would one not want to know, first, to what kinds and quantities of air pollution the respondents are directly exposed, and second, what they do to contribute to it? Actual measures of various aspects of air pollution are required. However, with the possible exception of Liechtenstein, a national estimate of levels of air pollution would tell us little about a particular respondent's exposure to air pollution. It would be quite meaningless except for the purpose of comparing countries' aggregate contributions to a mean global pollution level. It is local

---

[7]With strong models, sample sizes can be reduced, but there is little reason to believe that the social sciences will have such strong models anytime soon. The basic point is that we need an adequate representative sample of any area about which we want to draw generalizations.

air pollution that most affects the lives of respondents to a survey; it is that which must be measured[8].

We are then in the situation of needing local measurements in the social sciences as well as in the natural sciences. Does this mean that our research cannot be done because we need to have representative samples of every locality in a country, and we know that we cannot afford that? Not necessarily; it depends upon our choice of the units for which we will model. If we want only to describe the relationships of human activities to environmental change at the national level, then we need a nationally representative sample with local measurements attached to individual observations. However, if we want to discuss how environmental conditions and interactions will differ in the Great Plains and the Atlantic coast, we do need representative samples from those areas.

10 The nation-state as the presumptive modelling unit

As Harold K. Jacobson makes clear in his chapter in this volume, modelling at the level of the nation-state often provides the critical information that will be needed for policy formulation and evaluation. In much of the world, it is the nation-state that defines the territory occupied by a society, establishes economic policies, sanctions international trade, exercises police and military powers, and regulates human interactions with the environment. In some states the core energy, transportation, extraction, and heavy industrial economic sectors are all in the hands of the government. It is states that are involved in the international negotiations on climate change, ozone depletion, and loss of biodiversity, through the international organizations created by them. The concept of national sovereignty is still fiercely protected, at least at the level of rhetoric, by every government. One cannot imagine a successful climate change policy that bypasses the nation-states of the world and attempts to reach people through entities such as local governments, non-governmental organizations, or corporations – although all of these entities will also be important actors. So too will be the supra-national governments, such as the European Union, and trade and cooperation institutions embodying agreements among governments, such as GATT (General Agreement on Tariffs and Trade), NAFTA (North American Free Trade Agreement), and ASEAN (Association of South-East Asian Nations)[9].

That said, nation-states will prove, I believe, to be unwieldy units for the task of linking social science models with natural science models. Natural science models are rarely bounded by the nation-state, and natural science data

[8]People are also affected by distant air pollution, in the form of acid precipitation and possible global warming.

[9]For a fuller discussion of the difficulties of the concept of the nation-state, see Richard H. *Moss* and Richard C. *Rockwell*, 1990 [11].

(particularly data collected by remote sensing) are often not even keyed to the nation-state from which an observation was derived. Presenting a single national value of something like exposure to ultraviolet radiation would be nonsensical. In recognition of this, natural science analysis tends to be much more finely-grained than is social science analysis. In some cases, natural scientists model small ecosystems such as forests; in other cases, they model regions of continents, continents themselves, or the entire globe.

The political boundaries decreed by humans apparently have little modelling interest for natural scientists tracking air pollution, populations of migratory birds, or the processes of global environmental changes – except as explicit objects of study themselves in such areas as trans-boundary air pollution research. When the boundary between East and West Germany disappeared, natural science models of climate change did not have to change, but sociopolitical models of how to mitigate climate change and how to deal with its impacts – had such models existed – would have had to change dramatically. The concept of the ecosystem lies at the heart of much modelling in the natural sciences and could well find its way into the social sciences. Large nation-states may contain more than one major ecosystem, and thus the nation-state is further a problem for the natural sciences. In terms of the division of the world into major analytical regions used in the START program[10], the United States contains parts of three major regions (four, if Hawaii is included). These regions differ in climate and in plant and animal life, and somewhat in human habitation and economic activities. It is likely that the impacts of global warming will differ across these regions. To produce a single estimate for the entire U. S. of the potential loss of agricultural production caused by global warming of a certain amount would be to obscure what could be enormous variations by region and by latitude. Such a national point estimate would have little value for policy-makers: what they need to know is, for example, what might happen to wheat farming in the lower Great Plains. To be sure, it is appropriate also to know that U. S. agricultural productivity is unlikely to decline in a world warmed by about 2 °C, but that does not mean that the U. S. government will face no agricultural policy challenges[11]. Modelling the various ecosystems in the country will thus be important for a policy reason as well as for scientific reasons.

Attention to local behavior will be important for another reason of policy. Although it is national governments that sign international treaties, the act

[10] START is an acronym for Global Change System for Analysis, Research, and Training. It is a project of the International Geosphere-Biosphere Programme (IGBP), the World Climate Research Programme (WCRP), and the Human Dimensions Programme (HDP). START provides a framework for establishing regional networks for research and training on global environmental change and its causes and effects.

[11] Averages of any sort can be misleading, as the proverbial statistician learned to his dismay when attempting to walk across the Platte River, with its average depth of one foot.

of signing a treaty has no environmental importance in itself until the treaty has been implemented. Implementation always means people taking action at the local level. Even when the policy instrument is national (such as for a carbon tax), the environment will be affected only if people change their local behaviors (their uses of fuels). There will be implementors and non-implementors of any policy, whose behavior must be understood[12].

There are social scientific reasons as well for looking beneath the nation-state for units of social science modelling. For one thing, the late 1980s should have taught us that nation-states are sometimes transitory and fragile entities. The political shapes of Europe and parts of Asia changed in just a few years with the collapse of many centrally planned economies. Political shapes are still changing. The year 1992 has come and gone, and from the U.S. side of the Atlantic, European unity does not yet seem a reality. Nation-states may be more stable, on the whole, than any other human political creation[13], but they are recent inventions and do not necessarily foretell the shape of political organization in the 21st century. Even today, there are legitimate questions as to whether the institution that claims to be a national government in some countries in fact exercises any of the traditional powers of such governments.

Many nation-states are, in fact, relatively new. The United States did not attain its present territorial configuration until 1959, and even then was less than 200 years old. Germany did not come into existence as a governmental entity until the Franco-Prussian War (1870-71), the Austro-Prussian War (1866), or the Congress of Vienna (1814-15) depending upon which confederation one accepts as signifying the beginning of the German nation. The 1990s have seen a proliferation of small states as countries have broken apart.

Scientific interest in human impacts on the global environment extends back before many of today's nation-states were created, often to the beginnings of the Industrial Revolution in the 18th century. Models restricted to today's nation-states simply will not suffice for historical analyses. What is constant across time is territory – territory often governed by changing entities and possibly containing a changing ecosystem.

Many of the socioeconomic processes in which social scientists are most interested do not primarily occur at the level of the nation-state. There is every reason, for example, to suspect strong cultural influences on environmental actions and impacts. By "culture" I mean a pattern of attitudes,

[12] I am indebted to Detlef Sprinz for the stimulus for this observation.

[13] An argument could be made that city government is, historically, the most stable form of political organization. Empires have collapsed or appeared while the cities continued, even before the era of nation-states. Damascus was inhabited before 2000 B.C., and has been under the rule of Egyptians, Israelites, Assyrians, Persians, Alexander the Great, Seleucids, Romans, Saracens, Ottoman Turks, British, the League of Nations, and since 1941, Syria. Damascus was continuously occupied as a city throughout that period while all of its one-time rulers have undergone fundamental changes and most have disappeared.

values, laws or norms, and knowledge that is shared, to some extent, within a population. Culture is not bounded by national boundaries: a state may contain several cultures, and a culture may be spread across several states. Some aspects of culture are becoming global, but the beliefs that drive people's daily lives may be deeply rooted in the cultures of tribes, ethnicities, or regions of a country. In much of the world the nation-state simply does not have the cultural salience that it presently does in Europe and North America (and even there, there have been civil wars, terrorism, and rumors of secession, with cultural issues the occasion for the conflicts).

The task of linking social science models that exist solely at the level of the nation-state and natural science models in which nation-states play no role would be, at best, difficult. The incongruities of measurement and modelling are simply too large. There is little prospect of natural science organizing its observations and modelling along the lines of nation-states for the benefit of social science and little prospect of social science organizing its thinking around classical ecosystems. However, it must be, I believe, social science that accommodates the needs of the natural sciences. This can be read as implying that social science must bend to the will of natural science, but I hold that social science already has concepts in place and at work that will permit us to meet the natural sciences while sustaining the capacity to do all the social science that we need to do. The same is not true of the natural sciences.

11 Modelling of land-use/land-cover changes

Consider the topic of land-use/land-cover changes as an example of models that can profitably be constructed at levels below that of the nation-state. National policies do affect land use, but state and local policies – taxation, assessment, zoning, restrictive covenants, water rights – also constrain or channel the uses of the land. A model of land use that ignored varying sub-national conditions would be incomplete. This means that data must be collected at the sub-national level to provide the needed policy context. In addition, the impact of changes in land use on land cover is sure to differ strongly across ecosystems.

The authors of the proposal for a IGBP-HDP Core Project on Land-use/Land-cover Change remarked:

> "Comparative studies suggest that the relationships among human causes, land-use change, and land-cover change vary considerably across space and time. This sub-global complexity is currently masked in global-scale approaches that rely on global aggregate data and relationships. The loss of variability in relationships through aggregation diminishes the accuracy of models and projections of land- cover change and lessens their meaning for regional and local

assessments. The dynamics of land-use/cover change are better understood at sub-global levels, and global-scale models and projections will be improved if they are made regionally sensitive" (*Turner* II et al., 1993 [15]).

The authors propose middle-level units of study called "regions," which they seek to classify in terms of human driving forces, land uses, and land covers. Each region would possess "common patterns of relationships between human driving forces and land-use/cover change" (*Turner* II et al., 1993 [14]). Regions would be grouped into "situations" with common "cause-to-cover relationships". They note that modelling must take account of the varying scale of resolution in data collection, from the square-kilometer resolution of AVHRR (Advanced Very-High Resolution Radiometer – a form of remote sensing data) measurement of land cover to the political-boundaries resolution of data collection on human driving forces. They also note that square-kilometer resolution is too gross for detailed studies of land-use change. They plan to resolve the discrepancies between scales of resolution by proposing to code all data into Geographic Information Systems (GIS), which require precise geographical coordinates for each observation. Once those discrepancies are resolved, there is still the issue of modelling units: are land-use/land-cover changes studied at the national level or at some other level? Is it national taxation policy that is coded in the GIS or do we also code local laws?

The most powerful analytical unit for land-use/land-change modelling may well be what has been known in the social sciences as the "community," analogous in some ways to the geographers' "landscape". In the field of human ecology (a sub-discipline of sociology in most universities), the community has historically been the smallest geographical unit studied, and much research in that field has concentrated upon comparative studies of several communities. A community is defined as a unit having a population located in a common environment that is tied together by ties of economic interdependence. In this concept, a community could be a city, a metropolitan area, or a farming region. The community was traditionally seen as relatively closed, so that it could be treated as a system. However, as economic relations and social organization became more complex in the twentieth century, the idea that communities could be analyzed as closed systems became less and less tenable, to the point that today much analysis focuses on the interactions of communities with other communities or with regions of countries, with entire societies, or even with the entire globe.

Despite this increased complexity, human ecology retains its prime focus on specific populations living in common environments. I will argue that human ecology provides the sampling design and the units of modelling that we need for research on human interactions with the environment.

12 The POETic ecosystem

The most articulate brief statement of the perspective and concepts of a field of human ecology in which the natural environment plays a significant role is still that of Otis Dudley Duncan writing in 1959 (*Hauser* and *Duncan*, 1959 [5]). Duncan wrote:

> "A concrete human population exists not in limbo but in an environment. Moreover, to continue to exist, it must cope with the problems posed by an environment which is indifferent to its survival but offering (in varying degrees) resources potentially useful for the maintenance of life. By mere occupancy of an environment, as well as by the exploitation of its resources, a human population modifies its environment to a greater or lesser degree, introducing environmental changes additional to those produced by other organisms, geological processes, and the like" (*Hauser* and *Duncan*, 1959 [5]).

Duncan points out that human ecology, together with demography, differs from other social sciences in the unit of analysis that it chooses. Political science chooses a nation-state or some other governmental unit; anthropology, a culture; economics, a market; and psychology, a personality. To Duncan, "The unit of ecological analysis is a human population, more or less circumscribed territorially" (*Hauser* and *Duncan*, 1959 [5]). This unit is often taken to be the community, as the smallest unit of ecological analysis possessing properties approaching those of a system. This human population is described by a framework consisting of four interacting elements:

(P) Population – not simply its size but also its composition, spatial distribution, and vectors of its change.
(O) Organization – the social adaptation to the fact that individuals are interdependent and must cope collectively with their inability to survive in isolation, to which they respond with formal and informal expectations of each other, rules, laws, group formation, and divisions of labor.
(E) Environment.
(T) Technology – the set of techniques employed by a population to gain sustenance for itself and to organize itself to provide that sustenance[14].

This set of four interacting elements came to be known as the POET framework[15]. Duncan called this inter-related set "the ecological complex," initially avoiding what he saw as the assumption of its systemic character

[14] In a subsequent article, Duncan emphasized that in human populations the flow of information was a crucial aspect of an ecosystem (*Duncan*, 1964 [1]).

[15] Myth on the University of Michigan campus has it that Duncan initially presented this framework to his colleagues and students as the PETO framework, perhaps reflecting some implicit ordering, but that he quickly changed to POET when he saw the didactic advantages.

that was implicitly embedded in the term "ecosystem". To Duncan, systems were equilibrium-seeking or equilibrium-maintaining, and it was clear that static equilibria are not commonly observed in human interactions with the environment. However, by 1964 Duncan had adopted "ecosystem" to describe the interacting POET elements. Social scientists might do well to study what could be called "POETic ecosystems" in the search for a global modelling capacity, rather than pursuing the fiction of a global model.

13 Conclusions and a postscript

Social scientists and natural scientists can usefully converge, for many models, on the meso-scale POETic ecosystem. This could be a city, a region of a state or province, a river basin, a nation, or even a continent. Where possible, we should employ the definition of ecosystem or "community" found in human ecology: a unit having a population located in a common environment that is tied together by ties of interdependence. It is, however, critical that these be studied as POETic ecosystems – humans cannot be excluded from them and the environment must be taken into account.

The definitions of ecosystems used in both biology and human ecology have been partial rather than POETic. To biologists, the human role in the ecosystem has been somewhat of a bother: humans have been regarded as external to nature and have been systematically excluded from study, often by selecting areas that seem unaffected by human action. To the human ecologists working in the context of sociology departments, humans and their creations of technology and organization have taken primacy, and the environment has largely been disregarded as not constraining to human action (*Young*, 1994 [19]). The natural environment has thus disappeared from much recent work in human ecology. Human ecologists did not really know how to bring the environment into their analyses, just as ecologists had difficulty bringing humans into their analyses.

To biologists, the question of incorporation of an explicit concern with human populations into the emerging field of ecology was a debate at the founding of the Ecological Society of America around 1920 (*Duncan*, 1964 [1]). As Duncan observes, the idea that humans were the dominant species of the twentieth century must have proved uncomfortable in its implications. Duncan quotes F. F. Darling as reporting: "When some bold spirit began talking of human ecology he was promptly excommunicated". The separation of biological and social perspectives on ecosystems largely persisted until the rapprochement engendered by shared concern with global environmental changes. The willingness of ecologists participating in international scientific programs to invite social scientists to collaborate with them is a major forward step – or perhaps a step back in history to the views of some of the founders of ecology.

The rapprochement between social and natural scientists now offers the possibility of defining "ecosystem" so as to provide the sampling designs and units for modelling of human interactions with their environments. In this context, it makes no sense for biologists to draw ecosystem boundaries so as to assiduously exclude or ignore human populations, and it makes no sense for human ecologists to ignore the fact that the communities they study are located in river valleys, on plains, or upon the mountains. We can work together to define units that will serve both our needs, and social science can bring its concept of the community or the POETic ecosystem to the table.

If we draw our samples so as to be representative of POETic ecosystems, we will have the potential of linking work in social science with work in natural science. Further, with proper preparations, we can aggregate samples of ecosystems to samples of regions and even of nation-states, giving us purchase on the analytical problems addressed by Jacobson. We can put human organization, technology, and culture into environmental context at the local ecosystem level. This suggestion, admittedly, does not fully resolve the problem with which this chapter began: it does not definitively guide us to a particular scale for modelling. What it does imply is that small is better, for aggregation is easier than disaggregation. Put another way, it tells us to design our samples so as to afford us the greatest flexibility in choosing modelling units. In addition, it clearly urges us to treat humans as an element of ecosystems, if not as the element that is today so dominant worldwide that it now endangers the life support system on which humans depend.

Since this article was written and presented, research on global environmental changes has strongly taken the direction of emphasizing regional scales and integrated modelling. The International Geosphere-Biosphere Programme has defined a number of "transects" – systematic slices across mesoscale ecosystems – in which IGBP core projects will be conducting coordinated data collection and modelling around the globe. The International Institute for Applied Systems Analysis has finished the current phase of its research on the interactions of population growth and environmental change on the island of Mauritius (*Lutz*, 1994 [7]), producing an exemplar of how the context-sensitive, regionally-specific modelling advocated in this paper can be conducted. The most sophisticated treatment of the carrying capacity of the Earth to appear in decades was presented in 1995 by Joel E. *Cohen*. The Intergovernmental Panel on Climate Change is emphasizing integrated assessment at the national or sub-national level and has elevated the importance of the concept of "vulnerability" – which involves all the factors identified as needed in a model of the carrying capacity of the African continent. The Susquehanna River Basin Experiment (a joint project of Pennsylvania State University and the Marshall Space Flight Center) has successfully linked global General Circulation Models with regional climate models. The author can only applaud these and similar initiatives and looks forward to increased recognition that the proper choice of spatial scales is crucial to advancing knowledge of how humans influence the environment and are affected by it.

References

1. O. T. Duncan. Social Organization and the Ecosytem. In R. E. L. Faris, editor, *Handbook of Modern Sociology*, pages 36–82, Rand McNally, Chicago, 1964.
2. R. A. Eblen and W. R. Eblen, editors. *The Encyclopedia of the Environment.* Houghton Mifflin Company, Boston, 1994.
3. R. Freedman. Asias's recent fertility decline and prospects for future demographic change. Asia-Pacific Population Research Reports 1, East-West Center Program on Population, 1995.
4. P. M. Groffman and G. E. Likens, editors. *Integrated Regional Models: Interactions between Humans and their Environment.* Chapman & Hall, New York, 1994.
5. Ph. M. Hauser and O. D. Duncan. *The Study of Population: An Inventory and Appraisal.* University of Chicago Press, Chicago, 1959.
6. The International Geosphere-Biosphere Programme. *Global Change, Report No. 12, A Study of Global Change, The Initial Core Projects*, Stockholm, 1990.
7. W. Lutz. *Population-Development-Environment: Understanding Their Interactions in Mauritius.* Springer-Verlag, Heidelberg, 1994.
8. D. H. Meadows, D. L. Meadows, and J. Randers. *Beyond the Limits.* Chelsea Green Publishing Co., Post Mills, 1992.
9. D. H. Meadows, D. L. Meadows, J. Randers, and W. W. Behrens III. *The Limits To Growth: A Report for the Club of Rome's Project on the Predicament of Mankind.* Universe Books, New York, 1972.
10. D. H. Meadows, D. L. Meadows, J. Randers, and W. W. Behrens III. *The Limits To Growth: A Report for the Club of Rome's Project on the Predicament of Mankind.* Universe Books, New York, 1972.
11. R. H. Moss and R. C. Rockwell. Reconceptualizing security: a note about research. In S. Aguayo and Bruce Bagley, editors, *En Busca de la Seduridad Perdida: aproximaciones a la seguridad nacional mexicana*, Siglio Veintiuno Editores, Mexico City, 1990. In Spanish translation.
12. A. Palloni. The relation between population and deforestation: methods for drawing causal inferences from macro and micro studies. In L. Arizpe et al., editor, *Population and Environment: Rethinking the Debate*, pages 125–165. Westview Press, Boulder, 1994.
13. B. L. Turner II, W. C. Clark, R. W. Kates, J. F. Richards, J. Tuchman-Mathews, and W. B. Meyer, editors. *The Earth as Transformed by Human Action: Global and Regional Changes in the Biosphere over the Past 300 Years.* Cambridge University Press, Cambridge, 1990.
14. B. L. Turner II, R. H. Moss, and D. L. Skole, editors. *HDP Report No. 5, Relating Land Use and Global Land Change.* HDP, Barcelona, 1993.
15. B. L. Turner II, R. H. Moss, and D. L. Skole, editors. *Relating Land Use and Global Land-Cover Change: A Proposal for an IGBP-HDP Core Project.* International Geosphere-Biosphere Programme (IGBP), Stockholm, 1993.
16. U. S. Department of Commerce. *Our Increasingly Populated World*, Washington, 1992. Government Printing Office. Bureau of the Census, Statistical Brief SB/92-9.

17. I. M. Wallerstein. *The Modern World-System.* Academic Press, New York, 1974.
18. World Wide Web, Project Link: An Introduction, 1997. http://www.chass.utoronto.ca/link.
19. H. P. Young. Equity in Theory and Practice. Princeton, 1994.

Commentary on "Modelling Human Interactions with the Environment"

Lenelis Kruse-Graumann

1 Introductory remarks

Professor Rockwell pleads for a bottom-up approach. He is critical of so-called global entities as the basis for modelling and, furthermore, he is also critical of some of the bottom-up approaches that take the nation-states as the basic modelling units.

As appropriate basic modelling units he suggests "territories" which not only can be defined in spatial terms but also should reflect something like POET, i.e., the combination or intersection of population, organisation, environment, and technology.

If I understand his approach correctly, then he actually considers these models as a kind of complement to top-down models, hoping that these two roads of global change analysis eventually meet.

2 Social aspects: an indispensable facet of bottom-up approaches

Although I would hesitate to use the term "territory" as a significant word for what it is meant to signify, I am very much in favour of developing bottom-up approaches. I deliberately use the plural term "approaches" because I think there is no unique or single bottom from which to start and which then eventually determines what kind of data have to be collected and processed. What is at the bottom depends on what kind of questions one asks and what kind of problems one wants to solve and, furthermore, it depends on the assumptions and hypotheses that one has about critical variables and the causal linkages between these variables.

So let me suggest a further variant of a bottom-up approach. If we conceive of global environmental changes as results of people-environment interactions, we have to take people more seriously than we have done so far. Environmental changes are direct or indirect effects of human activities. Human activities, however, are not only causes of environmental changes, they are also affected by these changes. In addition, behaviour is at stake if people have to serve as potential managers and mediators of global changes. The objective of management may be prevention or mitigation of further changes and/or adaptation to already existing changes and their impact on human and natural systems. Management and engineering have to take place mainly on the local level, but also on the global or international level.

When I use the term behaviour I subsume a whole range of experiential and behavioural aspects as well as types and levels of action, i.e., how people perceive environmental changes, how they categorize and evaluate environmental changes, how they emotionally react to such perceived changes and are motivated to act and, eventually, how people really act in an ecologically relevant, destructive or ameliorating way.

When discussing actions as causes and effects of global environmental change it is useful to distinguish between different types of behaviour. A one-time act, such as investing in a better insulation of housing, or the recurrent modification of daily routines or complex patterns of life styles are not only differently effective for various global environmental problems but they also involve different "costs" for individuals and groups. Furthermore, we have to take into account different levels of behaviour, starting from individual behaviour through various social levels, e.g., families, companies, communities, to nation-states and international organisations. We also have to consider that people hold different roles and functions within environmentally relevant processes. They act as consumers or producers, as policymakers or administrators, and very important, as multipliers of opinions and values, such as journalists and conference contributors.

Hence, the essence of what I want to convey is that people and even researchers cannot live on POETry alone; they need some behavioural prose, social scientist's daily diet. To spell it out more concretely: bottom-up models can be constructed on various levels using different scales and degrees of resolution for individual and social actions and societal processes. The development of such models requires the cooperation of various disciplines. The question is, however, which level and which degree of resolution is appropriate of environmental politics. If we think of a bottom-up approach in terms of a model, then the character of this model should be heuristic rather than deterministic.

Commentary on "Modelling Human Interactions with the Environment"

Tillmann Mohr

1 Introductory remarks

My comment is mainly concerned with the idea of Professor Rockwell to establish territories from which the required data and the relevant input can be derived for his integrated models.

The problem Professor Rockwell is referring to is an old one, well known in natural science, a problem meteorologists have been dealing with for more than a century.

2 Dimensions of data requirement for a "set of models"

In order to define requirements for data necessary for the establishment, the validation, the running and the verification of the set of models Professor Rockwell is concerned with, a detailed analysis of the data requirements of each individual model has to be undertaken. The requirements depend on the models themselves, the equations governing the models, the numerical scheme used and the parameterization applied. Another factor is the scale of the models, whether they are national, regional or global models. The scale of the individual phenomenon described by the model has to be taken into account, and problems like data representativeness, data accuracy, etc. have to be addressed. Thus the requirements for time and space resolution and in general all requirements for the individual data will differ very much. It might be possible to group together these requirements and to arrive at the very end at territories. It is not so simple as it might look and it might be not possible at all. (In meteorology we have never tried to achieve this. Our answer to the problem has been the development of very sophisticated data assimilation schemes.)

Hence it is absolutely necessary to start with the establishment of the data requirements for the integrated models. The next step is to design an "observing" network – and this could be composed of the proposed territories – which will provide the data sets. In designing a network, common standards and formats for taking, collecting and exchanging the data will have to be developed not only on a regional but also on a global basis.

The establishment and the operation of the relevant "observing" network are ambitious undertakings, not just from an organisational point of view. More important are the funds necessary to run such a network.

However, if we believe that earth system analysis is a prerequisite to provide decision makers with the required information for a sustainable development, the funds will have to be provided not only for the relevant research and development of the integrated models but also for the routine operation of the "observing" network and the running of the models.

Chapter 4: Is Global Modelling Feasible?

Hans-Peter Dürr

Abstract On the microscopic level nature is not deterministic but at most allows the statement of probabilitites for the occurrence of future events (quantum physics). On the macroscopic level (ensembles with a large number of quantum states) these uncertainties may get averaged out in simple systems and hence become irrelevant (ergodic systems near thermal equlibrium). In the general case of complex systems, however, because of intrinsic instabilities, microscopic fluctuations will escalate and lead to unpredictable macroscopic behaviour (bifurcations, "chaotic order") which makes modelling extremeley difficult and quite often rather useless. Models in physics are constructed with the goal of approximating a natural system with regard to features relevant to our special questions and purposes such as to allow sufficiently reliable predictions for the future evolution of the system. Because of the basic non-deterministic structure of the physical laws this will only be successful for suitably aggregated properties and/or sufficiently simple systems. Because of its high complexity the system Earth (including not only physical quantities but also human factors) certainly does not meet the latter conditions. This does not render global modelling altogether unfeasible. By exploiting regularities of certain aggregated properties it may still be possible to learn something about the general dynamics, the stability and intrinsic correlations of the system, and hence give important guidance to our understanding (including possible consequences of our manipulations) of the system.

1 Introductory remarks

I am not an expert in the field of "Earth System Analysis" and hence my contribution to this Symposium will be that of an outsider. Professionally I am a nuclear physicist or more precisely an elementary particle physicist who has spent quite some time on the formulation of a Fundamental Quantum Field theory. Dealing with nonlinear field equations in this context I have learned that subjects belonging to a set defined by the prefix "non" (like in nonlinear) do not have to show any similarities, at all. Since the earth system is a highly complicated nonlinear system I am therefore fully aware of the difficulty of making general statements about the feasibility of modelling in such cases and would actually be well advised to rather abstain from making any comments on this subject. On the other hand I am convinced of the basic importance of the issues discussed at this Symposium and in particular as expressed by the subtitle: Integrating Science for Sustainability. Indeed the problem of appropriately taming human activities such as not to destroy the sustainability and resilience of our natural life-support system appears to be basic for the prospects of long-term survival of humankind and there is no question in my mind that science has to play an important rôle in this process.

I, therefore, very much welcome the establishment of the transdisciplinary Potsdam Institute for Climate Impact Research as an important and necessary adventure, and I wish to congratulate all of you and in particular H.J. Schellnhuber for your engagement and courage.

As I have already indicated, the question I am asked to address: "Is Global Modelling feasible?" in the context of the earth system cannot be easily answered. The general answer to this question, of course, is: yes! which however does not imply very much, because the important question is: How good can such models be? Or if we expand the meaning of the word "good": How adequate and accurate can we make such models? At present only very little can be said about this because an analysis and description of the earth system is – despite many decades of research – still at the beginning, because of its high complexity, of presumably long development.

I intend to deal with my topic in four steps:

Firstly, I will consider the basic (microscopic) structure of reality. Models are formal, mostly highly simplified images of what we consider as "reality". To admit such simplifications, at all, such a "reality" has to have a certain structure and to possess certain properties. As a nuclear physicist I have to realize that "reality" on the microscopic level does not, in general, exhibit a structure which allows fragmentation without severe losses. There are holistic features which prevent reduction and hence accurate modelling in the classical sense.

Secondly, we will concentrate on the effective structure of macroscopic systems. A macroscopic system is defined here as the (actually only mesoscopic) "world" in which we live, the earth, in contrast to the really macroscopic world as represented by planetary systems, gallaxies, the universe and as studied in astrophysics. According to our knowledge there is an important dependence on length scale in the description. The world has no "Matryoschka" (Russian doll) – structure or "fractal" structure in the sense that the microcosmos looks like a scaled-down macrocosmos. There occur qualitative changes by going up and down in size. Modelling of macroscopic systems hence will be distinctly different from modelling microscopic systems.

Thirdly, we will turn to the main question: Is global modelling feasible? "Global" is interpreted here in the sense of referring to the "globe", the earth, i.e. the question is asked for the feasibility of modelling the earth's system. The earth system is not only macroscopic but is very special because of the presence of life and even more so because of the presence of man and his capabilities.

Fourthly and finally, we will reflect on the pertinent question: Can there be working models for the earth such as to assist planetary engineering? Generally, modelling is not only a way to "grasp" and understand a complicated system in simpler terms as an agglomeration of certain parts and processes but to use successful models as guides for manipulating these systems in various ways. Engineering requires sufficiently accurate manipulation to achieve an intended goal. Because of the irreducible complexity of the system, however, engineering capabilities will be rather limited except in some special cases.

2 The basic structure of reality

Commonly reality is perceived as a complicated sum of interacting smallest objects. The smallest objects are identified as atoms, elementary particles, quarks or alike (they are analogous to Schellnhuber's "hardware"). The interactions of these fundamental constitutent particles (analogous to Schellnhuber's "software" or operational content) determine their dynamics and lead to well-prescribed changes in time which allows to predict future configurations. Formally everything can be exactly described in the frame work of the Lagrangian formalism as the motion of points in a high-dimensional phase space (spanned by the space-coordinate and momenta of all particles involved). This is what we call classical physics.

Modern physics, quantum physics, however tells us there are actually no objects and the dynamical laws are not deterministic but only of a statistical character. "Reality" is replaced by "potentiality", a potential for various possible realisations. In a way you can say there is no hardware anymore but only software, i.e. operational context, and this in an "as well as" fashion. Strictly speaking a system cannot be taken apart without partial destruction of the context. All systems should be regarded as being complex where everything is connected to everything in a highly non-trivial sense. Reduction of the system to simpler systems is only approximately possible. There are differerent ways of approximately reducing a system and each reduction will exhibit a different "reality" of the system. Depending on the observational approach an electron may look like a "particle" or like a "wave", i.e. as two different and actually incompatible realities.

To illustrate this strange situation I always like to recall a parable by the famous English astrophysicist Sir Arthur Eddington. He compared the scientist with an ichthyologist, a fish-expert, who investigates the world by catching fish in the ocean with a net. After years of fishing he arrives at a fundamental law of ichthyology: "All fishes are larger than two inches." He calls this finding a law because there was not a single catch where this observtion was not valid. He tells this exciting discovery to his best friend, the metaphysicist, but he is not surprised, at all. He points out to the ichthyologist the fact that the spacing of his net is just two inches and that he therefore will not be able to catch anything smaller. However, the ichthyologist is not impressed at all by this argument and responds: "What I cannot catch is, in principle, beyond ichthylogical knowledge, it does not refer to any object defined in ichthyology. What I cannot catch, therefore, is not a fish. In addition: The common sense people in the market place to whom I sell the fish do not care about fishes I cannot catch."

The analogy to scientists and the scientific method is obvious. The net refers not only to the particular scientific method but also to our special rational and fragmenting way of thinking. By limiting "reality" to what we can catch we have the enormous advantage to make "reality" objective and even tradeable. This properties, however, in general will not be sufficient to exhaust the "full reality", a "potential" which I can more extensively tap by, for example, using other nets or even more exotic extraction devices. There may be even something which I cannot pull out at all because it cannot be separated in principle. This indicates principle limits to knowledge in the usual sense and to the scientific method in particular to gain knowledge.

The question, however, arises: Who cares about this strange microscopic world? The world at larger scales, involving 10^{24} of such exotic atomic entities does not exhibit these features because they are completely averaged out to extremely high accuracy. The macroscopic world is the world of classical physics with real objects obeying deterministic dynamical laws. A statistical description is still required but only because of our ignorance or our incomplete knowledge of the particular contingent configurations.

3 The effective structure of macroscopic systems

For large ensembles of quantum states the uncertainty of space locations and momenta, and the undetermined change of configurations in time are, in general, averaged out statistically to a very high degree. However, this is not true in all cases. Indeed, quantum indeterminism can surface to the macroscopic level in systems with instability points which frequently occur if the system is far away from thermodynamical equilibrium. There microscopic fluctuations can initiate drastically different macroscopic behaviour. The initial conditions hereby are not stochastic in the classical sense but have a complex valued quantum statistical form which is not only characterized by probabilities (amplitudes) but also by coherence properties (phase relations). For "ordered" quantum systems, the superpositions of a large number of quantum states, therefore, will not necessarily lead to an averaging out of all quantum features. It appears very likely that the occurrence of animated matter is connected with this phenomenon. The biosphere including man hence may reflect more closely the basic character of "matter" than matter, the inanimate matter, itself. Macroscopic systems occur as mixtures of rigid structures obeying strict (deterministic) dynamical laws and ergodic (statistically averaged) behaviour near thermodynamical equilibrium and ordered open structures far away from thermodynamical equilibrium supported by a through-flux of energy and an influx of syntropy (negative entropy). In the latter case highly ordered systems can evolve by self-organisation.

Let me expand a bit on the concept of "order". I am employing in this context and exemplify it with a poem "Grenzen der Menschheit" (Limits of Humankind) by the German poet Johann Wolfgang von Goethe (Fig. 1).

Without doubt it is endowed with a highly sophisticated structure on a num-

Grenzen Der Menſchheit
Johann Wolfgang von Goethe

Wenn der uralte,
Heilige Vater
Mit gelaſſener Hand
Auf rollenden Wolken
Segnende Blitze
über die Erde ſät
Küß ich den letzten
Saum ſeineſ Kleideſ,
Kindliche Schauer
Treu in der Bruſt.

Denn mit Göttern
Soll ſich nicht meſſen
Irgendein Menſch.
Hebt er ſich aufwärtſ
Und berührt
Mit dem Scheitel die Sterne,
Nirgendſ haften dann
Die unſicheren Sohlen,
Und mit ihm ſpielen
Wolken und Winde.

Steht er mit feſten,
Markigen Knochen
Auf der wohlgegründeten
Dauernden Erde
Reicht er nicht auf,
Nur mit der Eiche
Oder der Rebe
Sich zu vergleichen.

Waſ unterſcheidet
Götter von Menſchen?
Daß viele Wellen
Vor jenen wandeln,
Ein ewiger Strom:
Unſ hebt die Welle,
Verſchlingt die Welle,
Und wir verſinken.

Ein kleiner Ring
Begrenzt unſer Leben,
Und viele Geſchlechter
Reihen ſich dauernd
An ihreſ Daſeinſ
Unendliche Kette.

Figure 1.

ber of different levels which we designate as a highly "ordered" system in the sense adopted here. Only a minor part of this "ordered" structure is objective, e.g. the occurrence of different (small and capital) letters, punctuation-signs, intervals, even correlations in the statistical distributions of the various letters, the syntax etc. The major part, however, is not objective, e.g. the arrangements of letters in words and of words in sentences, the semantics, the "beauty", the emotional impact etc. which decisively depend on the observer, his reading ability, his knowledge of the language (German in this case) and the literature and culture connected with it. You can strip off the latter features and put everybody on the same footing, for example, if you mirror the letters in the alphabet, i.e. replace $a \leftrightarrow z, b \leftrightarrow y$ etc. (Fig. 2). In this "approximation" all the objective features of the poem remain but

Tivmavm Wvi Nvmhxssvrg
Qlszmm Dloutzmt elm Tlvgsv

Dvmm wvi fizogv,
Svrortv Ezgvi
Nrg tvozhhvmvi Szmw
Zfh iloovmwvm Dlopvm
Hvtmvmwv Yorgav
:Fyvi wrv Viwv h:zg
P:fhh rxs wvm ovgagvm
Hzfn hvrmvh Povrwvh,
Prmworxsv Hxszfvi
Givf rm wvi Yifhg.

Wvmm nrg T:lggvim
Hloo hrxs mrxsg nvhhvm
Ritvmwvrm Nvmhxs.
Svyg vi hrxs zfud:zigh
Fmw yvi:fsig
Nrg wvn Hxsvrgvo wrv Hgvimv,
Mritvmwh szugvm wzmm
Wrv fmhrxsvivm Hlsovm,
Fmw nrg rsn hkrvovm
Dlopvm fmw Drmwv.

Hgvsg vi nrg uvhgvm,
Nziprtvm Pmlxsvm
Zfu wvi dlsotvti:fmwvgvm
Wzfvimwvm Viwv
Ivrxsg vi mrxsg zfu,
Mfi nrg wvi Vrxsv
Lwvi wvi Ivyv
Hrxs af evitovrxsvm.

Dzh fmgvihxsvrwvg
T:lggvi elm Nvmhxsvm?
Wzhh ervov Dvoovm
Eli qvmvm dzmwvom,
Vrm vdrtvi Hgiln:
Fmh svyg wrv Dvoov,
Evihxsormtg wrv Dvoov,
Fmw dri evihrmpvm.

Vrm povrmvi Irmt
Yvtivmag fmhvi Ovyvm,
Fmw ervov Tvhxsovxsgvi
Ivrsvm hrxs wzfvimw
Zm rsivh Wzhvrmh
Fmvmworxsv Pvggv.

Figure 2.

all the "order" on the higher levels – and this is actually the relevant part which makes us appreciate the poem – is completely lost. We may even go one step further in the approximation by arranging the poem according to the different symbols (Fig. 3) to make the objective part more explicit and easier for communicating the objective content. We may also try to improve somewhat on the objective features by calculating the improbability of the special arrangement (or the syntropy, its natural logarithm) of the 836 symbols or its inverse (the probability or its natural logarithms connected with the entropy) to find the extremely small number of $1 : 10^{1136}$ for this highly

AAAaaaaaaaaaaaaaaaaaaa
BBBbbbbbb
Cccccccccccccccccccccccc
DDDDDDdddddddddddddddddddddddddddddddddddd
EEEEEEEEEEeeeeeeeeeeeeeeeeeeeeeeeeeeeeeeeeeeeeeeeeeeeeeeee
eeeeeeeeeeeeeeeeeeeeeeeeeeeeeeeeeeeeeeeeeeeeeeeeeeeeeeeeee
eeeeeeeeeeeeeeeeeeeeeeee
ffffff
GGGGGggggggggggggggg
HHHHHhhhhhhhhhhhhhhhhhhhhhhhhhhhhhhhhh
IIiiiiiiiiiiiiiiiiiiiiiiiiiiiiiiiiiiiiiiiiiiiiiiiiiiiii
Jj
KKKKKkkkkk
Llllllllllllllllllllllllllllllllll
MMMMMMmmmmmmmmm
NNNNNnnnnnnnnnnnnnnnnnnnnnnnnnnnnnnnnnnnnnnnnnnnnnnnnnnnnn
nnnnnnnnnnnnnnnnnnnnnnnnnn
Oooooooooooooooo
P
RRRRRRrrrrrrrrrrrrrrrrrrrrrrrrrrrrrrrrrrrrrrrrrrrrrr
SSSSSSSSSSSsssssssssssssssssssssssssssssss
TTttttttttttttttttttttttttttttttttttttttttt
UUUUUUuuuuuuuuuuuuuuuuu
VVVvvvvvv
WWWWWWWWWwwwww
Zzzzz
ää
öö
üüüü
ßß
,,,,,,,,,,
......
:
?

Figure 3.

differentiated system (compare this number e.g. with the number 10^{90} of atoms in the Universe, or 10^{30} of seconds elapsed since the start of our Universe with the Big Bang). How in the world was Goethe able during his short life-time to generate such an unlikely structure, such a highly distinguished system? The answer, of course, is that he did not create it by "throwing dice" the 836 symbols in a single line, but he started with making full use of certain fixed substructes, the letters, words, syntax of sentences, cultural habits etc. limiting severely his possible options. Still it took the creative mind of a poet, his ordering hand, to generate the highly unlikely.

There is an important law in physics, the so-called Second Law of Thermodynamics (or Entropy Law) which states: All isolated (and strictly speaking ergodic) sytems change automatically with time from less likely to more likely configurations. If we connect the concept of "order" with the improbability of a configuration this implies, in the long run any special, any distinguished and differentiated structure will get destroyed and transformed into complete disorder. The syntropy as a measure of order will always decrease (or the entropy as a measure of disorder will always increase) in time. The opposite trend on earth exhibited by the evolution of life from primitive to highly differentiated living forms is not a violation of this law but due to the fact that there is a constant syntropy influx from the sun – estimated by Juri Svireshev, as we just learned from his lecture, to about 2.4×10^{20} J/K×years which acts as an "ordering hand" and keeps the earth system away from thermodynamical equilibrium. One should realize, however, that these syntropy increasing processes only occur in a small subsystem of the earth system, primarily in the biosphere and there in the plants, against the overwhelmingly dominating ("natural") entropy increasing processes. Any acceleration of the processes will favour the destructive tendency. Evolution looks like a rock climber. Because of the gravitational downward forces the tendency to fall is overwhelming. Still the climber can successfully overcome this "natural law" if he takes full advantage of the uneven rock and by carefully choosing and patiently securing each step.

It may be quite surprising to learn that the stability of a rather simple mechanical system like the planetary system to which our earth belongs cannot be demonstrated straightforwardly as anybody might believe. Because after all the sunrise every morning at the expected hour is one of the predictions nobody would ever doubt. Still the French mathematician Henri Poincaré demonstrated in 1889 that already the classical 3-body system (e.g. the system of sun, earth and moon) was an intrinsically non-integrable dynamical system, i.e. that its solution cannot be derived by mathematical analysis. Fortunately in the late 1950s and early 1960s Kolmogorov, Moser and Arnold demonstrated that for certain ranges of the relevant parameters (positions and momenta) such 3- and more-body-systems can, fortunately for us, still behave in a regular fashion. Hence our planetary system may be viewed as an originally dynamically rather turbulent system where the present planets with their particular orbits emerged as survivors in a billion-year process.

In a similar way the earth system is a very peculiar, highly nonlinear and statistically extremely unlikely system which is the result of a 4.5 billion year evolutionary process. The relative robustness of the ecosphere reflected by its high sustainability and resilience developed by a trial-and-error process and a subsequent selection of the most appropriate configuration. Under the continuous influx of solar radiation but changing physical parameters of the earth (partially as a consequence of the biological evolution) a high flexibility was generally more favored than high specialized performance. As a result it leads to the immense diversity of biological systems which are dynamically stabilized by countless interlacing regulatory circuits of forces and counterforces. The complexity of the biosphere is absolutly stupendous and its particular expression extremely non-unique if we consider the indetermined path through the innumerable bifurcations in the historical evolution.

4 Is global modelling possible?

In view of the high complexity and non-uniqueness of the earth system the answer to the question: Is modelling of the earth system possible? must be definitely: No! if we insist on any precision. This is obvious from a mathematical point of view because of the strongly nonlinear and intricately correlated character of the system. After all it took nature 4.5 billion years to establish this particular solution and this on the basis of "maximum knowledge" of the system. Modelling, however, never aims at precision but rather at a description of the "relevant" features, where relevance relates to different objectives whatever we believe is important to us. In the neighborhood of an equilibrium, also of a dynamical equilibrium, usually a simpler parameterization and hence a corresponding approximation of the system are possible. Strongly linked chaotic processes under syntropy influx may exhibit regularities which are not expressed in definite configurations but rather in certain patterns. Such patterns will only show up for certain ranges of the system variables. This means that there will only be answers for certain (and not all possible) questions and these answers can only be given in a "fuzzy" way. Such ranges of variables will exist in particular in the neighborhood of the dynamical equilibirum state. Modelling, therefore, should in particular be possible for sufficiently small deviations from the presently time-honored well-established earth system, i.e. for sufficiently small "impacts", measured not only by intensity (amplitude) but also by velocity of change, since a dynamical equilibrium is composed of a large number of regulatory circuits with characteristic response times. As a consequence there will be critical limits on intensity and time constants which determine what we mean by "sufficiently small" and hence on the validity range of our models. These critical limits have some similiarity with the critical points of material systems where phase transitions occur. How can we get an estimate on these critical values?

Since the earth system is the result of an evolutionary process it appears reasonable to obtain such values by looking at the magnitude and the velocity of these changes during the historical development of the earth. The rationale for this supposition is the hope that the ecosystem of the earth has generated over the billions of years of its evolution a sufficiently diverse "immune system" to successfully buffer these impacts. An important parameter in this context will be the syntropy influx from the sun. Strong deviations from this influx induced by tapping fossile energy resources – syntropy reservoirs collected from sun energy over hundreds of millions of years – may indeed cause severe shifts from the present equilibrium state. Other changes which overstrain the resilience of the earth system and upset its "immune system" may be connected with new substances, i.e. substances more or less uncommon in the present earth crust, like radioactive materials, new chemical compounds, materials from far inside the earth (heavy metals), but also large deviations from present concentrations of various already occurring substances. It will not be easy to assess the virulence of these dangers similarly as it is difficult to predict what infections of our body may cause serious sickness or even death.

To construct models of the earth system is only one problem; the validation and verification of these models are an even more difficult task, because of the multitude of parameters and the slowness of changes of the system as compared with human standards. Agreement of data with model predictions is not sufficient for proving their adequacy or correctness. One should search for two types of models:

1. "Engineering models" or functional models of sufficient mathematical complexity and with a sufficiently large number of parameters to roughly describe the situation for a selected set of questions.
2. "Physics models" which may not be empirically as good as engineering models but allow a physical interpretation by approximately simulating the causal structure and the physical correlations of the earth system. The goal in this case is to improve our understanding of the system and gain some rough long-term predictive power. The hope in this attempt would be, to quote Francis Bacon: "Truth emerges more readily from error than from confusion."

Although the predictive power of such models may be rather limited they may very well give important hints to some pertinent questions like closeness or remoteness of critical points (phase transitions) which should be avoided, if possible, because they may mean catastrophies. The question of course, is whether, after knowing this, we will ever have the capability and the instruments to choose a different course. This brings us to our last question.

5 Can there be working models such as to assist planetary engineering?

In envisaging planetary engineering in the sense of global engineering or engineering of the earth's system in one way or the other we should make the distinction regarding the goal of such attempts, namely

1. To avoid imminent disasters (in particular brought about by our own foolish behaviour) which may threaten the present coevolution of man and the ecosphere, i.e. to set up an ecologically sustainable development scheme for human civilization.
   It may look rather artificial to separate man from the ecosphere and emphasize our importance in this way. However, that is how we feel in our

anthropocentric attitude and it is not inappropriate to adopt this position, since if we disregard this point of view we have difficulties talking about planetary disasters and, even in the extreme case where we choose to saw off the branch on which we all are sitting, the disaster would be primarily only one of the human species.

2. To shape the future of the earth according to some clever superplan to reach certain goals.
   Here, of course, the difficult question arises: What do we want? And who are we? Clearly there will be different opinions. What is in some people's minds – or perhaps the minds of the global engineers – is to effect a different split between man and his environment, the rest of the ecosphere, or more concretely, a new partition which favours man (or only a certain subset of man) even more in comparison with the other forms of life on earth.

To my mind, an attempt of type 1 may be feasible if approached appropriately and cautiously. It does not really require any profound understanding of the earth system but simply the recognition that there does exist a possible solution, namely the state of the earth *now* or actually *before the strong anthropogenic impact* occurred. The goal hence would consist in an attempt to reestablish the original time-honoured situation within the tolerance of the earth's immune system as far as possible. To achieve this we simply have to strengthen negative feedbacks by applying defensive strategies which in particular require *moderation and deceleration of all anthropogenic processes.* Since work-capable energy is the main driving force for all human activities, a very effective access to this strategy would be what I like to designate as "the 1.5 Kilowatt Society", i.e. aiming at setting upper bounds on the average through-flow of anthropogenic, technically applied energy (i.e. essentially bounds on the consumption of primary energy from non-renewable sources) such that the energy consumption does not exceed 1.5 kWh per person and hour or 13 000 kWh per person and year (for comparison: USA: 11 kW, Europe: 6 kW, China: 0.8 kW, Bangladesh: 0.08 kW; present world average: 2.3 kW). The idea of this approach is that prevention is far better and much less risky than end-of-the-pipe corrections and repairs. This is very similar to health care where lowering the chances of infections and relying on or strengthening of the immune system are preferable to applying strong pharmaceutics with unknown side effects. Even if we do not succeed in fully reaching the "original state of health" of the earth we at least can reduce the size of human catastrophies in the process of adaptation.

The case 2 of planetary engineering constitutes to my mind, a very dangerous undertaking. To avoid major disasters it would, indeed, require models of the earth system of fantastically high quality which, in particular, do not break down by crossing boundaries to different "phases". I cannot see how this should be possible, at all, in view of the "chaotic" indeterminism of nonlinear, complex systems. Furthermore, an attitude which simply ignores these risks and optimistically strides ahead appears, to my mind, quite unacceptable because there is no good reason to expect that the present complex regulatory system will still work in a different phase or new dynamic equilibrium state if insufficient time is alloted for gradual adjustment through an evolutionary process.

6 Closing remarks

In summary I should emphasize that planetary engineering of the first type directed towards a sustainable development of human civilisation should not be even termed "engineering", because all engineering starts with setting a definite goal. The evolutionary process, however, does not proceed on the basis of a superplan designed for reaching a definite goal, but rather by trial and error and by approaching optima in a plus-sum-game. Sustainable development, therefore, similarly does not call for a definite plan and certain well-prescribed actions but actually demands an attitude which is open-minded, alert and circumspective and continously searches for win-win-strategies. In particular we should try to spot "critical points" as soon as possible because they may herald major catastrophies for us. To achieve this sensitivity studies of various parameters may be important. It would be extremely helpful if direct, model-independent indicators for such "critical points" could be found. I think of something like the sizzling noise of the water shortly before it reaches the boiling point which arises from the fact that the corresponding phase transition starts to occur locally causing large fluctuations. Of course, it may very well be that such definite signals come all too late for any possible counteraction further aggravated by the rather common situation that the destabilizing consequences may appear highly retarded relative to their initiating cause.

In closing let me stress that I definitely agree with H.J. Schellnhuber's final statement: "It is easier to shape the development of the earth than to predict its future!", but I have to add:

"It is easier to wreck the earth in an attempt to shape it than to predict this outcome."

Commentary on "Is Global Modelling Feasible?"

Lennart Bengtsson

1 Introductory remarks

Professor Dürr has given a most intellectually stimulating lecture on the many principal problems in the modelling of complex non-linear systems. I would like to continue along similar lines but try to address more specifically the climate prediction problem.

It is now just over 30 years ago that Edward Lorenz published his now famous paper on deterministic non-periodic flow where he, with mathematical elegance, demonstrated some interesting properties of the numerical solution of Rayleigh's convective equations. The result of this study and several similar investigations by Lorenz was the fundamental notion that although a deterministic system is predictable *in principle* it is *not so in practice* since neither the inital state nor the physics of the system are known exactly. Small errors in the initial state, which cannot be observed, grow very rapidly and have the effect, for example, that weather forcasts have a limited predictability of the order of two weeks.

Another fundamental result of Lorenz's studies was that at least simple non-linear versions of the hydrodynamical equations at times have a tendency to remain in comparatively closed domains where the error growth is small and at other times quickly move away to another part of the phase space domain corresponding to a rapid error growth.

The areas of stability or high predictability had the character of an attractor. It was in fact also later called a *stable attractor*. It is of considerable interest that even very complex flow as presently being studied by the European Centre for Medium-Range Weather Forecasts (ECMWF) using an atmospheric model with some 105 degrees of freedom shows up rather similar properties.

Consequently, atmospheric flow has not only a *limited predictability* but also a *highly variable predictability* in space and time. ECMWF presently provides probability forecasts based on a large number of parallel predictions by the same model but all based on different inital states. The inital states are provided with different error patterns corresponding to unstable eigenmodes of the large scale flow and with the amplitudes of the errors adjusted to the accuracy of the meteorological observations.

2 What about climate prediction?

It is useful to consider two kinds of climate prediction. In the first we intend to predict climate events per se in very much the same way as we predict the weather. A prediction for the average temperature during the next winter for northern Germany, for example, is a *prediction of the first kind.*

For the *second kind of prediction* we restrict ourselves to predicting the statistical structure of climate, that is we do not indicate when specific climate events will take place. A calculation of climate change due to the increase in the atmospheric concentration of greenhouse gases is an example of such a prediction.

3 Climate prediction of the first kind

The large-scale quasi-regular variations in the surface temperature of the tropical Pacific Ocean have been found to be associated with characteristic climate anomalies around the world. When there is a warm anomaly (El-Niño) there is excessive precipitation over northern Chile, Peru and Colombia but drought conditions over northeast Brazil. Indonesia and Australia are dryer than normal but there is high precipitation over California and southern United States. Warmer than normal conditions occur over Canada and northern United States. Although the climate signal is global, the strength of the climate signal varies from area to area due to other influences mostly of a chaotic nature. The climate response over northern Europe in particular is highly irregular but there is a weak signal for more continental conditions of the climate during an El Niño event. Atmospheric general circulation models reproduce the global response pattern quite realistically including the strength of the signal. For this reason climate forecasting systems, which now are under development, consist of a series of forecasts starting from slightly different initial states. This so-called Monte Carlo prediction is required in order to provide a measure of the reliability of the forecast and to extract the climate signal from the chaotic variability of the system.

4 Climate prediction of the second kind

For climate prediction of the second kind a Monte Carlo approach is also required. The reason is that also the average climate may have an unpredicted component due to the possibility, say, of different modes of the ocean circulation not presently possible to observe. It is therefore not satisfactory to undertake one climate simulation experiment only but instead several different ones related to different aspects of the ocean circulation should be undertaken. Such studies indicate that climate change due to the increase in the concentration of greenhouse gases in the atmosphere may well take several decades before being convincingly detected – alternatively the change may be detected much sooner.

The "Monte Carlo" aspect of climate prediction is of profound importance and should be properly considered both by modellers and users of climate model simulation.

Commentary on "Is Global Modelling Feasible?"

Wolf-Dieter Eberwein

1 Introductory remarks

At first glance, the question whether global modelling is feasible, does not make much sense because the answer is yes! Global simulation models worth that name have been built in the 1970s and 1980s, they have been discussed at great length within and without the scientific community, and numerous studies have been undertaken for research as well as policy purposes. Yet global modelling had vanished from the research stage by the end of the 1980s. Thus, the question that needs an answer first is whether any such research activity is indeed necessary and not just an academic fad which might explain why it did not survive the 1990s. Only then will I come back to the feasibility issue before addressing the question of research strategy. These comments are complementary to the presentation by Hans-Peter Dürr rather than a comment to his well taken arguments on the properties of complex systems and the ability to represent them mathematically.

2 Four interrelated global issues

Global modelling is an expensive enterprise, Karl Deutsch, the famous political scientist from Harvard and former director at the Science Center Berlin once stated. However, he added that not building them is even more expensive. *Meadows* et al. [4] state in beautifully simple terms the necessity for such an investment: "as long as there are global problems there will be a need for global models." Deutsch identified four global problems or problem areas. They are global in scale because they threaten the survival of humanity at large: population (growth), hunger, the environment, and war. The disturbing fact is that these problems are interdependent. The complex dynamics that determine their evolution over time require an adequate methodology: global modelling or computer simulation.

These statements have nontrivial implications of which three are of major importance. First, global modelling is a normatively grounded enterprise given its focus on global problems. Each single discipline needs to contribute to their conceptualization but in substantive terms it is a truly interdisciplinary venture evading disciplinary monopolization. Second, global modelling has both a theoretical as well as a practical focus. Theoretically, it is an experimental approach intended to better understand long term dynamics of the complex global system. In practical terms global models are intended to contribute to problem solving on a global scale. This naturally implies the possibly heroic belief that solutions can be found to resolve them. Third, global problems result from the interaction between the social and the physical worlds.

This last statement needs further elaboration. If global problems originated exclusively in the physical world, there is little we could do to resolve them. The laws of nature can not be changed. If global problems originating in the physical environment would not affect the social world, there would be no need to care about them, even less to resolve them. If these two conditions were met, global modelling would be a trivial exercise with respect to the problem solving dimension notwithstanding its unquestioned relevance from the basic research perspective. However, obviously the concrete situation is a different one. As of today environmental issues are highly salient. All over the world people begin to be seriously concerned about and affected by the consequences of the complex interaction between the physical and the social worlds. Ignoring the laws of nature by interfering continuously with the physical environment as we please inevitably creates global problems. This is certainly not a new phenomenon but it is new given its scale. We begin to realize that all of mankind will sooner or later have to pay the bill in terms of hunger, overcrowding, air pollution, water depletion etc. unless fundamental changes with respect to our exploitative approach to nature occur.

This last argument leads us to reconsider the feasibility issue of global modelling, its necessity having been established. In particular: can we develop models capable of representing the complex interaction between the physical and the social worlds? Furthermore, can models be developed which are of potential use for problem solving on a global scale? As far as the interaction dimension is concerned the conference participants all share the consensus that sustainability of the earth system is the appropriate substantive focus for such an enterprise even though the concept definitely needs further elaboration.

Whether this choice will finally lead to a model or a whole set of models contributing to problem solving is another matter. I do not have any answer at this stage. Before looking into the future and therefore possibly "reinventing the wheel" let me briefly look back and ask to what extent past global modelling efforts have indeed contributed to a better understanding of the sustainability issue and to problem solving (for an overview see *Meadows* et. al, 1982 [4]; *Eberwein* and *Siegmann*, 1987 [3]).

3 The evolution of global modelling

Global modelling was virtually invented in the late 1960s. The first generation of models, identified by some critics as models of doom, addressed the interaction between population growth, production, resource depletion and environmental degradation. The models, of which "Limits to Growth" became world famous, revealed that the world would collapse in the long run if human behavior remained virtually unchanged. This "alarmist" type of model made the point that sustainability, to put it this way, is endangered if the world continues to evolve as in the past.

The second generation of global models tried to improve the original constructs by disaggregating the world into regions and by developing much more specific model representations, primarily of the economic sector. The *Bariloche* group showed that the basic needs in the Third World could be satisfied within certain limits in several decades. The MOIRA-model showed that world hunger could be overcome if the prevailing distribution patterns were changed on a global scale. Or, as the Leontieff model made it clear, the goals set for the United Nations developmental decade could be achieved if the allocation mechanisms were adjusted globally.

Whereas the first generation of global models were critical in that they sensitized the public to global sustainability in a very general sense, the second generation of models did reveal that specific global issues can indeed at least partially be resolved if a change in behavior would take place. However, whether and how the appropriate policies could be implemented were left open. All these models had excluded politics in general, the nation state in particular. The nation states are not only responsible for the creation of the problems addressed, but also core players in resolving them. None of these models even come close to what Schellnhuber has called managerial models, that is models that may serve as decision tools.

The third generation of global models came closer to this end by introducing the states, GLOBUS on the one hand, SIMPEST on the other. But this improvement was offset by their exclusion of the environmental dimension which is germane to sustainability. Experiments with these models revealed that the best intentions which may induce specific governments to change their policies may not necessarily produce the expected outcomes, but rather generate counterproductive if not perverse effects (cf. *Eberwein* and *Groenen*, 1990 [2]; *Bremer* and *Hughes*, 1990 [1]).

Progress in global modelling has indeed been achieved over the past 25 years. These efforts have produced valuable insights if not building blocks for the fourth generation of global models, in particular the economic and political sectors. This new generation of models belongs to the class of sustainability models representing the longer term consequences of the interaction between the physical and the social worlds. In the presentations and discussions I have taken part in this conference, no reference has ever been made to these prior activities. I can only hope that this error of omission is purely accidental.

4 What can we learn from past experience?

Based upon my own experience in global modelling I want to mention four points.

4.1 Formal aspects

In addition to Dürr I just want to add one point with respect to the mathematical representation of complex systems. Even though the construction of very large models is possible given the hardware and software at our disposal today, very large and complex models will not necessarily enhance our understanding of the problems under study. There is – at least from some point on - an inverse relationship between model size and our ability to analyze and understand model behavior. There is an inescapable trade-off between size and intellectual manageability.

4.2 The adequate level(s) of analysis

Global modelling is problem oriented. The ultimate goal is what Schellnhuber calls "managerial systems" which means that the model(s) must include variables that can – in principle – be manipulated by policy makers in the real world in order to resolve specific problems. It has been argued that the nation-state no longer is the central locus of power. I would argue to the contrary. The nation-state is still the most important actor in global politics, certainly one of the most important ones. The real dilemma is that the nation-state is characterized by an increasing loss of control, squeezed between international, supranational and transnational actors on the one hand, subnational or transregional actors on the other. For practical purposes models should use the nation-state as the point of reference both in terms of a bottom-up strategy (leading to the global level) or a top-down strategy (in terms of a subnational or regional level). This means that the model or models need to include variables that policy makers can – in principle – manipulate in order to achieve certain goals. This implies as well that the modeller has to take into account who the potential recipient of the analyses and prescriptions derived from global models will be!

4.3 Time horizon

Critical with respect to modelling sustainability is the choice of the time horizon. Long-term for climatologists refers to thousands of years. Long-term for political scientists means at best twenty or thirty years. One just needs to think about the succession of major events in the 20th century, beginning with World War I and ending with the breakdown of Communism and the Soviet Union. If, say, climate can be changed systematically within a period of 200 years, such a model is very likely of little if any relevance at all to policy-making at the present time. What is then the longer time horizon if both domains are coupled? In other words, the commensurability of the physical sciences and the social sciences needs to be identified first before even building a model or models including both domains.

4.4 Human behavior

At the core of sustainability is the problem of change in politics and society. First of all, sustainability has next to its physical (or natural world) dimension a social one. With respect to the latter acceptability of a policy of sustainability is central. What may be necessary from an objective observer's point of view may be impossible to implement politically. Sustainability is fundamentally a political problem which relates to the global redistribution of wealth, the functioning of our economies and the structure of the global political system. Given the present level of inequality at a global scale sustainability as a long-term goal may not be feasible because of the lack of political and social acceptability.

5 Looking ahead

These are but a few aspects that need to be addressed in the development of this fourth generation of global models. Before concluding let me briefly raise five issues that are much more operational in nature.

1. Sustainability is a promising core concept for modelling. Before even starting with model construction, however, the variety of interdisciplinary linkages and related issues needs to be clarified. A modelling strategy that begins at the disciplinary level will severely impede if not preclude interdisciplinary model development at a later stage. Modellers love their products. As we all know love is strongly exclusionary.
2. Starting with smaller models enhances the understanding of model behavior. Starting with smaller models also greatly enhances the likelihood that interdisciplinary cooperation will be successful. The behavior of smaller models is easier to explain and therefore likely to increase the confidence in the results they produce. The conceptual components of smaller models are easier to transmit across disciplinary boundaries and therefore supportive for the creation of an interdisciplinary research culture than would be the case with very complex theoretical structures. In sum, shooting for one unified big model from the outset is too risky.
3. As a corollary, the strategy of developing a family of models should be choosen. This is the option PIK has selected and which makes perfect sense, not only for the reasons just mentioned above but also because sustainability implies conceptually a variety of substantive dimensions on the

one hand, different levels of aggregation (i.e. local, national and global, to mention but three) on the other. Such a multi-model strategy, however, is only productive if formal and/or substantive linkages are clearly identified a priori. By formal linkages I mean the identification of the specific set of variables which later on will serve to couple two or more models. By substantive linkages I mean the identification of one or more output variables from one model which another model may at a later stage need as input without ever formally coupling the two.

4. Global modelling is a social enterprise. It is not an institutionalized area of scientific activity and thus not a highly attractive career avenue for an academic. Career patterns are defined within disciplinary boundaries. This career uncertainty is therefore a potential source of disruption for such a longer-term enterprise. Ways and means of how to institutionalize an internal career system that offsets at least partially this potential shortcoming need to be envisaged by PIK from the outset. Otherwise this institution may at worst fail to develop into an attractive research site for dedicated and bright younger scholars which guarantees continuity; at best it may evolve into a departmentally disconnected site for modelling activities along disciplinary lines. Either option would preclude the achievement of the original goal.
5. Finally, model development should be pursued in cooperation with the potential users and consumers from as early on as possible. In particular, including political and social dimensions is a guaranteed source of disagreement and conflict with potential consumers inside and outside academia. Unless this is done too much time may be lost later on for marketing the product.

6 A short summary

The development of a fourth generation of global models focusing on the sustainability of the earth system is not only necessary given the nature of the global problems but also feasible from the scientific point of view. In order to achieve this very ambitious goal, one needs to think big and thus, one needs to think PIK!

References

1. Stuart A. Bremer and Barry B. Hughes. *Disarmament and Development: A Design for the Future.* Prentice Hall, Englewood Cliffs, 1990.
2. W.-D. Eberwein and W. Groenen. Long-term impact of alternative western policy options: Foreign policy change and disarmament. In H. Chestnut, P. Kopacek,

and T. Vamos, editors, *International Conflict Resolution Using System Engineering*, IFAC Workshop Series, No. 1, pages 113– 118. Pergamon Press, New York, 1990.
3. W.-D. Eberwein and H. Siegmann. Evaluating long-term developments by using global models: Studies and documents. Report BEP/GPI/4, Bureau of Studies and Programming (BEP), UNESCO, Paris, 1987.
4. D. Meadows, J. Richardson, and G. Bruckmann. *Groping in the Dark - the First Decade of Global Modeling.* John Wiley & Sons, New York, 1982.

Glossary

AGENDA 21 GC mitigation programme adopted at the UNCED in Rio 1992
AQUA Global Water Model in TARGETS
BAHC Biological Aspects of the Hydrological Cycle
BAU Business As Usual
BIOME Global Biosphere Model
BLS Basic Linked System
C_3, C_4 Plant Classes
CFC Chlor Fluor Carbon
CGCM Coupled Atmosphere-Ocean Model
CLUAM Climate Land Use Allocation Model
DGVM Dynamic Global Vegetation Model
ENSO El Niño Southern Oscillation
E & D Environment and Development
FCCC (UN) Framework Convention for Climate Change
GAIM Global Analysis, Interpretation, and Modelling
GC Global Change
GCM Global Circulation Model
GCTE Global Change and Terrestrial Ecosystems
GDP Gross Domestic Product
GES Global Environment and Society
GFDL Geophysical Fluid Dynamics Laboratory
GIS Geographical Information System
GISS Goddard Institute for Space Studies
GLOBUS Generated Longterm Optimization By Using Simulation
GNP Gross National Product
HDI Human Dimensions Index
IAM Integrated Assessment Model
IGBP International Geosphere-Biosphere Programme
IHDP International Human Dimensions Programme
IIASA International Institute for Applied Systems Analysis
IMAGE Integrated Model to Assess the Greenhouse Effect
IPCC Intergovernmental Panel of Climate Change
LAI Leaf Area Index
LCS Land Use Classification System

LGM Last Glacial Maximum
MAC Maximal Allowable Concentration
MFM Martian Fixed Mode
MOIRA Model for International Relations in Agriculture
NPP Net Primary Production
OECD Organization for Economic Cooperation and Development
PAR Photosynthetic Activ Radiation
PFT Plant Functional Type
POEM Potsdam Earth System Models
RAINS Regional Acidification, Information, and Simulation
SD Sustainable Development
SIAM Structural Integrated Assessment Model
SIMPEST Simulating Political, Economic, and Social Trends
TARGETS Tool to Assess Regional and Global Environmental and Health Targets for Sustainability
TBM Terrestrial Biogeochemical Model
UKMO United Kingdom Meteorological Office
UNCED United Nations Conference for Environment and Development
VFM Venerian Fixed Mode
WBGU Wissenschaftlicher Beirat (der Bundesregierung) Globale Umweltveränderu gen
WCRP World Climate Research Programme
WORLD3 World model #3

Register of Persons

Subject Index

Entries into this index are generally sorted with pagenumber as they appear in the text. Page numbers that are marked in **bold face** indicate that the entry appears in a title or subheading. Page numbers that are marked in *italics* indicate that the entry appears in the caption of a table or figure.

D

E

MIX
Papier aus verantwortungsvollen Quellen
Paper from responsible sources
FSC® C105338

If you have any concerns about our products,
you can contact us on
ProductSafety@springernature.com

In case Publisher is established outside the EU,
the EU authorized representative is:
Springer Nature Customer Service Center GmbH
Europaplatz 3, 69115 Heidelberg, Germany

Printed by Libri Plureos GmbH
in Hamburg, Germany